D0329177

ANNUAL REVIEW OF EARTH AND PLANETARY SCIENCES

ANNUAL REVIEW OF EARTH AND PLANETARY SCIENCES

VOLUME 27, 1999

RAYMOND JEANLOZ, *Editor*
University of California at Berkeley

ARDEN L. ALBEE, *Associate Editor*
California Institute of Technology

KEVIN C. BURKE, *Associate Editor*
University of Houston

http://www.AnnualReviews.org science@annurev.org 650-493-4400

ANNUAL REVIEWS 4139 EL CAMINO WAY P.O. BOX 10139 PALO ALTO, CALIFORNIA 94303-0139

ANNUAL REVIEWS
Palo Alto, California, USA

International Standard Serial Number: 0084-6597
International Standard Book Number: 0-8243-2027-1
Library of Congress Catalog Card Number: 72-82137

∞ The paper used in this publication meets the minimum requirements of American National Standard for Information Sciences—Permanence of Paper for Printed Library Materials, ANSI Z39.48-1992.

TYPESET BY TECHBOOKS, FAIRFAX, VA
PRINTED AND BOUND IN THE UNITED STATES OF AMERICA

PREFACE

This volume of the *Annual Review of Earth and Planetary Sciences* spans a wider range of disciplines than ever before, reflecting the Editorial Committee's efforts to enhance coverage of the "Fluid Earth" (e.g., oceans and atmosphere) while maintaining the series' traditional strengths in geology, geochemistry, geophysics, paleontology, and planetary science.

The nature of the Earth's crust is described on a wide range of length scales, from crystal-structural to continental (articles by J. Środoń, C.-F. Tsang, J. Revenaugh, and L. Sonder & C. H. Jones), including temporal constraints on surface processes (P. M. Vasconcelos). Moving beyond simplifying traditional assumptions, the hallmark of current research is to face head-on the actual heterogeneity present over the full spectrum of length and time scales in the Earth. The evidence for complexity emerging from studies of present-day or recent ocean circulation and climate changes (A. J. Weaver et al, and R. B. Alley & P. U. Clark), is complemented by the fossil record of past changes in environment and ecosystems (B. Van Valkenburgh, and J. P. Grotzinger & A. H. Knoll); the result is a strong linkage between the planetary, geological, and biological sciences.

Impact processes (H. J. Melosh), as illustrated by the K-T event (J. Smit), play a special role in Earth's history and, more generally, in the planetary sciences. Indeed, telescope observations of other planets (S. W. Lee & P. B. James) and of a recently discovered class of comets (D. Jewitt) illustrate the seamless connection between astronomy and geology by way of planetary science.

There is no better indication of this link than through the prefatory chapter by Caroline Shoemaker, who gives an account of a remarkable—and still active—career in both planetary astronomy and field geology. Her writing offers a tribute to the life of Gene Shoemaker, whose untimely death was a deep loss to the entire Earth and planetary science community. We acknowledge this by dedicating the volume to his memory.

RAYMOND JEANLOZ
EDITOR

Photo courtesy of the American Geophysical Union

IN MEMORY OF EUGENE M. SHOEMAKER
1928–1997

Annual Review of Earth and Planetary Sciences
Volume 27 (1999)

CONTENTS

 (continued)

SOME RELATED ARTICLES IN OTHER *ANNUAL REVIEWS*

From the *Annual Review of Astronomy and Astrophysics*, Volume 36, 1998:

Detection of Extrasolar Giant Planets, Geoffrey W. Marcy and R. Paul Butler

Astronomical Searches for Earth-Like Planets and Signs of Life, Neville Woolf and J. Roger Angel

Radio Emission from Solar Flares, T. S. Bastian, A. O. Benz, and D. E. Gary

The Lyman Alpha Forest in the Spectra of Quasistellar Objects, Michael Rauch

Chemical Evolution of Star-Forming Regions, Ewine F. van Dishoeck and Geoffrey A. Blake

From the *Annual Review of Ecology and Systematics*, Volume 29, 1998:

The Ecological Evolution of Reefs, Rachel Wood

Early Evolution of Land Plants: Phylogeny, Physiology, and Ecology of the Primary Terrestrial Radiation, Richard M. Bateman, Peter R. Crane, William A. DiMichele, Paul R. Kenrick, Nick P. Rowe, Thomas Speck, and William E. Stein

Possible Largest-Scale Trends in Organismal Evolution: Eight "Live Hypotheses," Daniel W. McShea

Carbon and Carbonate Metabolism in Coastal Aquatic Ecosystems, J.-P. Gattuso, M. Frankignoulle, and R. Wollast

The Chemical Cycle and Bioaccumulation of Mercury, François M. M. Morel, Anne M. L. Kraepiel, and Marc Amyot

From the *Annual Review of Energy and the Environment*, Volume 23, 1998:

Rewards and Penalties of Monitoring the Earth, Charles D. Keeling

The O_2 Balance of the Atmosphere: A Tool for Studying the Fate of Fossil Fuel CO_2, Michael L. Bender, Mark Battle, and Ralph F. Keeling

Climate Change Mitigation in the Energy and Forestry Sectors of Developing Countries, Jayant A. Sathaye and N. H. Ravindranath

From the *Annual Review of Fluid Mechanics*, Volume 31, 1999:

Transport by Coherent Barotropic Vortices, Antonello Provenzale

Nuclear Magnetic Resonance as a Tool to Study Flow, Eiichi Fukushima

 (continued)

Liquid and Vapor Flow in Superheated Rock, Andrew W. Woods

Planetary-Entry Gas Dynamics, Peter A. Gnoffo

A New View of Nonlinear Water Waves: The Hilbert Spectrum, Norden E. Huang, Zheng Shen, and Steven R. Long

From the *Annual Review of Materials Science*, Volume 28, 1998:

Jahn-Teller Phenomena in Solids, J. B. Goodenough

Isotropic Negative Thermal Expansion, Arthur W. Sleight

The Material Bone: Structure—Mechanical Function Relations, S. Weiner and H. D. Wagner

In Situ Studies of the Properties of Materials Under High-Pressure and Temperature Conditions Using Multi-Anvil Apparatus and Synchrotron X-Rays, J. B. Parise, D. J. Weidner, J. Chen, R. C. Liebermann, and G. Chen

High-Pressure Synthesis, Characterization, and Tuning of Solid State Materials, J. V. Badding

From the *Annual Review of Physical Chemistry*, Volume 49, 1998:

The Shuttle Glow Phenomenon, Edmond Murad

ANNUAL REVIEWS is a nonprofit scientific publisher established to promote the advancement of the sciences. Beginning in 1932 with the *Annual Review of Biochemistry*, the Company has pursued as its principal function the publication of high-quality, reasonably priced *Annual Review* volumes. The volumes are organized by Editors and Editorial Committees who invite qualified authors to contribute critical articles reviewing significant developments within each major discipline. The Editor-in-Chief invites those interested in serving as future Editorial Committee members to communicate directly with him. Annual Reviews is administered by a Board of Directors, whose members serve without compensation.

ANNUAL REVIEWS OF

Anthropology
Astronomy and Astrophysics
Biochemistry
Biomedical Engineering
Biophysics and Biomolecular Structure
Cell and Developmental Biology
Computer Science
Earth and Planetary Sciences
Ecology and Systematics
Energy and the Environment
Entomology
Fluid Mechanics
Genetics
Immunology
Materials Science
Medicine
Microbiology
Neuroscience
Nuclear and Particle Science
Nutrition
Pharmacology and Toxicology
Physical Chemistry
Physiology
Phytopathology
Plant Physiology and Plant Molecular Biology
Political Science
Psychology
Public Health
Sociology

SPECIAL PUBLICATIONS

Excitement and Fascination of Science, Vols. 1, 2, 3, and 4
Intelligence and Affectivity, by Jean Piaget

For the convenience of readers, a detachable order form/envelope is bound into the back of this volume.

Carolyn S Shoemaker

Annu. Rev. Earth Planet. Sci. 1999. 27:1–17

UPS AND DOWNS IN PLANETARY SCIENCE

Carolyn S. Shoemaker
US Geological Survey and Lowell Observatory, Flagstaff, Arizona 86001;
e-mail: css@lowell.edu

KEY WORDS: Gene Shoemaker, moon, impact, comets, Australia

ABSTRACT

The field of planetary science as it developed during the lifetimes of Gene and Carolyn Shoemaker has sustained a period of exciting growth. Surveying the skies for planet-crossing asteroids and comets and studying the results of their impact upon the planets, especially the Earth, was for Gene and Carolyn an intense and satisfying quest for knowledge. It all started when Gene envisioned man going to the Moon, especially himself. After that, one thing led to another: the study of nuclear craters and a comparison with Meteor Crater, Arizona; the Apollo project and a succession of unmanned space missions to the inner and outer planets; an awareness of cratering throughout our solar system; the search for near-Earth asteroids and comets; a study of ancient craters in Australia; and the impact of Shoemaker-Levy 9 on Jupiter. The new paradigm of impact cratering as a cause for mass extinction and the opening of space for the development of new life forms have been causes to champion.

INTRODUCTION

It is an honor to be the first woman asked to write a prefatory chapter for the *Annual Review of Earth and Planetary Sciences*. I am intrigued by the idea of addressing a period of scientific development in terrestrial and planetary science along with my own scientific growth. Planetary science as a comprehensive field of endeavor did not exist when I was in the midst of my college studies, but I have seen it emerge as a field that includes areas of every physical science and touches upon some social sciences. It requires an ability to think theoretically and to use skills found in mathematics, computer science, photography, and

0084-6597/99/0515-0001$08.00

mechanics. It challenges the imagination and provides the "stuff" of dreams. To work in planetary science is to work in an area that takes us both back to the origin of our solar system and beyond it into the future.

GENE SHOEMAKER

It is impossible to write about this part of my life without also discussing Gene Shoemaker in a major way. My successes were his successes; what I accomplished was the result of a close and happy partnership for the 46 years we were married. Gene was not only my husband, he was also my scientific teacher, mentor, and collaborator. He was an enthusiastic teacher, who conveyed the excitement he felt in geology, and who instilled that excitement in others. His love of science, of geology and astronomy, was contagious. He was one of a rare breed—a true Renaissance man.

I did not set out to work in science, either as an astronomer or a geologist; I have not had a formal education in either. When I went through college, few were thinking about going to the Moon or even into space. If they did, they wouldn't admit it. Comets and asteroids were not part of our world, and certainly they did not fall out of the sky. World War II was just over, and veterans flooded the colleges. They were an earnest, sober group, eager to get on with life, to get in and out of school in short order, and my outlook was similar. I was eager to get out and see the world after attending college in my hometown. Four years after I started my classes at Chico State, now known as California State University, Chico, I left school with a master's degree in history and political science, a secondary teaching credential for high school, and a contract as a seventh grade teacher in Petaluma, California. Then fate took a hand.

Just before I left home to assume my teaching position, my brother married one of my good friends from college, and the best man at his wedding was his roommate at Caltech, a young geologist named Eugene Shoemaker. The first things that impressed me about Gene were his sense of humor and ability to laugh at himself, his enthusiasm for virtually everything, and the energy that seemed to flow from him. Through the following year, while Gene attended Princeton University to work on his doctorate and I taught seventh grade, we corresponded with long letters each month. At the end of this year, I took a two-week camping trip with Gene, who was eager to show me the Colorado Plateau and took me to the most scenic places he knew—there were a lot of them. After the first week, I apparently passed the "love of camping" test. Gene asked me to marry him, and our wedding took place in 1951, exactly one year after my brother's wedding.

Our Early Years

For the first five years of our marriage, home was Grand Junction, Colorado. Gene, working for the U.S. Geological Survey (USGS), was participating in an exploration for uranium. After the first two nuclear bombs had been dropped on Japan, ending World War II, the Manhattan Project ended. The country had exhausted its supply of uranium, and more had to be found if atomic power was to become a reality. The Survey had contracted with the Atomic Energy Commission to find new sources on the Colorado Plateau, and Gene was involved in running a drilling project and in mapping during his spare time. Individual miners were searching everywhere and "digs" appeared on every hillside. As in all booms, some made it big and some "lost their shirts." Land scams were the order of the day. It was fun to be on the scene of the last great mining boom in our country.

When Gene went out to map, I went along. Women did not often go in the field in those days, but neither Gene nor I could see a reason why I shouldn't. For the rest of his life, Gene would champion the causes of women, an attitude he took from his very independent mother. Gradually I began to learn the stratigraphy of the Colorado Plateau and to perceive the many meaningful differences in the shades of orange and red that distinguished the various beds of that area. I was taught all about layer-cake geology in one of the choice regions of the world. I had never thought it necessary to pick something apart to appreciate it, but I was suddenly learning about faults and thrusts, discontinuities and structure; about the difference between sand, pebbles, rocks, and boulders; about what the land had looked like through different ages. Under Gene's enthusiastic teaching, many of the things that had had no meaning for me in my one geology class in college were becoming fascinating.

In the summer of 1952, Gene and I headed for Nevada and a summer field camp with James Gilluly, a long-time geologist with a heady reputation. To work with him necessitated an ability to be half mountain goat, have the legs and endurance of an Everest climber, have the speed of a greyhound, and have a determination to not be outdone by anyone, least of all by Gilluly. What a learning experience it was for both of us. Newly pregnant with our first child, I could not tag along with the men. Instead, I was camp cook for eight on a regular basis and often many more as they arrived to visit Jim. The stories that were told around the evening campfires opened a whole new world to me.

Three years after we were married, Gene and I headed east to Princeton with our baby daughter, Christy. The chance to run his own program in the search for uranium had been an opportunity too good to pass up, and Gene had taken a two-year break. When it ended, we headed back so he could finish the requirements for his PhD. He completed the Princeton requirements, and we returned to Grand Junction. Gene had been on leave from the USGS and it was time to get back to work. In fact, he was so involved in various projects that he had no time left to write his thesis. It was not until some years later, when the deadline for completing it was running out, that he took the subject that was topmost on his pile of work and turned in one of the shortest and most concise theses in Princeton's history—his work on Meteor Crater, Arizona, which became a landmark paper in cratering studies.

GOING TO THE MOON

In Grand Junction in 1953, Gene first confided to me his dream of going to the Moon. Not just going to the Moon, but being the first geologist, the first scientist, to go to the Moon and—why not?—being one of the first to set foot there. One day in 1948, as a young bachelor new to his job, he had been driving down a canyon road thinking about rocket development during World War II and after, when some of the best German rocket scientists came to work in the United States. Suddenly it dawned on him that during his lifetime man would go to the Moon. He planned to be ready, to be the most knowledgeable and best-prepared person when that time came. I was truly startled, for when he told me this it was a time when *no one* had such aspirations, because no one envisioned such a thing happening. I knew this wouldn't happen immediately, and like Gene I didn't discuss the subject with anyone else—but I wondered if I could go along. From the day he told me of his desire, I started looking at the night skies with much more attention.

Christy, Patrick, and Linda, our three children, went out in the field with Gene and me from very young ages. Field life was a regular, enjoyable part of our existence. Gene, above all, was a field geologist who enjoyed the challenge of reading the rocks and understanding the history they told. Field work to him was more a vacation than a job because he enjoyed it so much. He shared this passion with me, explaining all he could wherever we went.

In the fall of 1957, our family of five went to the Hopi Buttes for some mapping with two other geologists, Frank Byers and Carl Roach. Gene had found uranium in the Hopi Buttes, and further geological mapping of the Navajo and Hopi country was needed, with attention to diatremes (exploded volcanoes). One evening, upon returning to our camp on the Navajo Reservation near Indian Wells, we were told by Frank and Carl that the Russians had put up a spacecraft. Sputnik triggered a major change in our lives. Gene was dismayed by the Russian achievement; he knew that it would spur the United States to start its own space program before he himself felt ready to participate. He had his work on Meteor Crater and also the nuclear craters at the Nevada test site to finish first. Nevertheless, henceforth, he never deviated from that ultimate goal of participating in a space program that would fulfill his dream of lunar exploration.

From 1957 to 1960, Gene's research was directed toward the structure and mechanics of meteorite impact and nuclear explosion craters. In studying the latter, he could see the close relationship between the two types of craters. As he pointed out to me when we looked through his alidade at the Moon, a study of craters was necessary to understanding what man would be looking at there. His work on Meteor Crater, Arizona, led to the discovery with E.C.T Chao of coesite, a high-pressure form of silica not usually found in natural features on Earth. Coesite was a tool that indicated an extraterrestrial impact. I was impressed when we went to the Ries Basin in Germany, before the 1960 meeting of the 21st International Geological Congress in Copenhagen, to see Gene descend briefly into a quarry and return with samples he expected to contain coesite. He dispatched them by air to Ed Chao, who promptly analyzed them in the USGS lab in Washington, D.C. and found coesite. Sure enough, this diagnostic system using coesite was proving true to Gene's hopes for it!

Our family was living in Los Altos, California, by 1959, having moved there when the USGS program with the Atomic Energy Commission for finding uranium came to an end after the discovery of vast new resources in New Mexico. Gene elected to join the Menlo Park office of the USGS instead of moving to Denver. With his thesis completed, Gene began studying the Moon in earnest, and at last we could "come out of the closet" and talk openly with a few other young geologists there about his goals for going to the Moon. In 1961, President Kennedy made his famous pronouncement about going to the Moon and back. Gene and I followed each advance of man rocketing into space with the Mercury program, and we avidly consumed all the details released by the press.

The year of 1960 was eventful. Gene traveled constantly, pursuing the establishment of an Astrogeologic Study Group within the USGS. The USGS is an old conservative organization and to establish such a group was not easy. The Survey's charter was to study and investigate the geology of the United States and sometimes aid studies in other countries; "old-timers" were not eager to risk, as they perceived it, their careers on a truly "lunatic" idea and transfer into astrogeology. I yearned to hear more about the program, but with young children to tend to, I had to be content with following its progress whenever Gene was home to tell me about it.

Planetary science was in its infancy. Astronomers had long since left the solar system to study the universe, but geologists, literally, could hardly get off the ground. After the first group of seven Mercury astronauts had been selected, followed by a second group of nine Apollo astronauts, a third group was to be chosen. Gene was not so naive as to think there might be a scientist among them. Military pilots were the choice of NASA. In the fall of 1962, we moved to Washingtion, DC, so that Gene could work with NASA as Acting Director of the Manned Space Sciences Division. The move plunged us into the frenetic scene of the political arena. With the development of manned space flight, NASA was growing rapidly into a large organization. Gene could not, would not, imagine that the American public would be satisfied to go to the Moon just to beat the Russians there. The time had come to think about a real purpose for man to go into space; the selection of the science astronauts was about to occur.

Any other year, Gene would have applied to be an astronaut, but in 1962, he gradually became very ill. As the year advanced he saw one doctor after another, hoping for a diagnosis. After almost nine months, we finally learned that his adrenal glands were not functioning (a condition known as Addison's disease), and his bodily functions were shutting down. Once the diagnosis was made and supplementary cortisone prescribed, Gene made, within a couple of days, a miraculous recovery. Because blackouts were a common symptom associated with Addison's and because most people with the disease died before cortisone was discovered, he knew that he could never pass the stringent physical examination required of astronauts. Instead, he accepted the position of Chairman of the National Academy's Science Astronaut Selection Committee. It was now necessary for him to put aside his earlier dreams of going to the Moon; he had to do his best to make the space program work for other astronauts.

In 1963, with the organization of the Branch of Astrogeology, the family moved to Flagstaff, Arizona, where the space age would come to fruition for us. Flagstaff, which was a proven site for astronomical observing, was an ideal place for a small telescope (31 in) dedicated to studying the Moon, and in addition, nearby Meteor Crater and the craters of the San Francisco Peak volcanic field afforded valuable comparisons with the lunar landscape. There was much to learn before the astronauts arrived on the Moon.

Now came the years of extensive travel for Gene, as he developed a lunar geological time scale and suggested methods of geological mapping of the Moon. He became involved in the application of television systems to the investigation of extraterrestrial geology, taught courses in astrogeology at the California Institute of Technology, and served as coinvestigator of the television experiment on Project Ranger, principal investigator of the television experiment on Project Surveyor, and principal investigator of the geological field investigations in Apollo lunar landings from 1965–1970. He was Chief of the USGS Branch of Astrogeology until 1966 and then, in a shift from administrative work to scientific studies, became its Chief Scientist until 1968. One of his many projects was to institute the geologic field training of the astronauts who would go to the Moon and to take them to various sites throughout the world.

Another project was to learn to fly an airplane, on the premise that he could demonstrate how much more easily a scientist could learn to fly than a pilot could learn science. When he could not pass the medical exam required of pilots because of his Addison's, he encouraged me to get my own pilot's license, and we flew together.

A SHIFT IN COURSE

During all this time, as our children grew to their teens, I was a homemaker and entertained the seemingly infinite number of scientists who passed through our doors. Gene kept me abreast of developments in the space program as much as possible, and the many visitors also helped to maintain my interest in that program.

Then, in 1969, our family shifted course. Gene had become tired of the infighting involved in the space program and felt that the scientific goals would never come to pass. Only one scientist, geologist Harrison Schmitt, had been selected to go to the Moon. Gene was away from home three quarters of the year and was exhausted. Also, as our children entered high school he knew the family would need him. Leaving the Geological Survey, he took up the position of Chairman of the Division of Geological and Planetary Science at Caltech, and we moved to Pasadena, California. Unfortunately, the Apollo program had been set back one year after the fire in Apollo I, and Gene was still committed to work on another Apollo flight even as he took up his new responsibilities.

Caltech was a new experience for me. I had never encountered such a concentration of scientific, intelligent, single-minded prima donnas! I was fascinated to realize that not only were the professors, researchers, and students extremely capable and dedicated, they also loved their research as much as Gene did his. I enjoyed this atmosphere, although while our children were still at home finishing high school and going off to college I was not yet ready to take part in it.

At Caltech, Gene went on to become a coinvestigator with the television experiment on Project Voyager. This occupied much of his attention from 1978 to 1990. Now there was a mission! I shall never forget walking over to the Caltech campus, where, on a large screen in Beckman Auditorium, images of the Jupiter flyby were being displayed as they came in. Then and there, I fell in love with planetary science, the skies, our solar system, and the universe. I was excited and captivated. Exploration of the Moon during Apollo days had been vicariously, for me, a heady experience, but the idea that *I* could look at Jupiter and its moons in something like real time left me awestruck.

The First Palomar NEO (Near-Earth Object) Search

In 1973, Gene initiated the Palomar Planet-Crossing Asteroid Survey with Eleanor Helin. After studying craters on Earth and the Moon, then on Mercury and Mars, he became convinced that cratering was a fundamental process in the formation of our solar system. A look at all the craters on some of the moons of Jupiter—Europa, Ganymede, and Callisto—helped to confirm his belief that cratering by asteroids and comets had occurred on all the solid bodies of the solar system. Yet there were almost no statistics on the numbers of planet-crossing asteroids and comets, much less Earth-crossing bodies. He would have to obtain the statistics at the telescope, and Eleanor Helin was interested in taking part. Eventually a search program on the 18-in Schmidt telescope at Palomar Observatory, owned and operated by Caltech, was established. As it turned out, this telescope (really a camera), the oldest and smallest at Palomar, was the best telescope in the world for the search. With its wide-field view, one round, 6-in-diameter film covered 60 square degrees of sky. In the ten years of that survey, with the help of many students, techniques of observing were gradually developed, and the team began to find asteroids that crossed Earth's orbit. Still, the discovery rate was not high, and Gene hoped to develop "cheaper, better, and faster" methods, to quote a phrase current today.

BECOMING INVOLVED

By the late 1970s, our children had grown and left home. I felt a need to do something interesting. I wanted to find some area of strong compelling interest—something that would absorb my time and energy and stimulate the thinking process. When Gene proposed the idea of working with him on either his paleomagnetic projects or the search for Earth-crossing asteroids, I replied that I "had done the paleomagnetic bit" but would like to try working on the asteroid search.

In the beginning of this work, I went with Gene to his office at Caltech and started helping there. Then, on a somewhat casual basis in 1980, I started to search for asteroids on the large plates, 14 × 14 in, that were taken on the UK Schmidt telescope at the Siding Spring Observatory in Australia. Bobby Bus, at that time a Caltech student, was my teacher first in scanning and later at the telescope. He taught me how to scan the large glass plates on a blink comparator and to measure the positions of asteroids in the sky. He was meticulous in this work and showed me the importance of doing each step carefully. Because he was patient with me, I found the asteroid work enjoyable.

Gene, in the meantime, had concluded that the discovery process would be easier and faster with a stereomicroscope. This was a procedure that was originally developed by Max Wolf in Germany during the 1800s, but Wolf could not see stereo and so did not use the new system for long. Now Gene hoped that a stereomicroscope built especially for use with the films from the 18-in Palomar telescope would speed things up. Together with Helin, he had one made to their specifications by a company in California, which completed it in 1980. Up to that point, Helin had used a monocular microscope with success and was apparently not eager to try this system when it arrived. Gene, by then, was spending half the year at the USGS in Flagstaff and half the year at Caltech. When we returned to Pasadena for the winter quarter, he found that Helin was not using the microscope. A student who had been working with her had tried it out and was not very successful in finding any asteroids. Dismayed, Gene tried looking at films himself and then asked me to try. Neither of us had difficulty in seeing the asteroids, and we decided to experiment with the system further. When we returned to Flagstaff, we took the stereomicroscope and some 18-in films with us, set the microscope up at home, and commenced a search. We had fun as we sorted through thousands of stars, dust, scratches, and other marks on the films. In short order, once he knew that I could find asteroids, Gene left the scanning of films to me. As the years passed, the search for asteroids, and later comets, would increasingly become my particular project.

Back at Caltech, I learned to observe at Palomar Mountain Observatory. Because I am a morning person, I had real doubts about my ability to observe through the night; I had never stayed up all night in my life. I thought that I might have to be content with scanning the films others had taken for me. Nevertheless, I resolved to try observing. Sleep deprivation while observing was, in fact, a real problem for me in those first years. We would finish a night's observing and go to bed when the birds were beginning their early morning songfest; two or three hours later, I would wake up. However, as I learned the ropes—taking 10- and 20-min exposures and guiding on stars with an often cranky telescope; learning the method of developing the films and the procedures of recording information on the margin of films, bagging the films, and recording the observations in a notebook for Palomar and another notebook for us; and learning methods of planning nightly observations—and as I had more practice in scanning the films with the stereomicroscope, I found a real joy in going to the mountain. I would sit guiding at the telescope and find myself marveling that *I* was actually observing at *Palomar*. This was an observatory I had heard of during much of my life; I had visited it with my parents as a child at the time the 200-in telescope was newly completed and installed. To be sure, the 18-in was not a glamorous telescope in a world where "bigger" equates with "better," but I had a growing fondness for that instrument.

The Second Palomar NEO Search

In 1982, Gene and I started the Palomar Asteroid and Comet Survey (PACS) and started observing together for the first time. Helin had developed her own program separately and the two projects parted company. In the ensuing years, both projects used the 18-in at one end or the other of the "dark run," the two-week dark-of-the-moon period each month centered on the new moon. Gene and I decided to observe for seven nights each run, figuring that we should have at least four nights of good weather, which would enable us to take lots of fields. Soon, using a spectrographic film called Kodak IIa-D, we started taking 4-min as opposed to the earlier 10-min exposures. That was something to keep me awake! With only the two of us observing together, there was no time for anything else. We traded jobs on the telescope, one of us taking the first half of a set of four or five fields, and then the other taking the second half of identical fields. Whoever was not actually guiding the telescope helped first to position it on a field by providing the coordinates and recording times on the log sheets; then that person would rush downstairs to the dark room, change the film to a new one, and run back upstairs in time to start the process all over again. If we wanted a snack or a cup of coffee in the night to keep up the energy level, we would have to grab it on the fly. Our interest lay in photographing as much of the sky as possible; one long night we managed to expose 96 films between evening twilight and dawn, but we were really dragging by the end of the night. Throughout our program, we would take two exposures of the same field about 45 min apart. Placing a pair of films side by side on the stereomicroscope made it possible to detect the asteroids and comets by stereopsis as "floating" objects against a background of stars. The parallax, caused by the movement of the Earth during the time difference in the taking of the films, allowed us to see such nearby objects without the necessity of blinking the films, which was the more commonly used method of search.

One day in 1987, Alain Maury, then at Palomar, stopped by the dome. He told us about a new film, Kodak's Technical Pan 4415, which we could use if we would hypersensitize it to speed up the emulsion. This would allow us to take 8- or 10-min exposures with the Technical Pan rather than the 1-h exposures that this film would otherwise require to obtain images. We would no longer be able to take 4-min exposures, but we would improve our ability to find asteroids or comets because there was virtually no grain observable. We would be able to see the really faint (18 to 19 magnitude) asteroids. So we learned to hypersensitize our film for 6 h at 65°C with forming gas (a mixture of hydrogen and nitrogen) flowing through the boxes in the oven. We found more asteroids, took fewer films, and relaxed slightly. Use of the new film allowed us to develop most of the exposed films during the night, leaving more time in the day for "cutting the cookies," as we called the round pieces of film, hypering, and doing all the other things that were a regular part of observing.

I found that scanning films became easier and faster the more I practiced. Not everyone can see in stereo, as I learned when I worked with Caltech students. Fortunately, I had no difficulty with this and soon could find most of the interesting objects on a pair of films within 20 min. That is not to say that I didn't miss some, but they were usually faint, out of focus, or only on one film, with the second image mostly hidden by a nearby star. Sometimes I would become sleepy while scanning and realize when I finished a pair that I couldn't remember marking any asteroids. That was a signal to get up, move around, develop films, observe, or go outside to walk around the dome and just look at the sky. I gradually stopped seeing the artifacts (dust, scratches, other marks) in favor of seeing asteroids and comets, but more time passed before I started to see displaced images. Because we looked at negatives, most asteroids appeared as a single floating point of light in the form of a little black ball. The faster motion of some of these objects gave them a separate appearance, so that I saw two black balls, or two hyphens if the object moved even faster within the 8-min exposure. I learned about retrograde and prograde objects, which under the microscope would either float or sink, respectively. At first I saw only the bright objects, but with time I saw fainter and fainter objects. The search was fun, and when I discovered my first near-Earth asteroid, 1982 RA, I was thrilled. For the first time I noticed little hyphens, widely separated, instead of little floating balls, and I added short and long hyphens to my repertoire of things to look for.

Comets sounded exotic to me, and I wondered if I would be able to see any. The first comet I saw on our films was one I deliberately searched for according to the predicted position; it was Comet Bowell (1982 I). It was a unique comet because its coma was the result of one outburst of dust. It was well beyond the orbit of Jupiter when I saw it, a faint, fuzzy image on one film and scarcely brighter on the other. After that I was discouraged because it seemed impossible that I would ever discover a new comet. By the fall of 1983, however, I had seen a few other comets on our films.

A NEW COMET

One day in late 1983, I was scanning films at the USGS in Flagstaff in an effort to catch up with all that we had taken on a fairly good observing run (I was never able to scan all the films at the observatory). Suddenly, there it was! The image of a comet that I thought might be new. I ran from the little room where I scanned films at the back of the library down the hall to a phone in Gene's office and called Brian Marsden. He is Director of the Central Bureau for Astronomical Telegrams (to which all new discoveries are reported from around the world). Giving him approximate positions for the comet, I waited anxiously for him to tell me if it was new. Marsden checked his files and reported to me that this was, indeed, a new comet. A feeling of elation swept over me, one which I was to feel later with each new comet discovery. Each comet has a unique appearance on our films, so each new discovery is a new experience.

From the moment I found Comet Shoemaker 1983p, as it was designated, my interest in comets increased rapidly. There was a lot to learn about them, and one of the particular areas of study was a growing awareness of the role they play in impact throughout our solar system. Gene felt strongly that comets are responsible for the abundance of water and other elements necessary for life as we know it, that we may truly be made of comet "stuff." Comets were added to our search program, and by the time PACS came to an end in December, 1994, I had found 32 of them, 27 with Gene.

After we accidentally found our first Trojan asteroid, we started a special search for them. Trojan asteroids have been captured into the 1:1 resonant areas known as the Lagrangian equilibrium points of the Sun-Jupiter system; they comprise two swarms on either side of the giant planet, where they are held at a distance safe from capture by Jupiter and from which only a few have been observed in the process of slowly escaping. Gene calculated the numbers of Trojan asteroids in the two swarms and concluded that there should be many sufficiently bright enough for us to discover. The search for Trojans presented a new challenge to me. They are sufficiently far away from us that their parallactic displacement is small. That means that on our films, taken with fairly short intervals between exposures, Trojan asteroids hardly appeared to move at all or, in so many words, would barely "float" in stereo above the background stars. I had to develop a different perception of depth as I looked for them.

With us on that first search for Trojan asteroids, in 1985, was Richard Preston, a young, aspiring author who had come to Palomar to write a book about the 200-in telescope. We first met him at the dining table of the Monastery, which housed most of the astronomers during their observing runs at Palomar. We suggested that he come to the 18-in telescope some night to see how "real" or "old-fashioned" astronomy was done. This resulted in a nightly visit on that observing run and a request from him to come again. He was intrigued with our proposal to search for Trojans, and it was through his persistent urging that I finally took a hard look at the films for these objects. In succeeding years, we continued to search for them and ultimately discovered 47 new Trojans. From this search, we learned that the two swarms together probably contained as many as half the number of asteroids in the main asteroid belt between Mars and Jupiter.

By 1987, I was searching our films for "high floaters" (near-Earth asteroids) and "low floaters," those that seemed to be below the plane of the background stars, as though in a hole (very near asteroids), "displaced hyphens" (Earth or Mars crossers), fuzzy images with or without tails (comets), and, lastly, the "very low floating images" of Trojans. I was on the trail of unusual moving objects. Sorting all of these out from the thousands of images of stars on the films, along with measuring their positions in the sky with evolving techniques, kept me busy—I could work on the PACS project every waking moment, if I wanted, and enjoy it.

Looking at films was a little like looking out of a spaceship window; I was transported out into the universe where I could see galaxies, star clusters, and constellations. Because we used the same fields in successive years, many of these became very familiar; it seemed that the sky offered an order, peace, and constancy that made our world affairs seem petty. But in time, I realized that the universe is a rather wild place, full of collisions of asteroids, comets, stars, and molecular clouds. Worlds come and go there on a timescale so great as to be almost incomprehensible to us, and distances between most objects are so far as to be meaningless. Astronomy is a time machine that takes us back to beginnings; looking into space on our films gave me a sense of wonderment.

In 1987, Henry Holt, a retired planetary geologist with the USGS, joined Gene and me as a volunteer, part-time member of our observing team. When he expressed an interest in observing with us and came along for the first time, we found him to be a hard worker and congenial companion. What a luxury it was to have a third person with us on whom we could rely. At last we were able to develop our films throughout the night. We took turns guiding the telescope, changing film, and developing the exposures. I continued to be a part of those operations and also did all the scanning. Eventually, Holt was to take the summer runs on his own with students while Gene and I were away. Working with staff at Lowell Observatory in Flagstaff, Holt has found thousands of asteroids.

In 1989, we were joined by David Levy, who became a member of the PACS team for half of every year. Levy is a well-known amateur astronomer, author, lecturer, and educator. He became a reliable, capable, and loyal volunteer in our program. His infectious enthusiasm and sense of humor carried us through many a long night's observing. There is no doubt that our greatest discovery together, and in the whole program, was the discovery of Comet Shoemaker-Levy 9, the comet which had been pulled apart into 21 fragments by the gravity of Jupiter on a close pass in 1992, was discovered by us in 1993, and impacted Jupiter in 1994. What an adventure we had with that comet!

AUSTRALIAN CRATERS

Gene and I started another survey together in 1984, not astronomical but geological, on Australian craters, and we continued this program until his death in Australia in 1997. This survey was closely related to our "Mom and Pop" observing program, so called because the monetary support was very small and for a while the two of us were the only ones involved. Much of the year we looked up to find those objects that could collide with Earth in the future.

Then we took about three months in Australia to look down at the Earth to see where such objects had struck. The Australian continent has been stable, without a lot of mountain building and tectonic activity, which makes it perfect for the study of ancient rocks and impact structures. There are more meteorite craters (those with associated meteorites) in Australia than in any other country. We decided to look for undiscovered craters in Australia, as well as to do the work (mapping, gravity and magnetic studies, finding of meteorites and impact glass) on all the known craters where it had not been done before. During the years we were doing our crater work in Australia, most of the geologists there were involved with mapping and the search for mineral resources. This left an unoccupied niche in which we could work without treading on anyone's toes.

In order to do our field work, we purchased, after the first two years, a small Toyota Hilux 4-wheel drive pickup with sides and tailgate that let down for easy loading or unloading. We had two large plywood boxes constructed to hold all our gear: camping equipment and food; library; scientific instruments, including the alidade and tripod for mapping, theodolite for obtaining star positions and triangulating to get our own, magnetometer, and gravity meter; portable typewriter, stationery, and drafting supplies; and extra car parts, come-along and "roo" jack, and chains to get us out of difficult places. Two 50-gallon fuel drums were in front of the boxes and alongside, and at the end of the boxes, were ten 5-gallon water containers. We also carried, alongside, eight rubber doormats, ideal for getting us over sand dunes and out of creek beds. Eventually an electric air pump was installed under the hood of the Hilux for use in repairing the numerous flat tires we had each year (one year we had 23 within three months). Gene was an expert on changing and repairing tires when we were miles from anywhere. At the end of the field season, we would arrive back in the city of our departure (Perth, Canberra, or Adelaide), frequently with only four rather bad tires and no spare, despite having left for the field with at least three spares. For the first three years, while in the outback of Western Australia, we had a radio, installed in the pickup by Australia's Geological Survey. In 1996 and 1997, we carried a satellite phone and a GPS location finder. Needless to say, our pickup was one of the heaviest in the country, but our equipment and supplies made it safe and possible for us to spend three or four weeks deep in the outback with no one else around. We were down to basics, working from dawn to dark, with no regular telephone (on the satellite phone we could call out but no one could call us), no fax, no computer or word processor. We loved it.

Our trips to Australia provided some of our happiest and most rewarding times together. We were excited by all the exotic animals and the very different biota; the spring wildflower bloom must be one of the loveliest anywhere and we traveled among them for hours. We were delighted by the people, who seemed to us so very laid back, warm, and generous; we were enchanted by the beautiful, dark, starlit skies and would lie in our sleeping bags with binoculars to look at them together.

As the years progressed with our work in Australia, we had the opportunity to work on approximately 17 different craters and impact structures and checked out a large number of false hopes, which either were on the maps or were called to our attention by others. We came to know the country like the backs of our hands, and knew each and every track in most of Western Australia, Northern Territory, and South Australia, as well as many in Queensland and New South Wales. The friendships we made, with scientists, business people, professors, and station (ranch) people, were deep and lasting; even casual acquaintances would recognize us when we met them on a safari trip a long way from home. The appearance of our pickup certainly helped, but hardly any Australians would go out as we did, alone without several other vehicles along. People have often asked if we weren't worried by the dangers, but we always felt that if we were prepared for everything, there was nothing we could not handle together. There are no dangerous wild animals, we went in the winter when venomous snakes were not about, and we were well prepared to handle difficulties.

Tektite Studies

Another of our Australian projects involved the study of tektites. These small glassy objects are the result of an impact that throws the crater melt high into the atmosphere to come down across huge distances. Tektites and microtektites are found, for example, in relation to the huge Chicxulub crater event. They are found in many countries and in the ocean as the result of impacts at great distances. Australites are strewn across most of Australia and into the southern sea, having come from a structure, yet to be discovered, in or near Indochina. Gene and I spent a number of months with Ralph Uhlherr gathering information on tektites along the southern coast of Victoria, where Ralph had made a very large collection over time. This is a spectacular stretch of coastline, which draws numerous tourists every year. Personally, I was enchanted to work on something near the coast because I have always been drawn to the ocean, with its ever-changing moods, its crashing waves against the cliffs, its long sandy beaches. The cliffs are very high on this coast. Surely, this is one of the most spectacular places to do geologic field work, but it is also, to me, rather scary because we were working on slopes that plunged over cliffs a hundred feet high. Tektites were weathering out of a formation and working their way down the slopes.

THE FUTURE

In July of 1997, a head-on collision with another vehicle on a remote stretch of road, where commonly one might see only one or two other vehicles in a day, brought our adventures to an end with Gene's death and my injuries. Since the time of the accident, I have had to take stock of the direction of my life without Gene. Much of the Australian work remains for me to write up, along with the maps to complete. This work represents a huge investment of time and effort that should not have to be repeated but rather built upon by others. My big interest involving the discovery of planet-crossing asteroids and comets continues. At Lowell Observatory I am becoming involved with LONEOS (Lowell Near-Earth Object Survey), a project on a dedicated telescope near Flagstaff. Many objects on the films taken at Palomar in our PACS project remain to be worked on. And David Levy and I are continuing our comet search with his telescopes, both in Flagstaff in summer and in Tucson in winter.

I have watched delightedly the growing enthusiasm for asteroid and comet research, which has developed along with an increasing awareness of the hazards of impact on Earth. Today there are sky surveys capable of finding many thousands of asteroids and comets—something Gene and I saw the need of many years ago. I view these with some awe as I see the rates of discovery increase steadily. What a change this is from the years of the Palomar projects! An efficient vacuuming of the skies is underway, but I wonder if it is as much fun to those involved as the old-time astronomy I enjoyed so much. I rather doubt it. However, the feeling of accomplishment in a search that may someday save our home planet from great destruction will make up for it.

Impact remains an important focus of my attention with its many facets. Since the original hypothesis offered by Walter and Luis Alvarez in 1980 on asteroid or comet impact as a cause of the Cretaceous-Tertiary boundary event, there has been a gradual acceptance of impact cratering as new paradigm; our Earth bears the consequence of impact and some of the great extinctions of life have followed these. Large rocks and veritable mountains do fall out of the sky, a difficult concept for many geologists to accept; the effects of their impact on life are inescapable. The number of newly recognized impact craters increases by several each year. Some of them have been found as buried structures in the course of drilling programs. Gene and I found that although some characteristics of these structures are typical and almost always present, every structure is different and provides new knowledge about the complicated geologic effects of an impact. We've come a long way since the origin of Meteor Crater, Arizona was established and the whole study of impact began.

Planetary science as a field of knowledge, likewise, is growing steadily, and the more we know, the more questions arise to be answered. I have indeed been fortunate in adding a passion for science to my early enthusiasms. The learning process for both Gene and me was continual over our years together, and the challenge of trying to keep pace with developments on all fronts was a joy. Without Gene, I would never have known the excitement of planetary science nor have had the opportunities I did to work in that area; without me, he often said, his search for asteroids and comets and then the Australian cratering work would never have been attempted. Together, we could do more than either of us alone.

What more could I ask for as I go into the future? That question leads me to one more thought. Since the time of the accident in which I lost Gene, the awareness of our human need for others has dominated much of my thinking. Family and friends have been very important in my own recovery, and the concept of their importance, I discover, is as essential as the need for knowledge of our physical world. Without the human relationships we cherish, knowledge would count for naught; both are to be nourished. Henceforth, I'll continue my scientific exploration, knowing that I must not neglect the other side of living.

ACKNOWLEDGMENTS

I am indebted to the U.S. Geological Survey, Caltech and Palomar Observatory, NASA, and Lowell Observatory for providing institutional support to Gene's and my efforts over the years. A large number of friends from these institutions and throughout the world, as well as my family, immediate and extended, have helped make possible the memorable and happy life that Gene and I shared together.

Annu. Rev. Earth Planet. Sci. 1999. 27:19–53

NATURE OF MIXED-LAYER CLAYS AND MECHANISMS OF THEIR FORMATION AND ALTERATION

Jan Środoń
Institute of Geological Sciences PAN, Senacka 1, 31–002 Krakow, Poland;
e-mail: ndsrodon@cyf-kr.edu.pl

KEY WORDS: chlorite/smectite, illite/smectite, kaolinite/smectite, serpentine/chlorite, talc/smectite

ABSTRACT

Mixed-layer clay minerals are intermediate products of reactions involving pure end-member clays. They come from natural environments ranging from surface to low-grade metamorphic and hydrothermal conditions. Most often mixed layering is essentially two component, but more complicated interstratifications have also been documented. Variable tendency to form regular 1:1 interstratifications has been observed and explanations of this phenomenon have been proposed. Mixed-layer clays are either di- or trioctahedral; di/trioctahedral interstratifications are rare. Most mixed-layer clays contain smectite or vermiculte as a swelling component. Exceptions are all trioctahedral: serpentine/chlorite in low-temperature environments, and mica/chlorite and talc/chlorite at high temperatures. Solid state transformation and dissolution/crystallization are the two mechanisms responsible for the formation of different mixed-layer clays. In general, the weathering reactions that produce mixed layering are reversals of the corresponding high-temperature reactions, but the reaction paths are quite different. Weathering reactions alter smectite into kaolinite via mixed-layer kaolinite/smectite. Illite, chlorite, and micas react into mixed-layer clays involving vermiculite layer, then into vermiculite, and finally smectite. Interstratifications of smectite and glauconite, serpentine and chlorite, and smectite and talc are characteristic of early diagenesis and indicative of sedimentary environments. Three reactions involving mixed-layer clays—smectite to illite, smectite to chlorite, and serpentine/chlorite to chlorite—proceed gradually during burial diagenesis and are used for reconstructing maximum burial conditions, illite/smectite being the most useful tool. Rectorite, tosudite, talc/chlorite, and mica/chlorite are mixed-layer minerals

0084-6597/99/0515-0019$08.00

indicative of temperatures higher than diagenetic, characteristic of low-temperature metamorphism or hydrothermal alteration.

MIXED-LAYER VERSUS END-MEMBER CLAY MINERALS

Definitions and Nomenclature

Clay minerals are layer silicates: layers composed either of one octahedral and one tetrahedral sheet (1:1 layer) or of one octahedral sheet sandwiched between two tetrahedral sheets (2:1 layer) spread over tens to thousands of nanometers in the a* and b* directions. In the c* direction, the layers are stacked on top of each other, forming clay crystals a few to tens of layers thick. Layers are either electrically neutral or they have a negative charge resulting from isomorphic substitutions within the tetrahedral and/or octahedral sheet. The charged layers are bound by interlayer cations or sheets of interlayer hydroxide. The chemical bonds within layers are much stronger than between layers, which results in the plate or the lath habit of clay crystals. Variation among clay minerals, presented in Table 1 in the form of a classification scheme, reflects (*a*) the composition of layers (1:1 vs. 2:1 minerals), (*b*) the occupation of two or three out of three available cationic positions of the octahedral sheet (dioctahedral vs. trioctahedral minerals), (*c*) the layer charge and nature of the interlayer (talc + pyrophyllite vs. smectite + vermiculite + illite vs. chlorites), (*d*) isomorphic substitutions (e.g. montmorillonite versus beidellite), and (*e*) the mutual orientation of identical layers (polytypes, e.g. kaolinite versus dickite versus nacrite).

If crystals contain identical layers and interlayers, the strict periodicity of the structure in c* direction is preserved and we deal with a pure (end-member, discrete) clay mineral or a mixture of discrete clay minerals. If the strict periodicity is violated due to the differences in layer type or in the interlayer material, we deal with the phenomenon known as mixed layering (interstratification). Mixed layering is defined as random and characterized by the ordering parameter (Reichweite) equaling 0 (R0) if there is no preferred sequence in stacking of layers. If some sequences are privileged, such mixed layering is called ordered. Privileged alteration of single unit cells, i.e. ABABAB..., is known as R1 ordering; alteration of two cells of one type with one of another type, i.e. AABAAB..., produces R2 ordering, etc. See Moore & Reynolds (1997:174) for details of the ordering concept.

Theoretically, the ordered mixed layering of each type has a special case, when the characteristic pattern of interstratification is not only privileged but unique: only ABABAB... sequences are present in R1 clay crystals. Such

Table 1 Classification of layer silicates[a]

Layer type	Group	Subgroup	Species
1:1	Serpentine-kaolin $z \cong 0$	Serpentines (Tri)	Chrysotile, antigorite, lizardite, berthierine, odinite
		Kaolins (Di)	Kaolinite, dickite, nacrite, halloysite
2:1	Talc-pyrophyllite $z \cong 0$	Talc (Tri) Pyrophyllite (Di)	
2:1	Smectite $z \cong 0.2$–0.6	Trioctahedral smectites	Saponite, stevensite, hectorite
		Dioctahedral smectites	Montmorillonite, beidellite, nontronite
	Vermiculite $z \cong 0.6$–0.9	Trioctahedral vermiculites Dioctahedral vermiculites	
	Illite $z \cong 0.6$–0.9	Trioctahedral illites (?)	Ledikite (?)
		Dioctahedral illites	Illite, Fe-illite, glauconite, NH_4-illite, brammalite (?)
	Mica $z \cong 1.0$	Trioctahedral micas	Biotite, phlogopite, wonesite
		Dioctahedral micas	Muscovite, paragonite, phengite, celadonite
	Brittle mica $z \cong 2.0$	Dioctahedral brittle mica	Margarite
2:1	Chlorite z variable	Tri, Tri chlorites	Chamosite, clinochlore, etc.
		Di, Di chlorites	Donbassite
		Di, Tri chlorites	Sudoite, cookeite
		Tri, Di chlorites (?)	

[a]Modified from Moore & Reynolds (1997). z, charge/half-unit cell; Tri, trioctahedral; Di, dioctahedral; (?), minerals not sufficiently well documented to be recognized by AIPEA classification.

interstratification is called regular; such minerals (of R1 type only) exist in nature and are given separate names. Other names of mixed-layer clays are combinations of the components' names, e.g. illite/smectite (by informal convention we put first the component with smaller d_{001} spacing).

Identification and Quantification of Mixed Layering

Mixed layering in clays was discovered by means of X-ray diffraction (XRD) in the 1930s, soon after the crystalline nature of discrete clays was revealed by the XRD technique. Since then, XRD, supplemented by chemical analysis, has remained the major tool of mixed-layer clay research. Mixed-layer clay identification and quantification techniques, nomenclature, and the concept of ordering resulted from these studies. The subject is covered in detail by Moore & Reynolds (1997).

In the past 15 years, major new contributions to this field, including the concept of so-called fundamental particles, were provided by transmission electron microscopy (TEM). High-resolution TEM (HRTEM) of clay crystals, analyzed at parallel orientation to the electron beam, produces images of layers. If the experimental conditions are properly selected (Guthrie & Veblen 1989), the nature of the layers can be identified by measuring the layer spacings and obtaining the analytical electron microscopy (AEM) analyses of the bulk crystal. The major technical problem is posed by hydrated layers, particularly smectite, which dehydrate in the vacuum of the electron microscope and become indistinguishable from illite. Chemical treatment techniques for achieving stable, distinct layer spacings of smectitic clays under vacuum have been proposed (Środoń et al 1990, Vali & Hesse 1990). Unless such chemicals are harmless to the other component of the mixed-layer clay, artifacts can be expected (e.g. the opening of illite layers by the standard alkylammonium procedure).

TEM has also been used to attempt quantification of mixed layering. In contrast to XRD, which averages billions of crystals, TEM allows observation of individual crystals; thus the major problem is obtaining statistically valid numbers. It has been shown for illite/smectites (I/S) that if this problem is solved properly, layer ratios that closely match XRD data can be obtained from the analyses of approximately 100 mixed-layer crystals (Środoń et al 1990, Elsass et al 1997).

The Special Case of Clays Containing Expandable (Swelling) Interlayers

Clays containing expandable interlayers pose two additional identification and quantification problems. These problems have been identified for I/S, but the findings are applicable as well to other minerals with swelling interlayers.

OPERATIONAL VERSUS STRUCTURAL DEFINITIONS The first problem is created by the lack of full consistency between the structural (i.e. based on the nature of layers) and operational (i.e. based on the XRD $d_{(001)}$ spacing under specified conditions) definitions of smectite, vermiculite, and illite. Only the operational definitions can be used in XRD identification and quantification procedures. For example, Moore & Hower (1986) investigated Wyoming montmorillonite in Na form at 35% relative humidity and obtained the pattern of ordered interstratification of dehydrated and one water-layer smectite (9.6/12.4 Å). At these specific conditions, the clay widely used as a smectite standard gave characteristics of a mixed-layer clay. High-charge smectite, if K exchanged and in glycolated form, produces an XRD pattern indicating a three-component illite/vermiculite/smectite (10 Å/14 Å/17 Å) interstratification (Eberl et al 1986). At <10% relative humidity, I/S glycolated in Na form produces the patterns of illite/vermiculite (10 Å/14 Å) or illite/vermiculite/smectite, whereas Ca and Sr clays give illite/smectite characteristics (Eberl et al 1987). These examples indicate that the characteristics of a clay containing expandable layers are valid only for the specified exchangeable cation and relative humidity. In most cases, the interstratification can be kept simple (i.e. two-component) if divalent cations and high relative humidity are used.

MIXED-LAYER CRYSTALS VERSUS FUNDAMENTAL PARTICLES A different problem arises from the lability of mixed-layer crystals containing swelling interlayers, the phenomenon known for a long time as osmotic swelling. Frey & Lagaly (1979) showed that when Na is the exchangeable cation and extremely diluted suspensions of <0.2 μm fractions are used, both low- and high-charge smectite crystals can be delaminated into free individual monolayers and recombined by mixing the suspensions into a mixed-layer high-charge/low-charge smectite. In an analogous experiment, Nadeau et al (1984) produced from R0 and R1 clay one mixed-layer I/S of intermediate composition. They showed by TEM platinum-shadowing measurements that I/S crystals delaminate into particles that are either monolayers or sets of layers bound by interlayer K. The layers in such sets are oriented with respect to each other according to the hexagonal symmetry (single selected area diffraction pattern). Monolayers were found only in R0 clays, and they were called by the authors smectite fundamental particles. Sets of layers were called illite fundamental particles. Each I/S sample can be characterized by a specific distribution of fundamental particle thickness. The mean thickness increases with the percentage of illite layers measured by XRD.

Additional evidence of the lability of I/S crystals was provided by Clauer et al (1997), who, by supercentrifugation of dilute suspensions of I/S in Na form, were able to delaminate mixed-layer crystals and concentrate fundamental particles of different thickness in different fractions. Finally, Eberl et al (1998) succeeded in complete delamination of I/S in solid state. Such material, in which fundamental particles are all free and not associated into mixed-layer crystals, produces XRD characteristics of a fine-grained illite (Figure 1).

XRD measurements of smectite percentage are fully reproducible and characteristic of the air-dry state of the sample if the standard separation procedures of Jackson (1975) are used and the grain-size cut is not finer than 0.2 μm (Clauer et al 1997). Note, however, that the percentage of smectite obtained from this measurement corresponds to the limited size of the mixed-layer crystals, and thus is underestimated (Figure 2). If the crystals were infinitely thick, the edge effect would not interfere and the layer ratio measured by XRD would

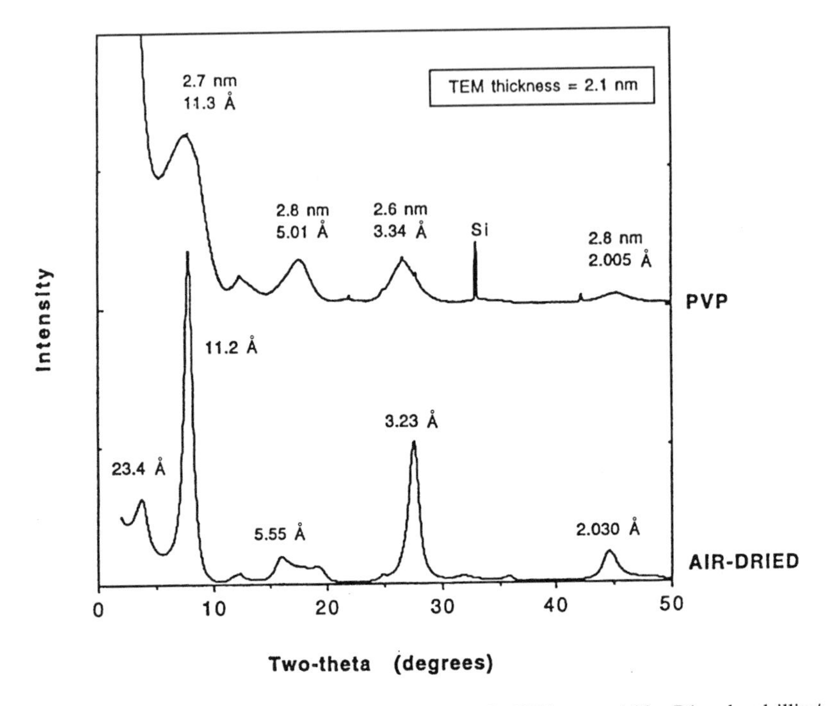

Figure 1 X-ray diffraction patterns of the same sample (45% expandable, R1 ordered illite/smectite) recorded intercalated with polyvinylpyrrolidone (characteristics of fine-grained illite: *upper curve*) and in air-dry state (characteristics of a mixed-layer clay with thick crystals: *lower curve*). This is an extreme case of lability of mixed-layer clays containing swelling interlayers (compare Figure 2). (From Eberl et al 1998.)

correspond precisely to the chemical and structural differences between smectitic and illitic components of the sample. This point is illustrated in Figure 3 by plotting fixed K content of I/S versus the percentage of smectite measured by XRD and calculated from the mean fundamental particle thickness (this calculation accounts for edge effects; see Środoń et al 1992). The nonlinearity of the first relationship is an artifact of the percentage-of-smectite measurement.

Layer Composition and Classification

During the 55 years of mixed-layer clay studies, seven regular 1:1 interstratifications have been discovered and given specific mineral names (Table 2). Five of them are trioctahedral and two are dioctahedral. One abundant mixed-layer clay, kaolinite/smectite (K/S), is not known to exist as a regular species. Several mixed-layer minerals with more than two types of component layers have also been documented.

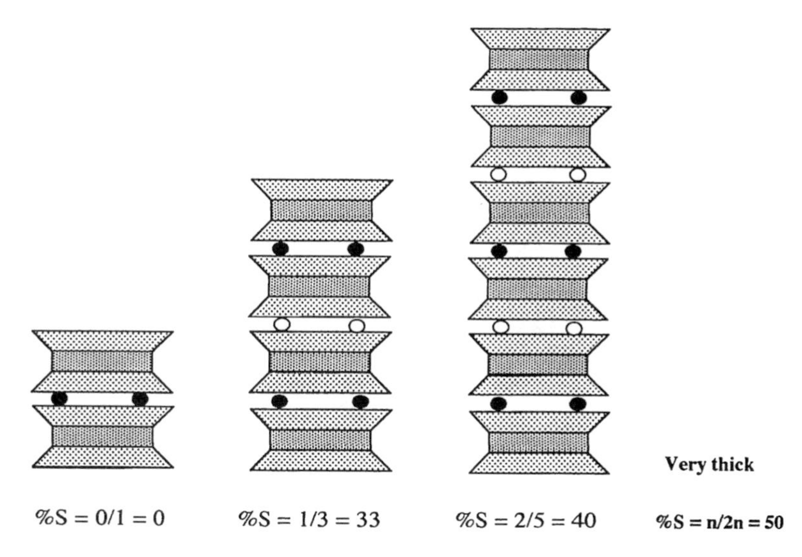

Figure 2 Effect of the limited thickness of mixed-layer crystals on the percentage of smectite interlayers (%S) measured by X-ray diffraction. (*Open circles*) smectitic interlayers; (*filled circles*) collapsed interlayers. Fundamental particles used in the model are two layers thick. In the extreme case of one fundamental particle/crystal, the material is fine-grained illite, as illustrated in Figure 1.

By far the most common in nature is I/S. Other mixed-layer clays are less ubiquitous, representing more exotic environments. Their origin and evolution in natural conditions are discussed below.

ORIGIN AND EVOLUTION OF MIXED-LAYER CLAYS IN NATURAL ENVIRONMENTS

Dioctahedral Mixed-Layer Clays

KAOLINITE/SMECTITE von Engelhardt & Goldschmidt (1955) were the first to observe interstratification of kaolinite with longer-spacing layers, and Sudo & Hayashi (1956) identified similar material as K/S. Altschuler et al (1963) described a weathering sequence of smectite into K/S. Hughes et al (1993) review most of the literature on K/S.

Natural occurrences of K/S are limited. The mineral is known from Recent soils developed in temperate to tropical climates and from the corresponding paleosols (Hughes et al 1993). A few papers report K/S from hydrothermally altered rocks (Sakharov & Drits 1973, Wiewióra 1973). If present in sedimentary rocks, K/S is most often a detrital component, as shown by Thiry (1981) in his study of Eocene clays from the Paris Basin. However, telogenetic (after basinal

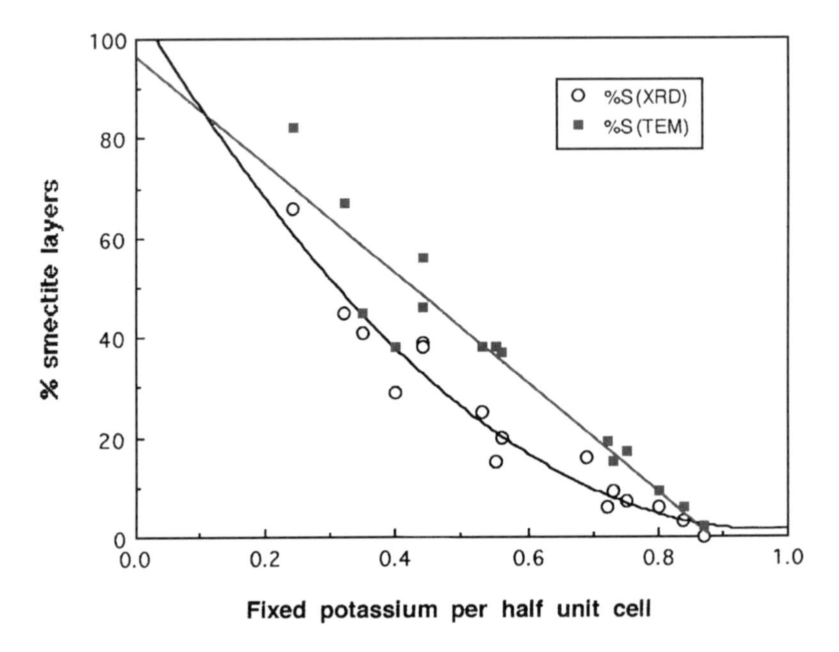

Figure 3 Trends in chemical composition of illite/smectites (fixed potassium) obtained using the percentage of smectite (%S) measured by XRD and TEM (data from Środoń et al 1992). The concave shape of the former curve is an artifact of the underestimation of %S by XRD (compare Figure 2).

cooling) origin of an exceptional ordered K/S has been indicated (Thomas 1989). Most probably, the abundance of K/S is highly underestimated because it is difficult to detect, in particular as a minor component, and has often been misidentified as halloysite (Delvaux et al 1990, Hughes et al 1993, Cuadros et al 1994). The data of Calvert & Pevear (1983) from a Gulf Coast well, as well as of Robinson & Wright (1987) and Wang et al (1996), who reported K/S associated with R1-ordered I/S, indicate that the mineral survives burial temperatures of at least 100°C.

Kaolinite/smectite forms in soils and weathering crusts developed on parent rocks of variable lithology, thus in a range of chemical environments from acidic (acid clays of Japan) to alkaline (calcrete of Robinson & Wright 1987). From the geological evidence, pH does not seem to be a controlling factor.

While Wiewióra (1973) and Sakharov & Drits (1973) propose crystallization of K/S from solution, most authors writing on the subject agree that K/S always evolves from smectite. K/S originating from kaolinite has not been

Table 2 Classification of regular (ABABAB...) interstratifications[a]

Type of Octohedral Sheet	Name	Component Layers
Trioctahedral	Aliettite	Talc/smectite
	Corrensite, low charge	Chlorite/smectite
	Corrensite, high charge	Chlorite/vermiculite
	Dozyite	Serpentine/chlorite
	Hydrobiotite	Biotite/vermiculite
	Kulkeite	Talc/chlorite
Dioctahedral	Rectorite K	Illite/smectite
	Rectorite Na	Paragonite/smectite
	Tosudite	Chlorite/smectite

[a]Modified from Moore & Reynolds (1997).

reported. Thus, if smectite is not present in the parent rock, chemical conditions favoring its formation must precede the conditions favoring kaolinite formation. Often, it means poor drainage followed by better drainage.

Środoń (1980) asked why smectite sometimes changes directly into discrete kaolinite and why the process sometimes proceeds via a K/S stage. Direct alteration is well documented, for example, from the bentonites and tonsteins. Both processes have been reproduced in the laboratory (Środoń 1980 and references therein; Delvaux et al 1989), but the experiments concerning the factors that control the reaction pathway are inconclusive.

Środoń (1980) suggested that K/S forms by the dissolution of smectite layers and the crystallization of kaolinite layers. Thus the two pathways of kaolinization of smectite differ by the site of kaolinite nucleation: within versus outside smectite crystals. A dissolution-crystallization mechanism was supported against layer transformation mechanisms by Delvaux et al (1989). Using electron spin resonance spectroscopy, Delvaux et al discovered that kaolinite layers in K/S originating from nontronite (both natural and synthetic K/S) have very low iron content of approximately 1% Fe_2O_3. Earlier, two versions of the layer transformation mechanism were proposed: stripping of one of the tetrahedral sheets of the 2:1 layer (Altschuler et al 1963, Hughes et al 1993), and inversion of one of the tetrahedral sheets in Al-hydroxy interlayered smectite (Schultz et al 1971, Brindley et al 1983a). Hydroxy-interlayered smectite has not been observed in association with natural K/S (Hughes et al 1993), but in some hydrothermal experiments such a starting material reacts into K/S (Środoń 1980, Delvaux et al 1989). If K/S is hydrothermally treated with excess Al present as discrete kaolinite, the reaction proceeds in the opposite direction, i.e. toward a more smectitic composition (Proust et al 1990). The role of Al seems crucial for the formation of K/S. If excess Al is not supplied and all other parameters of the hydrothermal experiment are kept the same, smectite reacts not into K/S but into I/S (Środoń 1980).

Hughes et al (1993) speculated on the role of structural Fe (iron-rich parent smectite) in promoting the K/S pathway of kaolinization. However, it seems that the correlation is accidental, resulting from the fact that smectites formed in soils are most often Fe-beidellites. At least four of the reported K/S series contain smectite layers of low or negligible Fe content (Shinoyama et al 1969, Schultz et al 1971, Wiewióra 1973, Brindley et al 1983a).

Schultz et al (1971) and Brindley et al (1983a) observed an intriguing deficiency of OH in K/S when they analyzed their thermogravimetric data against the kaolinite:smectite ratio established by XRD. The problem still remains unsolved, and at least three hypotheses can be offered: (*a*) kaolinite layers are defective; (*b*) undetected discrete smectite was present in the sample; or (*c*) the 17 Å/7 Å random model (glycolated smectite/kaolinite) used for XRD quantification was incorrect.

All reports on K/S, except that of Thomas (1989), indicate random mixed layering. All 2:1 layers other than smectitic, if present, are a minority. K/S clays developed on illitic bedrock exhibit swelling characteristics, though large heterogeneity of the 2:1 component is suspected (Hughes et al 1993). Kaolinization of micas proceeds by topotaxial growth of thick kaolinite crystals, as observed frequently by optical petrographers and confirmed by HRTEM (Jiang & Peacor 1991). The opposite process, diagenetic illitization of kaolinite, produces not mixed layering but discrete illite crystals. In K-deficient low-grade metamorphic environments, kaolinite reacts with quartz into pyrophyllite. Mixed-layer intermediate phases have not been reported.

ILLITE/SMECTITE The term illite/smectite (I/S) is used in this review in the strict sense: for interstratification of potassium illite with smectite. Other dioctahedral interstratifications of smectite with a 10 Å component are covered separately.

In the first period of investigation, I/S was identified and quantified as randomly interstratified clay by means of the mixing functions of Hendricks & Teller (1942). The clay was found in bentonites, hydrothermally altered rocks, soils, and sediments. Two former lithologies, providing monomineral samples, were studied extensively, and ordering was recognized and quantified using the direct Fourier technique of MacEwan (review of this early literature in Cole 1966). The chemistry of I/S was studied systematically by Hower & Mowatt (1966). After Weaver's study of mixed-layer clays in sedimentary rocks (1956), it became evident that illitization of smectite progressing with depth in sedimentary basins is a universal phenomenon. The process was related to petroleum occurrences (Burst 1969) that generated the interest of the oil industry. Perspectives of dating diagenesis via K-Ar measurements of I/S were provided by Weaver & Wampler (1970). Reynolds (1967) presented the technique of quantifying both layer ratio and ordering by computer simulation of the complete diffraction function. Shutov et al (1969) and Perry & Hower (1970) discovered a simultaneous increase of ordering and illite percentage with depth, and numerous studies of other basins followed. Eberl (1978) simulated in the laboratory all paths of dioctahedral smectite alteration. Środoń & Eberl (1984) provide a review of this period of research.

Bulk chemical evolution and the nature of the component layers These aspects were studied in detail for materials originated by diagenetic and hydrothermal alteration of pyroclastics, which provide pure monomineral samples. Meunier & Velde (1989) and Środoń et al (1992) analyzed several sets of published data by different techniques and came to the same conclusion: the charge of illitic interlayers is stable—approximately $0.9/O_{10}(OH)_2$—and different from that of muscovite. Thus the K_2O content can be used for calculating the mean thickness of illite fundamental particles. In some I/S, substitution of K^+ by NH_4^+ (Nadeau & Bain 1986) must be accounted for in such calculations.

Both studies indicated that the charge of smectitic interlayers of I/S varies for different data sets from about 0.3 to 0.6 per $O_{10}(OH)_2$ but seems not to evolve in the course of illitization. This result is surprising and deserves further verification because at the R0/R1 transition smectitic monolayers disappear and only contacts between illite fundamental particles remain as expandable (i.e. retaining the smectitic character) interlayers.

The bulk composition of silicate layers evolves during illitization: Al^{IV} and Al^{VI} increase and Fe and Mg decrease (Środoń et al 1992).

These chemical characteristics of I/S from altered volcanoclastics are not fully compatible with the conclusions drawn by the students of shale diagenesis, who are not able to separate their I/S from other minerals. Different lines of evidence, such as XRD pattern reaction to various chemical treatments (Sato et al 1996), direct Fourier transform (Foscolos & Kodama 1974), precise computer modeling of XRD patterns (Drits et al 1997), or catalytic activity (Johns & McKallip 1989), are used to argue that expandable layer charge evolves in the course of illitization from smectitic to vermiculitic. Thus I/S samples are really three-component systems. However, this opinion is incompatible with linear correlation between nonexchangeable K_2O and cation exchange capacity (CEC) observed by Foscolos & Kodama (1974). The problem has practical consequences. The percentage of swelling interlayers in a clay sample is an important property for geotechnics and borehole geophysics, and it could be calculated from CEC measurement if the expandable layer charge density evolution during diagenesis was predictable.

Interstratification pattern From the current perspective of TEM results, it is clear that by direct Fourier transform and then by modeling XRD patterns, Drits and his colleagues correctly identified the gradual evolution from random to long-range ordering without any privileged steps (Shutov et al 1969, Drits & Sakharov 1976). For a long time, however, the transition from random into R1 and then R3 pattern of ordering, with omission of R2 (Reynolds & Hower 1970), has been widely acknowledged because it was consistent with the accepted transformation mechanism of illitization (see below). Nadeau et al (1984) pictured mixed-layer crystals of I/S as stacks of fundamental particles of different thickness with swelling interfaces. They showed that in the course of illitization, distributions of thicknesses of fundamental particles evolve gradually without any privileged steps. Thus the evolution of ordering must be a continuous process: R1 ordering is dominant if 2-nm particles dominate the population, R2 ordering if 3-nm particles are the most frequent, etc. This viewpoint has been fully supported by the analysis of a large TEM data set (Środoń et al, in press). Regular interstratification (K-rectorite) is rare.

Tridimensional organization It was established early that the bulk organization of I/S evolves from turbostratic (random rotations and translations of layers) smectite structure, via 1Md (1M disordered) with decreasing amounts of turbostratic disorder, into 1M and $2M_1$ polytypes of pure illite (review in Środoń & Eberl 1984). Veblen et al (1990) interpreted cross fringes in their HRTEM photos as indicating that tridimensional order is characteristic for mixed-layer crystals, i.e. it continues across the expanding interlayers. In a carefully designed XRD experiment supplemented by computer modeling, Reynolds (1992) demonstrated convincingly that only illite fundamental particles are characterized by tridimensional order (1M polytype plus rotational faults n120° and n60°), whereas swelling interlayers display random faults. Thus the percentage of turbostratic character is proportional to the percentage of smectitic layers. An additional complication results from two types of vacancies: *trans* and *cis*, discovered in the octahedral sheet of illite. Both the nature and density of rotational faults and vacancies have been studied, but so far no clear correlations with the illitization process have appeared (Drits et al 1996, Moore & Reynolds 1997 and references therein).

Smectite illitization mechanism The problem of the smectite illitization mechanism has been vigorously debated from several standpoints over the past 30 years. Because of the abundance of this mineral, the illitization mechanism is helpful for understanding the evolution of pore water chemistry and sandstone cementation during diagenesis (Boles & Franks 1979). Other major fields of application are the interpretation of K-Ar dates and the use of percentage of expandability as a paleotemperature indicator.

The first aspect of the problem is purely chemical. Hower et al (1976) wrote the illitization reaction as:

$$K + Al + \text{smectite} \rightarrow \text{illite} + Na + Ca + Fe + Mg + Si + H_2O. \quad (1)$$

This reaction is nearly isovolumetric (approximately the same amount of elements lost and gained) and would lower the pH of the pore fluid. An alternative proposed by Boles & Franks (1979) assumes cannibalization of smectite:

$$K + \text{smectite} \rightarrow \text{illite} + Na + Ca + Fe + Mg + Si + O + OH + H_2O. \quad (2)$$

This reaction decreases the mass of clay by more than 30%, produces large amounts of quartz, and raises pH.

Many have argued since in favor of one or the other mechanism, but a convincing conclusion has not been reached. The difficulty comes from the unsolved problem of precise quantification of an I/S component in shales, which would be the most straightforward approach to test the alternative. In addition, the problem posed by Weaver & Beck (1971), i.e. whether potassium comes from within the shale or whether it migrates into it, is still open to debate.

The second aspect of the illitization mechanism is physicochemical: how do illite crystals form from smectite crystals? The first idea, promoted by Hower et al (1976) (but originating in earlier laboratory experiments and field observations of mica weathering by soil scientists), was a solid-state layer-by-layer transformation mechanism: the charge of the smectite interlayer increases due to Al substitution in the neighboring tetrahedra, potassium is fixed, and smectite converts to illite. The polarization effect on neigboring interlayers prevents them from following the lead, and R1 ordering emerges by this mechanism. After completion of the R1 sequence, if the remaining smectite interlayers begin collapsing, the R1 structure (ISIS) evolves directly into R3 (IIIS). R2 (IIS) is not allowed by this model.

This mechanism was challenged by Nadeau et al (1985). To explain the continuous evolution of fundamental particle thickness distributions, they assumed that smectite particles dissolve and illite particles nucleate and grow. Nadeau (1985) presented clear evidence of crystal growth, a positive correlation between fundamental particle thickness and ⟨ab⟩ area, and speculated on the possibilities of one-, two-, and three-dimensional growth mechanisms.

Inoue (1986) analyzed the ⟨ab⟩ dimensions of fundamental particles from a complete sequence of hydrothermal I/S and observed a morphological evolution: irregular flakes at the R0 stage of illitization, followed by development of laths and then hexagons. He explained his data assuming a layer transformation mechanism at the R0 stage followed by dissolution/precipitation plus growth at the R > 0 stage.

A decade ago, Eberl & Środoń (1988) and Inoue et al (1988) suggested from apparent steady state shapes of reduced distributions of fundamental particle dimensions that Ostwald ripening is the mechanism driving illitization, but it was soon realized that the distribution shapes are lognormal and thus different from known shapes resulting from Ostwald ripening (Eberl et al 1990). All conflicting ideas were recently discussed in detail by Inoue & Kitagawa (1994) and Altaner & Ylagan (1997).

Recently it was found that lognormal distributions of fundamental particle thickness evolve in a unique way in the course of illitization (Środoń et al, in press). Detailed modeling excludes Ostwald ripening as a plausible hypothesis and indicates the following sequence of events (Figure 4): nucleation of 2-nm illite particles at the incipient stage of the process, continuing nucleation and simultaneous one-dimensional growth until all smectite layers are consumed (about 40% smectite), and pure tridimensional growth at the subsequent stage. Smectite layers dissolve, providing nutrients for illite fundamental particles and the templates for their crystallization, but they do not serve as nuclei of the illite crystals. The process can be even more complicated if the nucleation and growth of illite also takes place outside of smectite crystals: a mixture of I/S and discrete illite should be the product, and such material has been characterized by HRTEM (Amouric & Olives 1991).

Controls of the illitization reaction and its use as a paleogeothermometer
The problem of the controls of the illitization reaction was posed by Eberl & Hower (1976), who interpreted their laboratory data on illitization of smectite cautiously, stating that they do not exclude neither kinetic nor thermodynamic control of the reaction steps. Hoffman & Hower (1979) discussed different factors that may affect the reaction product (temperature, time, pressure, availability of K, and other elements) and concluded that for shales, temperature is by far the most important factor. They calibrated an I/S paleogeothermometer for the Tertiary and Late Mesozoic basins: onset of illitization at 60°C; R0/R1 transition at approximately 100°C; and appearance of long-range ordering at 165°C, of illite at 210°C, and of $2M_1$ mica at 275°C. Weaver (1979) also did not detect the effect of time on the illite:smectite ratio, but McCubbin & Patton (1981) presented a kinetic interpretation of the available field data, stressing the time factor.

Since then, new arguments have been added to both approaches. Pytte & Reynolds (1989), Velde & Vasseur (1992), and Huang et al (1993) published alternative formulations of the kinetic equation, making it more flexible by introducing parameters such as reaction order, inorganic fluid chemistry, and

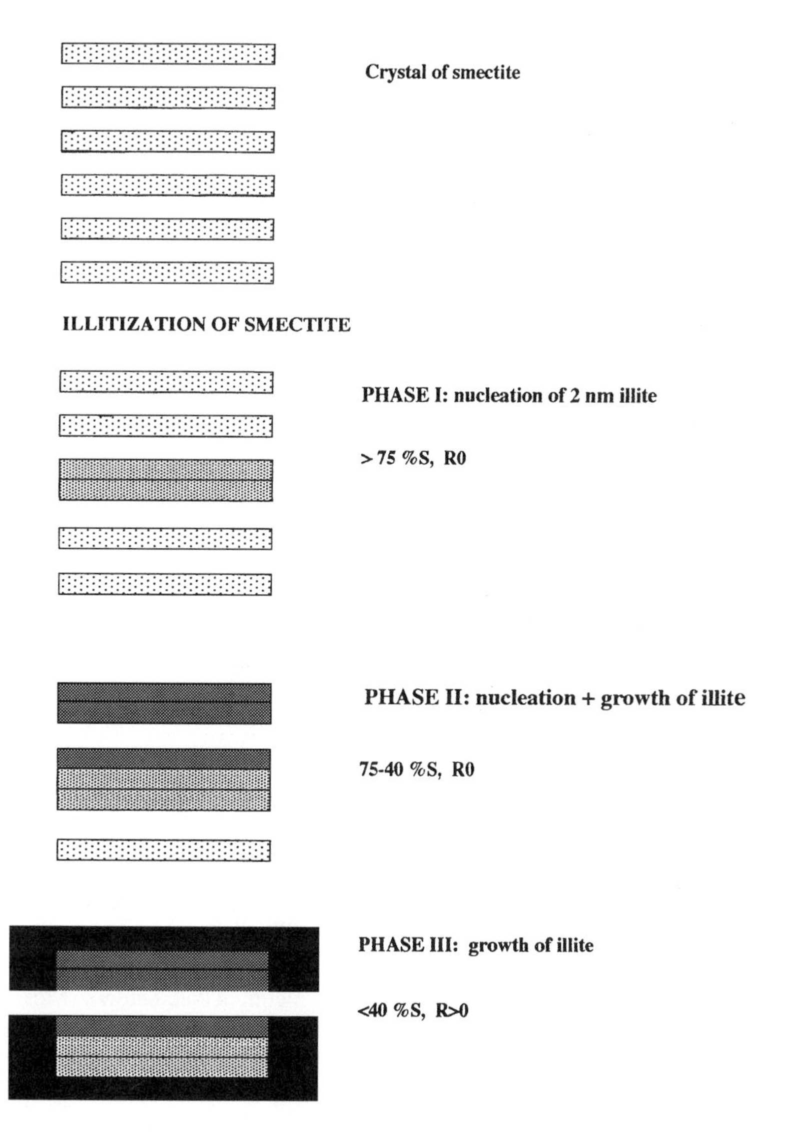

Figure 4 Model of the mechanism of illitization of smectite resulting from the analysis of fundamental particle thickness distributions (Środoń et al, in press). %S, percentage of smectite.

variable activation energies of random and ordered clays. Further complications emerged from the work of Small (1994), who found in laboratory experiments that the rate of illitization strongly depends on the presence of organic acids buffering the H/K ratio in the mica stability field. The extreme formulation is that of Wei et al (1996), which uses a distribution of activation energies for the starting smectite. In this approach, I/S is treated analogically to vitrinite and used as another technique in basin modeling.

Pollastro (1993) and Clauer et al (1997) summarized data from several sedimentary basins of well-known burial histories, ranging from Tertiary to Precambrian, and found consistent relationships between the illite:smectite ratio and maximum paleotemperatures, very close to the original scheme of Hoffman & Hower (1979).

Both sides of this ongoing dispute agree that illitization during burial is a long-lasting process. Direct evidence was provided by K-Ar dating of I/S separated into classes of fundamental particles (Clauer et al 1997).

Low-temperature illitization of smectite It has been recognized in the soil literature since the 1930s that smectites can fix limited amounts of potassium or ammonium irreversibly when submitted to alternating wetting and drying cycles. Eberl et al (1986) showed that the product of this reaction is mixed-layer I/S. At low and neutral pH, fixation involves only rearrangement of layers; it does not involve chemical reaction. Thus perfect juxtaposition of octahedral cavities across the interlayer space is achieved and potassium cations are locked. At elevated pH, silicate dissolution takes place and illitization advances. Eberl et al (1993) produced ordered I/S using this technique.

Numerous authors (e.g. Berkgaut et al 1994, Righi et al 1995) used a wetting/drying mechanism to explain their data from surface environments, but no undisputable evidence has been offered so far. The impact of this process on the mineral composition of sediments remains to be evaluated.

Reaction of illite toward smectite In 1937, Bray suggested that the illite reaction toward smectite takes place in soils. Several authors tried to reproduce it in the laboratory using different techniques of extracting potassium from illite, by analogy to the alteration of biotites to vermiculites (see below). Most often, the K-depleted layers displayed vermiculitic characteristics, but Tomita & Dozono (1972) produced ordered I/S very close to K-rectorite by treating dehydroxylated $2M_1$ sericite with sodium tetraphenylboron at room temperature.

The evidence offered by soil studies remains unclear. What is the sequence of steps leading from illite to smectite? Is I/S mixed layering an intermediate step of this reaction or does the reaction lead first to soil vermiculite and then in some instances to smectite? The answer to these questions is of primary importance for evaluating detrital input into sedimentary basins.

GLAUCONITE/SMECTITE Thompson & Hower (1975) showed that minerals in monomineral glauconite pellets are iron-rich analogues of I/S. A glauconitization reaction proceeds through the same steps of random and ordered interstratifications as illitization. The 10 Å layer draws its charge from octahedral Mg and tetrahedral Al substitution (as does illite) but it differs by a high percentage of octahedral Fe prevailing over Al. The end product of this process is nonexpandable clay mineral glauconite, again like illite differing from corresponding mica composition (celadonite) by lower K content. The glauconite layer seems to have a lower K occupancy than does illite, which Thompson & Hower (1975) attributed to stronger bonding due to higher tilt angle of the O-H, produced by high Fe content.

Excess occupancy of the octahedral sheet, common in glauconite/smectite (G/S), was attributed to hydroxy interlayers (Thompson & Hower 1975) based on the XRD characteristics on heating. It follows from the HRTEM study of Amouric et al (1995) that this chemical characteristic of some samples may result from interlayering with berthierine (iron serpentine). HRTEM observations indicate that, as during illitization, alteration of nontronite via mixed-layer G/S may be accompanied by crystallization of discrete glauconite crystals (Buatier et al 1993). A dissolution/precipitation mechanism of the alteration was inferred from Mössbauer studies (Buatier et al 1993).

Complete information on the mechanism of the glauconitization process and its geological implications is presented in the monograph edited by Odin (1988). Glauconitization is a low-temperature diagenetic phenomenon that takes place at the sediment-water interface. Iron and K content increase during this process, which requires approximately 1 Ma to reach the stage of pure nonexpandable glauconite. Glauconitization is accompanied by the dissolution of preexisting silicate phases, which become undetectable by K-Ar dating when the level of approximately 7% K_2O is reached.

Burial stops the glauconitization process. Data are lacking on the evolution of G/S during burial diagenesis.

Weathering of glauconite seems to reverse the glauconitization process: Fe and K are preferentially lost and more smectitic mixed layering appears before the material reacts into smectite, kaolinite, and iron oxides (Courbe et al 1981). In the laboratory, potassium can be removed from glauconites more easily than it can be removed from illites, producing expandable minerals with overall smectitic or vermiculitic characteristics (Robert 1973). More detailed studies of the process seem to be lacking.

NH_4-ILLITE/SMECTITE Pure ammonium illite has been described from numerous locations in close association with coals or organic-rich rocks of anthracitic grade, which were the source of NH_4 (Šucha et al 1994 and references therein). A unique occurrence of NH_4-I/S in Harghita Mountain volcanics of the Carpathians was characterized (Bobos et al 1995). The clays were found in pneumatolytic chimneys, associated with tourmaline. A complete range of mixed layering, with ordering analogous to potassium clays, was documented. The problem of the origin of NH_4 in this occurrence has not been solved.

NA(CA)-ILLITE SMECTITE Na(Ca)-I/S is reported in literature only as a regular variety, rectorite (alternatively as allevardite in older literature). Ordered 20% smectite Na-I/S in association with Na-illite was documented recently in the products of submarine hydrothermal alteration of basalts (B Jacquemont, J Honnorez, J Środoń, unpublished observations). The component layers of rectorite were identified as beidellite and paragonite (Brown & Weir 1963). The term brammalite seems more appropriate for the illitic component because the fixed-cation content is close to that of illite (0.9 per $O_{10}(OH)_2$) and not mica. Matsuda et al (1997) characterized samples containing dominant proportions of fixed Ca (margarite-like layers). Rectorite crystals are always very thick, as evident from sharp diffraction peaks and HRTEM photos (Matsuda et al 1997).

The geological occurrence of rectorite is restricted to hydrothermal alteration zones and low-grade metamorphic rocks. In both cases, it is associated with pyrophyllite, which indicates crystallization temperatures of approximately 280°C (Paradis et al 1983, Daniels & Altaner 1990), and occasionally with cookeite (Miser & Milton 1964, Nishiyama et al 1975, Daniels & Altaner 1990).

Natural occurrences of Na-I/S seem to correspond closely to the laboratory results of Eberl (1978). He observed that Na-smectite and glasses of this composition start reacting at much higher temperatures than those at which K equivalents react, that the rectorite composition is stable, and that rectorite-pyrophyllite or rectorite-tosudite assemblages are formed.

PARAGONITE/PHENGITE Frey (1969) suggested the presence of paragonite/phengite mixed layering in low-grade metamorphic shales from Alpine nappes. Using HRTEM and XRD to reexamine the samples, Livi et al (1997) were able to confirm only the intergrowth of thick paragonite and phengite crystals, which were detectable by AEM. They concluded that both XRD and HRTEM are incapable of distinguishing solid solution from interlayering of two 10 Å minerals. Other investigators of low-temperature metamorphic facies report only submicroscopic intergrowths of paragonite and phengite (Shau et al 1991 and references therein).

DIOCTAHEDRAL CHLORITE/SMECTITE Mixed-layer clay of dioctahedral chlorite/smectite composition is known from hydrothermal, pneumatolytic, and low-temperature metamorphic alteration zones. It was discovered by Sudo (1954), later named tosudite in his honor, and most of the early work was done in Japan (reviewed in Merceron et al 1988). The clay is always a regular interstratification of beidellite and di-dioctahedral chlorite (donbassite) or di-trioctahedral chlorite: sudoite (Mg rich) or cookeite (Li rich). In the laboratory hydrothermal experiments, only regular interstratifications were produced (Eberl 1978).

Tosudite has been found in associations with dioctahedral chlorite, serpentine, rectorite, and pyrophyllite, which suggests formation during acid hydrothermal alteration, when the temperature decreases well below 350°C (Merceron et al 1988). An exceptional report is that of Morrison & Parry (1986), who found tosudite in Permian redbeds and suggested its formation at approximately 100°C from percolating acid pore waters.

SOIL MIXED-LAYER CLAYS WITH AL-HYDROXY INTERLAYERS Mixed-layer clays with Al-hydroxy interlayers are a complex group of soil minerals recognized by MacEwan (1950) as containing incomplete dioctahedral Al chlorite layers. The identification was based on incomplete collapse to 10 Å during heating tests and was later confirmed by extraction of the interlayer Al-hydroxy material. Upon swelling tests, such layers display characteristics of vermiculite or smectite. Apparently, incompletely developed Al-hydroxy interlayer may not prevent the layer from swelling, although it increases the thermal stability of the material. The clays of this type were called soil vermiculites, soil chlorites, chlorite intergrades, or swelling chlorites. The first period of research is well summarized in a review by Rich (1968), and more complete information is provided by Barnhisel & Bertsch (1989).

Most often, the behavior of only the strongest XRD reflection (at approximately 14 Å in air-dry state) was tested; thus the data do not support a distinction between mixed layering and a homogenous mineral with the interlayer properties intermediate between chlorite and vermiculite (or smectite). In a more recent paper, Bain et al (1990) reported a careful XRD study of numerous samples from acid soils of Scotland. They identified mixed-layer illite/Al-hydroxy vermiculite derived from dioctahedral micas. The mineral is ordered, with clearly developed superlattice reflection, and its quantity increases toward the top of the soil profile, but the layer ratio seems stable. The amount of Al-hydroxy interlayering seems to be proportional to the pH of soil solution.

Al-hydroxy interlayering is believed to form at the level of the soil profile (B horizon) where organic complexes transporting Al decompose and pH increases to moderate values, causing hydroxyaluminium precipitation (April et al 1986). Olson (1988) suggested that soil vermiculites may play an important role in buffering Al in soil solutions.

Formation of soil vermiculites is a massive phenomenon in temperate climates, and this material has been identified in fresh sediments (Rich 1968). Its further evolution during diagenesis remains obscure.

PYROPHYLLITE/SMECTITE To my knowledge, pyrophyllite/smectite has not been reported in nature, but a series of compositions was produced by a hydrothermal treatment at >300°C of a montmorillonite + $AlCl_3$ (Eberl 1979). The reaction proceeded via ordered interstratifications, but a 1:1 mineral did not appear.

Trioctahedral Mixed-Layer Clays

SERPENTINE/CHLORITE Interstratification of 7 Å and 14 Å layers has been difficult to detect by XRD because of the coincidence of the peak positions. Dean (1983) interpreted the broadening of odd chlorite reflections as an indication of a 7 Å interstratified component, and Ahn & Peacor (1985) presented HRTEM evidence of 7 Å layers interstratified in chlorite from the Gulf Coast. Since then, serpentine/chlorite (Sp/C) has been recognized more often, and XRD techniques for the quantification of layer ratios were developed (Hillier & Velde 1992, Reynolds et al 1992). American literature on the topic was reviewed by Moore & Reynolds (1997).

Most of the present information on Sp/C comes from sedimentary rocks and, in particular, from the sandstone grain coating studies. Invariably, Sp/C is recognized as an iron-rich mineral. The 14 Å component is identified as chamosite and the 7 Å component is probably most often berthierine (Hillier 1994), though odinite (Reynolds et al 1992) and kaolinite (Hillier & Velde 1992) were also suspected. To complicate things, Amouric et al (1995) identified material similar to odinite, i.e. di-trioctahedral Fe-serpentine defined by Bailey (1988), as mixed-layer Fe-kaolinite/berthierine. Bailey et al (1995) recognized regularly interstratified Sp/C and named it dozyite. Ryan & Reynolds (1996) found that Sp/C interstratification is accompanied by interstratification of polytypes (1b0/1a).

Sandstone coating studies indicate a gradual decrease of the 7 Å component in chlorite with depth (Reynolds et al 1992, Ryan & Reynolds 1996, Hillier 1994). According to Hillier (1994), 7 Å layers disappear completely at 150°–220°C. Lee & Peacor (1983) observed that 7 Å layers disappear at the transition zone between shales and slates, and interpreted the reaction mechanism from their HRTEM observations as the lateral layer replacement of two serpentine layers by one chlorite layer. Amouric et al (1988) confirmed Sp/C interlayering in low-grade metamorphic shales but warned that lateral transition of 14 Å into 7 Å layers may be an imaging artifact.

A complete series of mixed layering from serpentine to chlorite has not been reported; thus it is currently unclear whether pure serpentine or Sp/C is a starting composition. Odin (1988) identified both 7 Å and 14 Å iron-rich minerals as components of Recent verdine facies sediments in tropical shallow seas; either they crystallize simultaneously or chloritization starts at the sea bottom. Hornibrook & Longstaffe (1996) interpreted their stable isotope results as indicating a berthierine-chamosite reaction in brackish-freshwaters at $<70^\circ$C.

Amouric et al (1995) documented an alternative evolution of green pellets in the verdine facies: Fe-kaolinite → Fe-serpentine → glauconite. According to the data of Odin (1988), it indicates deepening of the sea because the glauconite facies is always found in deeper sediments than the verdine facies is found. These relationships seem to indicate that the presence of Sp/C interstratification in sedimentary rocks is indicative of shallow tropical-water sedimentary conditions. However, retrograde formation of Sp/C from chlorite at high temperatures in a massive sulfide deposit has been inferred from HRTEM observations (Jiang et al 1992).

CHLORITE/SMECTITE (VERMICULITE) Chlorite/smectite(vermiculite) (C/S) was first identified as a regular species and called corrensite (Lippman 1954). It was soon recognized as a common mineral of hypersaline facies of Central Europe and North America (Bodine & Madsen 1985 and references therein), in hydrothermally altered basalts and ultrabasic rocks, in burial diagenetic sequences of volcanoclastics and graywackes, and in contact metamorphic zones (Reynolds 1988 and references therein; Drits & Kossovskaya 1990 and Russian references therein). Kübler (1964) first realized that the process operating during burial is the alteration of corrensite into chlorite, and Hoffman & Hower (1979) proposed to use the appearence of corrensite as a paleotemperature indicator. Jackson (1963) suggested the opposite reaction: removal of every second brucite sheet from chlorite during weathering. Johnson (1964) reported corrensite in a soil profile developed on chlorite-bearing rocks. Reynolds (1988) presented a review of C/S including details of its XRD identification.

Interstratification pattern Kimbara & Sudo (1973) realized early that chloritization is not fully analogous to illitization of smectite: instead of a continuous process, overlapping succession of three phases was observed. Inoue et al (1984) presented excellent documentation of the phenomenon. From the slight migration of peak position they concluded that the three coexisting phases are C/S with $<20\%$C, corrensite, and C/S $> 85\%$C. This interpretation is pursued further with the argument that both low-chlorite and high-chlorite phases may indeed be pure end-member minerals mixed with dominant corrensite (Reynolds 1988).

Careful modeling of XRD patterns coupled with HRTEM observations will be needed to verify these speculations. The HRTEM data published so far do not seem fully conclusive: while they confirm the presence of thick crystals of corrensite often intergrown with thick crystals of chlorite, Shau & Peacor (1992) and Bettison-Varga & Mackinnon (1997) produced clear HRTEM evidence of irregular (including random) chlorite/smectite interstratification. Whether this is an exception or the rule remains to be seen.

Bulk chemical evolution and the nature of the component layers Many authors examined chemical aspects of the smectite-to-chlorite evolution, recasting chemical analyses into structural formulae by different techniques. This procedure leads to contradictory conclusions concerning octahedral Al. Inoue (1985, 1987) reported opposite trends when recasting the same set of data by two different techniques. The reproducible pattern that emerges from these studies is a decrease of Si and interlayer exchangeable cations, and an increase of tetrahedral Al in the course of chloritization. The trends are continuous, despite apparently noncontinuous mixed layering. Octahedral cation ratios do not seem to evolve in a systematic manner; they seem to reflect local bulk chemistry.

In burial and hydrothermal sequences, the starting component is identified as saponite (e.g. Inoue 1985, Schiffman & Fridleifsson 1991). Published XRD data seem to indicate that expandable layers preserve their smectitic character during chloritization, i.e. corrensite is low-charge corrensite according to AIPEA nomenclature (Table 2). However, Drits & Kossovskaya (1990) reported exceptions to this rule. The problem of possible evolution of smectitic layer charge and occupancy (incomplete brucite sheets) in the course of chloritization has not been studied in detail.

C/S as a paleogeothermometer The appearance and disappearance of corrensite (identified by the presence of superlattice reflections), which are the easiest points of the smectite chloritization reaction to identify, have been traced with respect to temperature. In active hydrothermal systems, both subaerial and submarine, the first appearence of corrensite was reported between 150° and 200°C, and its presence was noted up to temperatures of approximately 300°C (Schiffman & Fridleifsson 1991 and references therein; Shau & Peacor 1992). In a submarine hydrothermal vent site, where the complete reaction was observed within fewer than 100 m, a 300°C crystallization temperature of end-member chlorite was measured by oxygen isotopes (quartz-chlorite pair) (Buatier et al 1995).

The 150°–300°C range established by these data is coincident with the 130°–300°C range in which Cathelineau (1988) observed a linear increase of tetrahedral Al in chlorites from the Los Azufres and Salton Sea geothermal fields. Schiffman & Fridleifsson (1991) offer a convincing interpretation that a C/S series was misinterpreted as chlorite and warn that the exact value of Al^{IV} in C/S depends on the technique of calculating structural formulae.

Burial diagenetic studies indicate lower temperatures at the first appearance of corrensite: 60°–160°C (as summarized by Hillier 1993). It has not been reported from Recent sediments. All these data indicate that corrensite common in hypersaline facies is also a product of burial diagenesis of precursor smectites, as suggested by Bodine & Madsen (1985).

Smectite chloritization mechanism Inoue et al (1984) imagined the smectite chloritization mechanism as a transformation (hydroxide sheets precipitated into smectitic interlayers). The presence of corrensite as a distinct species was explained by the tendency to segregation of corrensite packets. Recently, chloritization by a dissolution-precipitation mechanism has also been considered (Bettison-Varga & Mackinnon 1997). Studies of the thickness distribution of chlorite fundamental particles may help solve this problem.

Chlorite reaction toward smectite First reported by Johnson (1964), it has been well documented since then that the weathering of chlorite often produces corrensite as an intermediate phase. Most often, this material is characterized as chlorite/vermiculite, i.e. the high-charge corrensite (Proust 1982, Proust et al 1986 and references therein), and the final product is vermiculite, which may evolve into smectite. In terms of interstratification pattern, the reaction seems similar to chloritization because of the privileged position of corrensite, but details are difficult to establish in soil assemblages.

The process involves loss of Fe and Mg to such an extent that the mineral gradually loses its trioctahedral character and the end-member vermiculite is strictly dioctahedral (Proust 1982, Proust et al 1986). These conclusions were confirmed by an extremely detailed HRTEM study by Banfield & Murakami (1998), who found that Fe and Mg are gradually lost from both octahedral sheets and hydroxy interlayers. They explained the development of corrensite by a stabilization effect of the expanded interlayer imposing structural rearrangement on the neighboring layers (a modern version of the old polarization concept). However, the end-member vermiculite was considered alternatively as a dissolution-reprecipitation product.

Ross & Kodama (1976) and Senkayi et al (1981) reproduced the vermiculitization of chlorite in the laboratory. They concluded that the location of ferrous iron plays an important role. If it is present in easily accessible sites (hydroxy interlayers), chlorite converts directly to vermiculite upon oxidation. If iron is located in the octahedral sheet, corrensite development is favored.

MICA/CHLORITE Mica/chlorite (M/C) minerals, close to the regular interlayering, were first discovered using XRD by Eroshchev-Shak (1970). Another occurrence of ordered, almost regular interstratification was documented by HRTEM and XRD (Maresch et al 1985). Veblen (1983), Veblen & Ferry (1983), and Olives & Amouric (1984) produced HRTEM evidence of random mixed layering. In all but one occurrence, the 10 Å component was biotite. The exception was wonesite/chlorite accompanying discrete wonesite (sodium trioctahedral mica) in the sample studied by Veblen (1983). All M/C minerals occur as big crystals producing HRTEM images of excellent quality. Most often individual layers continue all across the crystals, but occasionally layer terminations or lateral alterations of a 10 Å into a 14 Å layer are observed.

The majority of described M/C occurrences are from rocks altered hydrothermally or by low- to medium-grade regional metamorphism: granite, gneiss, metabasalt, or metapelites. In three occurrences, the reaction clearly represents the chloritization of biotite, but in metabasalts Maresch et al (1985) convincingly documented the opposite direction of the alteration and inferred from the literature data that the process may be more common.

Two models of the lateral, layer-by-layer solid-state alteration were inferred from the HRTEM observations of biotite → chlorite reaction. Olives & Amouric (1984) suggested that interlayer K in biotite is laterally replaced by precipitated brucite sheet, a reaction resulting in volume increase and a lowering of pH. Veblen & Ferry (1983) interpreted their data as indicating the removal of two tetrahedral sheets of a biotite layer, with the remaining octahedral sheet serving as the hydroxy interlayer of the new chlorite layer. This mechanism increases pH and decreases the volume of mineral. The first mechanism was supported by Amouric et al (1988) and recognized as a better explanation (in a reversed version) for the alteration of chlorite to biotite (Maresch et al 1985).

TALC/SMECTITE Brindley (1955) first suggested talc interstratification in trioctahedral smectite stevensite, and Dyakonov (1993) suggested stevensite interstratification in kerolite (disordered talc). A regular species has been recognized and named aliettite (Veniale & van der Marel 1969). Occurrences have been reported from weathered ophiolites and dolomites.

A nickel laterite developed on ultramafic rocks provided a broad range of talc/smectite (T/S) minerals (Wiewióra et al 1982). Random, R1, and $R > 1$ interstratifications were documented. R0 clays were found in veinlets and interpreted from texture and spatial relationships as a replacement for sepiolite. In the surrounding wall rocks, ordered T/S was found. Ultrasonic treatment could increase the percentage of swelling interlayers in these minerals.

Eberl et al (1982) reported a complete range of T/S from Plio-Pleistocene Amargosa Flat desert lake and spring deposits. The clays resemble the veinlet material of Wiewióra et al (1982): they are randomly interstratified and associated with sepiolite. End-member kerolite was also reported. The expandable component was identified as stevensite. No systematic spatial variation in composition was found, and Eberl et al (1982) suggested variations in the temperature of formation (hot springs) or solution chemistry as factors responsible for the variation in expandability. Whether the reaction proceeds from smectite to talc, from talc to smectite, or both remains to be seen.

A high-temperature occurrence of smectite interlayering in talc was reported by Shau & Peacor (1992): pseudomorphs after olivine in a submarine basalt alteration zone (200°–380°C).

TALC/CHLORITE A unique occurrence of talc/chlorite (T/C) was reported from metamorphosed dolomites of the evaporite sequence in Algeria (Schreyer et al 1982). This perfectly regular mineral named kulkeite occurs as single crystals. It is associated with discrete chlorite and talc and interpreted as an intermediate step of the reaction series at a temperature estimated at up to 400°C. As with accompanying discrete talc, the talc layers in kulkeite contain substitution of Na + Al for Si.

Discrete chlorite, talc, and irregular T/C were documented by HRTEM in pseudomorphs after olivine in a high-temperature (200°–380°C) alteration zone of submarine basalt (Shau & Peacor 1992). Comparing the occurrences of T/S and T/C, I would not be surprised if T/C turned out to be the product of high-temperature alteration of T/S.

MICA/VERMICULITE The product of trioctahedral weathering, called hydrobiotite since the end of the nineteenth century, was recognized by Gruner (1934) as mica/vermiculite (M/V) interstratification. Several examples of interstratification close to 1:1 were described. Brindley et al (1983b) found samples fulfilling the criteria of regular interstratification and reserved the name hydrobiotite for the regular 1:1 M/V. Weiss (1980) presented convincing XRD evidence that approximately 50% mica interstratification can be random (R0). Single-crystal studies of individual flakes revealed a range of compositions: 12%–60% mica layers. Pozzuoli et al (1992) documented a complete range of M/V compositions.

Meunier & Velde (1979) showed that fluid composition may control the reaction path: biotite in two weathering profiles on the same granite was altered directly into vermiculite or into hydrobiotite, depending on the level of Ca in the soil solutions.

The relations characteristic of natural systems were reproduced during room-temperature experiments by removing potassium from micas. Hoda & Hood (1972) discovered that alteration of phlogopite leads directly to vermiculite, whereas alteration of biotite proceeds via mixed-layer phases. The published XRD patterns suggest a weak tendency toward regular interstratification. Gilkes (1973) confirmed the development of mixed layering in the course of alteration of biotite. In his experiments, which were conducted using different reagents, a strong tendency to hydrobiotite-type interstratification was observed; in one experiment, however, direct development of vermiculite was also documented.

Dioctahedral/trioctahedral mixed-layer clays Dioctahedral/trioctahedral mixed-layer clays are rare. They have never been detected by XRD. The only evidence comes from HRTEM observations.

Ahn & Peacor (1987) presented convincing documentation of kaolinite interstratification in biotite. Kaolinite layers are always present as pairs, and they occasionally replace laterally one biotite layer. A dissolution-precipitation reaction implying the increase of volume was postulated. Another example involving a 7 Å layer is berthierine/glauconite, investigated by Amouric et al (1995).

Peacor and coworkers (Ahn et al 1988 and references therein) presented HRTEM evidence of 10 Å/14 Å interstratification, including zones of regular ordering. They identify it as illite/trioctahedral chlorite based on a chemical composition intermediate between the accompanying big crystals of illite and chlorite. Alternative explanations must be considered because interstratified chlorite and illite layers share the same octahedral sheet, which cannot be di- and trioctahedral at the same time. It is well established that under low-temperature metamorphic and hydrothermal conditions illite and trioctahedral smectite often form parallel intergrowths of big crystals, which are the products of alteration of biotite (Morad & Aldahan 1986). Chlorite of such origin could well preserve some biotite mixed layering. Another possibility is that the interstratified chlorite is not tri- but di-trioctahedral.

GENERAL CONCLUSIONS

Figure 5 presents a summary of the mineral reactions involving mixed-layer clays. Accumulated data on natural occurrences as well as laboratory evidence leave little doubt that mixed-layer clays are intermediate products of reactions starting from discrete clay minerals. Direct precipitation of mixed-layer clays from solution has been proposed, but convincing evidence for such a mechanism is lacking. Hydrothermal experiments indicate that also rectorite and tosudite form from a smectite (beidellite) precursor (Eberl 1978).

Two mechanisms have been invoked to explain the origin of mixed layering. Solid-state transformation assumes that a layer of new mineral is formed by alteration of the interlayer material, accompanied by necessary adjustments of the surrounding portions of layers. Dissolution/precipitation implies that layers of the daughter mineral nucleate and grow at the expense of the layers of the parent mineral. It seems that both mechanisms are well documented in some instances (transformation in weathering reactions of mica and chlorite,

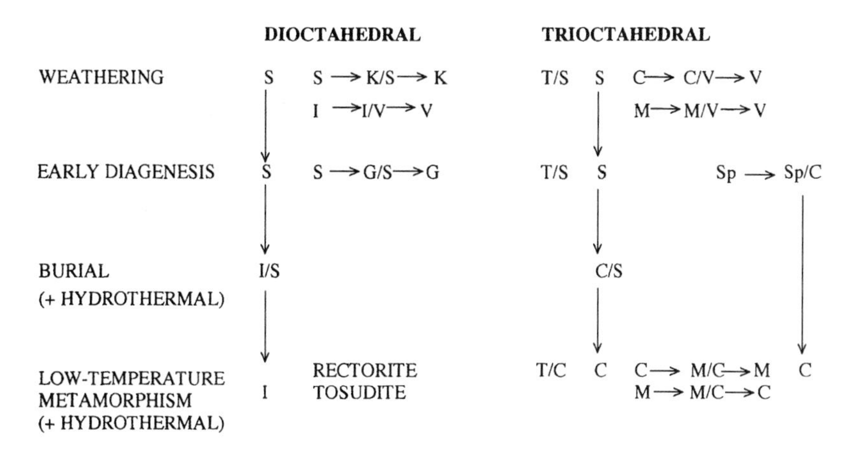

Figure 5 Summary of the occurrences of mixed-layer minerals at different stages of the rock cycle. S, smectite; K, kaolinite; V, vermiculite; T, talc; C, chlorite; I, illite; M, mica; G, glauconite; Sp, serpentine. (*Arrows*) The direction of the alteration process.

dissolution/precipitation in the illitization of smectite), but in most cases the actual mechanism is not well understood. Whatever the mechanism, layers of a given type continue most often throughout the crystal. Lateral changes from one to another type of layer within mixed-layer crystals are reported in HRTEM studies but are uncommon.

Mixed-layer minerals exhibit a variable tendency to form regular 1:1 interstratifications. In kaolinite/smectite it is very weak (mostly R0 clays); in illite/smectite, glauconite/smectite, and talc/smectite it is intermediate (R0, R1, and $R > 1$ clays); in chlorite/smectite (vermiculite) and mica/vermiculite it is strong (privileged occurrence of the regular clays); and in paragonite/smectite and dioctahedral chlorite/smectite it is very strong (almost exclusive occurrence of regular species rectorite and tosudite). This tendency has been explained (for a detailed discussion, see Sawhney 1972) by a so-called polarization effect: alteration of a given interlayer space affects the structure of surrounding layers and decreases the probability of the alteration of the next interlayer spaces. An ABABAB pattern results.

The polarization concept implies a solid-state transformation mechanism, but the tendency to form regular 1:1 interstratifications can also result from the dissolution/nucleation + growth mechanism. The nature of the product should depend in this case on the relative rate of nucleation and the growth of fundamental two-layer particles (illite or chlorite) and on the site of nucleation (Figure 6). Fast nucleation/slow growth inside smectite crystals would result in a strong tendency to 1:1 interstratification, and vice versa. If the nucleation and

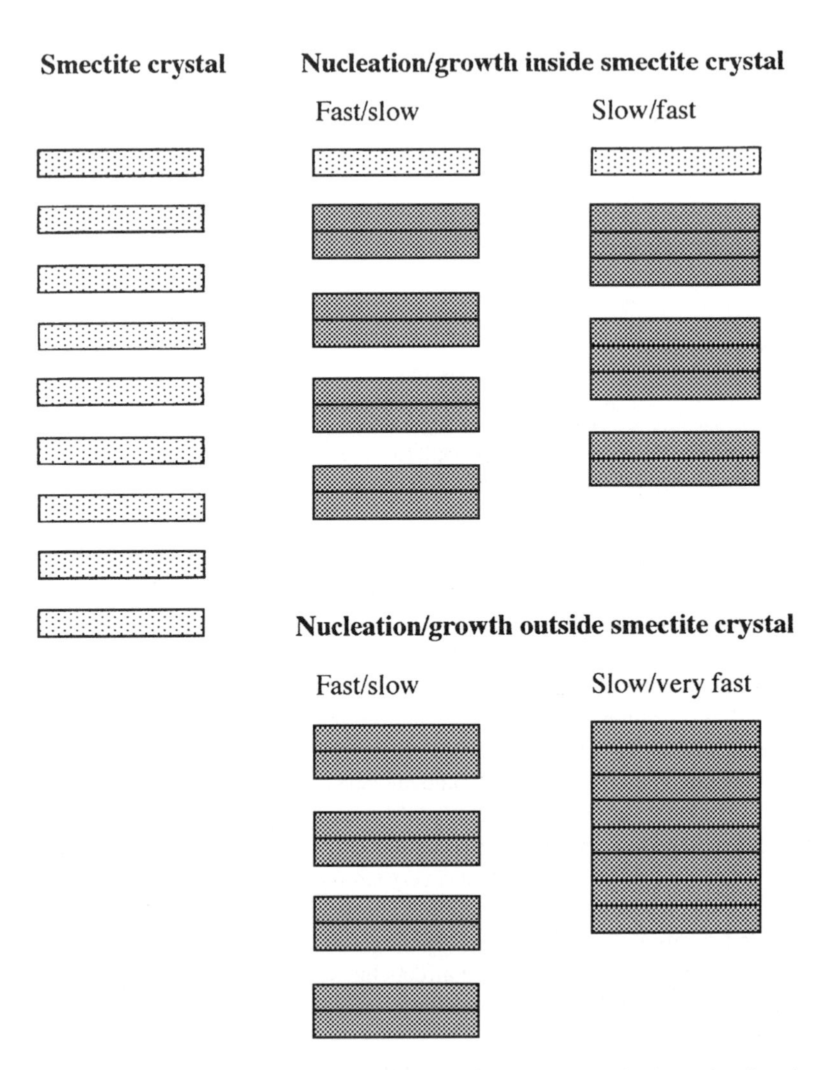

Figure 6 Conceptual model explaining variable tendency to 1:1 interstratification during alteration of smectite by a dissolution/nucleation + growth mechanism. Variable relative rates of nucleation versus growth of two-layer fundamental particles of illite or chlorite, combined with the site of nucleation inside or outside of smectite crystals, result in a range of potential products: regular or irregular interstratifications or discrete nonexpandable mineral.

growth occurred outside of the smectite crystal, the respective products would be a regular clay and ordered clay with <50% smectite or pure nonswelling end-member. It is also feasible that the outside nucleation path can produce only a discrete mineral if the smectitic external surfaces of two-layer nuclei are template effects inherited from smectite and the particle nucleated outside of smectite crystal is different. It appears that often the reaction proceeds by a combination of these reaction pathways, as shown by a common occurrence of authigenic mixtures of a mixed-layer clay and the corresponding nonexpandable discrete phase (kaolinite/smectite + kaolinite, illite/smectite + illite, glauconite/smectite + glauconite, corrensite + chlorite).

In a general sense, the weathering reactions leading toward kaolinite or smectite are the reversals of the diagenetic reactions producing illite and chlorite. In detail, however, the reaction series are quite different, and no reaction involving mixed layering can be reversed along the same path. Mixed layering involving vermiculite seems to be an intermediate product, specific for the weathering reactions proceeding by the transformation mechanism. It is not yet clear if the major portion of this material reacts all the way to vermiculite, and then dioctahedral smectite in the weathering environment, or if it is eroded and transported to sedimentary basins, where it undergoes diagenetic reactions.

ACKNOWLEDGMENT

A major portion of this review was written during my visit at the Universitè Louis Pasteur (PAST program, May–June 1998) in Strasbourg.

Literature Cited

Ahn JH, Peacor DR. 1985. Transmission electron microscopic study of diagenetic chlorite in Gulf Coast argillaceous sediments. *Clays Clay Miner.* 33:228–36

Ahn JH, Peacor DR. 1987. Kaolinization of biotite: TEM data and implications for an alteration mechanism. *Am. Mineral.* 72:353–56

Ahn JH, Peacor DR, Coombs DS. 1988. Formation mechanisms of illite, chlorite and mixed-layer illite-chlorite in Triassic volcanogenic sediments from the Southland Syncline, New Zealand. *Contrib. Mineral. Petrol.* 99:82–89

Altaner SP, Ylagan RF. 1997. Comparison of structural models of mixed-layer illite/smectite and reaction mechanisms of smectite illitization. *Clays Clay Miner.* 45:517–33

Altschuler ZS, Dwornik EJ, Kramer H. 1963. Transformation of montmorillonite to kaolinite during weathering. *Science* 141:148–52

Amouric M, Gianetto I, Proust D. 1988. 7, 10 and 14 Å mixed-layer phyllosilicates studied structurally by TEM in pelitic rocks of the Piemontese zone (Venezuela). *Bull. Mineral.* 111:29–37

Amouric M, Olives J. 1991. Illitization of smectite as seen by high-resolution electron microscopy. *Eur. J. Mineral.* 3:831–35

Amouric M, Parron C, Casalini L, Giresse P. 1995. A (1:1) 7-Å Fe phase and its transformation in recent sediments: an HRTEM and AEM study. *Clays Clay Miner.* 43:446–54

April RH, Hluchy MM, Newton RM. 1986. The nature of vermiculite in Adirondack soils and till. *Clays Clay Miner.* 34:549–56

Bailey SW, Banfield JF, Barker WW, Katchan G. 1995. Dozyite, a 1:1 regular interstratification of serpentine and chlorite. *Am. Mineral.* 80:65–77

Bailey SW. 1988. Odinite, a new dioctahedral-trioctahedral $Fe3^{+}$-rich 1:1 clay mineral. *Clay Miner.* 23:237–47

Bain DC, Mellor A, Wilson MJ. 1990. Nature and origin of an aluminous vermiculitic weathering product in acid soils from upland catchments in Scotland. *Clay Miner.* 25:467–76

Banfield JF, Murakami T. 1998. Atomic-resolution transmission electron microscope evidence for the mechanism by which chlorite weathers to 1:1 semi-regular chlorite-vermiculite. *Am. Mineral.* 83:348–57

Barnhisel RI, Bertsch PM. 1989. Chlorites and hydroxy-interlayered vermiculite and smectite. In *Minerals in Soil Environments*, ed. JB Dixon, SB Weed, pp. 729–88. Madison, WI: Soil Sci. Soc. Am.

Berkgaut V, Singer A, Stahr K. 1994. Palagonite reconsidered: paracrystalline illite-smectites from regoliths on basic pyroclastics. *Clays Clay Miner.* 42:582–92

Bettison-Varga L, Mackinnon IDR. 1997. The role of randomly mixed-layered chlorite/smectite in the transformation of smectite to chlorite. *Clays Clay Miner.* 44:506–16

Bobos I, Sucha V, Soboleva SV. 1995. Mixed-layer NH_4-illite in the hydrothermal area Harghita Bai (East Carpathians). *Abstr. Euroclay '95, Leuven, Belgium,* pp. 384–85

Bodine MW, Madsen BM. 1985. Mixed-layered chlorite/smectites from a Pennsylvanian evaporite cycle, Grand County, Utah. *Proc. Int. Clay Conf., Denver, 1985,* pp. 85–93

Boles JR, Franks SG. 1979. Clay diagenesis in Wilcox sandstones of southwest Texas: implications of smectite diagenesis on sandstone cementation. *J. Sediment. Petrol.* 49:55–70

Bray RH. 1937. Chemical and physical changes in soil colloids with advancing development in Illinois soils. *Soil Sci.* 43:1–14

Brindley GW, Suzuki T, Thiry M. 1983a. Interstratified kaolinite/smectites from the Paris Basin; correlations of layer proportions, chemical compositions and other data. *Bull. Mineral.* 106:403–10

Brindley GW. 1955. Stevensite, a montmorillonite-type mineral showing mixed-layer characteristics. *Am. Mineral.* 40:239–47

Brindley GW, Zalba PA, Bethke CM. 1983b. Hydrobiotite, a regular 1:1 interstratification in biotite and vermiculite layers. *Am. Mineral.* 68:420–25

Brown G, Weir AH. 1963. The identity of rectorite and allevardite. *Proc. Int. Clay Conf.* 1:27–37

Buatier MD, Früh-Green GL, Karpoff AM. 1995. Mechanisms of Mg-phyllosilicate formation in a hydrothermal system at a sedimented ridge (Middle Valley, Juan de Fuca). *Contrib. Mineral. Petrol.* 122:134–51

Buatier MD, Ouyang K, Sanchez JP. 1993. Iron in hydrothermal clays from the Galapagos Spreading Centre mounds: consequences for the clay transition mechanism. *Clay Miner.* 28:641–55

Burst JF. 1969. Diagenesis of Gulf Coast clayey sediments and its possible relation to petroleum migration. *Bull. Am. Assoc. Petrol. Geol.* 53:73–93

Calvert CS, Pevear DR. 1983. Paleopedogenic(?) mixed-layer kaolinite/smectite from 9000 ft in a Gulf Coast well. *Prog. Abstr. 20th Annu. Meet. Clay Miner. Soc., Buffalo, NY,* No. 50

Cathelineau M. 1988. Cation site occupancy in chlorites and illites as a function of temperature. *Clay Miner.* 237:471–85

Clauer N, Środoń J, Francu J, Sucha V. 1997. K-Ar dating of illite fundamental particles separated from illite/smectite. *Clay Miner.* 32:181–96

Cole WF. 1966. A study of a long-spacing mica-like mineral. *Clay Miner.* 6:261–81

Courbe C, Velde B, Meunier A. 1981. Weathering of glauconites: reversal of the glauconitization process in a soil profile in western France. *Clay Miner.* 16:231–43

Cuadros J, Delgado A, Cardenete A, Reyes E, Linares J. 1994. Kaolinite/montmorillonite resembles beidellite. *Clays Clay Miner.* 42:643–51

Daniels EJ, Altaner SP. 1990. Clay mineral authigenesis in coal and shale from anthracite region, Pennsylvania. *Am. Mineral.* 75:825–39

Dean RS. 1983. Authigenic trioctahedral clay minerals coating Clearwater Formation sand grains at Cold Lake, Alberta, Canada. *Prog. Abstr. 20th Annu. Meet. Clay Miner. Soc., Buffalo, NY.* p. 79

Delvaux B, Herbillon AJ, Vielvoye L, Mestdagh MM. 1990. Surface properties and clay mineralogy of hydrated halloysitic soil clays. II. Evidence for the presence of halloysite/smectite (H/Sm) mixed-layer clays. *Clay Miner.* 25:141–60

Delvaux B, Mestdagh MM, Vielvoye L, Herbillon AJ. 1989. XRD, IR and ESR study of experimental alteration of Al-nontronite into mixed-layer kaolinite/smectite. *Clay Miner.* 24:617–30

Drits VA, Kossovskaya AG. 1990. Clay minerals: smectites, mixed-layer silicates. *Transact. AS USSR 446*. 212 pp. (In Russian)
Drits VA, Sakharov BA. 1976. X-ray structural analysis of mixed-layer minerals. *Transact. AS USSR 295*. Moscow: 256 pp. (In Russian)
Drits VA, Sakharov BA, Lindgreen H, Salyn AL. 1997. Sequential structure transformation of illite-smectite-vermiculite during diagenesis of Upper Jurassic shales from the North Sea and Denmark. *Clay Miner.* 32:351–71
Drits VA, Salyn AL, Sucha V. 1996. Structural transformations of interstratified illite-smectites from Dolna Ves hydrothermal deposits: dynamics and mechanisms. *Clays Clay Miner.* 44:181–90
Dyakonov YS. 1963. The results of an X-ray study of kerolites. *Dokl. AS SSSR* 148:901–11 (In Russian)
Eberl DD. 1978. Reaction series for dioctahedral smectites. *Clays Clay Miner.* 26:327–40
Eberl DD. 1979. Reaction series for dioctahedral smectite: the synthesis of mixed-layer pyrophyllite/smectite. *Proc. Int. Clay Conf., Oxford, 1978,* 375–83
Eberl DD, Hower J. 1976. Kinetics of illite formation. *Geol. Soc. Am. Bull.* 187:1326–30
Eberl DD, Jones BF, Khoury HN. 1982. Mixed-layer kerolite/stevensite from Amargosa Desert, Nevada. *Clays Clay Miner.* 30: 321–26
Eberl DD, Nuesch R, Šucha V, Tsipursky S. 1998. Measurement of fundamental particle thicknesses by X-ray diffraction using PVP-10 intercalation. *Clays Clay Miner.* 46:89–97
Eberl DD, Środoń J. 1988. Ostwald ripening and interparticle diffraction effects for illite crystals. *Am. Mineral.* 73:1335–45
Eberl DD, Środoń J, Kralik M, Taylor B, Peterman ZE. 1990. Ostwald ripening of clays and metamorphic minerals. *Science* 248:474–77
Eberl DD, Środoń J, Lee M, Nadeau PH, Northrop HR. 1987. Sericite from the Silverton caldera, Colorado: correlation among structure, composition, origin, and particle thickness. *Am. Mineral.* 72:914–35
Eberl DD, Środoń J, Northrop HR. 1986. Potassium fixation in smectite by wetting and drying. In *Geochemical Processes at Mineral Surfaces. ACS Symposium Series No. 323*, ed. JA Davis, KF Hayes, pp. 296–326. Washington, DC: Am. Chem. Soc.
Eberl DD, Velde B, McCormick T. 1993. Synthesis of illite-smectite from smectite at Earth surface temperatures and high pH. *Clay Miner.* 28:49–60
Elsass F, Środoń J, Robert M. 1997. Illite-smectite alteration and accompanying reactions in a Pennsylvanian underclay studied by TEM. *Clays Clay Miner.* 45:390–403
Eroshchev-Shak VA. 1970. Mixed-layer biotite-chlorite formed in the course of local epigenesis in the weathering crust of a biotite gneiss. *Sedimentology* 15:115–21
Foscolos AE, Kodama H. 1974. Diagenesis of clay minerals from Lower Cretaceous shales of north eastern British Columbia. *Clays Clay Miner.* 22:319–35
Frey E, Lagaly G. 1979. Selective coagulation and mixed-layer formation from sodium smectite solutions. *Proc. Int. Clay Conf., Oxford, 1978.* pp. 131–40
Frey M. 1969. A mixed-layer paragonite/phengite of low-grade metamorphic origin. *Contrib. Mineral. Petrol.* 24:63–65
Gilkes RJ. 1973. The alteration products of potassium depleted oxybiotite. *Clays Clay Miner.* 21:303–13
Guthrie GD Jr, Veblen DR. 1989. High-resolution electron microscopy of mixed-layer illite/smectite: computer simulations. *Clays Clay Miner.* 37:1–11
Gruner JW. 1934. The structures of vermiculites and their collapse by dehydration. *Am. Mineral.* 19:557–75
Hendricks SB, Teller E. 1942. X-ray interference in partially ordered layer lattices. *J. Phys. Chem.* 10:147–67
Hillier S. 1993. Origin, diagenesis, and mineralogy of chlorite minerals in Devonian lacustrine mudrocks, Orcadian Basin, Scotland. *Clays Clay Miner.* 41:240–59
Hillier S. 1994. Pore-lining chlorites in siliciclastic reservoir sandstones: electron microprobe, SEM and XRD data, and implications for their origin. *Clay Miner.* 29:665–80
Hillier S, Velde B. 1992. Chlorite interstratified with a 7 Å mineral: an example from offshore Norway and possible implications for the interpretation of the composition of diagenetic chlorites. *Clay Miner.* 27:475–86
Hoda SN, Hood WC. 1972. Laboratory alteration of trioctahedral micas. *Clays Clay Miner.* 20:343–58
Hoffman J, Hower J. 1979. Clay mineral assemblages as low grade metamorphic geothermometers: application to the thrust faulted disturbed belt of Montana, USA. *Soc. Econ. Paleontol. Mineral. Spec. Publ.* 26:55–79
Hornibrook ERC, Longstaffe FJ. 1996. Berthierine from the Lower Cretaceous Clearwater Formation, Alberta, Canada. *Clays Clay Miner.* 44:1–21
Hower J, Eslinger E, Hower M, Perry E. 1976. Mechanism of burial metamorphism of argillaceous sediment. 1. Mineralogical and chemical evidence. *Geol. Soc. Am. Bull.* 87:725–37
Hower J, Mowatt TC. 1966. The mineralogy of illites and mixed-layer illite-montmorillonites. *Am. Mineral.* 51:825–54

Huang WL, Longo JM, Pevear DR. 1993. An experimentally derived kinetic model for smectite-to-illite conversion and its use as a geothermometer. *Clays Clay Miner.* 41:162–77
Hughes RE, Moore DM, Reynolds RC Jr. 1993. The nature, detection, occurrence, and origin of kaolinite/smectite. In *Kaolin Genesis and Utilization*, ed. H Murray, W Bundy, C Harvey, pp. 291–323. Boulder, CO: Clay Mineral. Soc.
Inoue A. 1985. Chemistry of corrensite: a trend in composition of trioctahedral chlorite/smectite during diagenesis. *J. Coll. Arts Sci. Chiba Univ.* B18:69–82
Inoue A. 1986. Morphological change in a continuous smectite-to-illite conversion series by scanning and transmission electron microscopies. *J. Coll. Arts Sci. Chiba Univ.* B19:23–33
Inoue A. 1987. Conversion of smectite to chlorite by hydrothermal and diagenetic alterations, Hokuroku Kuroko mineralization area, northeast Japan. *Proc. Int. Clay Conf., Denver, 1985.* 158–64
Inoue A, Kitagawa R. 1994. Morphological characteristics of illitic clay minerals from a hydrothermal system. *Am. Mineral.* 79:700–11
Inoue A, Utada M, Nagata H, Watanabe T. 1984. Conversion of trioctahedral smectite to interstratified chlorite/smectite in Pliocene acidic pyroclastic sediments of the Ohyu district, Akita Prefecture, Japan. *Clay Sci.* 6:103–16
Inoue A, Velde B, Meunier A, Touchard G. 1988. Mechanism of illite formation during smectite-to-illite conversion in a hydrothermal system. *Am. Mineral.* 73:1325–34
Jackson ML. 1963. Interlayering of expansible layer silicates in soils by chemical weathering. *Clays Clay Miner.* 11:29–46
Jackson ML. 1975. *Soil Chemical Analysis—Advanced Course.* Madison, WI: ML Jackson (self published)
Jiang WT, Peacor DR. 1991. Transmission electron microscopic study of the kaolinization of muscovite. *Clays Clay Miner.* 39:1–13
Jiang WT, Peacor DR, Slack JF. 1992. Microstructures, mixed layering, and polymorphism of chlorite and retrograde berthierine in the Kidd Creek massive sulfide deposit, Ontario. *Clays Clay Miner.* 40:501–14
Johns WD, McKallip TE. 1989. Burial diagenesis and specific catalytic activity of illite-smectite clays from the Vienna Basin, Austria. *Am. Assoc. Geol. Bull.* 73:472–82
Johnson LJ. 1964. Occurrence of regularly interstratified chlorite-vermiculite as a weathering product of chlorite in a soil. *Am. Mineral.* 49:556–72
Kimbara K, Sudo T. 1973. Chloritic clay minerals in tuffaceous sandstone of the Miocene Green Tuff formation, Yamanaka district, Ishikawa Prefecture, Japan. *J. Jpn Assoc. Mineral. Petrol. Econ. Geol.* 68:246–58
Kübler B. 1964. Les argiles, indicateurs de metamorphisme. *Rev. Inst. Franc. Petrole* 19:1093–112
Lee JH, Peacor DR. 1983. Interlayer transitions in phyllosilicates of Martinsburg shale. *Nature* 303:608–9
Lippman F. 1954. Uber einen Keuperton von Keiserweiher bei Maulbron. *Heidelberg Beitr. Mineral. Petrol.* 4:30–144
Livi KJT, Veblen DR, Ferry JM, Frey M. 1997. Evolution of 2:1 layered silicates in low-grade metamorphosed Liassic shales of Central Switzerland. *J. Metamorph. Geol.* 15:323–44
MacEwan DMC. 1950. Some notes on the recording and interpretation of X-ray diagrams of soil clays. *J. Soil Sci.* 1:90–103
Maresch WV, Massone HJ, Czank M. 1985. Ordered and disordered chlorite/biotite interstratifications as alteration products of chlorite. *Neues Jahrbuch für Mineralogie Abhandlungen* 152:79–100
Matsuda T, Kodama H, Yang AF. 1997. Carectorite from Sano Mine, Nagano Prefecture, Japan. *Clays Clay Miner.* 45:773–80
McCubbin DG, Patton JW. 1981. Burial diagenesis of illite/smectite: the kinetic model. *Abstr. Bull. Am. Assoc. Petrol. Geol.* 65:956
Merceron T, Inoue A, Bouchet A, Meunier A. 1988. Lithium-bearing donbassite and tosudite from Echassie'res, Massif Central, France. *Clays Clay Miner.* 36:39–46
Meunier A, Velde B. 1979. Biotite weathering in granites of western France. *Proc. Int. Clay Conf., Oxford, 1978.* 405–13
Meunier A, Velde B. 1989. Solid solutions in I/S mixed-layer minerals and illite. *Am. Mineral.* 74:1106–12
Miser HD, Milton C. 1964. Quartz, rectorite, and cookeite from the Jeffrey Quarry near North Little Rock, Pulaski County, Arkansas. *Ark. Geol. Comm. Bull.* 21:1–29
Moore D, Hower J. 1986. Ordered interstratification of dehydrated and hydrated Na-smectite. *Clays Clay Miner.* 34:379–84
Moore D, Reynolds RC. 1997. *X-Ray Diffraction and the Identification and Analysis of Clay Minerals.* Oxford: Oxford Univ. Press. 378 pp.
Morad S, Aldahan AA. 1986. Discussion and comments on the paper: electron-optical studies of phyllosilicate intergrowths in sedimentary and metamorphic rocks. *Mineral. Mag.* 50:340–43
Morrison SJ, Parry WT. 1986. Dioctahedral corrensite in Permian red beds, Lisbon Valley, Utah. *Clays Clay Miner.* 34:613–24

Nadeau PH. 1985. The physical dimensions of fundamental clay particles. *Clay Miner.* 23:499–514
Nadeau PH, Bain DC. 1986. Composition of some smectites and diagenetic illitic clays and implications for their origin. *Clays Clay Miner.* 34:455–64
Nadeau PH, Wilson MJ, McHardy WJ, Tait JM. 1984. Interstratified clays as fundamental particles. *Science* 225:923–35
Nadeau PH, Wilson MJ, McHardy WJ, Tait JM. 1985. The conversion of smectite to illite during diagenesis: evidence from some illitic clays from bentonites and sandstones. *Mineral. Mag.* 49:393–400
Nishiyama T, Shimoda S, Shimosaka K, Kanaoka S. 1975. Lithium-bearing tosudite. *Clays Clay Miner.* 23:337–42
Odin GS. 1988. *Green Marine Clays*. Amsterdam: Elsevier. 445 pp.
Olives J, Amouric M. 1984. Biotite chloritization by interlayer brucitization as seen by HRTEM. *Am. Mineral.* 69:869–71
Olson CG. 1988. Clay-mineral contribution to the weathering mechanisms in two contrasting watersheds. *J. Soil Sci.* 39:457–67
Paradis S, Velde B, Nicot E. 1983. Chloritoid-pyrophyllite-rectorite facies rocks from Brittany, France. *Contrib. Mineral. Petrol.* 83: 342–47
Perry E, Hower J. 1970. Burial diagenesis in Gulf Coast pelitic sediments. *Clays Clay Miner.* 18:165–77
Pollastro RM. 1993. Considerations and applications of the illite/smectite geothermometer in hydrocarbon-bearing rocks of Miocene to Mississippian age. *Clays Clay Miner.* 41: 119–33
Pozzuoli A, Vila E, Franco E, Ruiz-Amil A, de la Calle C. 1992. Weathering of biotite to vermiculite in Quaternary lahars from Monti Ernici, central Italy. *Clay Miner.* 27:175–84
Proust D. 1982. Supergene alteration of metamorphic chlorite in an amphibolite from Massif Central, France. *Clay Miner.* 17:159–73
Proust D, Eymery JP, Beaufort D. 1986. Supergene vermiculitization of a magnesian chlorite: iron and magnesium removal process. *Clays Clay Miner.* 34:572–80
Proust D, Lechelle J, Lajudie A, Meunier A. 1990. Hydrothermal reactivity of mixed-layer kaolinite/smectite: experimental transformation of high-charge to low-charge smectite. *Clays Clay Miner.* 38:415–25
Pytte AM, Reynolds RC Jr. 1989. The thermal transformation of smectite to illite. In *Thermal History of Sedimentary Basins*, ed. ND Naeser, TH McCulloh, pp. 133–40. New York: Springer-Verlag.
Reynolds RC Jr. 1967. Interstratified clay systems: calculation of the total one-dimensional diffraction function. *Am. Mineral.* 52:661–72
Reynolds RC Jr. 1988. Mixed layer chlorite minerals. In *Hydrous Phyllosilicates (Exclusive of Micas). Reviews in Mineralogy 19*, ed. SW Bailey, pp. 601–29. Washington, DC: Mineral. Soc. Am.
Reynolds RC Jr. 1992. X-ray diffraction studies of illite/smectite from rocks, <1 μm randomly oriented powders, and <1 μm oriented powder aggregates: the absence of laboratory-induced artifacts. *Clays Clay Miner.* 40: 387–96
Reynolds RC Jr, DiStefano MP, Lahann RW. 1992. Randomly interstratified serpentine/chlorite: its detection and quantification by powder X-ray diffraction methods. *Clays Clay Miner.* 40:262–67
Reynolds RC Jr, Hower J. 1970. The nature of interlayering in mixed-layer illite-montmorillonites. *Clays Clay Miner.* 18:25–36
Rich CI. 1968. Hydroxy interlayers in expansible layer silicates. *Clays Clay Miner.* 16:15–30
Righi D, Velde B, Meunier A. 1995. Clay stability in clay-dominated soil systems. *Clay Miner.* 30:45–54
Robert M. 1973. The experimental transformation of mica toward smectite; relative importance of total charge and tetrahedral substitution. *Clays Clay Miner.* 21:167–74
Robinson D, Wright VP. 1987. Ordered illite-smectite and kaolinite-smectite: pedogenic minerals in a Lower Carboniferous paleosol sequence, South Wales. *Clay Miner.* 22:109–18
Ross GJ, Kodama H. 1976. Experimental alteration of a chlorite into a regularly interstratified chlorite-vermiculite by chemical oxidation. *Clays Clay Miner.* 24:183–90
Ryan PC, Reynolds RC Jr. 1996. The origin and diagenesis of grain-coating serpentine/chlorite, Tuscaloosa Formation. *Am. Mineral.* 81:213–25
Sakharov BA, Drits VA. 1973. Mixed-layer kaolinite-montmorillonite: a comparison of observed and calculated diffraction patterns. *Clays Clay Miner.* 21:15–17
Sato T, Murakami T, Watanabe T. 1996. Change in layer charge of smectites and smectite layers in illite/smectite during diagenetic alteration. *Clays Clay Miner.* 44:460–69
Sawhney BL. 1972. Selective sorption and fixation of cations by clay minerals: a review. *Clays Clay Miner.* 20:93–100
Schiffman P, Fridleifsson GO. 1991. The smectite to chlorite transition in Drillhole NJ-15, Nesjavellir Geothermal Field, Iceland: XRD, BSE and electron microprobe investigations. *J. Metamorph. Geol.* 9:679–96

Schreyer W, Medenbach O, Abraham K, Gebert W, Muller WF. 1982. Kulkeite, a new metamorphic phyllosilicate mineral: ordered 1:1 chlorite/talc mixed-layer. *Contrib. Mineral. Petrol.* 80:103–9

Schultz LG, Shepard AO, Blackmon PD, Starkey HC. 1971. Mixed-layer kaolinite-montmorillonite from the Yucatan Peninsula, Mexico. *Clays Clay Miner.* 19:137–50

Senkayi AL, Dixon JB, Hossner LR. 1981. Transformation of chlorite to smectite through regularly interstratified intermediates. *Soil Sci. Soc. Am. J.* 45:650–56

Shau YH, Feather ME, Essene EJ, Peacor DR. 1991. Genesis and solvus relations of submicroscopically intergrown paragonite and phengite in a blueshist from northern California. *Contrib. Mineral. Petrol* 106:367–78

Shau YH, Peacor DR. 1992. Phyllosilicates in hydrothermally altered basalts from DSDP Hole 504B, Leg 83—a TEM and AEM study. *Contrib. Mineral. Petrol.* 112:119–33

Shinoyama A, Johns WD, Sudo T. 1969. Montmorillonite-kaolinite clay in acid clay deposits from Japan. *Proc. Int. Clay Conf., Tokyo, 1969,* 1:225–31

Shutov VD, Drits VA, Sakharov BA. 1969. On the mechanism of a postsedimentary transformation of montmorillonite into hydromica. *Proc. Int. Clay Conf., Tokyo, 1969.* 1:523–32

Small JS. 1994. Fluid composition, mineralogy and morphological changes associated with the smectite-to-illite reaction: an experimental investigation of the effect of organic acid anions. *Clay Miner.* 29:539–54

Środoń J. 1980. Synthesis of mixed-layer kaolinite/smectite. *Clays Clay Miner.* 28:419–24

Środoń J, Andreolli C, Elsass F, Robert M. 1990. Direct high-resolution transmission electron microscopic measurement of expandability of mixed-layer illite/smectite in bentonite rock. *Clays Clay Miner.* 38:373–79

Środoń J, Eberl DD. 1984. Illite. In *Micas. Reviews in Mineralogy 13*, ed. SW Bailey, pp. 495–544. Washington, DC: Mineral. Soc. Am.

Środoń J, Eberl DD, Drits VA. 1999. Evolution of fundamental particle size during illitization of smectite and implications for the illitization mechanism. *Clays Clay Miner.* In press

Środoń J, Elsass F, McHardy WJ, Morgan DJ. 1992. Chemistry of illite-smectite inferred from TEM measurements of fundamental particles. *Clay Miner.* 27:137–58

Šucha V, Kraus I, Madejova J. 1994. Ammonium illite from anchimetamorphic shales associated with anthracite in the Zemplinicum of the western Carpathians. *Clay Miner.* 29:369–77

Sudo T. 1954. Long spacing of about 30 Å in Japanese clays. *Clay Miner.* 2:193–203

Sudo T, Hayashi H. 1956. Types of mixed layer minerals from Japan. *Proc. 4th Natl. Conf. Clays Clay Miner., Univ. Park, PA, 1956.* 389–412

Thiry M. 1981. Sédimentation continentale et altérations associées: calcitisations, ferruginisations, et silicifications. Les argiles plastiques du Sparnacien du Basin de Paris. *Sci. Geol. Mem.* 64:173

Thomas AR. 1989. A new mixed-layer mineral—regular 1:1 mixed layer kaolinite/smectite. *Prog. Abstr. 26th Annu. Meet. Clay Miner. Soc., Sacramento, CA.* p. 69

Thompson GR, Hower J. 1975. The mineralogy of glauconite. *Clays Clay Miner.* 23:289–300

Tomita K, Dozono M. 1972. Formation of an interstratified mineral by extraction of potassium from mica with sodium tetraphenylboron. *Clays Clay Miner.* 20:225–31

Vali H, Hesse R. 1990. Alkylammonium ion treatment of clay minerals in ultrathin section: a new method for HRTEM examination of expandable layers. *Am. Mineral.* 75:1443–46

Veblen DR. 1983. Microstructures and mixed layering in intergrown wonesite, chlorite, talc, biotite, and kaolinite. *Am. Mineral.* 68:566–80

Veblen DR, Ferry JM. 1983. A TEM study of the biotite-chlorite reaction and comparison with petrologic observations. *Am. Mineral.* 68:1160–68

Veblen DR, Guthrie GD, Livi KJT, Reynolds RC Jr. 1990. High-resolution transmission electron microscopy and electron diffraction of mixed-layer illite/smectite: experimental results. *Clays Clay Miner.* 38:1–13

Velde B, Vasseur G. 1992. Estimation of the diagenetic smectite-to-illite transformation in time-temperature space. *Am. Mineral.* 77:967–76

Veniale F, van der Marel HW. 1969. Identification of some 1:1 regular interstratified trioctahedral clay minerals. *Proc. Int. Clay Conf., Tokyo, 1969.* 1:233–44

von Engelhardt W, Goldschmidt H. 1955. Ein Tonmineral der Kaolinit-Halloysit-gruppe von Provins (Frankreich). *Heidelberg Beitr. Mineral. Petrogr.* 4:319–24

Wang H, Frey M, Stern W. 1996. Diagenesis and metamorphism of clay minerals in the Helvetic Alps of Eastern Switzerland. *Clays Clay Miner.* 44:96–112

Weaver CE. 1956. Distribution and identification of mixed-layer clays in sedimentary rocks. *Am. Mineral.* 41:202–21

Weaver CE. 1979. Geothermal alteration of clay minerals and shales: diagenesis. *Tech. Rep.*

Off. Nucl. Waste Isol. Atlanta: Georgia Inst. Technol. 176 pp.

Weaver CE, Beck KC. 1971. Clay water diagenesis during burial: how mud becomes gneiss. *Geol Soc. Am. Spec. Pap.* 134:1–96

Weaver CE, Wampler JM. 1970. K, Ar, illite, burial. *Bull. Geol. Soc. Am.* 81:3423–30

Wei H, Roaldset E, Bjoroy M. 1996. Parallel reaction kinetics of smectite to illite conversion. *Clay Miner.* 31:365–76

Weiss Z. 1980. Single-crystal X-ray study of mixed structures of vermiculite and biotite (hydrobiotites). *Clay Miner.* 15:275–81

Wiewióra A. 1973. Mixed-layer kaolinite-smectite from Lower Silesia, Poland: final report. *Proc. Int. Clay Conf., Madrid, 1972.* 2:75–88

Wiewióra A, Dubińska E, Iwasińska I. 1982. Mixed-layering in Ni-containing talc-like minerals from Szklary, Lower Silesia, Poland. *Proc. Int. Clay Conf., Bologna-Pavia, Italy, 1981.* pp. 111–25

Annu. Rev. Earth Planet. Sci. 1999. 27:55–73

GEOLOGIC APPLICATIONS OF SEISMIC SCATTERING

Justin Revenaugh
Earth Sciences, University of California, Santa Cruz, CA 95064; jsr@monk.ucsc.edu

KEY WORDS: seismic scattering, continental crust, seismogenesis, lithosphere, volcanoes

ABSTRACT

Once disregarded as noise, scattered seismic waves are finding increasing application in subsurface imaging. This sea change is driven by the increasing density and quality of seismic recordings and advances in waveform modeling which, together, are allowing seismologists to exploit their unique properties. In addition to extensive application in the energy exploration industry, seismic scattering is now used to characterize heterogeneity in the lower continental crust and subcrustal lithosphere, to examine the relationship between crustal structure and seismogenesis, and to probe the plumbing of active volcanoes. In each application, the study of seismic scattering brings wavelength-scale structure into sharper focus and characterizes the short scale-length fabric of geology.

INTRODUCTION

The notion of a horizontally stratified, or layer-cake, Earth pervaded seismology for many years, its prevalence as much a reflection of seismologists' unwillingness to surrender analytical and numerical simplicity as of gravity's dominance in geologic processes. With time, the intersection of need and increasing computing power produced successive generalizations of seismic structure, each seeking to explain a greater percentage of the recorded wave field and to extract more information from the same wiggles. The products of the latest stage of this evolution—three-dimensional reflection profiling and tomography—are spectacular and very much a part of geology.

Only a few holdovers of the layer-cake paradigm remain. The most important is scale-length separation. The subsurface is treated not as a true three-dimensional continuum but rather as a collection of short-wavelength, laterally

0084-6597/99/0515-0055$08.00

continuous features (reflectors) superposed on a smoothly varying background, splitting the heterogeneity spectrum in two and forcing different, but quite limited, forms of regularity on each half. The efficacy of this approach in everyday practice is sufficient validation. Nonetheless, the evidence of its failure appears on every seismogram, manifest, for instance, as the slowly decaying coda trailing "primary" arrivals in time. Created by a multiple-scale, laterally varying seismic structure that does not lend itself to the former parameterization, the scattered arrivals composing coda are unfairly treated as "noise," that is, as an unavoidable nuisance. That attitude began to give way in the 1970s with the pioneering work of K. Aki (e.g. Aki 1969, Aki & Chouet 1975). Recognizing coda of regional S waves as the sum of diffusely scattered body and surface waves, Aki's work (Aki 1969) introduced an alternative parameterization of Earth structure, one based on the statistics of velocity and density fluctuations rather than on deterministic description (see also Chernov 1960).

Statistical characterization of the subsurface is extremely powerful, describing aspects of the subsurface that, if usually not dominant, are at least always present. For many wave propagation regimes (e.g. turbulent media) it is the only natural parameterization. Information about this portion of the heterogeneity spectrum comes almost entirely from scattered waves, that is, from the waves produced by primary wave interaction with it. This review focuses on recent geologic applications of seismic scattering with specific emphasis on the structural fabric of the lower continental crust and subcrustal lithosphere, regions of active volcanism, and the relationship of crustal heterogeneity to the seismogenic process.

THEORETICAL BACKGROUND

Throughout much of the Earth, short scale-length fluctuations of velocity and density about the mean are small, typically a few percent or less. For such a weakly inhomogeneous medium, it is possible to separate the primary wave field (waves that would exist in the medium without any fluctuations) from the scattered wave field (waves generated by primary wave interaction with fluctuations). The ability to separate wave field components is largely responsible for the success of the deterministic parameterization which explains only the primary wave field.

Properties of the scattered wave field depend on both the primary wave field and the heterogeneity. Scattering is most efficient when the wavelength of primary waves, λ ($= 2\pi/k$, where k is wavenumber), is commensurate with the scale length of heterogeneity, a, that is, when $ka \approx 1$. For $ka \gg 1$, velocity varies little over a wavelength, and the medium appears essentially homogeneous. The travel time and amplitude of primary arrivals vary, but scattered waves carry little energy. Conversely, for $ka \ll 1$, velocity varies so frequently over a wavelength that wave energy is held entirely by the scattered wave field, in which case long-range propagation is modeled accurately by primary wave propagation in an equivalent homogeneous medium. Between these two extremes, the scattered wave field can be very complex.

It is common to imagine scattering as wave field interaction with a distribution of discrete scatterers, allowing a separation of the scattered wave field into singly and multiply scattered components, distinguishing between scattering of the primary and scattered wave fields. For weakly inhomogeneous media, singly scattered energy dominates, and multiple scattering becomes important only at long lapse times (arrival times much longer than that of the primary waves). Singly scattered waves are further classified as either forward scattered or back scattered. Forward-scattered waves travel subparallel to the primary waves and are the dominant component of early *P*-wave coda. Back-scattered waves propagate at high angles to the primary wave field and account for a larger percentage of *S*-wave coda. Phase conversion (e.g. *P* to *S* wave and body to surface wave) is an important process, producing a rich sequence of secondary arrivals sensitive to compressional and shear velocity, density, and roughness of internal boundaries and the free surface.

Figure 1 shows a typical regional array recording of a teleseism in which a slowly decaying coda continues long after passage of the primary *P* wave. Variation from station to station proves that this late-arriving energy is generated locally in the lithosphere beneath the stations, but even the best three-dimensional deterministic models predict little of this energy. Missing from those models are scale lengths of heterogeneity comparable to seismic wavelength: scale lengths that are averaged over or healed out of primary waves propagating many wavelengths (Gudmundsson 1996, Snieder & Lomax 1996) but that are the most effective scatterers.

CONTINENTAL LITHOSPHERE

A number of authors have attempted to relate teleseismic *P*-wave coda levels, as well as amplitude and phase fluctuations of the primary *P* wave, to random heterogeneity in the lithosphere beneath single stations or local arrays. The first such study (Aki 1973) used Chernov (1960) theory to explain amplitude and time delay variation of 2-s period *P* waves beneath the Montana Large-Aperture Seismic Array (LASA). In Chernov theory, heterogeneity has a Gaussian autocorrelation function characterized by a root-mean-square (RMS) velocity fluctuation ε^2 and correlation length-scale a:

$$\langle \nu(\mathbf{x})\nu(\mathbf{x}+\mathbf{H})\rangle = \varepsilon^2 \exp\left(\frac{-|\mathbf{h}|^2}{a^2}\right) \tag{1}$$

where the angle brackets imply integration over $\mathbf{x}$.

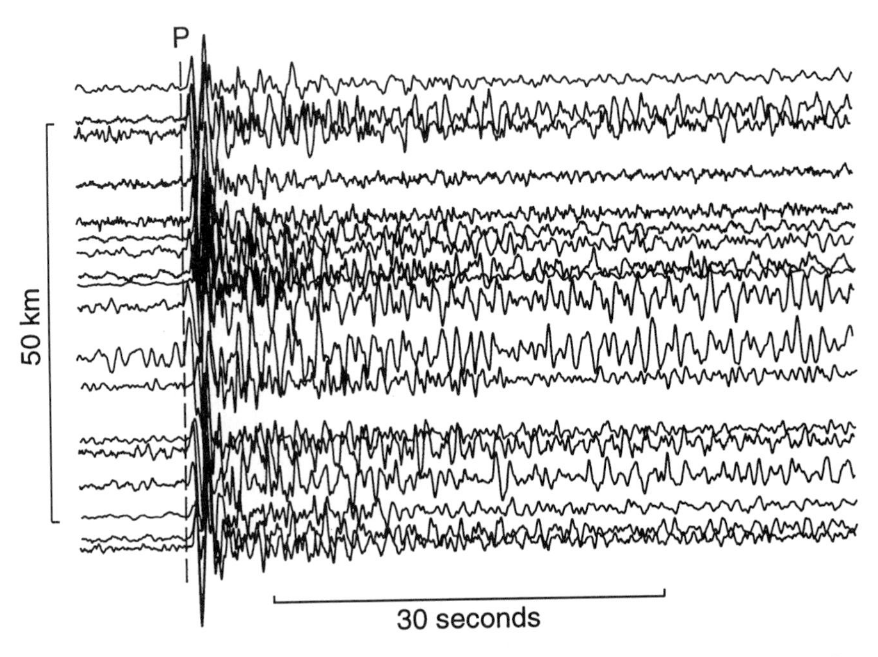

Figure 1 Record section of an earthquake in the Tonga-Fiji subduction complex (January 17, 1995; $M = 6.0$; hypocentral depth of 637 km) recorded by the Hawaiian Volcano Observatory seismic network. Almost all of the energy in the slowly decaying and spatially incoherent coda following the impulsive, short-duration *P*-wave is generated by scattering in the volcanic edifice and lithosphere below the array.

For LASA, a layer extending to ~60 km depth with $\varepsilon^2 = 4\%$ and $a = 10$ km explains the data well (Aki 1973). Other authors report longer correlation lengths for the Norwegian Seismic Array ($a = 15$–60 km) (Berteussen et al 1975), the southern California seismic array ($a = 25$ km) (Powell & Meltzer 1984), and the Gauribidanaur array in southern India ($a = 20$ km) (Berteussen et al 1977), indicating variation in the fundamental scale of lithospheric structure beneath the four sites. However, the single-layer, Gaussian autocorrelation representation of random inhomogeneity in these studies is extremely limited by poor data resolution. Flatté & Wu (1988) introduced a new measure of amplitude and phase coherence, enabling them to significantly expand the representation of lithospheric heterogeneity. Working with Norwegian Seismic Array (NORSAR) data, they obtained a multilayer power law model of compressional velocity fluctuation, extending to ~250 km depth (Flatté et al 1991). In the upper crust, above ~15 km depth, the heterogeneity spectrum is white between 5.5 and 110 km wavelength (the spectrum is unconstrained outside this band). Between 15 and 175 km depth, the lithosphere is enriched in long-wavelength variability while retaining greater short-wavelength variation than the subjacent asthenosphere. This is indicative of diffusive processes erasing shorter-scale-length heterogeneity in the asthenosphere. Retention of significant short-wavelength variability in the lithosphere is consistent with the work of Silver and co-workers (e.g. Silver & Chan 1991), documenting whole lithospheric strain fabrics persisting over hundreds of millions of years. Clearly, the Mohorovicic discontinuity is not the base of geology (Silver 1996).

The Russian peaceful nuclear explosion refraction profiles document a *Pn* wave propagating within the shallow subcrustal lithosphere to distances of several thousand kilometers (Ryberg et al 1995) and followed by a high-frequency coda. Conventional layered Earth models do not predict *Pn* at these distances, nor do they produce significant coda energy. Tittgemeyer et al (1996), however, show that multiple scattering within a highly heterogeneous mantle lid produces teleseismic *Pn* phases and high-frequency coda. Superposition of velocity fluctuations with $\varepsilon^2 = 4\%$ and a vertical correlation length scale of 2 km onto a positive upper mantle velocity gradient reduces the effective velocity of the upper mantle *P* wave at low frequencies and results in multiple scattering of the higher-frequency components. This highly heterogeneous waveguide must extend to at least 100 km depth. The short correlation length scale may be an artifact of modeling, which restricted velocity fluctuations to one dimension, but it is more likely a true property of cold Russian platform lithosphere reflecting differences in the correlation length scale of heterogeneity in the horizontal (long) and vertical (short) directions (e.g. Wagner 1996, 1997).

Although the long-range Russian profiles are unique, European controlled source profiles display similar levels of *Pn* coda and there are abundant single-station observations of *Sn* and *Pn* at teleseismic distances (Molnar & Oliver 1969). Enderle et al (1997) take this as evidence of an extensive upper mantle *Pn* waveguide. What is responsible for the velocity fluctuations is a question of considerable interest. Without access to lithospheric mantle it is difficult to associate scale lengths and RMS velocity fluctuations to specific geologic structures or the causative processes. In recent years, there have been several attempts to image localized scatterers in continental lithosphere by using local array recordings of regional and teleseismic waves. These studies use a hybrid parameterization of heterogeneity, specifying scattering strength (or some associated parameter) as a function of position. We refer collectively to the resulting techniques as scattered-wave imaging. The two most common approaches, semblance (Lay 1987, Lynnes & Lay 1989) and stochastic Kirchhoff migration (Revenaugh 1995b), both rely on travel time information to map arrivals in coda to specific scatterer positions but differ in how they detect these arrivals. Semblance uses a measure of waveform similarity that assumes scattered waves radiate in all directions with equal magnitude and unchanging polarity. Stochastic Kirchhoff migration uses fluctuations of coda energy, rather than waveform similarity, eliminating the constraint on scattered wave polarity. By design, both are better able to detect isolated features with scale lengths comparable to seismic wavelength ("point" scatterers) than extended bodies or surfaces. Neither is capable of estimating the absolute strength of scattering; each provides a measure of relative variation.

Using the semblance method, Mohan & Rai (1992) image a strong scatterer in the lower crust and upper mantle west of the Gauribidanaur (India) array. They associate this feature, localized between 35 and 55 km depth, with a granitic intrusion thought to be a Precambrian suture zone between the east and west Dharwar craton.

Revenaugh (1995b) uses stochastic Kirchhoff migration to image low-angle forward scatterers in the lithosphere beneath southern California. As applied to a data set of short-period recordings of teleseismic *P* waves, the method reveals a "curtain" of high scattering strength paralleling the southern flank of the Transverse Ranges (Figure 2). This feature coincides with a slab of

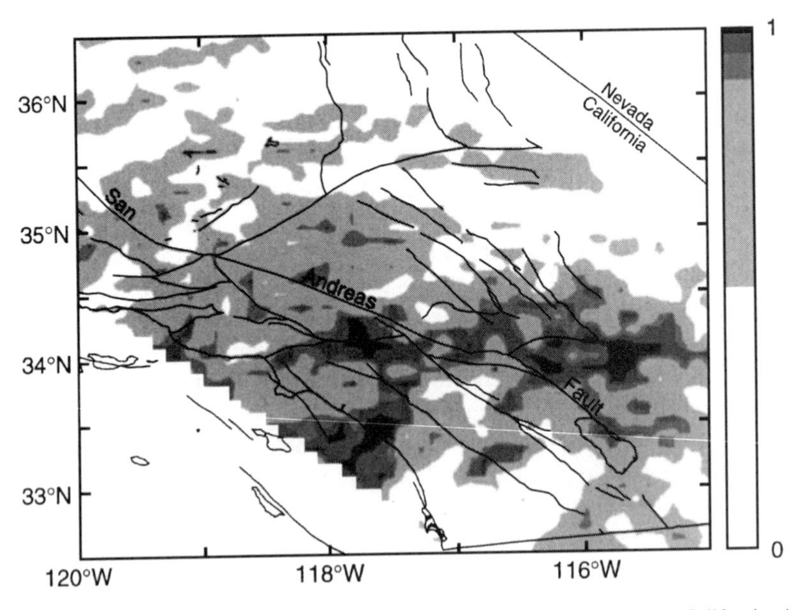

Figure 2 Image of scattering strength for a depth slice at 150 km beneath southern California. An east-striking sheet of strong scattering dips almost vertically to greater than 250 km depth. This marks the southern flank of a high-velocity slab that has been independently imaged by seismic tomography.

high *P*-wave velocity in the upper mantle seen in tomographic images (e.g. Humphreys & Clayton 1990). Associated with subcrustal lithospheric subduction of both the North American and Pacific plates through the Big Bend region of the San Andreas fault system, this primarily thermal anomaly should not be an efficient scatterer of short-period (~1 s) *P* waves. Revenaugh (1995b) reconciles this discrepancy by appeal to progressive tearing of the downgoing slab, juxtaposing cool slab interior with warmer mantle rocks. Localized highs in the scattering pattern underlie along-strike steps in the Transverse Ranges, suggesting a dynamic tie between slab tears and shallower orogenic processes. Away from this structure, scattering variability in the upper mantle is minimal. Using a similar technique applied to regional *S*-wave coda, Nishigami (1991) also finds low levels of lateral variation in upper-mantle scattering strength.

Common to the scattered-wave imaging studies is an absence of strong lateral variations in scattering strength in the subcrustal lithosphere away from structural boundaries. Heterogeneous over tens of kilometers but homogeneous over hundreds of kilometers, uppermost continental mantle apparently consists of large blocks of statistically uniform material. Variation is dominantly vertical, most likely reflecting thermal-activation processes and differentiation but potentially imparted or affected by growth processes such as subhorizontal subduction accretion (Bostock 1997).

REFLECTIVITY OF CONTINENTAL CRUST

The deterministic model of Earth structure, in particular splitting of the heterogeneity spectrum into long-wavelength background variations and short-wavelength reflectors, was developed to model the subhorizontal layering and strong impedance contrasts of shallow, sedimentary sections. Seismic reflection profiling is premised on this description of the subsurface, beginning with acquisition and continuing through to processing and interpretation. Its success in exploration is a testament to the power of the deterministic description in the shallow environment. As applied to crystalline basement rocks, however, this description is a signal failure. Laterally coherent reflectors are almost entirely absent, replaced by short, anastomosing reflections and a much more diffuse "coda." Common processing and interpretation techniques assuming single scattering and a horizontal heterogeneity spectrum depleted in high wavenumber energy produce false coherence, biasing the result toward the deterministic model (Gibson & Levander 1990, Emmerlich et al 1993). Progress in this environment requires three things: (*a*) a statistical model of the subsurface capable of explaining relevant aspects of the backscattered wave field, (*b*) sufficient theory and numerical ability to accurately model wave propagation, and (*c*) the means to robustly extract information from data. In a series of papers, Levander, Holliger, Goff, and other workers have addressed these three fronts.

Initial development of a statistical model of crystalline crust relied on models of continuous Gaussian or self-similar media, similar to those used in studies of subcrustal lithosphere. Although it is simple to use, continuous inhomogeneity does a poor job of representing exposures of former mid- and lower-crustal rocks that tend to be structurally complex assemblages of distinct lithology, rather than continuously variable fields. This discrepancy led to the introduction of "modal" fields: two- and three-dimensional fields that assume only a small number of discrete values or lithologies (Goff et al 1994).

The spatial variation of modal fields is described by an autocorrelation function. This is typically the von Karman autocorrelation function developed for turbulent media, which describes a self-similar or self-affine fractal field with variability at all length scales and no periodicities (a self-affine field requires an anisotropic rescaling to appear self similar). The von Karman autocorrelation constrains only the amplitude spectrum of the heterogeneity, saying nothing about phase, emphasizing its inherently statistical nature. The success of this description in modeling field observations of exposed basement rocks (Holliger et al 1993, Levander et al 1994) suggests that the modal field description is a sufficient characterization of the deep seismic structure of continental crust.

Unless the RMS variation of seismic heterogeneity is small (several percent or less), multiple scattering will be important. The effects of data processing steps assuming dominant single scattering, or predictable multiple scattering as in the case of reflection "multiples," must be understood. These effects are manifest both in deterministic imaging of the subsurface (e.g. migration) and in statistical inference. For example, one must understand the effects of tuning and detuning of seismic waves by various length scales of heterogeneity before attempting to interpret amplitudes in terms of RMS velocity variation.

Much work has gone into the analytical description of multiply scattered wave fields in inhomogeneous media. For elastic (vector) waves this is enormously difficult, but progress is being made (e.g. Li & Hudson 1995, 1996; Shapiro & Treitel 1997). At present, the most efficient means of understanding wave interaction with random inhomogeneity is through Monte Carlo modeling (e.g. Wagner & Langston 1992, Wagner 1996). A number of investigators have followed this approach, investigating the effects of standard seismic processing on synthetic data for random, two-dimensional acoustic heterogeneity. As a summary of the primary results, I note that (*a*) common-midpoint stacking, acting as a dip filter, artificially enhances the horizontal coherency of images of random inhomogeneity (Gibson 1991); (*b*) migration does not eliminate this smearing and produces erroneous horizontal coherence, even for a scattering medium with little or no lateral continuity (Emmerlich et al 1993); and (*c*) unmigrated images bear little resemblance to the underlying structure and individual reflection sequences may represent interference more than structure (Holliger et al 1994). These effects worsen as the importance of multiple scattering increases. To date, most active-source studies have synthesized data in only two spatial dimensions, have coupled P-wave velocity with density, and have ignored S-wave velocity variability. Although the latter appears to be relatively unimportant for near-vertical incidence (Emmerlich 1993), it is very likely that these studies underestimate the amount of image distortion induced by multiple scattering. More powerful prestack migration and three-dimensional imaging are likely to improve the situation, but these expensive techniques find only rare application in deep crustal imaging where the deleterious effects of multiple scattering are worst.

This somewhat grim reality beckons the question: what information can we meaningfully extract from deep crustal reflection profiles? The answer, of course, depends on the setting, both geological and experimental. Deterministic structure, which satisfies the single scattering approximation, may be evaluated, although the focusing and defocusing effects of random inhomogeneities will introduce amplitude and phase distortions. Of the latter, only a statistical description is possible. The aspects of a statistical description that can be determined depend on the inhomogeneity itself, the accuracy of the assumed description, and the temporal and spatial resolution of the data. For example, Gibson (1991) relates the lateral coherence of active-source reflection data to lateral coherence of the inhomogeneity for a smooth (nonmodal) field with little multiple scattering. On the other hand, Holliger et al (1992) show that no similar estimate is possible for a model of deep crustal inhomogeneity based on the Ivrea Zone. In this case, homogenization of the wave field due to multiple scattering largely desensitizes the wave field to lateral coherence of the medium, and it is difficult to relate the remaining sensitivity to true lateral coherence of the medium (Hurich 1996).

Pullammanappallil et al (1997) discuss a Monte Carlo based optimization scheme for estimating the horizontal characteristic length scale a_x and Hurst number ν (a parameter of the von Karman autocorrelation related to fractal dimension) from seismic exploration data. They apply this method to two deep crustal seismic profiles in the Basin and Range Province. For depths of 12 to 21 km, a_x is ~240 m and $\upsilon = 0.5$ (fractal dimension of ~2.8). The latter is equivalent to the value determined from field studies of the Ivrea Zone, Italy (Holliger et al 1993) and upper- and middle-crustal rocks of the Lewisian gneiss complex, Scotland (Levander et al 1994), but much larger than values for the shallow upper crust where small cracks dominate the scattering inhomogeneity (Holliger 1996, 1997; Leary & Abercrombie 1994). The characteristic length scale is roughly a quarter of the first Fresnel zone at depth and much shorter than the length scales estimated for the subcrustal lithosphere. Along these lines, there is a clear break in reflectivity structure (e.g. Mooney & Meissner 1992, Enderle et al 1997) and a change in scattering character (Ritter et al 1997) associated with the Mohorovicic discontinuity, suggesting different length scales of heterogeneity above and below. As estimates of this type become more common, it should be possible to infer dominant processes involved in the production of inhomogeneity.

SCATTERING AND SEISMOGENESIS

There is a growing body of literature suggesting that structural features within the crust strongly influence the seismogenic process. Some of the best indications draw from travel time tomography, showing clear correlations of velocity anomalies with centers of main shock slip and aftershocks (e.g. Michael & Eberhart-Phillips 1991, Foxall et al 1993, Zhao et al 1996). To many, this will not seem surprising—certainly, crustal properties modulate slip, and some, if not much, of the richness of the distribution of seismicity must mirror complexities of the crust. However, numerical simulations of rupture on interacting faults show that it is quite possible to produce complex spatial and temporal patterns in a uniform crust (e.g. Rundle 1988). To what extent and in what ways is the seismogenic process influenced by properties of the surrounding crust? Along with comprehensive seismic and geodetic monitoring, an answer to this question requires that we thoroughly characterize the crust around active faults. This entails many geological and geophysical probes. Seismic scattering figures heavily.

An early observation in the study of regional *S*-wave coda is the pronounced variation in coda Q^{-1} between tectonically active and stable areas. Coda Q^{-1}, a measure of the attenuation rate of coda-wave amplitude with time, is highest (attenuation greatest) in active regions (e.g. Singh & Hermann 1983). Coda Q^{-1} reflects a complex mixture of intrinsic and scattering attenuation, the latter measuring the rate at which energy is scattered out of a wave front. Geographic variability in coda Q^{-1} is due largely to variation in intrinsic attenuation, which is greater in tectonically active (warm) regions. However, scattering attenuation also varies and is greatest in extensively faulted regions (e.g. Nishigami 1991, Jin et al 1994). This implies that the crust, and perhaps the uppermost mantle, surrounding active faults is highly "opaque" to seismic waves, that is, that it scatters seismic waves strongly.

Coda attenuation measurements sample a roughly ellipsoidal volume of crust and upper mantle containing the source and receiver. To minimize the impact of nonisotropic scattering and nonuniformly distributed scatterers, measurements are made at long lapse times and thus average over a large area. This averaging clouds the connection between scattering and faults, in particular, the relative roles of the fault zone versus distributed cracks.

Several recent studies attempt to constrain more tightly the geographic distribution of scattering strength proximal to faults. Nishigami (1991, 1997) uses the single scattering model of Sato (1977) as the basis of an inversion method for imaging variations of volumetric scattering strength in three dimensions. For regional *S*-wave coda data from the Hokuriku district, central Japan, scattering strength in the upper crust near mapped faults is 20% greater than regional background levels. Using power scaling rules (Aki and Richards 1980), Nishigami concludes that velocity fluctuations near the faults are 10% greater than elsewhere. The scattering highs do not clearly define the fault traces, however, and are spread over 10–30 km, suggesting that extensive cracking outside the fault zone proper may contribute significantly to scattering.

Matsumoto et al (1998) apply a migration-like operator to active-source reflection data sampling the source region of the 1995 Kobe, Japan ($M = 7.2$) earthquake. Their images of subsurface *P*-wave scattering strength reveal two patches of strong scattering. One, spread out between 10 and 25 km depth, lies southwest of the hypocenter in a complexly faulted region. The second lies just below the hypocenter, suggesting that rupture initiated in or near a highly inhomogeneous region characterized by structural scale lengths of order 500 m. The tomographic study of Zhao et al (1996) identifies the hypocenter as a low *P*-wave velocity, high Poisson's ratio region, which they attribute to a local concentration of fluid-filled cracks, a natural candidate for strong scattering.

Revenaugh (1995c) modified the stochastic Kirchhoff migration method of Revenaugh (1995b) to image scattering variations in the crust. This approach has the considerable advantage of not relying on local seismicity, allowing imaging of quiescent fault segments. As implemented, it estimates geographic variation of the statistical significance of scattering (or scattering "potential") in the upper crust ($<$12–15 km depth) rather than the strength of scattering, a much less stable quantity. Scattering potential is a measure of the likelihood that scattering strength locally exceeds the regional mean; scattering potentials near unity imply locally strong scattering, whereas potentials near zero mark the weakest local scattering strengths. In other words, scattering potential is a relative index of scattering strength.

Color Plate 1 shows scattering potential for the region of southern California surrounding the June 28, 1992, Landers earthquake ($M_W = 7.3$). Scattering potential varies considerably over a variety of length scales. The highest scattering potentials occur close to the rupture area, but they do not delineate the fault, and there are high scattering potential patches far from mapped fault segments. There is, however, a noticeable tendency for aftershocks to cluster in regions of high scattering potential. This observation is clearer in Figure 3,

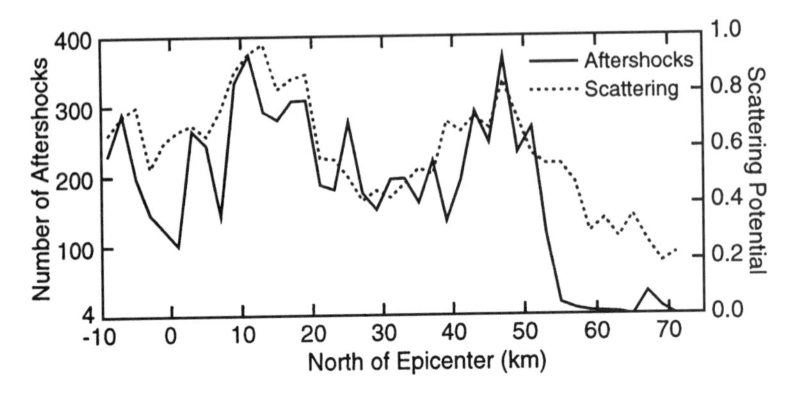

Figure 3 Along-fault comparison of mean scattering potential and $M_L = 2.0$ aftershock seismicity within 8 km of the rupture zone of the 1992 Landers earthquake (Figure 3). The excellent correlation implies that the seismic structure responsible for scattering modulates seismogenesis.

which compares along-fault variation in scattering potential and aftershock density. There are a number of implications of this clustering. Foremost, it requires that aftershock occurrence be tied, at least in part, to properties of the crust near the fault. This raises a question: Does repeated slip on the fault and/or near-fault stress variation induce scattering by increasing fracture density, for example, or does crustal structure predating fault development influence present-day seismogenic processes? The answer may lie anywhere in between these two extremes, but the correlation of a measurable property of the crust with aftershock density and coseismic slip (Revenaugh 1995c) implies some predictability of the earthquake process, at least in terms of along-fault slip variability.

To further elucidate the relation of crustal scattering and seismogenesis, Revenaugh & Reasoner (1997) image the San Andreas fault system through central and southern California by using stochastic Kirchhoff migration. The paucity of seismicity along many stretches of the San Andreas precludes meaningful correlation of scattering potential and aftershock seismicity. At present, the instrumental record of seismicity is too short to allow such a comparison. Revenaugh (1995c) notes that transitions from high to low scattering correlate geographically with fault jumps and bends along the Landers rupture zone, prompting Revenaugh & Reasoner (1997) to look for a similar correlation between segment bounds and scattering along the San Andreas.

Segment bounds delimit fault stretches that rupture singly or multiply in large earthquakes. They are determined from historic and paleoseismicity and geologic constraints, making them more representative of long-term fault behavior than the short instrumental record alone. Revenaugh & Reasoner (1997)

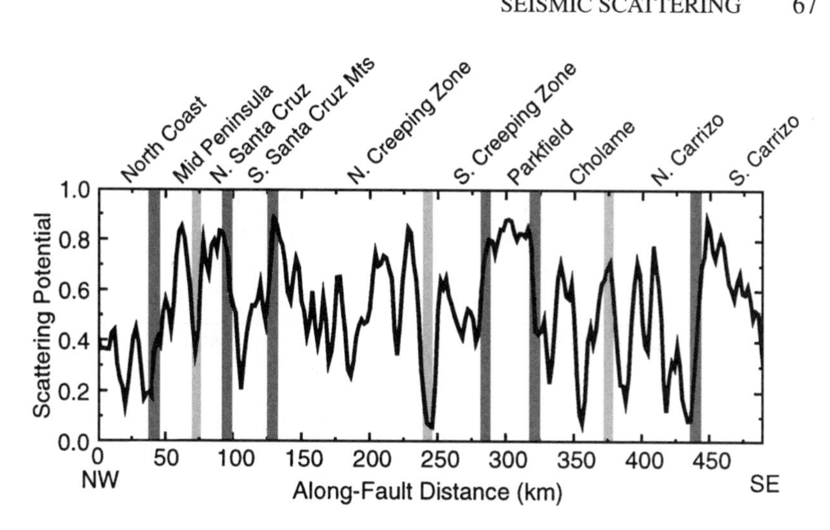

Figure 4 Along-fault gradient in scattering potential and segment bounds (vertical bars) for a ~500 km stretch of the San Andreas Fault in central California. Six of nine segment bounds (Working Group on California Earthquake Probabilities 1988) lie in regions of high scattering gradient (gradients outside the one standard deviation level), an outcome with a <5% chance of random occurrence.

find that segment bounds coincide with high along-fault scattering gradients for the San Andreas fault in central California (Figure 4). The correlation of segment bounds and scattering gradients suggests either that segment bounds delimit distinct crustal blocks (with distinct scattering signatures) or that segment bounds mark transitions between distinct stress states that dynamically influence scattering, the latter potentially manifested as changes in crack density or crack orientation. Distinguishing these two hypotheses is the same question of precedence raised by Revenaugh (1995c) for scattering near the Landers earthquake. Revenaugh & Reasoner (1997) consider it in unique fashion, by searching for evidence of offset in the pattern of scattering potential along the San Andreas fault. They obtain optimal cross-fault correlation for 315 km of right-lateral offset (Figure 5), a value in excellent agreement with geologic estimates of cumulative offset. The fact that scattering offsets (that is, that some portion of the scattering field advects with the crust), suggests that preexisting structure imparts a strong influence on fault structure.

Temporal Change

The static offset of scattering potential does not exclude a dynamic component. Indeed, there is mounting evidence of a component of crustal and shallow

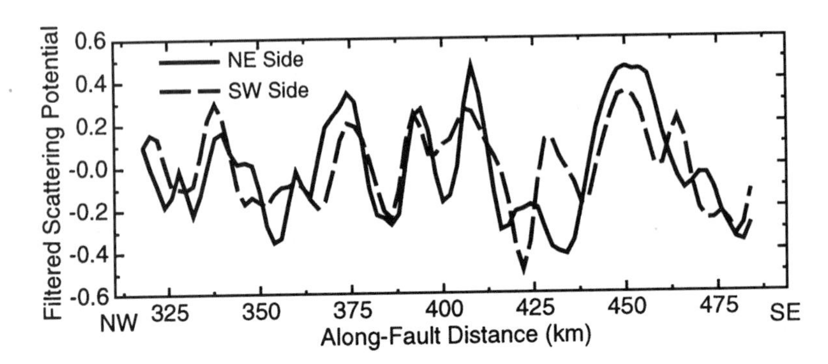

Figure 5 Cross-fault comparison of scattering potential along the San Andreas Fault in central California after removal of 315 km of right-lateral offset. Profiles have been filtered to remove long-wavelength variations. The excellent correlation implies scattering advects with the crust. Its relationship to aftershock seismicity (Figure 4) and segment bounds (Figure 5) implies seismogenesis is strongly influenced by heterogeneity predating the fault.

mantle scattering that responds dynamically to changes in the regional stress field. This comes from temporal variation in coda Q^{-1} of regional S waves.

Intrinsic attenuation is an essentially static property of the shallow lithosphere, such that any temporal variation in coda Q^{-1} must reflect changes in scattering attenuation. Significant temporal variation in coda Q^{-1} has been observed in several tectonic settings and over time scales ranging from weeks to decades (Chouet 1979, Jin & Aki 1989, Su & Aki 1990). More important is the apparent connection with seismicity: large temporal gradients in coda Q^{-1} often coincide with changes in seismic b value, a measure of the production rate of small earthquakes relative to large (e.g. Jin & Aki 1989). However, the sense of correlation is not consistent—an increase in b value may accompany an increase or a decrease in coda Q^{-1}—and there are coda Q^{-1} gradients with no obvious association to seismicity. Jin & Aki (1989) and Aki (1992) suggest that this behavior indicates changes in crack density in the semiductile lithosphere. Scattering strength and attenuation are highly sensitive to crack density, causing coda Q^{-1} to vary. The sense of variation, increase or decrease, depends on the scaling of dominant seismic wavelength to fracture size. Likewise, small earthquakes on these fractures may raise or lower seismic b value depending on the local stress state and the method of computing b value. The result is a coincidence in timing, but no consistency in sign. The spatial and temporal coarseness of coda Q^{-1} and seismic b-value measurements make it difficult to envision an effective monitoring program based on these observations, but the correlation is intriguing and may offer insight into the boundary conditions of the seismogenic process.

Other Scattered-Wave Imaging Studies

Chávez-Pérez & Louie (1998) use regional *P*-wave coda to investigate crustal structure near the 1994 Northridge earthquake. Combining aspects of controlled-source profiling and scattered-wave imaging, they map a north-dipping reflector, possibly corresponding to the Elysian Park Thrust (Davis & Namson 1994), and an offset subhorizontal reflector that they interpret as a midcrustal detachment. Innovative studies of this sort have the potential to greatly advance our knowledge of the deep configuration of faults mapped in the near surface.

Scattered waves are an important component of earthquake strong motion. Spudich & Iida (1993) use a source array to image scatterers contributing to the early *S*-wave coda. They find that laterally propagating waves scattered from a basin edge were prominent in the early coda. These are a mixture of *S* and surface waves (*Rg*). A number of other studies note basin-edge surface waves (e.g. Frankel & Vidale 1992), which appear to be a dominant component of early coda. Revenaugh (1995a) and Revenaugh & Mendoza (1996) demonstrate a regional correlation of surface topographic roughness and the efficiency of *P* to *Rg* scattering, which they use to formulate a parametric model of high-frequency *Rg* velocity. Knowledge of *Rg* producers (scatterers) and propagation velocity will vastly improve seismologists' ability to predict ground motion.

SCATTERING AND VOLCANOES

Seismic scattering is finding increasing application as a means of quantifying the short scale-length heterogeneity and magma plumbing of active volcanoes. Scattering is well suited to this task. First arrivals avoid low-velocity magma bodies, which are difficult to image tomographically unless they are large, but can be very strong scatterers. Likewise, faults and thin layers have little effect on travel time, but they do produce coda arrivals.

Nishigami (1997) applies his coda-envelope inversion (Nishigami 1991) to two active volcanoes in central Japan (Mount Ontake and Mount Nikko-Shirane). Strong scattering beneath Mount Ontake extends to a depth of $\sim$7 km, close to the maximum depth of microseismic activity. Strong scattering is confined to a narrow linear zone to 7 km depth, where it spreads out laterally. Interestingly, the zone of high scattering is nearly void of microseisms. A similar line of high scattering strength extends beneath Mount Nikko-Shirane to 20 km depth. In both cases the coincidence of the volcano and the vertical zone of high scattering suggests that a magma conduit is responsible.

Mikada et al (1997) adapt a diffraction tomography approach to image scattering beneath Izu-Oshima volcano, an island about 100 km south of Tokyo, Japan. Their detailed image shows several patches of strong scattering. Most conspicuous is a cloud of scatterers centered at $\sim$9 km depth directly beneath the volcano crater. Mikada et al associate this feature with the primary magma reservoir. They further associate smaller and shallower patches of high scattering strength with subchambers tapped during eruptive activity in 1986.

Nishimura et al (1997) apply an envelope inversion method to active-source recordings from the Jemez volcanic field, New Mexico. Their profiles of the depth dependence of scattering peak in the upper 5–7 km beneath the Valles Caldera at a level more than a factor of five greater than similar profiles for the Rio Grande Rift and Colorado Plateau. They conclude that two episodes of eruption and collapse of the caldera, the presence of rhyolite domes, pyroclastic material, and intrusions into caldera fill are responsible. Scattering beneath the Rio Grande Rift is stronger than that beneath the Colorado Plateau, again indicating a significant role of intrusion and volcanism in the production of scattering heterogeneity.

CONCLUSIONS

For many years, seismology had one way of parameterizing the Earth: as a layered structure colored by smooth velocity perturbations, much like a Mark Rothko painting except that the edges are sharp (reflectors). Capturing much of Earth's inhomogeneity, this representation has been enormously successful, but seismology, in particular, seismic scattering, has much more to tell us. Scattered waves feel the "fabric" of crust and mantle rock and carry that information to the surface. The pictures they paint vary with the style of representation. The purely statistical representation is like a Jackson Pollock drip painting: what matters is the fabric, not the individual arcs of color. To realize its full potential, the sophistication of representation and modeling must rise, giving full consideration to vector waves, realistic background structures, and three-dimensional wave propagation, as well as the connection between scattering heterogeneity and geology. The work of Levander, Holliger, and others demonstrates the tremendous promise of using geology to design statistical representations of seismic inhomogeneity, but there is much work left to do. The same is true of scattered-wave imaging, which sees the world through the eyes of a pointillist, as a distribution of individual scatterers. To continue making progress, the relation of scatterers to geology and other geophysical images of the subsurface must be better understood, and seismology must strive to make the scatterers more versatile, better able to capture the essence of complex larger structure. In these regards, scattered-wave seismology is still on the steep limb of the learning curve.

ACKNOWLEDGMENTS

The author received support from National Science Foundation Grant EAR-9614874 and Lawrence Livermore National Lab/Institute for Geophysics and Planetary Physics.

Literature Cited

Aki K. 1969. Analysis of the seismic coda of local earthquakes as scattered waves. *J. Geophys. Res.* 74:615–31

Aki K. 1973. Scattering of *P* waves under the Montana LASA. *J. Geophys. Res.* 78:1334–46

Aki K. 1992. Higher-order interrelations between seismogenic structures and earthquake processes. *Tectonophysics* 211:1–12

Aki K, Chouet B. 1975. Origin of coda waves: source, attenuation and scattering effects. *J. Geophys. Res.* 80:3322–42

Aki K, Richards P. 1980. *Quantitative Seismology*, San Francisco: Freeman. 932 pp.

Berteussen KA, Christoffersson A, Husebye ES, Dahle A. 1975. Wave scattering theory in analysis of *P* wave anomalies at NORSAR and LASA. *Geophys. J. R. Astron. Soc.* 42:402–17

Berteussen KA, Husebye E, Mereu RF, Ram A. 1977. Quantitative assessment of the crust-upper mantle heterogeneities beneath the Gauribidanaur seismic array in southern India. *Earth Planet. Sci. Lett.* 37:326–32

Bostock MG. 1997. Anisotropic upper-mantle stratigraphy and architecture of the Slave craton. *Nature* 390:392–95

Chávez-Pérez S, Louie JN. 1998. Crustal imaging in southern California using earthquake sequences. *Tectonophysics* 286:223–36

Chernov J. 1960. *Wave Propagation in a Random Medium*, New York: McGraw-Hill. 168 pp.

Chouet B. 1979. Temporal variation in the attenuation of earthquake coda near Stone Canyon, California. *Geophys. Res. Lett.* 6:143–46

Davis TL, Namson JS. 1994. A balanced cross-section of the 1994 Northridge earthquake, southern California. *Nature* 372:167–69

Emmerlich H, Zwielich J, Müller G. 1993. Migration of synthetic seismograms for crustal structures with random heterogeneities. *Geophys. J. Int.* 113:225–38

Enderle U, Tittgemeyer M, Itzin M, Prodehl C, Fuchs K. 1997. Scales of structure in the lithosphere-images of processes. *Tectonophysics* 275:165–98

Flatté SM, Wu RS. 1988. Small-scale structure in the lithosphere and asthenosphere deduced from arrival-time and amplitude fluctuations at NORSAR. *J. Geophys. Res.* 93:6601–14

Flatté SM, Wu RS, Shen ZK. 1991. Nonlinear inversion of phase and amplitude coherence functions at NORSAR for a model of nonuniform heterogeneities. *Geophys. Res. Lett.* 18:1269–72

Foxall W, Michelini A, McEvilly TV. 1993. Earthquake travel time tomography of the southern Santa Cruz Mountains: control of fault rupture by lithological heterogeneity of the San Andreas fault zone. *J. Geophys. Res.* 98:17,691–710

Frankel A, Vidale J. 1992. A three-dimensional simulation of seismic waves in the Santa Clara Valley, California, from a Loma Prieta aftershock. *Bull. Seismol. Soc. Am.* 82:2045–74

Gibson BS. 1991. Analysis of lateral coherency in wide-angle seismic images of heterogeneous targets. *J. Geophys. Res.* 96:10,261–73

Gibson BS, Levander AR. 1990. Apparent layering in common-midpoint stacked images of two-dimensionally heterogeneous targets. *Geophysics* 55:1466–77

Goff JA, Holliger K, Levander A. 1994. Modal fields: a new method for characterization of random seismic velocity heterogeneity. *Geophys. Res. Lett.* 21:493–96

Gudmundsson O. 1996. On the effect of diffraction on travel time measurements. *Geophys. J. Int.* 124:304–14

Holliger K. 1996. Upper-crustal seismic velocity heterogeneity as derived from a variety of *P*-wave sonic logs. *Geophys. J. Int.* 125:813–29

Holliger K. 1997. Seismic scattering in the upper crystalline crust based on evidence from sonic logs. *Geophys. J. Int.* 128:65–72

Holliger K, Carbonell R, Levander AR. 1992. Sensitivity of the lateral correlation function in deep seismic reflection data. *Geophys. Res. Lett.* 19:2263–66

Holliger K, Levander AR, Goff JA. 1993. Stochastic modeling of the reflective lower crust: petrophysical and geological evidence from the Ivrea Zone (northern Italy). *J. Geophys. Res.* 98:11,967–80

Holliger K, Levander A, Carbonell R, Hobbs R. 1994. Some attributes of wavefields scattered from Ivrea-type lower crust. *Tectonophysics* 232:267–79

Humphreys ED, Clayton RW. 1990. Tomographic image of the southern California mantle. *J. Geophys. Res.* 95:19,725–46

Hurich CA. 1996. Statistical description of seismic reflection wavefields: a step towards quantitative interpretation of deep seismic reflection profiles. *Geophys. J. Int.* 125:719–28

Jin A, Aki K. 1989. Spatial and temporal correlation between coda Q^{-1} and seismicity and its physical mechanism. *J. Geophys. Res.* 94:14,041–59

Jin A, Mayeda K, Adams D, Aki K. 1994. Separation of intrinsic and scattering attenuation in southern California. *J. Geophys. Res.* 99:17,385–848

Lay T. 1987. Analysis of near-source contributions to early *P*-wave coda for underground explosions, 3. Inversion for isotropic scatterers. *Bull. Seism. Soc. Am.* 77:1767–83

Leary P, Abercrombie R. 1994. Fractal fracture scattering origin of *S*-wave coda: spectral evidence from recordings at 2.5 km. *Geophys. Res. Lett.* 21:1683–86

Levander A, England RW, Smith SK, Hobbs RW, Goff JA, Holliger KA. 1994. Stochastic characterization and seismic response of upper and middle crustal rocks based on the Lewisian gneiss complex, Scotland. *Geophys. J. Int.* 199:243–59

Li X, Hudson JA. 1995. Elastic scattered waves from a continuous and heterogeneous layer. *Geophys. J. Int.* 121:82–102

Li X, Hudson JA. 1996. Multiple scattering of elastic waves from a continuous and heterogeneous region. *Geophys. J. Int.* 126:845–62

Lynnes CR, Lay T. 1989. Inversion of P coda for isotropic scatterers at the Yucca Flat test site. *Bull. Seism. Soc. Am.* 79:790–804

Matsumoto S, Obara K, Hasegawa A. 1998. Imaging *P*-wave scatterer distribution in the focal area of the 1995 *M*7.2 Hyogo-ken Nanbu (Kobe) earthquake. *Geophys. Res. Lett.* 25:1439–42

Michael AJ, Eberhart-Phillips D. 1991. Relations among fault behavior, subsurface geology, and three-dimensional velocity models. *Science* 253:651–54

Mikada H, Watanabe H, Sakashita S. 1997. Evidence for subsurface magma bodies beneath Izu-Oshima volcano inferred from a seismic scattering analysis and possible interpretation of the magma plumbing system of the 1986 eruptive activity. *Phys. Earth Planet. Inter.* 104:257–69

Mohan G, Rai SS. 1992. Imaging of seismic scatterers beneath the Gauribidanaur (GBA) array. *Phys. Earth Planet. Inter.* 71:36–45

Molnar P, Oliver J. 1969. Lateral variations of attenuation in the upper mantle and discontinuities in the lithosphere. *J. Geophys. Res.* 74:2648–82

Mooney WD, Meissner R. 1992. Multi-genetic origin of crustal reflectivity: a review of seismic reflection profiling of the continental lower crust and Moho. *Tectonophysics* 275:199–219

Nishigami K. 1991. A new inversion method of coda waveforms to determine spatial distribution of coda scatterers in the crust and uppermost mantle. *Geophys. Res. Lett.* 18:2225–28

Nishigami K. 1997. Spatial distribution of coda scatterers in the crust around two active volcanoes and one active fault system in central Japan: inversion analysis of coda envelope. *Phys. Earth Planet. Inter.* 104:75–89

Nishimura T, Fehler M, Baldridge WS, Roberts P, Steck L. 1997. Heterogeneous structure around the Jemez volcanic field, New Mexico, USA, as inferred from the envelope inversion of active-experiment seismic data. *Geophys. J. Int.* 131:667–81

Powell CA, Meltzer AS. 1984. Scattering of *P*-wave beneath SCARLET in southern California. *Geophys. Res. Lett.* 11:481–84

Pullammanappallil S, Levander A, Larkin SP. 1997. Estimation of crustal stochastic parameters from seismic exploration data. *J. Geophys. Res.* 102:15,269–86

Revenaugh J. 1995a. The contribution of topographic scattering to teleseismic coda in southern California. *Geophys. Res. Lett.* 22:543–46

Revenaugh J. 1995b. A scattered-wave image of subduction beneath the Transverse Ranges, California. *Science* 268:1888–92

Revenaugh J. 1995c. Relationship of the 1992 Landers, California, earthquake sequence to seismic scattering. *Science* 270:1344–47

Revenaugh J, Mendoza H. 1996. Mapping shallow heterogeneity with teleseismic *P* to *Rg* scattered waves. *Bull. Seismol. Soc. Am.* 86:1194–99

Revenaugh J, Reasoner C. 1997. Cumulative offset of the San Andreas fault in central California: a seismic approach. *Geology* 25:123–26

Ritter JRR, Mai PM, Stoll G, Fuchs K. 1997. Scattering of teleseismic waves in the lower crust: observations in the Massif Central, France. *Phys. Earth Planet. Inter.* 104:127–46

Rundle JB. 1988. A physical model for earthquakes. 1. Fluctuations and interactions. *J. Geophys. Res.* 93:6237–54

Ryberg T, Wenzel F, Mechie J, Egorkin AV, Fuchs K. 1995. High-frequency teleseismic Pn wave observations beneath northern Eurasia. *J. Geophys. Res.* 100:18,151–63

Sato H. 1977. Energy propagation including scattering effects; single isotropic scattering approximation. *J. Phys. Earth* 25:27–41

Shapiro SA, Treitel S. 1997. Multiple scattering

of seismic waves in multilayered structures. *Phys. Earth Planet. Inter.* 104:147–59
Silver PG. 1996. Seismic anisotropy beneath the continents: probing the depths of geology. *Annu. Rev. Earth Planet. Sci.* 24:385–432
Silver PG, Chan WW. 1991. Shear wave splitting and subcontinental mantle deformation. *J. Geophys. Res.* 96:16,429–54
Singh S, Herrmann RB. 1983. Regionalization of crustal coda Q in the continental United States. *J. Geophys. Res.* 88:527–38
Snieder R, Lomax A. 1996. Wavefield smoothing and the effect of rough velocity perturbations on arrival times and amplitudes. *Geophys. J. Int.* 125:796–8–12
Spudich P, Iida M. 1993. The seismic coda, site effects, and scattering in alluvial basins studied using aftershocks of the 1986 north Palm Springs, California, earthquake as source arrays. *Bull. Seismol. Soc. Am.* 83:1721–43
Su F, Aki K. 1990. Temporal and spatial variation of coda Q^{-1} associated with the north Palm Springs earthquake of July 8, 1986. *Pure Appl. Geophys.* 133:23–52
Tittgemeyer M, Wenzel F, Fuchs K, Ryberg T. 1996. Wave propagation in a multiple scattering upper mantle—observations and modelling. *Geophys. J. Int.* 127:492–502
Wagner GS. 1996. Numerical simulation of wave propagation in heterogeneous wave guides with implications for regional wave propagation and the nature of lithospheric heterogeneity. *Bull. Seismol. Soc. Am.* 96:1200–6
Wagner GS. 1997. Regional wave propagation in southern California and Nevada: observations from a three-component seismic array. *J. Geophys. Res.* 102:8285–311
Wagner GS, Langston CA. 1992. A numerical investigation of scattering effects for teleseismic plane wave propagation in a heterogeneous layer over a homogeneous half-space. *Geophys. J. Int.* 110:486–500
Working Group on California Earthquake Probabilities. 1988. Probabilities of large earthquakes occurring in California on the San Andreas fault. *US Geol. Surv. Open File Rep.* 88-398:1–62
Zhao D, Kanamori H, Negishi H, Wiens D. 1996. Tomography of the source area of the 1995 Kobe earthquake: evidence for fluids at the hypocenter? *Science* 274:1891–94

Annu. Rev. Earth Planet. Sci. 1999. 27:75–113

THE GLOBAL STRATIGRAPHY OF THE CRETACEOUS-TERTIARY BOUNDARY IMPACT EJECTA

J. Smit
Department of Sedimentary Geology, Vrije Universiteit, 1081HV Amsterdam, Netherlands; e-mail: smit@geo.vu.nl

KEY WORDS: K/T boundary, Cretaceous-Tertiary boundary, ejecta, tektites, microkrystites, Chicxulub crater, extinction, stratigraphy

ABSTRACT

The Chicxulub crater ejecta stratigraphy is reviewed, in the context of the stratigraphy of underlying and overlying rock sequences. The ejecta sequence is regionally grouped in (*a*) thick polymict and monomict breccia sequences inside the crater and within 300 km from the rim of the crater known from drill holes in and close to the breater, and exposures near the border of Yucatan and Belize; (*b*) Gulf of Mexico region, <2500 m from the crater, with up to 9 m thick, complex, tsunami-wave influenced, tektite-bearing sequences in shallow marine (<500 m deep) environments and tektite bearing, decimeter thick gravity-flow deposits in deep water sites; (*c*) an intermediate region between 2500 and 4000 km from the crater where centimeter thick, tektite-bearing layers occur, and (*d*) a global distal region with a millimeter thin ejecta layer. The distal ejecta layer is characterized by sub-millimeter sized microkrystites, often rich in Ni-rich spinels and (altered) clinopyroxene. Wherever present, the ejecta layers mark exactly the sudden mass-mortality horizon of the K/T boundary. What exactly caused the mass mortality is still uncertain, but it appears the main event leading to the K/T mass extinctions.

INTRODUCTION

It is often surprising how easily the Cretaceous/Tertiary (K/T) boundary can be identified in outcrops and drill cores. Whether in New Zealand, Spain, or the mid-Pacific, the boundary is often visible as a bedding plane. The only

0084-6597/99/0515-0075$08.00

exceptions are deep ocean clay cores and areas with strong bioturbation. This simple observation has far-reaching consequences. In shallow-water sequences, interruptions in sedimentation such as diastems, discontinuities, or disconformities often occur due to sea level changes and tectonic uplift, and a sharp bedding plane is not uncommon. Therefore, it is not surprising to encounter the K/T boundary as a gap in shallow-water sequences.

On the other hand, it is not obvious to expect gaps in deep-water sequences. Before the Deep-Sea Drilling Project (DSDP) and its successor the Ocean Drilling Program (ODP), it was widely believed that sedimentation on the deep ocean floor is more or less continuous. However, DSDP/ODP showed (Van Andel et al 1977, Worsley 1974) that most deep-sea records are also interrupted by omission surfaces ascribed to extremely low sedimentation rates, deepening of the calcium carbonate compensation depth, slumping and contour or turbidity currents. The K/T bedding plane, which also occurs in the deep sea, is therefore often routinely dismissed as representing missing strata and hiatuses produced by sea level changes. However, the sea level chart (Haq et al 1988), which is based on seismic stratigraphy rather than biostratigraphy, does not reveal obvious sea level changes near the K/T boundary to which the bedding-plane discontinuity can be attributed. Also, it is not obvious that important gateways were opened or closed at the time (Figure 1).

A number of deep-water sequences, in particular those near the continental margins, offer continuous records and high sedimentation rates (2–8 cm/kyr). The upper Cretaceous and Paleocene in these sequences are represented by thick packages of strata, usually bedded in cyclic Milankovitch-style patterns (Kate & Sprenger 1993). Such thick packages allow evaluation of the completeness of the K/T boundary transition, through detailed magnetobiostratigraphy and cyclostratigraphy supplemented with other sedimentary evidence at the K/T boundary itself. However, even in these complete sequences, it is hard to prove positively that sedimentation has remained continuous across the bedding plane discontinuity of the K/T boundary.

Controversies exist regarding many aspects of the K/T boundary stratigraphy. It is claimed that the upper Cretaceous record shows evidence of a gradual trend or change toward the K/T boundary, likely the result of deteriorating climate leading to extinctions. If such a trend is real, it may be difficult to make a distinction between (mass) extinctions related to the gradual trend and mass extinctions related to the consequences of the impact event itself.

The K/T impact ejecta-layer stratigraphy is also controversial in several aspects. It is widely believed that this layer is the result of a large meteorite impact because of the global association with an iridium anomaly, shocked minerals, and impact spherules, but alternative views also exist, mostly related to volcanism (Officer & Page 1996). The conflicting evidence seems to result

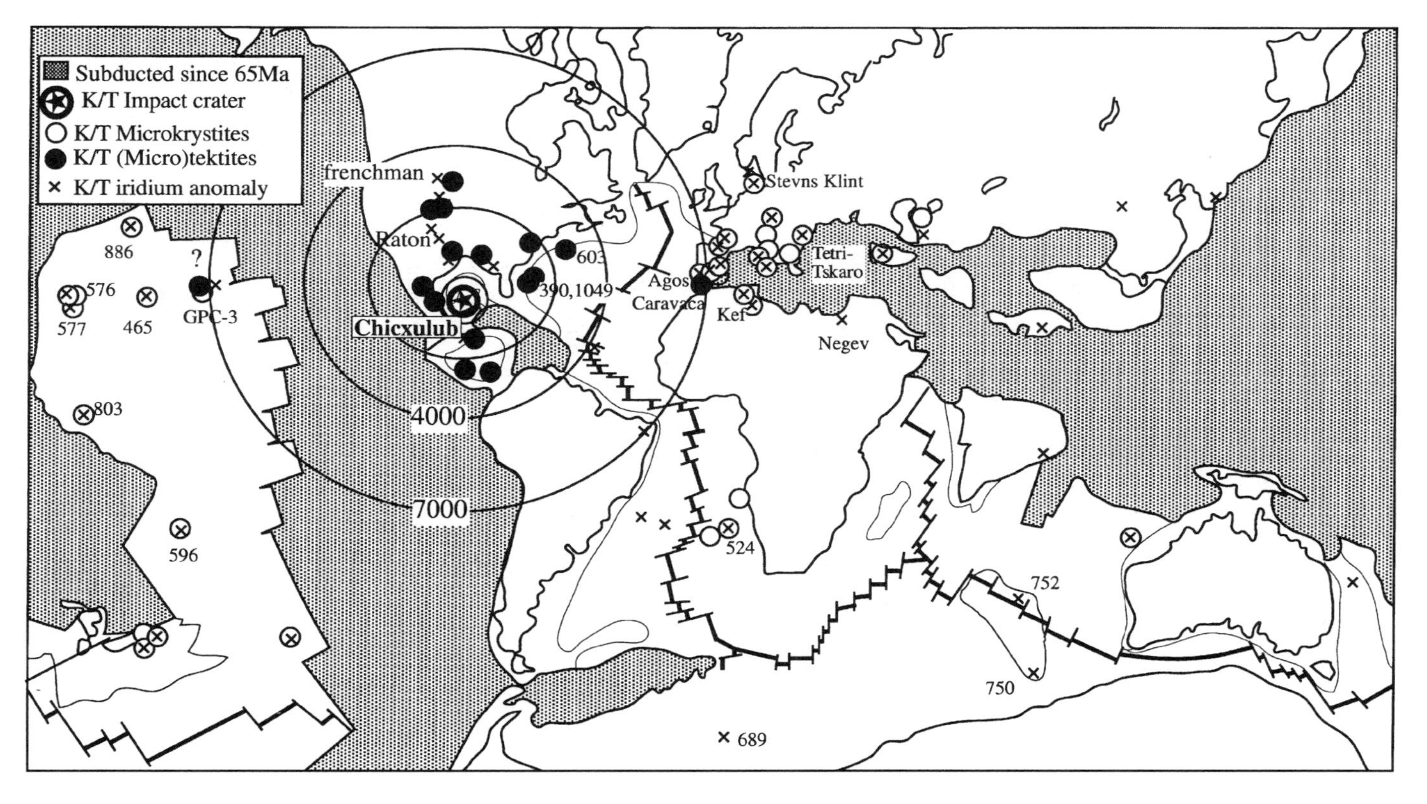

Figure 1 Distribution of K/T ejecta, projected on a paleogeographic map at 65 Ma. Distances of proximal, intermediate, and distal ejecta layers are 2500, 4000, and 7000 km, respectively. Numbers are ODP/DSDP sites.

from the complete alteration of the layer and the frequently strong bioturbation, which has led to different interpretations of the ejecta products.

Four types of ejecta deposits can be distinguished: (*a*) the distal settings, global in scale, where an ejecta layer that is a few millimeters thick has survived, (*b*) the continental Western Interior, with 1- to 2-cm-thick ejecta layers in coal-swamp deposits, (*c*) the Gulf of Mexico region, associated with high-energy clastic deposits, and (*d*) ejecta blanket deposits, up to 3.5 crater radii away from the crater rim.

In this paper I review the global stratigraphy of the K/T boundary ejecta deposits in chronological order in relation to the information provided by the sediments deposited before and after the ejecta layer. I first review the stratigraphy of the upper Maastrichtian. Next, I discuss the many aspects and disguises of the ejecta layer in each of the four regions mentioned above. Finally, I examine the first deposits of the basal Paleocene that provide the information on the biotic and environmental consequences of the Chicxulub impact.

K/T BOUNDARY DEFINITION

For a better understanding of the K/T boundary events and their controversies, it is essential to understand the historical and official definition of the K/T boundary.

Most geologists associate the K/T boundary with mass extinctions, and the definitions of the stratotypes of the Maastrichtian and the Danian stages are not in conflict with this view. The Danian, erected by Desor (1846), is the senior stage and was followed as much as possible in the definition of a new K/T boundary stratotype. The base of the Danian stratotype section is located at Stevns Klint in Denmark; at the base it includes the K/T boundary clay known as the "Fiskeler." The ejecta layer was later (Alvarez et al 1980) shown to be present at the base of the clay. The ejecta layer is part of the Fiskeler (Christensen et al 1973) and, therefore, the K/T boundary is located below the ejecta layer (Figure 2).

The International Commission on Stratigraphy (ICS) (Cowie et al 1989) has defined the new K/T boundary Global Stratotype Section and Point (GSSP) in the El Haria formation in outcrops west of the town of El Kef in Tunisia, based on the abundant presence of well-preserved calcareous and organic-walled microfossils and the unusually complete and expanded section. The upper Maastrichtian and Paleocene are more than 500 m thick and consist predominantly of drab hemipelagic calcareous marls with a few micritic limestone intervals. Obvious breaks in sedimentation or hard grounds are absent in the entire interval. The K/T boundary is marked by a reddish 2-mm-thick layer, the ejecta

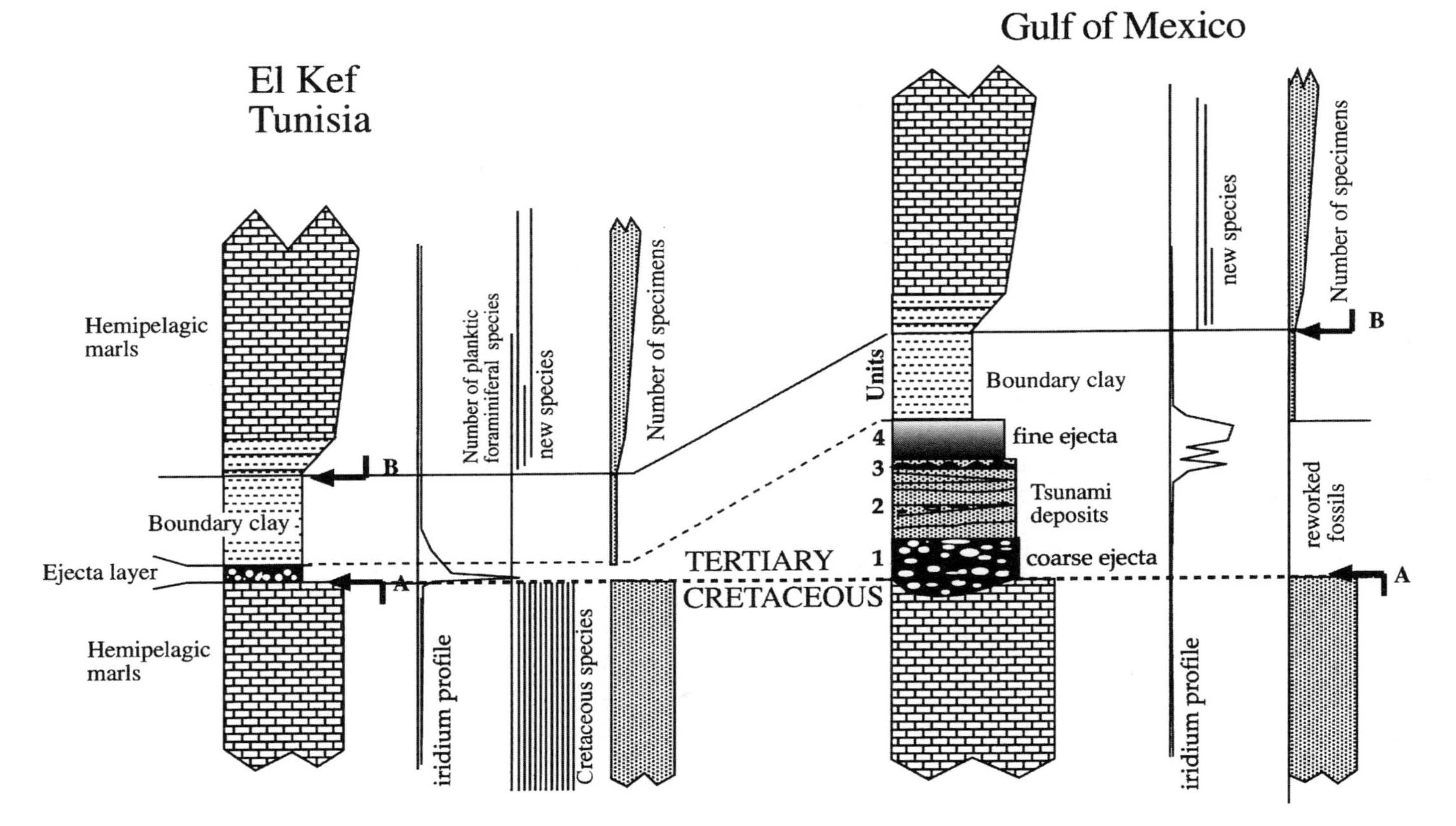

Figure 2 Schematic stratigraphy of the K/T boundary stratotype section at El Kef, compared with the proximal unit I-IV sequence in the Gulf of Mexico. In distal sites, iridium profile and other types of ejecta coincide; in proximal sites these are separated vertically. (*A*) Mass-extinction level. (*B*) First occurrence of new planktic foraminiferal species.

layer, immediately overlying calcareous marls with abundant Maastrichtian calcareous microfossils. The ejecta layer is enriched in iridium and contains shocked quartz and nickel-rich spinels (Robin et al 1991), as well as altered microkrystites (Smit & Romein 1985). The red ejecta layer is directly overlain by the K/T boundary clay, a 25-cm-thick clay interval, extremely poor in calcareous microfossils. In line with the Danian stage, the ICS commission placed the K/T boundary GSSP below the ejecta layer. The GSSP is an instant in time, represented by the bedding plane between the red ejecta layer and the fossiliferous marls below the boundary. All other criteria (such as the mass-extinction level, last occurrence of Maastrichtian fossils, first occurrence of Paleocene fossils, iridium anomaly, microkrystites, microtektites, soot, impact diamonds, amino acids, and negative shift of $\delta^{13}C$ and $\delta^{18}O$) are closely associated and can be used for correlation but are not part of the K/T boundary definition.

This definition is not inaccurate for the majority of distal K/T localities because all ejecta peak in the same thin layer. Problems occur in defining the K/T boundary around the Gulf of Mexico because the different types of ejecta (tektites, iridium anomaly) are at differenct stratigraphic levels (Gartner 1996, Keller & Stinnesback 1996, Keller et al 1994, Smit et al 1994) (Figure 2).

UPPERMOST CRETACEOUS (LATE MAASTRICHTIAN)

Open-ocean pelagic carbonate sequences with slow sedimentation rates are often incomplete or condensed. The most complete upper Maastrichtian sedimentary sequences occur in outer-shelf sections, in particular along the former Tethys ocean. Examples included in this review are the sections of Agost (Groot et al 1989), Caravaca, and the Bay of Biscay region (Zumaya) in Spain (Smit & Romein 1985); Bottaccione in Italy (Luterbacher & Premoli Silva 1964); and El Kef in Tunisia (Keller 1988, Nederbragt & Koning 1994). Those sections are in pelagic to hemipelagic facies and contain rich foraminiferal and nannofossil faunas and floras and insignificant amounts of macrofossils—usually echinoids, inoceramids, ammonites, and numerous trace fossils. There is no evidence for hard grounds or omission surfaces that abound in typical shallow-water sections, e.g. on the Gulf of Mexico coastal plain, Stevns Klint, and the Maastricht area.

There are a number of ways to determine the temporal sequence of events relevant to the K/T boundary. Because of the abundance of microfossils and the ease of use, traditional biostratigraphy is most frequently used but is often hampered by diachronous datum levels and facies depth dependence of faunal assemblages, which mask global biotic changes.

The late Maastrichtian is represented by the *Abathomphalus mayaroensis* foraminiferal biozone, which is subdivided by three datum levels near its termination (entry of *Pseudoguembelina hariaensis* and *Plummerita hantkeninoides* and exit of *Gansserina Gansseri*; Pardo et al 1996). The first appearance of *A. mayaroensis* is at about 2 Myr below the K/T boundary in low latitudes and, at up to 4.5 Myr in high latitudes (Huber 1992). Planktic foraminiferal faunas remain rich and diversified throughout this interval; the zone has the highest species richness of the Cretaceous. Considering the length of the zone, remarkably few evolutionary changes occur in this interval. Whether a number of large ornamented tropical species disappeared less than 15 kyr before the K/T boundary, thus before the Chicxulub impact event, as claimed by Keller (1988, 1996), is still debated. A blind-test effort at the K/T type section near El Kef did not yield a definitive answer (Keller 1997, Smit & Nederbragt 1997). However, all the disappearing species were still found just below the K/T boundary by at least one of the blind-test participants.

Nannofossil biostratigraphy is not better suited for zonation of the late Maastrichtian. Low-latitude marker species either are absent at high latitudes or are as diachronous as planktic foraminifers. The species traditionally used for defining the base of the Tertiary (*Biantholithus sparsus, Biscutum romeinii, B. parvulum, Criciplacolithus primus*) are shown to be already present in the Maastrichtian (Romein et al 1996), although their size is much smaller. The most reliable criterion for defining the K/T boundary using nannofossils is the relative abundance of so-called persistent or disaster species above the K/T boundary clay (Gartner 1996, Percival & Fischer 1977).

A magnetostratigraphy has been established for a few well-known sections, (Gubbio, Caravaca, Agost, Bjala), but a reliable magnetostratigraphy for the upper Cretaceous of the Biscay region or El Kef (Kate & Sprenger 1993; Moreau et al 1994, 1989; Roggenthen 1976; Lindinger 1989) could not be established. Yet magnetostratigraphy may provide a better estimate than biostratigraphy for completeness of sections (Kent 1977). The K/T boundary occurs in magnetochron C29R, with a duration presently estimated at 833 kyr (Berggren et al 1995). Earlier estimates are considerably less, 570 kyr (Berggren et al 1985). To assess the size of the gap, or amount of time involved in the K/T interval, sedimentation rates can be linearly extrapolated upward and likewise downward to the K/T boundary (Figure 3). The estimated time involved to deposit both Cretaceous and Tertiary intervals within C29R is more than the total time allowed for C29R and may indicate an increased sedimentation rate in the vicinity of the K/T boundary interval, rather than the time involved in the K/T interval. In another approach, the orbital cyclicities measured in the K/T interval could be used (Herbert & D'Hondt 1990, Kate & Sprenger 1993). Berger et al (1989) calculated the duration of the precession periods at 65 Myr ago at 18.7 and 22.5 ka, respectively, averaged at 20.65 ka. In C29r, at least 14 cycles are observed in the Cretaceous (Agost) and 15 in the Paleocene (Zumaya) (Kate

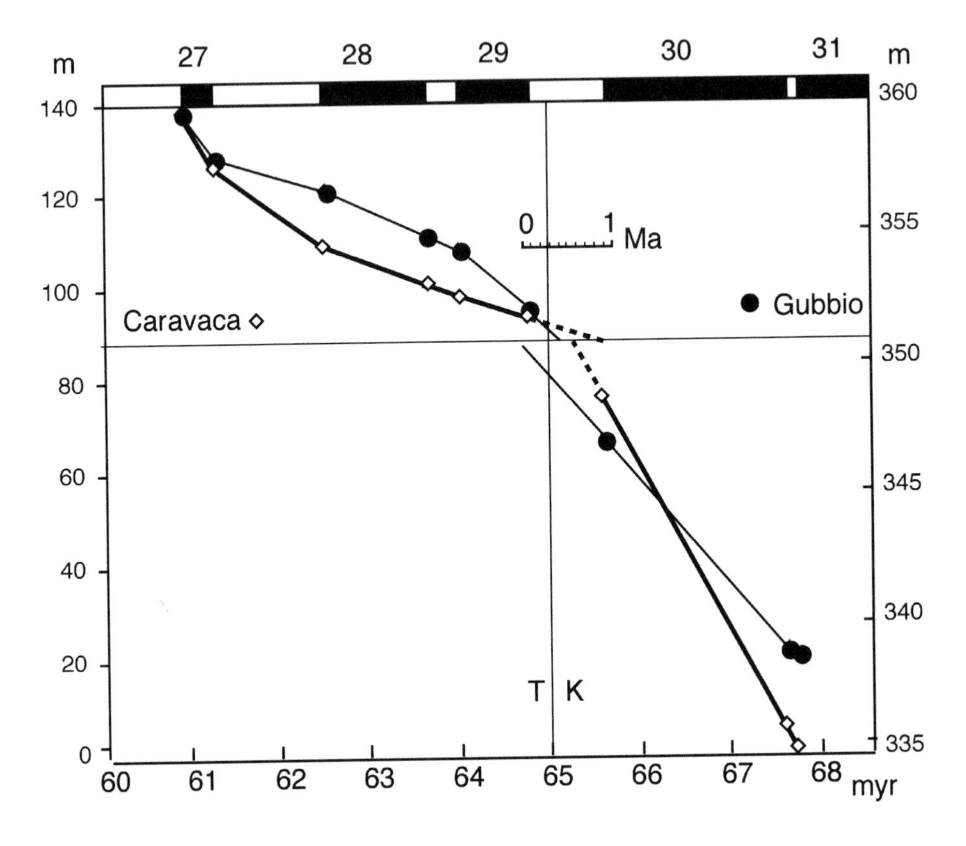

Figure 3 Accumulation rates of (*filled dots*) Gubbio and (*diamonds*) Caravaca, based on magnetostratigraphy (Chrons 27–31), extrapolated toward the K/T boundary. The misfit at the K/T boundary indicates (*a*) increased sedimentation rate in C29R, (*b*) an understimation of the duration of C29R. No significant gaps at the K/T boundary are expected.

& Sprenger 1993), using the above cycle lengths, accounting for 542.3 and 652.5 kyr, respectively, averaged at 594.5 kyr. This would better fit the earlier (Berggren et al 1985) estimate of 570 kyr. However, if the recent estimate of 833 kyr is valid, there may be a gap: at least eight precession cycles appear to be missing, contrary to the results from simple extrapolation. This contradiction raises questions about either the applicability of astronomical cycles or the length of C29r determined from seafloor magnetic anomalies.

The lithology of the complete continuous sections shows major cyclic lithological variations (0.5–2 Myr) in the Maastrichtian that are probably related to sea level change. The lithological succession at Zumaya, which is more or less the same in all the Bay of Biscay sections, compares well with the sections in Agost and Caravaca in Spain and El Kef in Tunisia, indicating that these changes are probably eustatic.

The more calcareous, lithified parts of the sections occur in the base of the upper Maastrichtian, in the middle, and near the top, about 13 cycles below the K/T boundary. Although the carbonate content in these limestone intervals is higher, the thickness of the observed precession cycles is less, showing a reduced flux of both clay and pelagic carbonate. These intervals probably represent sea level highstand to maximum-flooding stages, in good agreement with the studies on benthic foraminifers (Pardo et al 1996) (Figure 4).

Deformation of the uppermost Cretaceous strata is noticeable in regions close to the Chicxulub crater. The strata underlying the ejecta layers often show signs of soft-sediment deformation, whereas strata overlying the ejecta

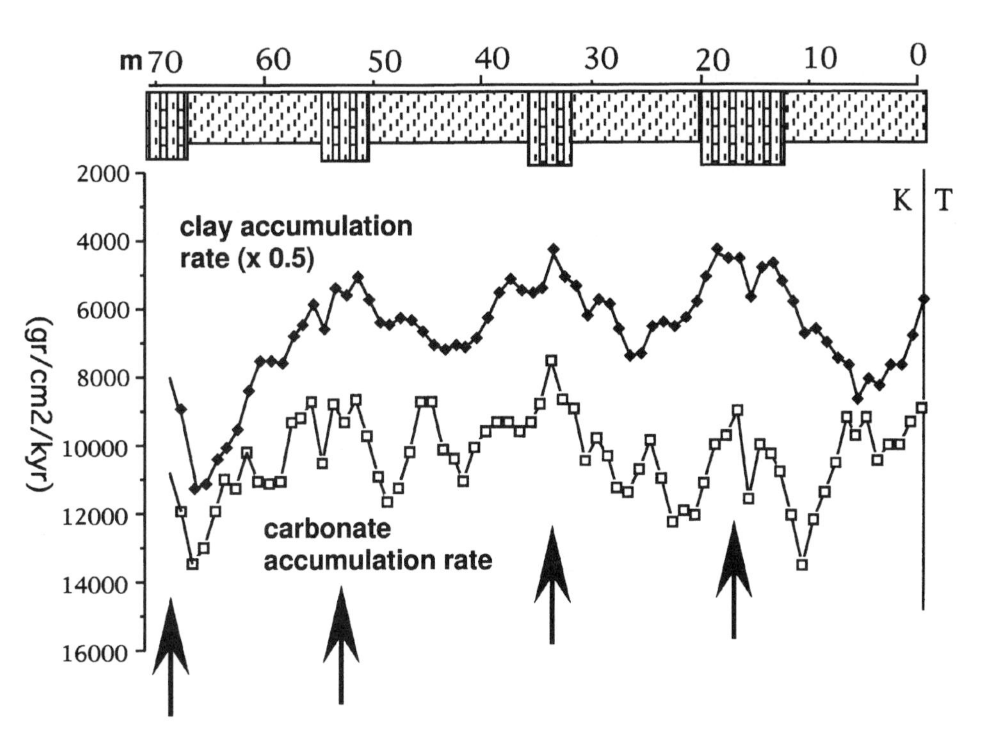

Figure 4 Accumulation rates of carbonate and clay, calculated on the average duration and thickness of the observed precession cycles (Kate & Sprenger 1993) for Zumaya section, Bay of Biscay region, Spain. Arrows indicate maximum flooding surfaces. The flooding surface at 70 m may correspond to the 67.5 ma maximum flooding surface from Haq et al (1988).

layers are undisturbed. This phenomenon is particularly striking on the Blake Nose, Florida. Sediments of ODP Site 1049 directly underlying the ejecta layer are tightly slump-folded, whereas the overlying stratigraphy is undisturbed and shows the normal biostratigraphic succession. This is also visible on the seismic facies interpreted from the seismic profiles across Blake Nose. The same large-scale slumping of the upper Cretaceous is inferred from the lithology at and seismic profiles across DSDP Sites 536 and 540 (Alvarez et al 1992b, Buffler et al 1984), but again the lower Paleocene is undisturbed. In outcrops around the Gulf Coast, upper Cretaceous soft-sediment deformation has been frequently observed (Smit et al 1996). An intriguing example is the Moscow Landing, Alabama outcrop. Normal faulting, with associated soft-sediment deformation of the Maastrichtian chalk, began slightly before the emplacement of ejecta but continued through deposition of the ejecta and associated clastic beds, which are presumably emplaced by tsunami waves. The slumping is confined to strata below the ejecta layer, but the faulting continued into Paleocene, as indicated by the offset of the overlying burrowed transgressive surface and thickness differences of the same lithologic units on either side of the fault planes. Slumping has also been observed in the Mimbral, Rancho Nuevo, and Cuauhtemoc outcrops in northeastern Mexico (Alvarez et al 1992a) before or during emplacement of the ejecta (tektites). Large-scale slope failures are inferred from the lithological successions in Bochil (Chiapas) and El Caribe in Guatamala. All these phenomena indicate deformation and large-scale slope failures related to seismic energy input from the Chicxulub impact itself, some of it induced before the emplacement of the ejecta from the same impact event (Figure 5).

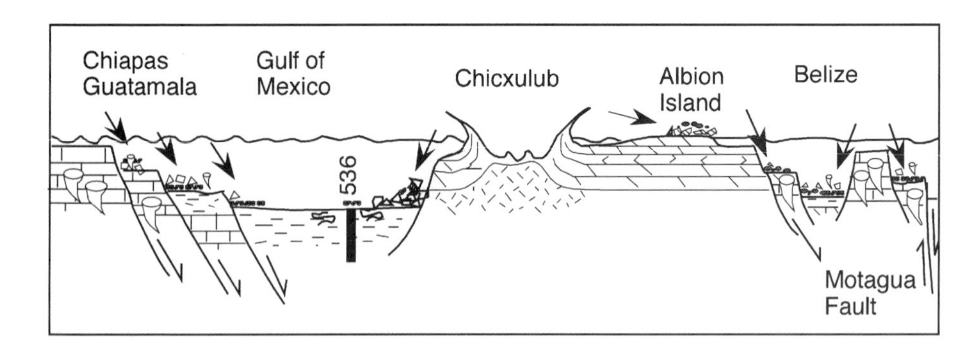

Figure 5 Schematic cross section through the Chicxulub crater and surrounding dolomite or rudist-bearing carbonate platform margins with possible impact triggered slope-failure features (not to scale).

Transition Upper Maastrichtian Ejecta Layer

In complete distal sections, the K/T transition is additionally marked by a characteristic burrowed, but not completely bioturbated, horizon in the top of the Maastrichtian. The burrows are conspicuous, typically filled with dark clayey material from the clay layer above the K/T boundary. In particular, the Tunisian sections (Kef, Elles) (Smit et al 1997), those in Spain (Agost, Caravaca) (Smit 1977), and the Furlo and Petriccio sections in Italy (Montanari et al 1983) contain a well-preserved trace-fossil fabric. Sediments above and below the K/T boundary are bioturbationally well mixed. The sequence of trace-fossil fabrics across the K/T shows a significant decrease precisely at the K/T boundary. This means, indirectly, that bottom life also decreased dramatically in these (outer) shelf sections at intermediate (200–1000 m) water depths (Figure 6). As premium for K/T boundary studies, the ejecta layer in these sections is therefore stratigraphically undisturbed. Nevertheless, burrow traces of *Chondrites*, *Planolites*, and *Zoophycos* have often scavenged the ejecta layer, because the burrows are filled with impact spherules (Kotake 1989). The scavenging of the ejecta layer obviously caused enhanced concentrations of iridium in the upper 10 cm of the Maastrichtian and a decrease in iridium in the ejecta layer

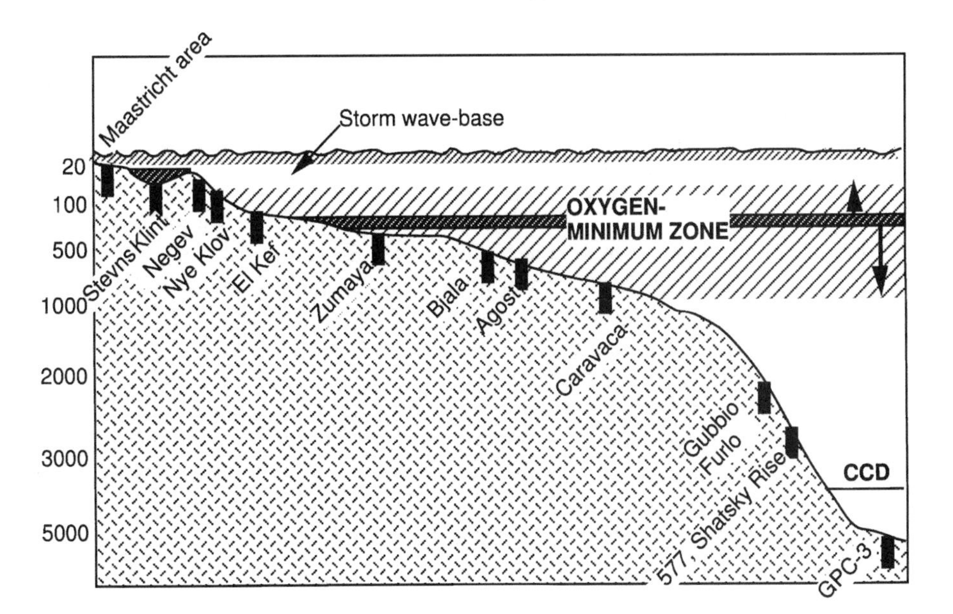

Figure 6 Depth of deposition in meters of distal K/T boundary sites, as estimated from various sources. Hatched areas and arrows indicate the presumed expansion of the oxygen minimum zone directly after the impact event.

(Montanari 1990). All ichnofossil types found are typical for burrowing of soft, muddy substrates. It is often mentioned that the discontinuity at the K/T boundary may indicate an erosion or hard ground surface, but evidence of these mud burrowers does not support this view.

The ejecta layer is not well preserved in open-ocean deep-water sections such as DSDP Site 577 (Michel et al 1985) and abyssal clay cores such as GPC-3 (Kyte et al 1996, 1993). Also, the K/T boundary interval in the Negev sections Ein Mor and HorHaHar is completely bioturbated, leading to a smearing that extends the iridium anomaly and first and last occurrences of foraminifers. Sediments across the K/T interval are saturated with ichnofossils, and numerous body fossils of crustaceans, notorious burrowers, occur below and above the boundary.

The top few centimeters of the Maastrichtian just below the ejecta layer are characterized by dissolution, creating the illusion of a period of transition toward the ejecta layer. At El Kef, only internal casts of foraminifers are preserved in this interval. In addition, in Agost and Caravaca the $CaCO_3$ content is lowered. In the Apennine sections (Gubbio, Furlo, Petriccio), the surfaces bounding the ejecta layer are obviously stylolithic, indicating diagenetic dissolution. At Stevns Klint, so-called dissolution horse-tail structures occur below the ejecta layer. Most of the dissolution has a diagenetic character, most likely the result of leaching by sulfuric acid produced by dissociation of the abundant pyrite framboids in the K/T boundary clay.

K/T EJECTA LAYER

The ejecta layer at the K/T boundary shows distinct characteristics related to distance from the Chicxulub crater. Those sites that possess a thick >3-cm layer containing (altered) tektites of impact glass, characterized by a distinct bubbly texture (Figure 1), are labeled proximal. These sites occur mainly in the Gulf of Mexico, the Caribbean, and the east coast of the United States up to New Jersey (Olsson et al 1997), DSDP Site 603 (Klaver et al 1986) up to 2500 km from the crater center. Continental North American sites, ranging in distance from 2200 km (Raton Basin) (Izett 1990) to 4000 km (Alberta) (Carlisle et al 1991) and ranging in thickness from 0.5 to 2 cm, are labeled intermediate and partially owe their characteristics to the coal-swamp depositional environment. Sites at a distance of 7000 km and greater are labeled distal and are characterized by an ejecta layer thickness of only a few millimeters that contains abundant microkrystites (Smit et al 1992a).

Distal Sites

At sites more than 7000 km from the Chicxulub crater, the thickness of the ejecta layer, when properly reconstructed, is fairly constant at not more than 2–3 mm

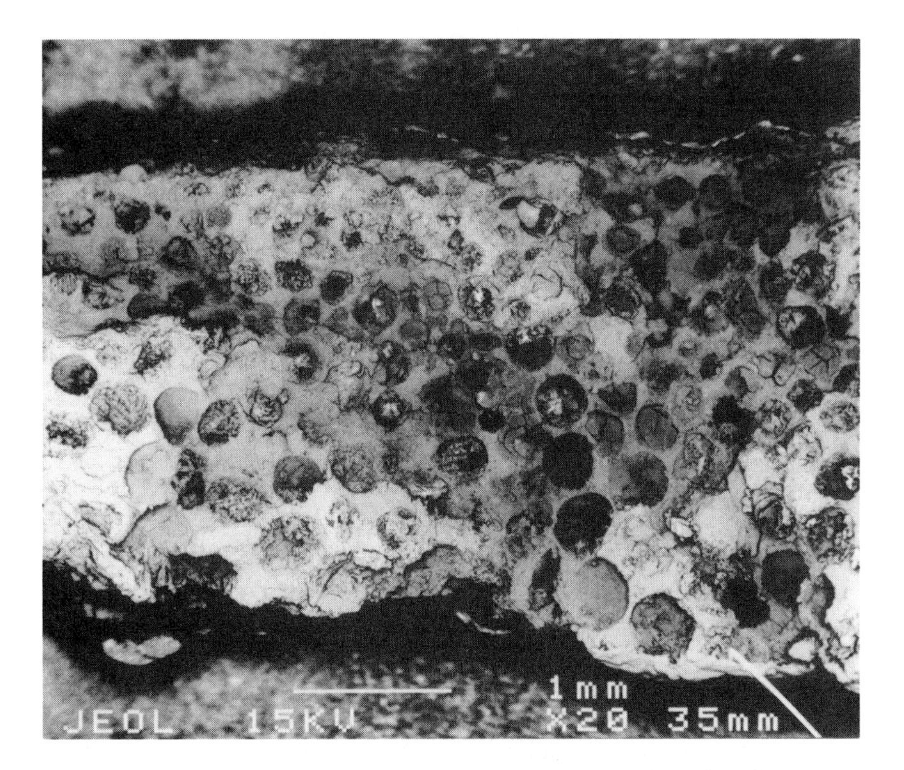

Figure 7 SEM image of the distal ejecta layer, showing graded spherule layer altered into an early diagenetic goethite concretion that is highly enriched in iridium (86 ng/g), compared to the ejecta non-concretionary clay adjacent to the concretion (2 ng/g). Tetri-Tskaro, near Tbilisi, Republic of Georgia.

(Figure 7). Owing to variable degrees of compaction and local resedimentation, it is not possible to reconstruct a reliable thickness distribution as a function of distance from the impact crater. The site with an undisturbed ejecta layer closest to the Chicxulub crater is Alamedilla in Spain, some 7000 km away, and the most distal site is Woodside Creek in New Zealand, now about 15,000 km away. The ejecta layer thickness at both sites is a few millimeters thick. The distal ejecta layer is invariably altered, and at some sites only components such as high-temperature magnesioferrite spinels, shocked minerals, impact diamonds, and, thus far only at DSDP site 577, Ca-rich clinopyroxene (augite) have survived diagenetic alteration (Figure 8) (Smit et al 1992a).

The stratigraphic distribution of the different components in the ejecta layers is determined mainly by differences in settling velocity and thus in grain size. In the Tetri-Tskaro section in Georgia the microkrystites are clearly graded (Figure 7). In a water column exceeding a few tens of meters, grain size

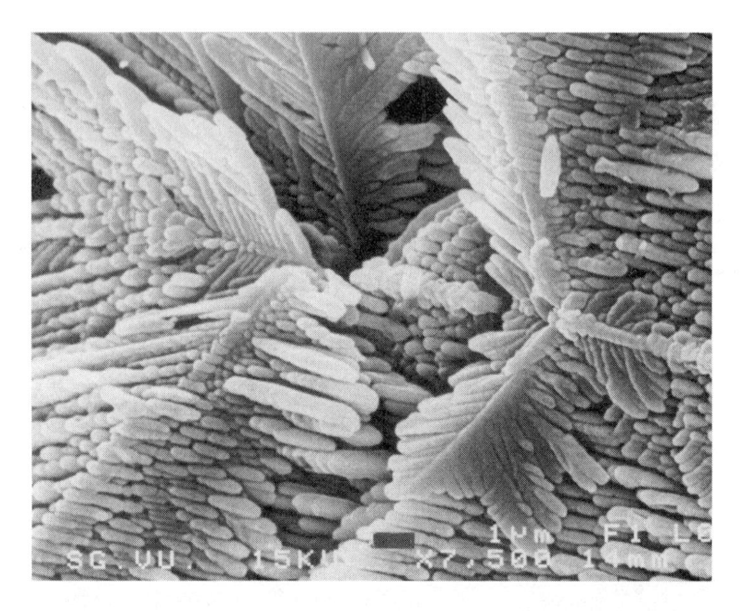

Figure 8 SEM image of clinopyroxene (augite) dendrites, from a microkrystite of the K/T boundary of DSDP Site 577b. Interstitial glass has been etched out by seawater.

distribution is determined by differential settling and does not yield information about arrival times of the different ejecta or help determine whether one or more simultaneous impacts occurred at the K/T boundary, as suggested by Robin et al (1993). In the Petriccio section (Montanari 1990), smaller-size shocked minerals are concentrated in the lower part of the ejecta layer, contrary to the distribution of shocked minerals in the ejecta layer in the western interior of the United States (Izett 1990). A position below the peak of the (larger) microkrystites is difficult to explain by settling of ejecta from one impact. However, as in the other Appennine K/T sections, the ejecta layer and overlying boundary clay in the Petriccio section appear to be well mixed, favoring redistribution by sedimentary processes over two impact events.

The most abundant, macroscopically visible components in the distal ejecta layer are spherules with a relict crystalline texture (Smit et al 1992a), designated microkrystites (Glass & Burns 1988). These microkrystites are restricted to the ejecta layer, except where dispersed by bioturbation. The microkrystites cannot be easily classified in terms of a simple mixing of different compositional end-members, as is the case for the more proximal tektites (Sigurdsson et al 1991), because there is no compositional data owing to ubiquitous alteration of the layer. A proxy for composition may be the mineralogy, inferred from the relict crystalline texture. It is possible to distinguish several groups of microkrystites within one ejecta layer at one site. Microkrystites are altered to goethite in Stevns Klint, Denmark, to As-rich pyrite in Zumaya, to potassium feldspar in Caravaca, and to glauconite in Fonte d'Olio, Italy (Montanari 1990, Smit et al 1992a), but owing to the different diagenetical and (micro) chemical environments, it is doubtful that the distinctions between sites reflect meaningful differences in original composition. Different groups in a single sample can be distinguished in Agost and Caravaca, Spain; Furlo and Petriccio, Italy; and DSDP Site 577 (Pacific). One microkrystite group is characterized by skeletal Ni-rich magnesioferrite spinels (Kyte & Bostwick 1995), embedded in a dark smectitic matrix containing ghosts of olivine crystals. Another group consists dominantly of K-spar, with skeletal-texture pseudomorphs after clinopyroxene and plagioclase (Montanari et al 1983, Smit et al 1992a). A third group of green glauconite microkrystites in some places contains skeletal K-spar crystals. The latter two groups do not contain magnesioferrite crystals and may represent mixing between two end members. Do the two groups (with and without spinels) represent different chemical compositions or a different cooling history? The spinel-bearing group is clearly more mafic in composition and significantly enriched in iridium (Montanari et al 1983); the other two groups, glauconite and K-spar rich, are not. Therefore, it is likely that the spinel-bearing microkrystites are formed from the hotter parts of the ejecta vapor cloud (Kyte & Bostwick 1995) containing a larger proportion of the bolide than the other, clinopyroxene-rich microkrystite type.

Several features associated with the composition and global distribution of the Ni-rich magnesioferrite spinels are not easily explained. Most K/T spinels located in the Pacific and a few in the South Atlantic (DSDP Site 524) and Indian Ocean (ODP Site 761) have significantly higher MgO, Al_2O_3, and Fe_2O_3/FeO content than do spinels from the Tethys and show myrmekitic intergrowth with, presumably, Ni-rich periclase (Mg, Ni, Fe) O, indicating formation at extremely high temperatures (Kyte & Bostwick 1995). At a finer regional scale, there are considerable differences, particularly in Cr content among the European sites. The asymmetric global distribution may support oblique impact. The high Fe_3/Fe_2 ratio in all magnesioferrite spinels, but in particular in the Pacific spinels, requires a relatively high oxygen fugacity during crystallization. Interaction with atmospheric oxygen is usually inferred: one hypothesis (Robin et al 1993) posits the formation by ablation of many bolides at relatively low altitudes, whereas another hypothesis (Kyte & Bostwick 1995) suggests atmospheric mixing of part of the impact vapor cloud. The Robin et al (1993) hypothesis seems inconsistent with the uniform composition over a large area and with the periclase inclusions, whereas the Kyte & Bostwick (1995) hypothesis seems inconsistent with the observation that impact ejecta (microtektites) are usually highly reduced. Probably related to this problem is the high oxidation state of the impact glass found at Beloc, Haiti, and Mimbral, Mexico (Jehanno et al 1992, Oskarsson et al 1996). Although atmospheric interaction may play a role, liberation of oxygen from the vaporized CO_2- and SO_3-rich evaporitic and carbonate sedimentary cover from the Chicxulub target area, as suggested by Oskarsson et al (1996), cannot be excluded.

The number of microkrystites can be estimated in some sections with an undisturbed ejecta layer (Agost, Caravaca, Tetri-Tskaro). In these sections the ejecta layer remains constant in thickness along the outcrop for at least 50 m. In other sections, diagenesis and bioturbation prevent an accurate estimate. Basically, the ejecta layer consists of a dense packing of spherules, 250 μm in mean size, yielding about 20,000 microkrystites per square centimeter. There is no reason to believe that the distribution elsewhere is much different. Woodside Creek, New Zealand; the Pacific sites GPC-3 and DSDP Site 577; South Atlantic (DSDP Site 524); and Stevns Klint yield hundreds of microkrystites per cubic centimeter. The total number of microkrystites, assuming a global coverage of comparable microkrystite density, is 1×10^{23} microkrystites or a volume of about 850 km^3 deposited in distal sites alone.

Between 4000 and 7000 km from the crater, there are very few data from K/T ejecta layers, although GPC-3 may occur in this range. Kyte et al (1996) and Smit et al (1992a) note the occurrence of smooth hollow clay spherules, sometimes with splash forms, without relict crystalline textures, in GPC-3 and ODP Site 886. Ruiz et al (1992) report goethite spherules and droplets with similar morphology from Agost, and recently I found droplets at Alamedilla, Spain (J Smit, unpublished data). Because of the similarity to the altered spherules from Dogie Creek and the Gulf Coast, these droplets are interpreted as altered tektites and indicate that in the North Pacific and Spain, tektites and microkrystites occur together in the same ejecta layer.

Intermediate Sites

The ejecta layer in intermediate sites, 2000–4000 km from the Chicxulub crater, is represented by a couplet of claystone layers in continental deposits, invariably coal-swamp deposits now turned into coal or lignite. At sites where the K/T transition, as determined by palynostratigraphy, occurs in overbank deposits, the ejecta layer has been spread out over several decimeters and is no longer visible as a layer. The reported thickness of the ejecta layer varies from 2 cm in Raton Basin (Izett 1990), Red Deer River (Lerbeckmo et al 1996), and Dogie Creek (Bohor et al 1987) to 0.5 cm in the Seven Blackfoot coulee in Montana and near Frenchman River in Saskatchewan (Lerbeckmo et al 1996). There is no clear relation of thickness to distance from Chicxulub, although, in general, the thickness in the Raton Basin is consistently around 2 cm. A thickness around

1.5–2 cm in Montana and Canada seems a local maximum, but 0.5 cm is the average.

The ejecta layer has a distinct, dual-layer stratigraphy, consisting of a thick, lower claystone layer, often sharply separated from a thinner upper layer. The lower claystone layer is often termed the kaolinitic layer (Bohor et al 1987) or K/T boundary claystone (Izett 1990), and the upper layer is often termed the fireball, magic, or K/T boundary impact layer because it contains high iridium concentrations and the bulk of the shocked minerals. The lower claystone layer is composed mainly of kaolinite, and the upper layer is mainly composed of a mixture of kaolinite and illite/smectite mixed-layer clays. The dual stratigraphy extends from Brownie Butte in Montana (Bohor et al 1984) to Raton Basin in New Mexico. Izett (1990) showed clearly that the lower claystone has signs of reworking because of the presence of detrital clastic grains and vitrinite particles and the discontinuous, lenticular nature of the layer. Additionally, the lower claystone layer at Dogie Creek, Wyoming, a site where the layer almost entirely consists of goyazite spherules, is clearly cross-bedded.

It has been argued that the dual-layer stratigraphy represents two different impact events. If so, these have to be almost simultaneous because both layers, though separate, are amalgamated. Not even a single season of fallen leaves separates the layers. Reported discontinuities, truncations of root traces, and cutoffs of pumice fragments of glass can be explained by sedimentary processes during emplacement of both layers. The evidence for reworking in the lower layer suggests that the emplacement of the lower layer was affected by disturbances in the shallow-water coal swamp, possibly induced by the blast loaded with tektites. On the other hand, because of its continuous thickness, the upper layer seems to result from relatively quiet settling of fine impact debris through the atmosphere.

Proximal Ejecta Sites

The proximal sites <2500 km from the Chicxulub crater (Figures 9, 10, and 11) can be roughly subdivided into five types: (*a*) sites near the Atlantic continental margin and Atlantic Coastal Plain that contain mainly graded tektite deposits, (*b*) shelf to outer shelf locations around the Gulf of Mexico at relatively shallow <500 m depths, (*c*) sites around the carbonate platform in Chiapas, Guatamala, and southern Belize, (*d*) deep-water sites in the Gulf of Mexico and the Caribbean, and (*e*) ejecta-curtain deposits in and close to the Chicxulub crater, extending to northern Belize.

ATLANTIC COASTAL PLAIN AND CONTINENTAL MARGIN The ejecta layer near the Atlantic coast consists of a single, 3- to 17–centimeter-thick, often graded layer of greenish spherules. The morphology of the spherules, including splash

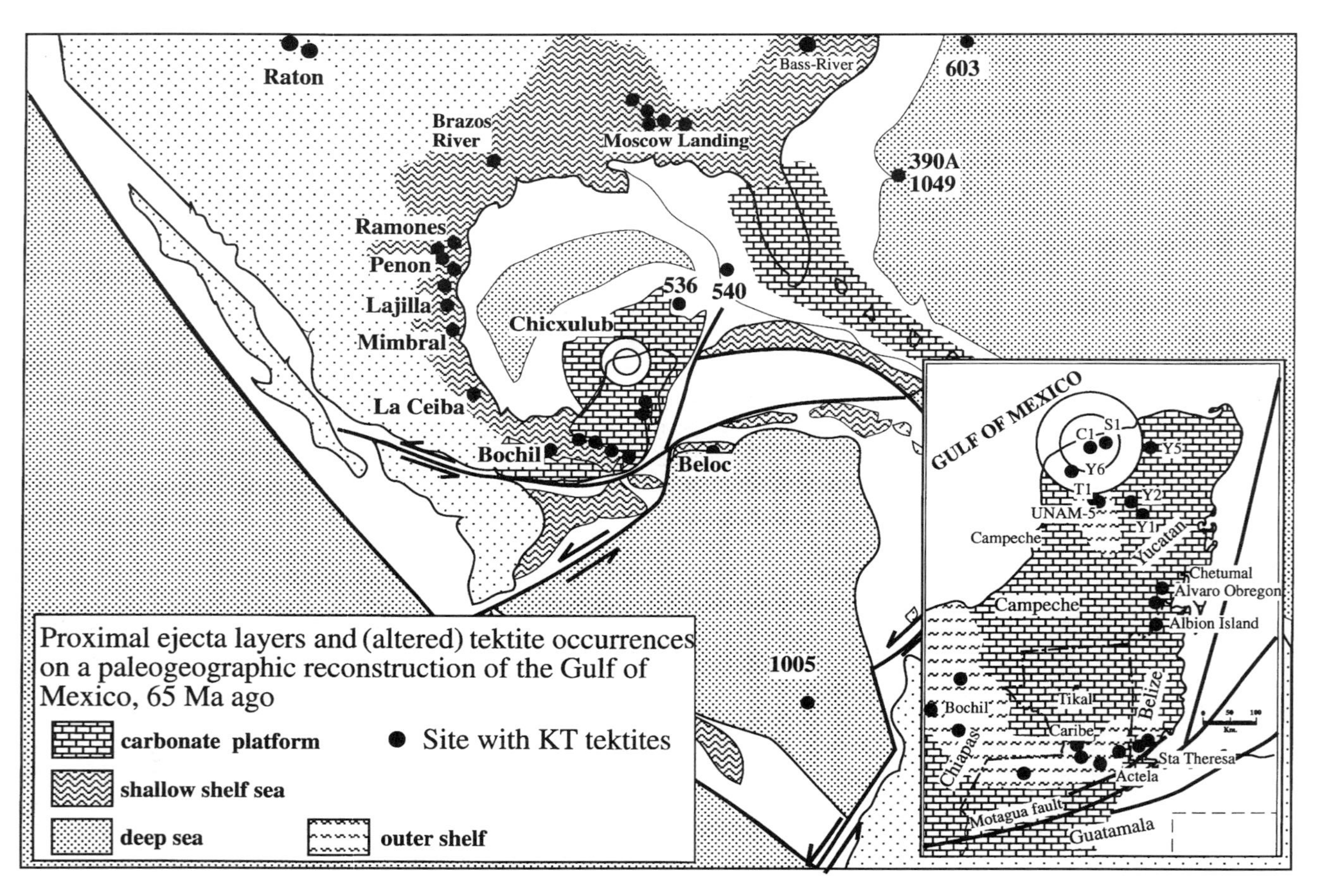

Figure 9 Paleogeographic map of the Gulf of Mexico and Yucatan with the proximal sites discussed in the text. Numbers are DSDP/ODP sites.

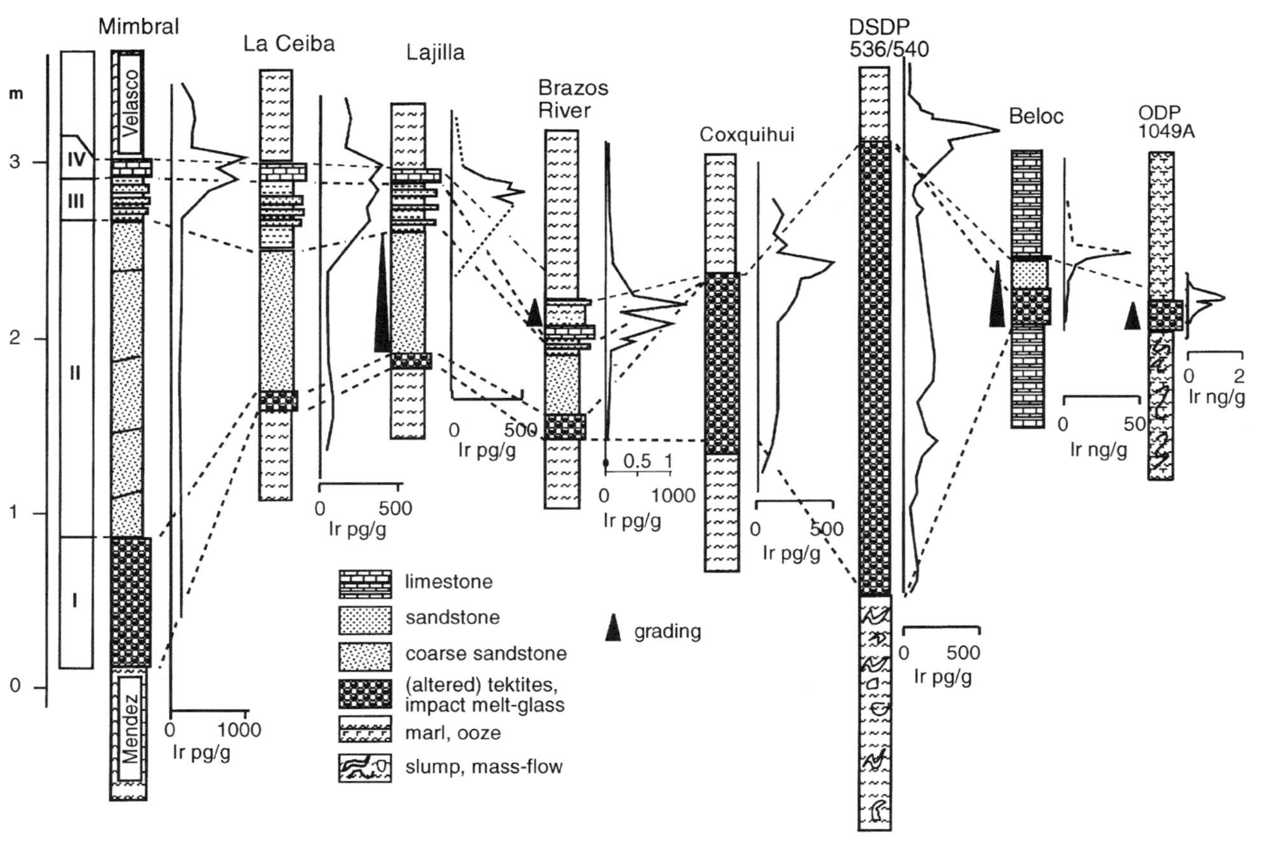

Figure 10 Scheme of the stratigraphy of the proximal sites with measured iridium profiles discussed by region in the text. The Ir data are from F Asaro and P Claeys (personal communication), except La Lajilla and ODP Site 1049 data are from R Rocchia (unpublished data). Roman numerals I to IV indicate the subdivisions of the clastic beds with evidence for tsunami waves (Smit et al 1996). Coarse ejecta, mostly altered tektites, are consistently separated stratigraphically from the fine ejecta indicated by Ir.

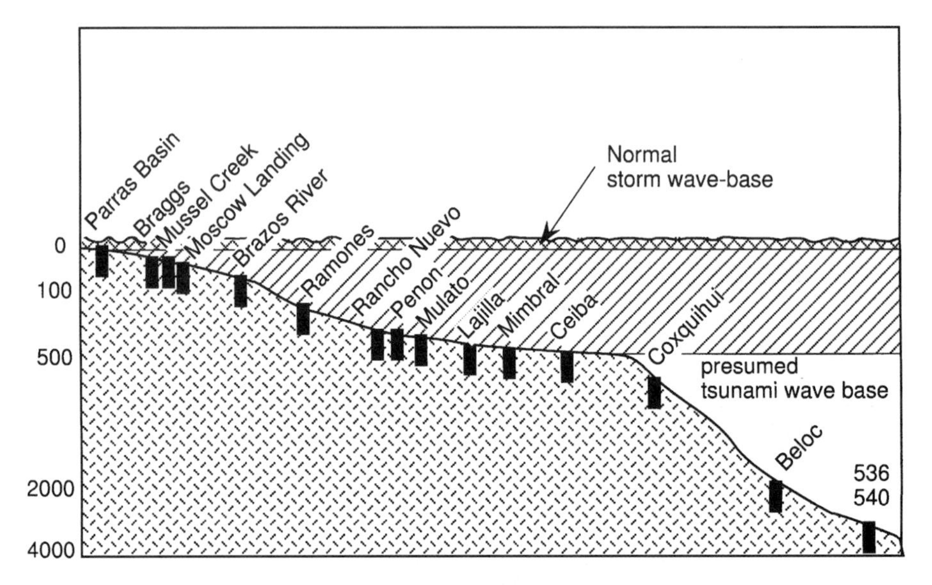

Figure 11 Depth of deposition of the proximal sites discussed in the text. *Hatched area* is the estimated tsunami-wave base. *Cross-hatched area* is the normal storm-wave base.

forms with flow banding and abundant internal vesicles (Klaver et al 1986), is identical to the spherules with a preserved tektite glass core from Haiti (Izett 1991, Sigurdsson et al 1991) and the goyazite spherules from Dogie Creek.

The distinction between the Atlantic and the intermediate sites at 2000–2500 km is somewhat arbitrary. Main differences are the deposition in a marine environment and the slightly greater thickness (>3 cm). A dual-layer stratigraphy is not observed but, as noted above, should not be expected here from a single impact. Mean spherule (tektite) size is about equal: 0.95 mm in Dogie Creek (Bohor et al 1987) and 1.1 mm in Bass Creek (Olsson et al 1997) and DSDP Site 603 near New Jersey (Klaver et al 1986). In ODP Site 1049 (a redrill of DSDP Site 390A), a single 7- to 17-centimeter–thick graded layer of green spherules (<2.3 mm) is observed, capped by a red layer. The abundantly present Cretaceous foraminifera in the graded ejecta layer suggests that the grading did not result from differential settling through the water column but from a gravity flow. Iridium concentrations are enhanced throughout the layer but, in particular, are enhanced just above the spherule layer (not in the red capping), again indicating that the spherule layer has been remobilized after deposition. Clear evidence for tsunami wave activity has not been found in any of the sites, although Olsson et al (1997) report large clasts in sediments overlying the ejecta layer.

GULF OF MEXICO COASTAL PLAIN These sites are invariably characterized by a complex set of clastic sandstones, containing coarse ejecta at the base and an iridium anomaly at the top. The sites extend from Moscow Landing in Alabama to La Ceiba, a site east of Mexico city and renamed Tlaxcalantongo by Stinnesbeck et al (1996). Smit et al (1992b, 1996), Stinnesbeck et al (1993), and Stinnesbeck & Keller (1996) subdivide the set from bottom to top into three macroscopically visible units (Figure 12). A fourth unit is discernible only after detailed grain size analysis (Smit et al 1996). Bohor (1996) compares the subdivision with a classical, idealized Bouma Ta-Te sequence.

The depositional environment of the U.S. Gulf Coast is in a shallow shelf sea. In eastern Mexico, it is in a deep outer shelf, according to the lithology and foraminiferal assemblages in the underlying Maastrichtian sediments (Figure 11).

The lowermost unit I is characterized by channeled, poorly sorted, coarse-grained pebbly sandstones containing flat rip-up clasts, abundant spherules interpreted as (altered) tektites, and limestone particles interpreted as unmelted limestone ejecta (Smit et al 1996). The unit displays low-angle cross bedding, often lateral accretion-type channel infill. The channels are mostly discontinuous. The matrix between the spherules and clasts consists primarily of planktic foraminifera that have been winnowed from the surrounding seafloor.

Unit II is composed of a series of upward thinning and fining lenses or extended sheets of finer-grained, well-sorted calcareous sandstone. The lenticular bodies are more extended than the channels from unit I, but lateral channel infill structures have not been recognized. Lithic clastic grains are more frequently found in unit II than in unit I, but the dominant component of the sandstones are planktic foraminifera. Locally, plant debris has accumulated in layers. Rounded, armored mud balls, the armoring consisting of tektite-like spherules, are frequently found at the base.

At Mimbral, up to six, stacked and imbricated lenticular sandstone layers were recognized, each with a clearly coarse basal part containing tektite-like spherules, probably reworked from unit I. Each layer has eroded into the top of the underlying sheet, producing disconformities. In La Lajilla and El Penon, where the outcrops can be traced over several hundreds of meters, about eight sheet-like, thinning-upward sandstone bodies were identified.

Unit II displays numerous sedimentary structures that are particularly well developed at La Lajilla (Figure 12). These structures include parallel laminations with associated primary current lineation, lunate, linguid, and climbing ripples. More than 200 individual current directions measured in several locations in unit II (Brazos River, Rancho Nuevo, El Peñon, La Lajilla, Mimbral, and La Ceiba) indicate bimodal current directions in each of the locations. Often the dominant directions differ by almost 180° (Smit et al 1996).

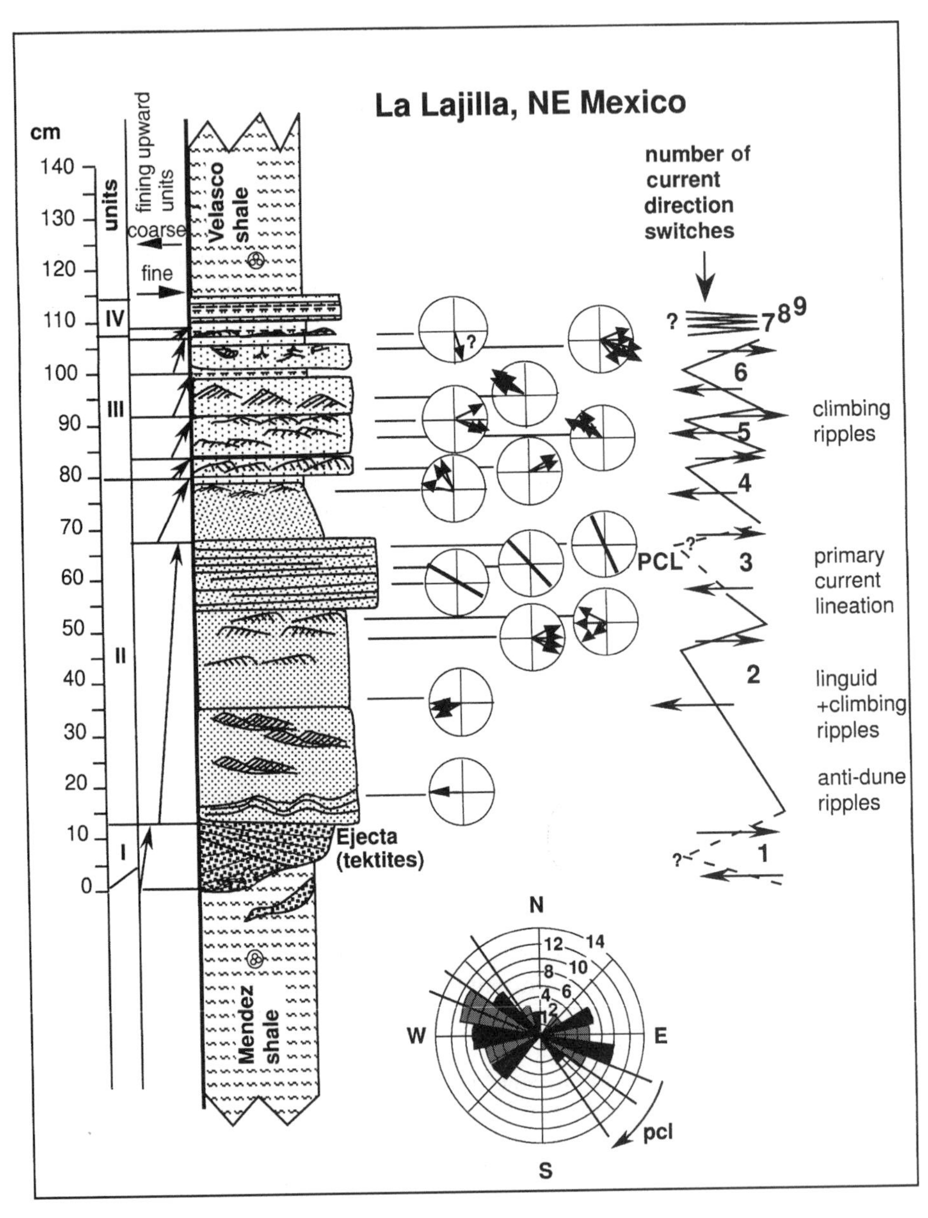

Figure 12 La Lajilla, northeast Mexico. Lithology of K/T clastic beds is subdivided in units I to IV. The fining-upward units, types of sedimentary structures, and measured current directions are indicated. Primary current lineations (PCL); 1–9 indicate the approximate number of current direction changes, assumed to indicate the passage of successive tsunami waves.

The transition between units II and III is defined by the first of a series of alternating soft, thin silt layers and thin, rippled, fine-grained sandstone layers. These silty layers are the lowermost layers enriched in iridium, although the highest iridium concentrations occur in the graded silt layers above the highest rippled sand layer.

The current directions measured in the ripples remain variable, again often differing by 180°. The rippled sandstone layers are sheet-like and continue over the entire outcrop; at La Lajilla and El Peñon, they continue over at least 150 m. The upper two to three layers contain diverse burrow traces, mostly confined to the surface of the sandstone layers. The ichnoassemblages are tiered, that is, a different assemblage is found on each successive level. *Chondrites* occupy the deeper levels; different types of *Zoophycos*, *Planolites*, and *Rhizocorallium* occupy the upper levels. One type of burrow, consisting of straight 1-cm–diameter tubes, extends down to the lowest silt layer from unit III, and approximately 1 m below the top of the highest ripple layer in straight vertical burrows, before spreading out horizontally into the silt layer in huge, 3-m-long, radiating, often bifurcating burrows. All these ichnofossils can be explained by colonization after deposition of the entire clastic unit and do not necessarily indicate a considerable length of time between deposition of sand layers. Keller et al (1997) mention burrows in layers from units I and II. However, a number of flame structures were found in the base of the same layers, where mud has been squeezed into the bottom of the sand layer by loading. These could have been mistaken for burrow traces.

Unit IV, the highest unit, is silty and varies in thickness on top of the last observed sandstone layer. Bralower et al (1998) label this unit the K/T boundary cocktail because of the abundance of a mixture of reworked microfossils and ejecta. Unit IV is often visible in outcrop because it forms a conspicuous, more lithified micritic layer of 5–10 cm thickness, which has been mistaken locally for a Cretaceous (Mendez formation) shale layer, probably because it consists largely of reworked material from the underlying Mendez shale (Keller et al 1997). A detailed grain size analysis (Smit et al 1996) has shown, although it is not apparent in outcrop, that this unit is size graded, and therefore originated from the settling of fine resuspended material. Iridium concentrations are highest in this fine-grained graded unit, indicating that the fine-grained iridium carrier settled at the same time.

Origin of the Clastic Layers

The clastic layers are interpreted differently. Smit et al (1996) interpreted the complex of layers as a result of the interaction of large tsunami waves, produced by the Chicxulub impact. Although the impact itself did not occur in the deep ocean basin of the Gulf of Mexico, a large part of the ejecta curtain, presumably thousands of cubic kilometers, fell in the Gulf, and slope failures along the Campeche escarpment may additionally have triggered tsunamis. Keller & Stinnesbeck (1996) and Stinnesbeck & Keller (1996) interpreted the layers as the result of a sea-level lowstand followed by transgression, or of repeated gravity-flow deposits related to eustatic sea level changes over thousands of years during the uppermost Maastrichtian. Bohor (1996) also interpreted the layers as a result of gravity flow deposits, not extended in time but as one coherent gravity-flow event triggered by the Chicxulub impact. He considered the lithological succession as comparable with a single idealized Bouma sequence. Stinnesbeck & Keller (1996) based their view on an alternative position of the K/T boundary, above the clastic layers, and the presumed presence of burrow fabrics and "normal hemipelagic layers" in, between, and above the sandstone layers. In their view, because thousands of years passed between deposition of the tektites in unit I and the iridium in unit IV, they proposed two impacts, the lower of the two related to the Chicxulub impact, the other at the K/T boundary. However, the grain size distribution of these normal hemipelagic layers is clearly different from, and coarser-grained than, undisputed upper Maastrichtian and Paleocene normal hemipelagic layers. Also, because the burrows in the lower units could not be confirmed, their interpretations are in question—particularly the claim that thousands of years would have passed between deposition of individual sand layers.

Bohor's (1996) view of the clastic units as a single turbidite is at odds with the frequent hydrodynamic jumps indicated by the grain size shifts at each successive layer and the reversed current directions. Moreover, the grain size distribution related to the specified Bouma intervals is different from that in a classical turbidite. Although hydrodynamic jumps may occur occasionally in composite turbidites, the complex K/T sequence is not comparable with any turbidite sequence in well-studied flysch basins. Also, repeated reversals of current directions are never observed in a single turbidite and are difficult to explain by gravity-flow mechanisms.

A problem with the interpretation of Smit et al (1996) is the great distance from the presumed coastline. It is difficult to imagine how large-scale oscillatory water movements, although triggered by exceptionally large waves, can transport material, including clastic sand grains and wood fragments, from a coastline at least tens of kilometers away. The oscillatory movements are necessary to explain the measured reversals in current direction, so reworking by tsunami waves is at least involved in part of the story. However, the transport of near-coastal material to the locations farther into the gulf may be partially a consequence of gravity flows.

Chiapas, Guatamala, and Southern Belize

The sites in Chiapas, Guatamala, and southern Belize all occur on or around the Yucatan-Chiapas carbonate platform and incorporate very coarse mass flows underlying an iridium anomaly, sometimes associated with altered tektite-like spherules (Montanari et al 1994, Stinnesbeck et al 1997). A typical sequence is the Bochil section in Chiapas, where a 70-m-thick graded mass flow, containing >5-m-sized platform limestone blocks often with rudists and miliolids, occurs at the K/T boundary. The matrix of the mass flow contains a mixture of deep (bathyal) and shallow water biota, such as planktic foraminifers, miliolids, orbitoids, and rudists. The absence of any dolomite or anhydrite blocks, which constitute the dominant components in the ejecta blanket/curtain sites, for example, in northern Belize and Yucatan, indicates that these blocks are not ejecta from the Chicxulub impact ejecta curtain but represent local mass wasting from the edge of the carbonate platform. The coarse, often conglomeratic, mass flows are overlain by a graded sandstone deposit, followed by a thin rusty silt/mudstone layer. The top of the sandstone and the thin, fine silt/mudstone contain an iridium anomaly and numerous shocked quartz grains. Shocked quartz (J Smit, unpublished data) and enhanced iridium concentrations (E Fourcade et al 1998) are also present in the El Caribe (Guatemala) section directly above the graded mass flow. Stinnesbeck et al (1997) describe four sections in Guatemala where platform limestone facies underlie a breccia unit with spherules, which is itself overlain by outer-neritic to upper bathyal pelagic sediments of basal (P1a) Paleocene age, indicating an abrupt subsidence of a few hundred meters.

Recently, a new site with tektite-like spherules and splash forms was found in southern Belize, near Santa Teresa. These features occur in green boulders, a (minor) component in a conglomerate of dominantly Cretaceous shallow-water limestone pebbles, rudists (*Barrettia* sp.) and boulders of brown pelagic marls with a well-preserved lower Paleocene P1a foraminiferal assemblage. The geological setting is complicated by the block faulting related to movements along the nearby Motagua fault, separating the Chortis and North American plates and active in late Cretaceous and early Paleocene times (Figure 5). The tectonic activity has repeatedly triggered mass flows in grabens adjacent to highs with platform limestones. Most of these mass flows are not related to the Chicxulub impact.

Beloc (Haiti), Coxquihui (Mexico), and DSDP Sites 536/540

Beloc (Haiti), Coxquihui (Mexico), and DSDP Sites 536/540 are located in deep water, between 600 and 3000 m (Figure 10), and do not contain evidence for tsunami waves in the coarse-grained K/T ejecta deposits.

In Coxquihui, the K/T ejecta deposits are a 90-cm-thick, poorly graded sandstone bed consisting of three cross-bedded sublayers. The components are almost exclusively vesicular spherules of calcite. Iridium contents are elevated throughout the layer but are most enriched in the top. This site, although only

30 km from La Ceiba, is either too deep to have been influenced by the tsunami waves or too far away from the coastline for clastic sands to reach the site. The overlying and underlying pelagic marls are reddish colored, suggesting deposition below the oxygen minimum zone (>600 m) in contrast to the other eastern Mexican outcrops, and contain a fauna with deepwater style planktic species (large *Globotruncana contusa*, *G. stuarti*).

The Beloc, Haiti site is famous for the content of vesicular glass cores inside spherules (Figure 13), droplets, and 1-2 cm blebs with rims of smectite. Beloc is still the only site known, besides the rare occurrence in Mimbral and La Lajilla in Mexico, where the impact glass is abundantly present.

The physical, isotopic, and chemical properties of the glass have been treated by Blum (1992), Izett (1991), Jehanno et al (1992), Koeberl (1992), Koeberl

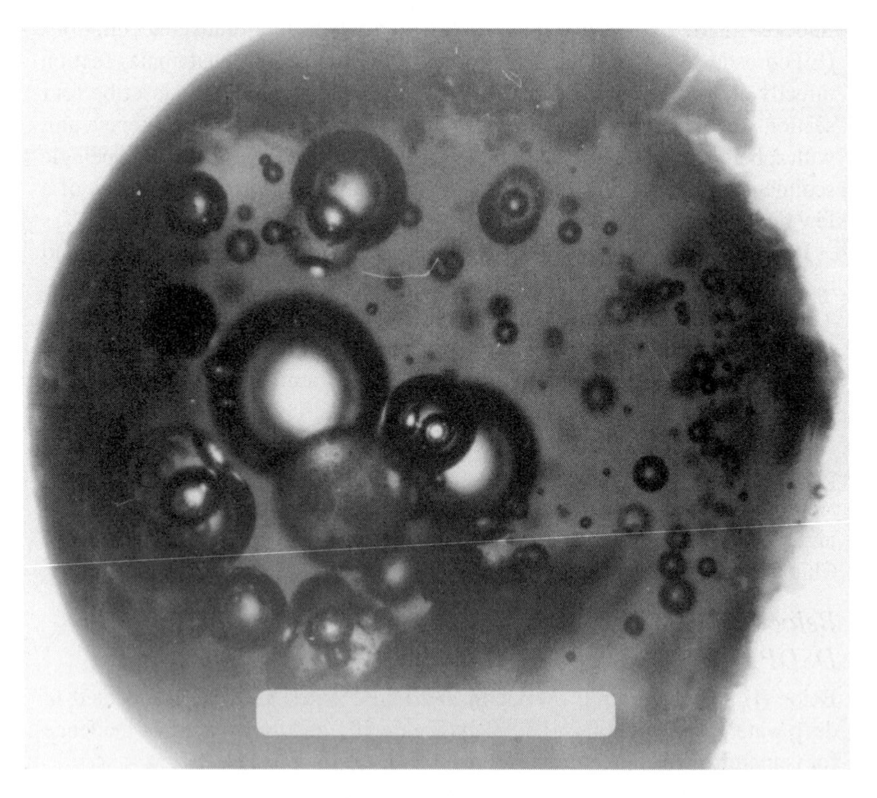

Figure 13 Thin section of round tektite consisting of brown, low-CaO glass, from the M site of Beloc, Haiti. The vesicles, when cracked and immersed in a liquid, appear to be almost at vacuum pressure. This vesicular texture is usually inherited through pseudomorphosis and provides a fingerprint for the recognition of Chicxulub ejecta. Scale is 1 mm.

& Sigurdsson (1992), Kring & Boynton (1991), Maurasse (1991), Oskarsson et al (1996), Sigurdsson et al (1991), and Swisher et al (1992), and I highlight only some of the more important features. The glass has two end-member compositions: a light, yellowish CaO (<31%) and MgO—rich to dark brown and relatively CaO (~5%) and MgO poor. These compositions probably reflect the amount of dolomite/limestone dissolved in the glass. Some CaO-rich glass contains sulfur, and some does not, reflecting different mixtures of anhydite and limestone. The trace-element composition of the brown glass is similar to the composition of the melt inside the crater (Y6-N17) (Figure 14). Some glasses show schlieren (Figure 15) of both end-member compositions in one single

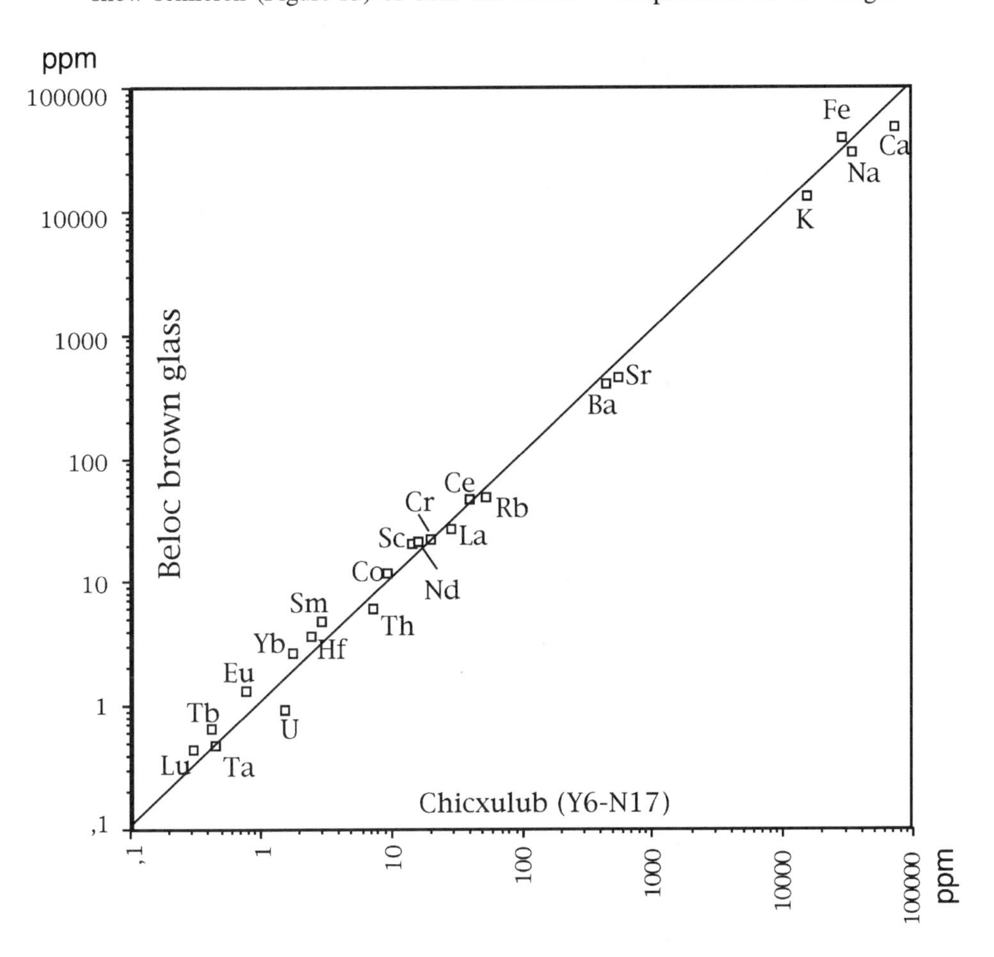

Figure 14 Comparison of trace-element compositions of the glass matrix of sample Y6-N17 from drill core Yucatan-6 in the Chicxulub crater, with the brown, low-CaO glass from Beloc, Haiti.

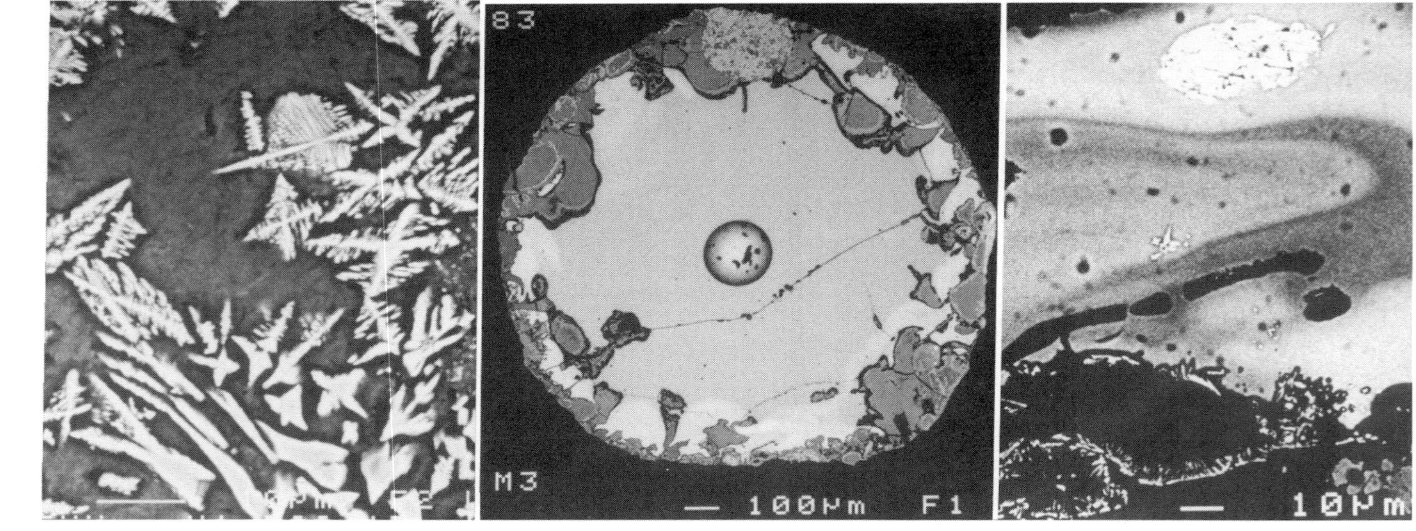

Figure 15 Backscatter SEM image of a complete glassy tektite (M site, Beloc, Haiti) with only marginal alteration to smectite at the rim (*center*). The rim is characterized by CaO-rich glass, in schlieren, that contains pockets of a melilite crystal meshwork (*left*) and unmelted inclusions of anhydrite (*right*). SEM scanning of a 2-cm large fragment of the ejecta layer with more than 60 individual tektites, where each tektite still has at least a glass core (*left*), shows that >50% of tektites have such a melilite and CaO-rich rim.

spherule. On the surface the schlieren are often differentially etched by weathering. Such schlieren are also visible on completely altered spherules of other proximal sites (DSDP 603, Dogie Creek, Mimbral) (Izett 1991). Lyons & Officer (1992) suggested that only 5% of the tektites are of the high-CaO type. However, at the M site (Maurasse 1991) which contains the least weathered spherules and where even the rim is preserved, 95% of the spherules contain a CaO-rich rim, with abundant skeletal melilite crystals. Thus the CaO-rich glass is at least as common as low-CaO dark glass. Water content is extremely low in the glass ($<0.02\%$), being comparable to other tektites and impact glasses but unlike any known volcanic glass. The gas pressure in the abundant vesicles inside the glass is almost zero, as determined in crushing experiments, indicating that the glass was solidified in ballistic flight outside the atmosphere. Sm/Nd isotope compositions indicate target basement-rock model ages between 400 and 1100 Myr (Izett 1991), which is compatible with the Pan-African basement of Yucatan. The $\delta^{18}O$ composition of the yellow glass is compatible with a mixture of basement and limestone (Izett 1991). The $^{40}Ar/^{39}Ar$ ages of both the tektite glass and the melt sheet sample C1-N10 in the crater are indistinguishable at 65 Myr (Swisher et al 1992). The age, chemical, and isotopic compositions all strongly indicate that the impact glass from Haiti is derived from the Chicxulub crater.

The Beloc ejecta layer occurs in the Beloc formation, a sequence of well-bedded siliceous pelagic limestones and marls, overlying basaltic pillow lavas. Maurasse (1991) has shown that the Beloc formation and underlying basalts are obducted from the Caribbean ocean floor, and the original depth of deposition is >2000 m. The ejecta layer has a complex stratigraphy and is in most places redeposited, as a mass flow, and slump-folded. The two southernmost outcrops are the only deposits where the spherules still contain relict glass. In some sites the spherule layer is graded and often cross-bedded. The thickness of the layer varies from 12 cm to more than 70 cm. In the B site (Jehanno et al 1992) the layer crops out over >30 m and has a constant thickness of about 12 cm. In this site and the D site, the layer is imbricated in places by package gliding. I believe that the 12-cm thickness is close to the original thickness of the ejecta layer, not the >50 cm as reported elsewhere.

Iridium and shocked minerals are concentrated in the top of the layer (Jehanno et al 1992) (Figure 10) except in those locations where the ejecta layer is redeposited. In the H-site the iridium anomaly and spherule layer seem separated by 20 cm of normal limestone (Jehanno et al 1992). However, this limestone interval is cross bedded and sandy, comparable to the Bouma Tc-d interval in a turbidite.

DSDP Sites 536 (2790 m) and 540 (2926 m) are 420 and 530 km, respectively, northeast of the crater rim, probably too far from the crater to contain ejecta blanket diamictite below the iridium-rich ejecta deposits (Alvarez et al

1992b). Site 540 contains an unusually thick (2.6 m) unit of dark smectitic, cross-bedded, melt-clast breccia, grading upward. The top is enriched in iridium and shocked minerals (Figure 10). The cross bedding indicates that the melt-clast breccia is not primary fallout but transported in mass flow. The distribution of the melt clasts and the small nonmelted clasts is almost identical to the top of the ejecta sequence of borehole UNAM-5. The melt breccia in Site 540 is underlain by 45-m matrix-supported pebbly mudstone and in Site 536 by a poorly recovered 100 m interval of shallow-water grain stone fragments. These units indicate mass wasting from the Campeche platform margin, most likely caused by slope failure induced by the seismic energy of the impact.

Northern Belize and Southern Yucatan

On both sides of the Rio Hondo, ejecta-blanket deposits, discovered during the 1998 Planetary Society expedition, crop out in Albion Island (Ocampo et al 1996) and along the road from Chetumal to La Union. These outcrops with ejecta deposits are the nearest known to the Chicxulub crater, only 340 km from the crater center. The ejecta sequence is almost identical to that at Albion Island, but the sequence is more complete and overlying Paleocene sediments were also found (Figure 16). The ejecta sequence overlies the Barton Creek dolomite, a thick-bedded coarse dolomite with fossiliferous layers containing carcineretid crabs and *Nerinea* sp. of Maastrichtian age. The contact is an undulating, irregular surface, showing evidence for subaerial exposure (Pope et al, submitted to *Geology*). The basal layer, the spheroid bed of Ocampo et al (1996), follows this surface closely, and, although it rests on the highs or lows of the undulating contact, maintains an unusual constant thickness of about 1 m. The matrix consists of pulverized dolomite containing rounded dolomite clasts, deformed greenish clay blebs that may be the remains of glassy ejecta, and centimeter-sized, concentric banded dolomite spheroids. Internal layering is discontinuous. Large boulders do not occur, in contrast to the overlying diamictite bed. The overlying diamictite bed also has a dolomite matrix and contains dolomite boulders up to 10 m in size, some of which contain the same fossils as the underlying Barton Creek dolomite. Most of the large boulders are concentrated in the lower part. Smaller boulders and cobbles, the vast majority of dolomite, occur throughout the diamictite. The diamictite further contains about 20% of up to 4-cm large green droplets and blebs with internal vesicles, identical to altered impact glass in other proximal locations in the Gulf of Mexico region. A few subhorizontal shear zones occur, similar to shear zones found in the Bunte Breccia of the Ries crater. The diamictite is at least 30 m thick. Near Alvaro Obregon the diamictite is capped by a 15-cm-thick calcrete soil, a caliche, in turn overlain by 0.5 m of thin-bedded, fine-grained

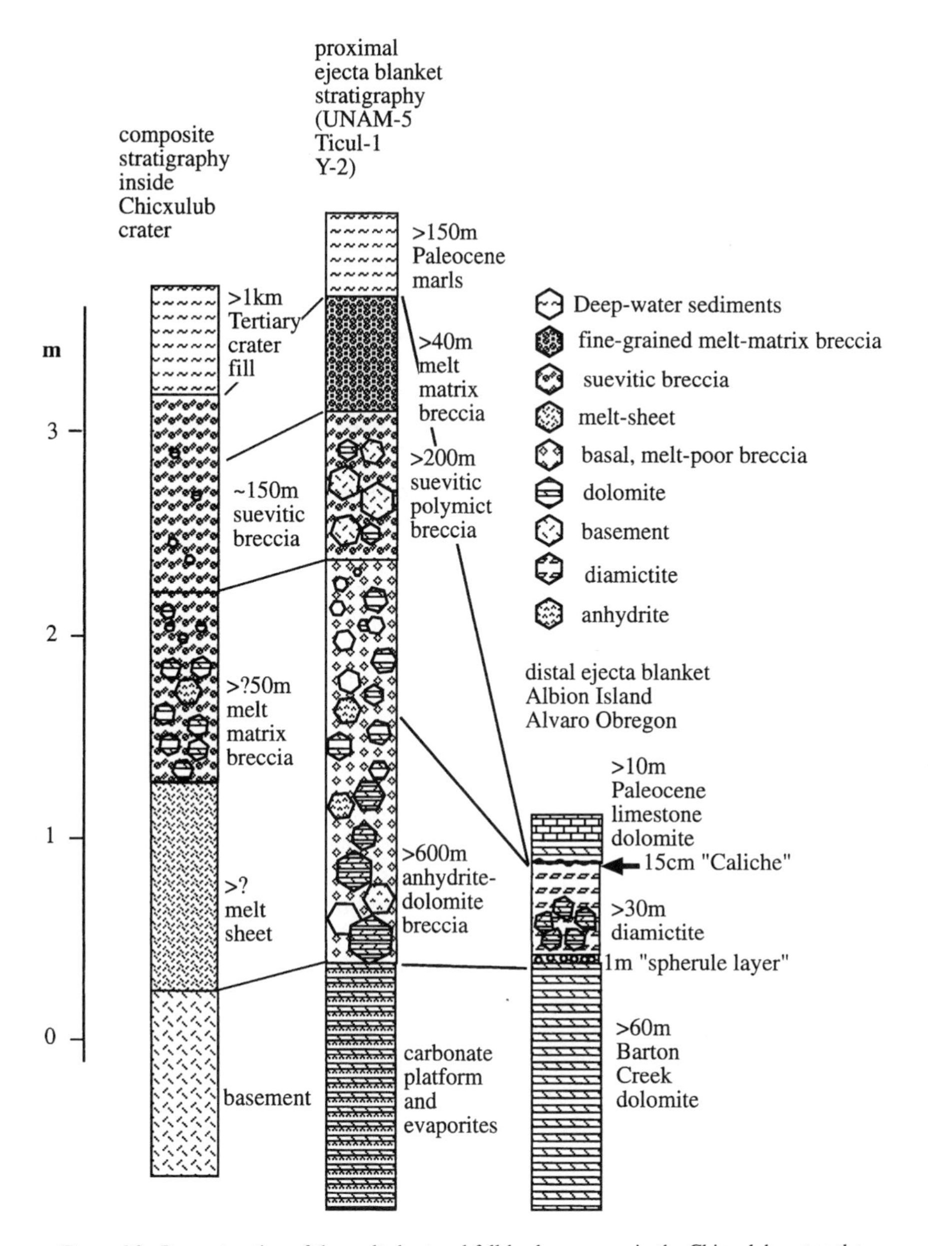

Figure 16 Reconstruction of the melt sheet and fall-back sequence in the Chicxulub crater, the proximal ejecta blanket sequence as known from drill holes Y-2, Ticul-1, and UNAM-5, and the more distal ejecta blanket sequence in Albion Island (Belize) and Alvaro Obregon in Mexico.

dolomite layers. The caliche layer is interpreted as a subaerial exposure of the ejecta, indicating that substantial erosion probably has removed the top of the ejecta sequence, prior to deposition of the dolomite beds. The thin-bedded dolomite grades into thin-bedded micritic limestones with a very small marine microfauna, ostracods, and foraminifers, probably representing a basal Paleocene transgressive sequence.

Pope et al (1998) suggest that the basal spheroid bed may represent the early blasts from the vapor plume. This may be true for some of the components, such as the vesicular clay blebs in the layer, but the presence of shear planes and slickensides, the constant thickness of the bed, the pulverized character of the matrix, and the layering in the bed indicate that this bed is produced by grinding during the horizontal movement of the ejecta curtain over the irregular karstic surface. The dolomite spheroids may have obtained their concentric banding by accretion within the layer, which is heated, and may be even partially melted through friction between the sliding ejecta curtain and authochtonous dolomite rather than from accretion in a turbulent, volatile-rich plume.

Ejecta In and Close to the Chicxulub Crater

Spot samples are recovered in the Chicxulub crater from drill cores Yucatan-6, Chicxulub-1, and Sacapuc-1 and just outside the crater in cores Y-1, Y-2, and Y-4. Shallow drill holes UNAM-5 and UNAM-6 (Marin & Sharpton 1994, Sharpton et al 1996, Hildebrand et al 1991) are continuously cored but penetrated only the top 300 m of the ejecta sequence. The stratigraphy of the ejecta is still preliminary, because the data are fragmentary and the results of further, continuously cored, deep drilling in the crater itself are awaited. Based on the seismic results of Morgan et al (1997), I believe the diameter of the crater to be 195 km.

The basal ejecta in the crater are holocrystalline igneous rocks, probably representing the almost clast-free impact melt sheet (C1-N9, 10). The $^{40}Ar/^{39}Ar$ age data from this melt sheet (Swisher et al 1992) show that the age of the melt is indistinguishable from the age obtained from impact glass at the K/T boundary in Beloc, Haiti, and Mimbral. Above the melt sheet is a melt-matrix breccia, dominated by melt clasts welded together and containing clasts of anhydrite and basement rocks up to 7 cm in size, often surrounded by coronas of augite. The overlying polymict suevitic breccias are graded from pebble to sand size, often containing reworked shallow-water biota of Cretaceous age and shocked minerals. The suevitic breccia is overlain by micritic pelagic sediments of Paleocene age. It is possible that between the melt-matrix breccia and the melt sheet a very coarse clastic unit occurs with >1-m-sized limestone, anhydrite, and dolomite boulders, considered by Meyerhoff et al (1994) to represent in situ anhydrite and limestone layers.

Just outside the crater, the stratigraphy of the ejecta curtain deposits (the basal part in particular) is poorly constrained. However, from the data from the T-1 and Y-2 wells from the base, and UNAM-5 and UNAM-6 from the top of the sequence, a stratigraphic sequence can be reconstructed (Figure 16).

The basal ejecta are presumably a coarse anhydrite and dolomite-clast breccia, with small amounts of melt clasts. The unit is >600 m thick, comparable and correlated to the diamictite bed of Albion Island (Sharpton et al 1996). The lack of anhydrite clasts in the Belizian diamictite and the similarity of the dolomite clasts to the underlying Barton creek dolomite suggest that the diamictite components in Belize may be derived from a local source, presumably by secondary cratering, rather than from the Chicxulub target rocks. These bunte breccia types of ejecta are overlain by a breccia unit rich in melt clasts and basement clasts that may be well over 300 m thick (UNAM-5). This dolomite/anhydrite breccia is overlain, with a sharp contact, by a >40 m (UNAM-5) sequence of ejecta that consists primarily (>90%) of melt clasts welded together and containing small <1-cm basement clasts and shocked grains. The melt clasts near the top display the familiar vesicles, and the size of the clasts decreases upward. The top of the sequence displays cross bedding, indicating that the ejecta clasts are now sand-sized, reworked, and transported by currents, presumably produced by seawater rushing back and filling the crater. A comparable cross-bedded melt clast unit occurs in the top of the ejecta sequence in DSDP Hole 540.

The ejecta blanket just south of the crater (UNAM-5) is overlain by micritic pelagic sediment, containing basal Paleocene nannofossils (J Pospichal, personal communication). On first sight, this might indicate that the area had subsided several hundred meters from near sea level on the evaporitic carbonate platform before the impact and suggests that the site would still be inside the topographic crater. However, its position in the graben, indicated by the gravity low extending south from the crater near the Ticul fault, indicates that this area may have been a deep-water trough before the impact (Figure 9). This pre-impact deep-water region may also be the source of the controversial deep-water Cretaceous planktonic foraminifers in the wash-back sediments above the ejecta in the crater (Lopez Ramos 1981, Meyerhoff et al 1994).

Although fragmentary, it appears that the stratigraphy of the proximal ejecta blanket is comparable to ejecta sequences observed near other impact craters, such as the Ries crater (Newsom et al 1990). Basically, the sequence represents the inverted stratigraphy of the impacted target rocks, with the melted basement rocks at the top of the sequence.

The ejecta blanket deposits extend somewhat beyond Albion Island (260 km, or 2.8 crater radii from the crater rim) but probably do not reach as far as middle

Belize (Pooks Hill) or El Caribe in Guatamala, 375 km and 430 km or 3.9 and 4.5 crater radii from the rim of the crater, respectively.

LOWERMOST PALEOCENE

The lowermost Paleocene sediments deposited directly above the ejecta layer record the first reactions of the biosphere to the consequences of the Chicxulub impact event. The first observation is the presence of a few-centimeters-thick detrital clay layer, the K/T boundary clay, directly on top of the ejecta. With the possible exception of the El Kef section, there is no transitional interval, that is, from carbonate-rich to carbonate-poor, indicating a gradual collapse of the oceanic ecosystems. The conclusion is that the primary production in the oceans was reduced suddenly and dramatically as a consequence of the impact event. This is supported by the negative $\delta^{13}C$ shift recorded globally in the first sediments above the ejecta layer. The thickness of the K/T clay layer corresponds to the sedimentation rate of clay in sediments deposited before and after the K/T boundary and ranges from 1 cm in the Apennines, 6.5 cm in Agost, and 10 cm in Caravaca, to 25 cm in El Kef. The length of time involved in deposition of the boundary clay layer, and by inference the length of the carbonate and primary production crisis in the oceans, can be extrapolated from the background clay sedimentation rates. Extrapolation from the Cretaceous, with higher clay flux than the basal Danian, yields a duration between 4000 and 8000 years, and extrapolation from the Danian yields a duration between 8000 and 16,000 years.

The boundary clay still has enhanced iridium concentrations, but these decrease exponentially upward. Diffusion, bioturbation, prolonged ocean residence times of iridium and redeposition from primary deposits are offered as explanations. The absence of any anomalous iridium in the boundary clay where the ejecta layer is missing, as in the Geulhemmerberg, Holland (Smit & Rocchia 1996), indicates no evidence for a prolonged supply of iridium.

The boundary clay contains a few other noteworthy clues. In the base of the clay, there is no confirmed record yet of new Paleocene species. The remaining biota are either reworked or surviving elements from the Cretaceous oceans.

Dinoflagellate cysts occur abundantly in the K/T boundary clay, and are so abundant that A Loeblich (personal communication) termed the Fish clay in Denmark a "dinoflagellate soup." Apparently, this group, which suffered no extinctions, flourished directly after the impact, and therefore its distribution and abundance patterns might yield clues about the environmental conditions of the oceans just after the impact, where groups such as nannoplankton and foraminifera have disappeared (Brinkhuis & Schioler 1996). One such clue is the immigration of *Palynodinium grallator*, a cool-water species from Denmark in the Maastrichtian, in the basal boundary clay of El Kef (Brinkhuis et al 1998). This is the first empirical support for dust-cloud-cooling scenarios in the geological record. Subsequently, still in the base of the boundary clay, warm tropical species (*Trithynodinium evitii*, *Manumiella druggii*) migrated from the tropics to Denmark, which supports the earlier reported negative shift (Romein & Smit 1981) in $\delta^{18}O$ in the boundary clay that indicated a substantial warming of ocean surface water during deposition of the remainder of the boundary clay.

Some organic molecular markers, exclusively found in the basalmost part of the boundary clay, indicate a short, intense period of fermentation of massive amounts of organic matter on the seafloor in El Kef and Geulhemmerberg sections (J de Leeuw et al, unpublished data). This is the best indication so far for mass mortality following the impact event. Evidence of mass mortality was earlier suspected on the basis of abundant framboidal pyrite in the boundary clay and the leaching and dissolution of the top of the Maastrichtian.

New incoming species of both the planktic foraminifera and nannofossils occur in the top of the boundary clay. It is unclear whether reports of new species already in the boundary clay (Keller 1997) are due to the frequent ichnofossils penetrating from above and forming traces filled with younger material. The boundary clay in the Geulhemmerberg, which has no overlying strata with basal Paleocene species, however, contains exclusively Cretaceous survivor species.

Beyond the clear conclusion that the mass extinctions and mass mortality are closely linked to the Chicxulub impact event, it is not yet clear how the impact event is linked to the extinctions. Many suggestions have been offered, but the original suggestion of Alvarez et al (1980), that a short period of photosynthesis shutdown followed by a food-chain collapse in combination with a prolonged period (>3 kyr) of global warming, would explain most of the biotic events.

ACKNOWLEDGMENTS

During my investigations of the mass extinctions at the K/T boundary, I have benefited greatly from discussions and other contributions of many colleagues: A. Montanari, W. Alvarez, P. Claeys, Th. B. Roep, J. Van Hinte, H. Vonhof, T. v. Eijden, T. v. Kempen, G. Ganssen, G. Klaver, H. Brinkhuis, J. Jagt, P. Willekes, M. Konert, R. van Elsas, and numerous others. I thank S. Kars for the many excellent SEM graphs. Financial support is acknowledged of the Vrije Universiteit, The Netherlands Organization for Scientific Research (NOW), and the Royal Dutch Academy of Sciences (KNAW). This is NSG contribution No. 981002.

Literature Cited

Alvarez LW, Alvarez W, Asaro F, Michel HV. 1980. Extraterrestrial cause for the Cretaceous-Tertiary extinction. *Science* 208: 1095–108

Alvarez W, Grajales NJM, Martinez SR, Romero MPR, Ruiz LE, et al. 1992a. The Cretaceous-Tertiary boundary impact-tsunami deposit in NE Mexico. *Geol. Soc. Am. Abstr. Programs* 24:A331

Alvarez W, Smit J, Lowrie W, Asaro F, Margolis SV, et al. 1992b. Proximal impact deposits at the Cretaceous-Tertiary boundary in the Gulf of Mexico: a restudy of DSDP Leg 77 Sites 536 and 540. *Geology* 20:697–700

Berger A, Loutre MF, Dehant V. 1989. Influence of the changing lunar orbit on the astronomical frequencies. *Paleoceanography* 4:555–64

Berggren WA, Kent DV, Flynn JJ, Van Couvering JA. 1985. Cenozoic geochronology. *Geol. Soc. Am. Bull.* 96:1407–18

Berggren WA, Kent DV, Swisher CCS, Aubry M-P. 1995. A revised Cenozoic geochronology and chronostratigraphy. In *Geochronology, Time Scales and Global Stratigraphic Correlation*, Spec. Publ. 54:129–212. Tulsa, OK: Soc. Econ. Paleontol. Mineral.

Blum JD. 1992. Oxygen isotope constraints on the origin of impact glasses from the Cretaceous-Tertiary boundary. *Science* 257:1104–7

Bohor BF. 1996. A sediment gravity flow hypothesis for siliciclastic units at the K/T boundary, northeastern Mexico. In *The Cretaceous-Tertiary Event and Other Catastrophes in Earth History*, ed. G Ryder, D Fastovski, S Gartner, Spec. Pap. 307:183–96. Boulder, CO: Geol. Soc. Am. 569 pp.

Bohor BF, Foord EE, Modreski PJ, Triplehorn DM. 1984. Mineralogic evidence for an impact event at the Cretaceous-Tertiary boundary. *Science* 224:867–69

Bohor BF, Triplehorn DM, Nichols DJ, Millard HT. 1987. Dinosaurs, spherules and the "magic" layer: a new K-T boundary clay site in Wyoming. *Geology* 15:896–99

Bralower TJ, Paull CK, Leckie RM. 1998. The Cretaceous-Tertiary boundary cocktail: Chicxulub impact triggers margin collapse and extensive sediment gravity flows. *Geology* 26:331–34

Brinkhuis H, Schioler P. 1996. Palynology of the Geulhemmerberg Cretaceous/Tertiary boundary section (Limburg, SE Netherlands). In *The Geulhemmerberg Cretaceous/Tertiary Boundary Section (Maastrichtian Type Area, SE Netherlands)*, ed. H Brinkhuis, J Smit, Spec. Issue Geol. Mijnbouw 75:193–213. Dordrecht, The Netherlands/Norwell, MA: Kluwer Academic

Buffler RT, Schlager W, Bowdler JL, Cotillon PH, Halley RB, et al. 1984. *Initial Reports of the Deep-Sea Drilling Project*, Washington, DC: US Govt. Print. Off. 365 pp.

Carlisle BB, Braman DR. 1991 Nanometre size diamonds in the Cretaceous/Tertiary boundary clay of Alberta. *Nature* 352:708–9

Christensen L, Fregerslev S, Simonsen A, Thiede J. 1973. Sedimentology and depositional environment of lower Danish Fish clay from Stevns Klint, Denmark. *Bull. Geol. Soc. Denmark* 22:193–212

Cowie JW, Zieger W, Remane J. 1989. Stratigraphic Commission accelerates progress, 1984–1989. *Episodes* 112:79–83

Desor E. 1846. Sur le Terrain Danien, nouvel étage de la craie. *Bull. Soc. Geol. Fr.* 4:181

Fourcade E, Rocchia R, Gardin S, Bellier JP, Debrabant P, et al. 1998. Age of the Guatemala breccias around the Cretaceous Tertiary boundary: relationships with the asteriod impact on the Yucatan. *C.R. Acad. Sci. Paris* 327:47–53

Gartner S. 1996. Calcareous nannofossils at the Cretaceous-Tertiary boundary. In *Cretaceous-Tertiary Mass Extinctions*, ed. N Macleod, G Keller, pp. 27–48. New York: Norton

Glass BP, Burns CA. 1988. Microkrystites: a new term for impact-produced glassy spherules containing primary crystallites. In *Lunar and Planetary Science Conference*, ed. S Ryder, New York: Pergamon. 18:455–8

Groot JJ, de Jonge RBG, Langereis CG, ten Kate WGHZ, Smit J. 1989. Magnetostratigraphy of the Cretaceous-Tertiary boundary at Agost (Spain). *Earth Planet. Sci. Lett.* 94:385–97

Haq BU, Hardenbol J, Vail PR. 1988. *Mesozoic and Cenozoic Chronostratigraphy and Cycles of Sea-level Change*. Spec. Publ. 42:71–108. Tulsa, OK: Soc. Econ. Paleontol. Mineral.

Herbert TD, D'Hondt SL. 1990. Precessional climate cyclicity in Late Cretaceous-Early Tertiary marine sediments: a high resolution chronometer of Cretaceous-Tertiary boundary events. *Earth Planet. Sci. Lett.* 99:263–75

Hildebrand AR, Penfield GT, Kring DA, Pilkington M, Camargo ZA, et al. 1991. Chicxulub crater: a possible Cretaceous/Tertiary boundary impact crater on the Yucatán Peninsula, Mexico. *Geology* 19:867–71

Huber BT. 1992. Upper Cretaceous planktic foraminiferal biozonation for the austral realm. *Mar. Micropaleontol.* 20:107–28

Izett GA. 1990. The Cretaceous/Tertiary boundary interval, Raton Basin, Colorado and New Mexico, and its content of shock-metamorphosed minerals: evidence relevant to the K/T boundary impact-extinction theory. *Geol. Soc. Am. Spec. Pap.* 249:1–100

Izett GA. 1991. Tektites in Cretaceous-Tertiary boundary rocks on Haiti and their bearing on the Alvarez impact extinction hypothesis. *J. Geophys. Res.* 96:20879–905

Jehanno C, Boclet D, Froget L, Lambert B, Robin E, et al. 1992. The Cretaceous Tertiary boundary at Beloc, Haiti: no evidence for an impact in the Caribbean area. *Earth Planet. Sci. Lett.* 109:229–41

Kate WGT, Sprenger A. 1993. Orbital cyclicities above and below the Cretaceous/ Paleogene boundary at Zumaya (N Spain), Agost and Relleu. *Sediment. Geol.* 87:69–101

Keller G. 1988. Extinction, survivorship and evolution of planktic foraminifera across the Cretaceous/Tertiary boundary at El Kef, Tunisia. *Mar. Micropaleontol.* 13:239–63

Keller G. 1996. The Cretaceous-Tertiary mass extinction in planktonic foraminifera: biotic constraints for catastrophic theories. See Gartner 1996, pp. 19–84

Keller G. 1997. Analysis of El Kef blind test I. *Mar. Micropaleontol.* 29:89–93

Keller G, Lopez-Oliva JG, Stinnesbeck W, Adatte T. 1997. Age, stratigraphy, and deposition of near-K/T siliciclastic deposits in Mexico: relation to bolide impact? *Geol. Soc. Am. Bull.* 109:410–28

Keller G, Stinnesbeck W. 1996. Sea-level changes, clastic deposits, and mega tsunamis across the Cretaceous Tertiary boundary. See Gartner 1996, pp. 415–50

Keller G, Stinnesbeck W, Lopez-Oliva JG. 1994. Age, deposition and biotic effects of the Cretaceous/Tertiary boundary event at Mimbral, NE Mexico. *Palaios* 9:144–57

Kent DV. 1977. An estimate of the duration of the faunal change at the Cretaceous-Tertiary boundary. *Geology* 5:769–71

Klaver GT, Kempen TMG, Bianchi FR, Gaast SJ. 1986. Green spherules as indicators of the Cretaceous Tertiary boundary in DSDP Hole 603b. In *Initial Reports of the Deep Sea Drilling Project*, ed. JE VanHinte, W. Wise, 93:1039–55. Washington, DC: US Govt. Print. Off. 1205 pp.

Koeberl C. 1992. Watercontent of glasses from the K/T boundary, Haiti: an indication of impact origin. *Geochim. Cosmochim. Acta* 56:4329–32

Koeberl C, Sigurdsson H. 1992. Geochemistry of impact glasses from the K/T boundary in Haiti: relation to smectites and a new type of glass. *Geochim. Cosmochim. Acta* 56:2113–29

Kotake N. 1989. Paleoecology of the Zoophycos producers. *Lethaia* 22:327–41

Kring DA, Boynton WV. 1991. Altered spherules of impact melt and associated relic glass from the K/T boundary sediments in Haiti. *Geochim. Cosmochim. Acta* 55:1737–42

Kyte FT, Bostwick JA. 1995. Magnesioferrite spinel in Cretaceous/Tertiary boundary sediments of the Pacific basin: remnants of hot, early ejecta from the Chicxulub impact? *Earth Planet. Sci. Lett.* 132:113–27

Kyte FT, Bostwick JA, Zhou L. 1996. The Cretaceous-Tertiary boundary on the Pacific plate: composition and distribution of impact debris. See Bohor 1996, pp. 389–402

Kyte FT, Leinen M, Heath GR, Zhou L. 1993. Cenozoic sedimentation history of the central North Pacific: Inferences from the elemental geochemistry of core LL44-GPC3 *Geochim. Cosmochim. Acta* 57:1719–40

Lerbeckmo JF, Sweet AR, Duke MJM. 1996. A normal polarity subchron that embraces the K/T boundary: a measure of sedimentary continuity across the boundary and synchroniety of boundary events. See Bohor 1996, pp. 465–76

Lindinger M. 1989. *The Cretaceous Tertiary boundary in the Caravaca and Kef sections.* PhD thesis. ETH Zurich. 213 pp.

Lopez Ramos E. 1981. *Geologia de Mexico.* Mexico City: UNAM. 446 pp.

Luterbacher HP, Premoli Silva I. 1964. Biostratigrafia del limite cretaceo-terziario nell'Appennino centrale. *Riv. Ital. Paleontol. Stratigr.* 70:67–128

Lyons JB, Officer CB. 1992. Mineralogy and petrology of the Haiti Cretaceous/Tertiary section. *Earth Planet. Sci. Lett.* 109:205–24

Marin LE, Sharpton V. 1994. Recent drilling and core recovery within the Chicxulub impact structure, northern Yucatan, Mexico. *Eos, Trans. Am. Geophys. Union, Fall Meet.* 75(44):408–9 (Abstr.)

Maurasse FJMR. 1991. Impacts, tsunamis, and the Haitian Cretaceous-Tertiary boundary layer. *Science* 252:1690–93

Meyerhoff AA, Lyons JB, Officer CB. 1994. Chicxulub structure: a volcanic sequence of Late Cretaceous age. *Geology* 22:3–4

Michel HV, Asaro F, Alvarez W, Alvarez LW. 1985. Elemental profile of iridium and other elements near the Cretaceous/Tertiary boundary in Hole 577B. In *Initial Reports of the Deep-Sea Drilling Project*, ed GR Heath, LH Burckle, 86:533–38. Washington DC: US Govt. Print. Off. 689 pp.

Montanari A. 1990. Authigenesis of impact spheroids in the K/T boundary clay from Italy: New constraints for high-resolution stratigraphy of terminal cretaceous events. *J. Sediment. Petrol.* 61:315–39

Montanari A, Claeys P, Asaro F, Bermudez J, Smit J. 1994. Preliminary stratigraphy and iridium and other geochemical anomalies across the KT boundary in the Bochil section (Chiapas, southeastern Mexico). In *New Developments Regarding the KT Event and Other Catastrophes in Earth History.* LPI Contrib. 825:84–85 (Abstr.) Houston: Lunar Planet. Inst.

Montanari A, Hay RL, Alvarez W, Asaro F, Michel HV, et al. 1983. Spheroids at the Cretaceous-Tertiary boundary are altered impact droplets of basaltic composition. *Geology* 11:668–71

Moreau MG, Cojan I, Ory J. 1994. Mechanisms of remanent magnetization in marl and limestone alternations. Case study: Upper Cretaceous (Chron 31-30), Sopelana, Basque Country. *Earth Planet. Sci. Lett.* 123:15–37

Moreau MG, Mary C, Orue Etxebarria X. 1989. Magnetostratigraphy of the Sopelana K/T boundary section. *Terra Abstr.* 1:254

Morgan J, Warner M, Brittan J, Buffler R, Camargo A, et al. 1997. Size and morphology of the Chicxulub impact crater. *Nature* 390:472–76

Nederbragt AJ, Koning JA. 1994. Morphologic variation in Turonian to Maastrichtian Heterohelix globulosa (Ehrenberg). *Proc. K. Ned. Akad. Wet.* 97:429–44

Newsom HE, Graup G, Iseri DA, Geissman JW, Keil K. 1990. The formation of the Ries crater, West Germany: evidence of atmospheric interactions during a large cratering event. In *Global Catastrophes in Earth History; An Interdisciplinary Conference on Impacts, Volcanism, and Mass Mortality*, ed. VL Sharpton, PD Ward, Spec. Pap. 247:195–206. Boulder, CO: Geol. Soc. Am. 631 pp.

Ocampo AC, Pope KO, Fisher AG. 1996. Ejecta blanket deposits of the Chicxulub crater from Albion Island, Belize. See Bohor 1996, pp. 75–88

Officer CB, Page J. 1996. *The Great Dinosaur Extinction Controversy.* Helix Books, Reading, MA: Addison-Wesley. 209 pp.

Olsson RK, Miller KG, Browning V, Habib D, Sugarman PJ. 1997. Ejecta layer at the Cretaceous-Tertiary boundary, Bass River, New Jersey (Ocean Drilling Program Leg 174AX). *Geology* 25:588–90

Oskarsson N, Helgason O, Sigurdsson H. 1996. Oxidation state of iron in tektite glasses from the Cretaceous Tertiary boundary. See Bohor 1996, pp. 445–52

Pardo A, Ortiz N, Keller G. 1996. Latest Maastrichtian and Cretaceous-Tertiary boundary foraminiferal turnover and environmental changes at Agost, Spain. See Gartner 1996, pp. 139–72

Percival SF, Fischer AG. 1977. Changes in calcareous nannoplankton in the Cretaceous-Tertiary biotic crisis at Zumaya, Spain. *Evol. Theory* 2:1–35

Pope KO, Ocampo AC, Fischer AG, Alvarez W, Fouke BW, et al. 1999. Proximal Chicxulub ejecta from Albion Island, Belize. *Geology.* (In press)

Robin E, Boclet D, Bonte P, Froget L, Jehanno C, et al. 1991. The stratigraphic distribution of Ni-rich spinels in Cretaceous-Tertiary boundary rocks at El Kef (Tunisia), Caravaca (Spain) and Hole 761C (Leg 122). *Earth Planet. Sci. Lett.* 107:715–21

Robin E, Froget L, Jehanno C, Rocchia R. 1993. Evidence for a K/T impact in the Pacific Ocean. *Nature* 363:615–18

Roggenthen WM. 1976. Magnetic stratigraphy of the Paleocene: a comparison between Spain and Italy. *Mem. Soc. Geol. Ital.* 15:73–82

Romein AJT, Smit J. 1981. Carbon-oxygen isotope stratigraphy of the Cretaceous-Tertiary boundary interval: data from the Biarritz section (SW France). *Geol. Mijnbouw* 60:541–44

Romein AJT, Willems H, Mai H. 1996. Calcareous nannoplankton of the Geulhemmerberg K/T boundary section, Maastrichtian type area, The Netherlands. See Brinkhuis & Schioler 1996, pp. 231–38

Ruiz FM, Huertas MO, Palomo I, Barbieri M. 1992. The geochemistry and mineralogy of the Cretaceous Tertiary boundary at Agost (southeast Spain). *Chem. Geol.* 95:265–81

Sharpton VL, Marin LE, Carney JL, Lee S, Ryder G, et al. 1996. A model of the Chicxulub impact basin based on evaluation of geophysical data, well logs and drill core samples. See Bohor 1996, pp. 55–74

Sigurdsson H, D'Hondt S, Arthur MA, Bralower TJ, Zachos JC, et al. 1991. Glass from the Cretaceous-Tertiary boundary in Haiti. *Nature* 349:482–87

Smit J. 1977. Discovery of a planktonic foraminiferal association between the Abathomphalus mayaroensis zone and the "Globigerina" eugubina zone at the Cretaceous/Tertiary boundary in the Barranco del Gredero (Caravaca, SE Spain): a preliminary report. *Proc. K. Ned. Akad. Wet.* 80:280–301

Smit J, Alvarez W, Claeys P, Montanari A, Roep TB. 1994. Misunderstandings regarding the KT boundary deposits in the Gulf of Mexico. See Montanari 1994, p. 116 (Abstr.)

Smit J, Alvarez W, Montanari A, Swinburne N, Kempen TM, et al. 1992a "Tektites" and microkrystites at the Cretaceous Tertiary boundary: two strewnfields, one crater? *Proc. Lunar Planet. Sci. Conf.* 22:87–100

Smit J, Keller G, Zargouni F, Razgallah S, Shimi M, et al. 1997. The El Kef sections and sampling procedures. *Mar. Micropaleontol.* 29:69–72

Smit J, Montanari A, Swinburne NHM, Alvarez W, Hildebrand AR, et al. 1992b Tektite-bearing, deep-water clastic unit at the Cretaceous-Tertiary boundary in northeastern Mexico. *Geology* 20:99–103

Smit J, Nederbragt AJ. 1997. Analysis of the El Kef blind test II. *Mar. Micropaleontol.* 29:94–100

Smit J, Rocchia R. 1996. Neutron activation analysis of trace elements in the Geulhemmerberg Cretaceous/Tertiary boundary section, SE Netherlands. See Brinhuis & Schioler 1996, pp. 269–74

Smit J, Roep TB, Alvarez W, Montanari A, Claeys P, et al. 1996. Coarse grained, clastic sandstone complex at the K/T boundary around the Gulf of Mexico: deposition by tsunami waves induced by the Chicxulub impact? See Bohor 1996, pp. 151–82

Smit J, Romein AJT. 1985. A sequence of events across the Cretaceous-Tertiary boundary. *Earth Planet. Sci. Lett.* 74:155–70

Stinnesbeck W, Barbarin JM, Keller G, Oliva JGL, Pivnik DA, et al. 1993. Deposition of channel deposits near the Cretaceous-Tertiary boundary in northeastern Mexico: catastrophic or "normal" sedimentary deposits? *Geology* 21:797–800

Stinnesbeck W, Keller G. 1996. K/T boundary coarse-grained siliciclastic deposits in northeastern Mexico and northeastern Brazil: evidence for mega-tsunami or sealevel changes? See Bohor 1996, pp. 197–210

Stinnesbeck W, Keller G, Adatte T, Oliva JGL, MacLeod N. 1996. Cretaceous-Tertiary boundary clastic deposits in northeastern Mexico: impact tsunami or sealevel lowstand? See Gartner 1996, pp. 471–518

Stinnesbeck W, Keller G, Cruz J, Leon C, MacLeod N, et al. 1997. The Cretaceous-Tertiary transition in Guatemala: limestone breccia deposits from the South Péten basin. *Geol. Rundsch.* 86:686–710

Swisher CC III, Nishimura JMG, Montanari A, Pardo EC, Margolis SV, et al. 1992. Coeval 40Ar/39Ar ages of 65.0 million years ago from Chicxulub crater melt-rock and Cretaceous-Tertiary boundary tektites. *Science* 257:954–58

Van Andel TH, Thiede J, Sclater JG, Hay WW. 1977. Depositional history of the South Atlantic Ocean during the last 125 million years. *J. Geol.* 85:651–98

Worsley T. 1974. The Cretaceous-Tertiary boundary event in the ocean. In *Studies in Paleoceanography*, ed. WW Hay, Spec. Publ. 20:94–125. Tulsa, OK: Soc. Econ. Paleontol. Mineral. 235 pp.

Annu. Rev. Earth. Planet. Sci. 1999. 27:115–48

HUBBLE SPACE TELESCOPE OBSERVATIONS OF PLANETS AND SATELLITES

Philip B. James
Department of Physics and Astronomy, University of Toledo, Toledo, Ohio 43606; e-mail: pbj@physics.utoledo.edu

Steven W. Lee
Laboratory for Atmospheric and Space Physics, University of Colorado, Boulder, Colorado 80309; e-mail: lee@syrtis.colorado.edu

KEY WORDS: solar system, Mars, Jupiter, Saturn, outer planets

ABSTRACT

Although exploration by remote, in situ spacecraft has been the primary tool used in scientific exploration of the planets and other solar system objects for the last three decades, the unique capabilities of the Hubble Space Telescope (HST) have made it an invaluable tool for solar system research and have led to many important discoveries. HST's extended lifetime has made it possible to continue observations of planets and satellites that were started by earlier space missions and has thereby provided new insight into dynamic surface and atmospheric phenomena on these bodies. The ultraviolet capability of HST has made it possible to study important time-variable phenomena such as the auroras on Jupiter and Saturn and the circulation of planetary atmospheres. This review provides an overview of HST observations of and discoveries relating to planets and satellites in the solar system.

INTRODUCTION

Hubble Space Telescope (HST) was launched in April of 1990. Shortly after launch it was discovered to have spherical aberration in the primary mirror, resulting in an undesirable point spread function (PSF) that threatened to negate the potential advantages of a space imaging system. During the year

0084-6597/99/0515-0115$08.00

following launch, great progress was made in using techniques of mathematical deconvolution to remove the smearing effects of the PSF. Although it would be inaccurate to portray the images as fully restored by deconvolution, various algorithms were successful in rendering the images scientifically viable. Mars, Jupiter, and Saturn were among the first targets studied by HST because the large signal-to-noise ratio obtainable with short exposures minimized jitter problems, making these targets ideal for deconvolution despite their extended nature and complexity (Beebe 1991).

The original HST had five instruments (not counting the Fine Guidance Sensors). The Wide Field and Planetary Camera (WFPC1) was a two-dimensional imaging camera covering wavelengths between Lyman α and about 1.1 μm. The WFPC1 had eight separate 800-by-800 CCDs in two groups: four Planetary Camera (PC) chips having contiguous fields of view and a pixel scale of .046 arcsecs/pixel, and four Wide Field (WF) chips having twice the scale.

The Faint Object Camera (FOC) is a long-focal-ratio, photon-counting imager capable of taking high-resolution images of the sky with scales as small as 0.014 arcsecs/pixel in the 1150 to 6500 Å wavelength range. The Faint Object Spectrograph (FOS) was used to make spectroscopic observations of astrophysical sources from the near ultraviolet to the near infrared (1150–8000 Å). The Goddard High Resolution Spectrometer (GHRS) obtained spectra from about 1150 to 3200 Å at resolving powers up to 80,000 depending on the gratings selected. The High Speed Photometer (HSP) was used to make very rapid photometric observations with a variety of passbands from the near ultraviolet to the visible.

In December 1993 the first shuttle servicing mission made extensive modifications in the observatory that, among other things, corrected the optical problems. A new instrument, COSTAR, that corrected the optics for the off-axis instruments, took the place of the HSP in the instrument bay. A new Wide Field and Planetary Camera with an optical path that compensates for the spherical aberration took the place of WFPC1. WFPC2 has one PC chip at .0455 arcsecs/pixel and three WF Camera chips at .0996 arcsecs/pixel that have contiguous fields of view.

The FOS and GHRS were removed from HST during the second servicing mission in February, 1997 and replaced by the Near Infrared Camera and Multi Object Spectrometer (NICMOS) and by the Space Telescope Imaging Spectrograph (STIS). NICMOS provides imaging capabilities in broad-, medium-, and narrow-band filters in the wavelength range 0.8–2.5 microns. NICMOS uses a 256-by-256 HgCdTe array and has three separate cameras with different scales of .043, .075, and .2 arcsecs/pixel. STIS provides both imaging and spectroscopic capabilities from 1150–11,000 Å in a CCD mode with .05 arcsec pixels or with the CS_2Te MAMA detector that has .024 arcsec pixels.

Table 1 Range of resolution of HST for various solar system bodies and number of HST orbits required for complete coverage of a diurnal cycle

Planet/satellite	Pixels/equatorial diameter	Kilometers/pixel (sub-Earth point)	HST orbits/ sidereal day
Mars	77–564	12–88	16
Jupiter	670–1100	130–215	7
Io	17–28		
Saturn	330–457	265–365	7
Titan	14–20		
Uranus	75–90	570–690	11
Neptune	48–52	950–1030	10
Triton	2–3		
Vesta	4–14	36–120	4

HST's diverse capabilities make it ideal for many types of solar system observations. In particular, HST acquires ultraviolet images with high spatial resolution and ultraviolet spectra with high resolving power, a capability that has had a significant impact on the study of planetary atmospheres. The good spatial resolution coupled with the broad range of wavelengths in which targets can be imaged makes HST an ideal instrument for monitoring the atmospheres of Mars and the outer planets at synoptic scale. Table 1 shows the range of spatial resolutions that can be obtained with WFPC2 for various planets and satellites. The table also includes the number of HST orbits required to monitor a complete diurnal cycle on each planet.

Using software developed by Jet Propulsion Laboratory, HST is able to point at and track solar system targets with sub-arcsecond accuracy between and during exposures. This software has been implemented as the mission has evolved; in the earliest observations planets were specified as fixed targets. Observations of solar system objects are affected by the 50° minimum elongation relative to the sun allowed for HST pointing. This characteristic has precluded studies of the inferior planets except for one set of observations of Venus near maximum elongation in January of 1995 (Na & Esposito 1995).

All the HST instruments mentioned above have been used for planetary observations. While the cameras and spectrographs have been most frequently used, HSP was used at least once to monitor an occultation of a twelfth-magnitude star by Saturn's rings (Elliot et al 1993). NICMOS and STIS are sufficiently new that studies involving these instruments are only now beginning to appear in publication. Several studies involved coordinated observations by HST and other observatories and spacecraft; these studies include simultaneous spectroscopy by the Hopkins Ultraviolet Telescope and imaging by HST to study the

Jovian aurora (Morrissey et al 1997); observations coincident with the Galileo probe entry (Beebe et al 1996); and a large number of observations related to the impact of Comet P/Shoemaker-Levy 9 (SL-9) on Jupiter.

The purpose of this review article is to give an overview of the important scientific discoveries relating to planets and satellites that have been made by Hubble Space Telescope. We focus on HST contributions and mention observations from other observatories only when absolutely necessary for establishing context. Constraints on the length of this review led us to focus on planets and satellites and to exclude the many excellent HST observations of comets. These observations include the many HST orbits dedicated to the collision of SL-9 with Jupiter. HST contributed greatly to the SL-9 campaign, and this was undoubtedly one of HST's finest hours with respect to solar system studies. The authors feel that the SL-9 campaign should be considered as an entity with the HST observations as one component; therefore we have not specifically referred to those observations in this article.

MARS

A set of WFPC1 observations of Mars in December, shortly after the 1990 opposition of the planet, was the first General Observer project executed by HST (James et al 1991); despite HST's problems the observations were approved as a Director's Discretionary project because of the critical timing of the opposition. Imaging and FOS spectral observations continued during 1991 until the elongation of Mars relative to the sun dropped below HST's 50° limit. Analysis of these images clearly showed that deconvolution of WFPC1 images restores a large measure of the images' interpretive value and that WFPC1 provides useful images of Mars even when its angular size is near the minimum (James et al 1994). Ratios of UV filter images at 230 nm and 336 nm, respectively within and outside the Hartley absorption band of ozone, revealed the amount of ozone in the atmosphere as a function of geographic position. The violet and UV images were used in conjunction with scattering models to constrain the opacities due to dust and condensates in the Martian atmosphere.

HST WFPC2 images acquired near the 1995 opposition of Mars revealed a zonal band of condensate clouds encompassing the planet between -10° and $+30^\circ$ latitude (James et al 1996a); a typical optical depth curve is shown in Figure 1. The maximum optical depth of these clouds was about 0.3 near $+20^\circ$, and observations of clouds beyond the morning terminator suggested altitudes of about 8 km. The season of the observations was spring in the northern hemisphere when Mars is near the aphelion of its orbit. Further study of these and other images acquired during 1994–95, as well as the same seasons in 1992–93 and 1996–97, showed that this belt of clouds occurred during all three years

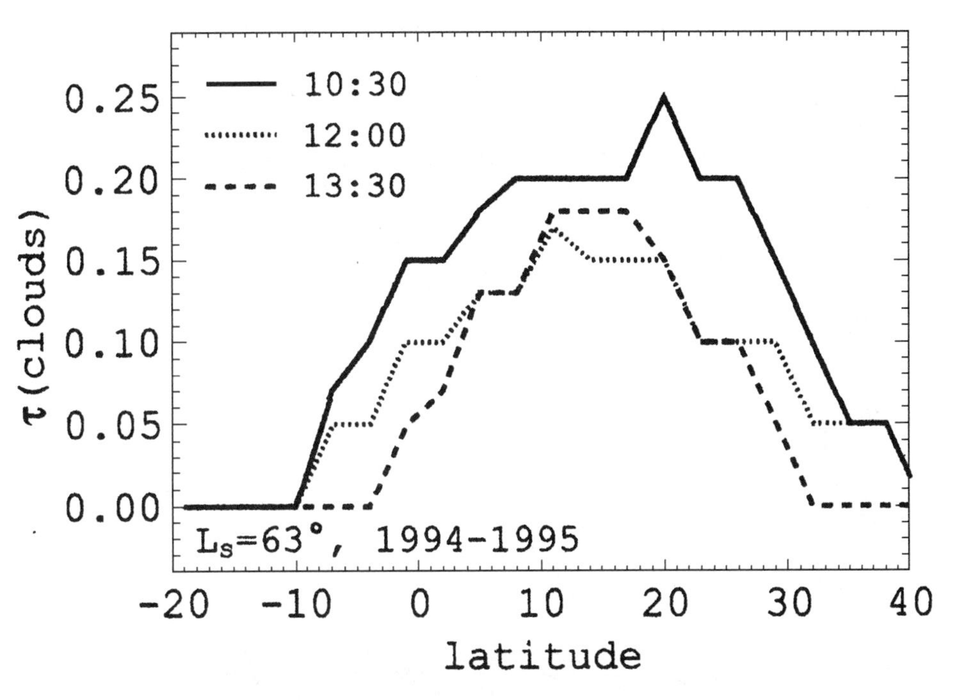

Figure 1 The optical thickness of condensate clouds in the vicinity of Syrtis Major during 1994–95 Hubble observations is plotted as a function of latitude. The season corresponding to this graph is midspring, and curves are shown for three values of the local time. The clouds over Syrtis are representative of the global cloud band observed during three Martian years of HST observations. The figure, originally published by Wolff et al (1998), is reproduced here with permission of *Journal of Geophysical Research*.

(Wolff et al 1998). The clouds are present from mid-spring through midsummer but are strongest between $L_s = 60°$ and $100°$ (L_s is the areocentric longitude of the sun relative to Mars measured from the vernal equinox). Clancy et al (1996a) pointed out that these clouds indicate low-level saturation and condensation of water vapor in the ascending branch of the seasonal Martian Hadley cell. This process is consistent with atmospheric temperatures determined from the intensity of microwave emission by CO molecules in the Martian atmosphere. Because such condensation could cap the expected north-to-south transfer of water vapor during this season, the clouds may be an important factor in confining water to the northern hemisphere.

Discrete condensate clouds were also observed extensively in all years of HST imaging (James et al 1994 and 1996a). Clouds in both the Tharsis and Elysium regions were prominent during the Martian seasons corresponding to

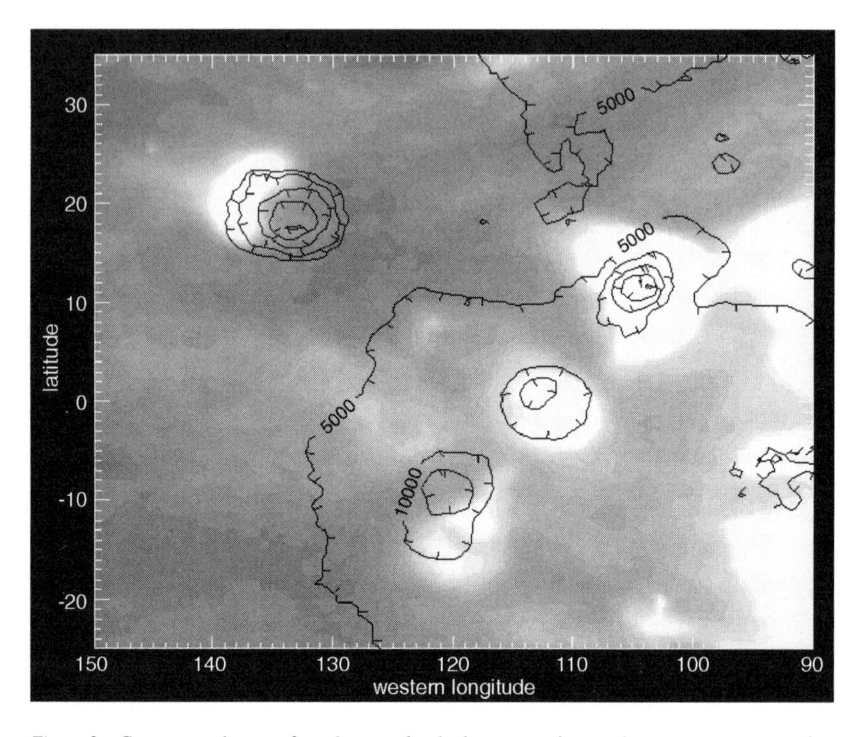

Figure 2 Concentrated areas of condensate clouds shown superimposed on contours representing topography in the Tharsis region of Mars. Olympus Mons, the largest Martian volcano, is in the upper right, and the three large volcanoes of the Tharsis ridge run diagonally across the image. The optical depths of these discrete clouds, viewed in February 1995, is substantially larger than the background opacity of the global cloud belt shown in Figure 1. The figure, originally published by James et al (1996), is reproduced here with permission of *Journal of Geophysical Research.*

most HST observations, namely spring and summer in the northern hemisphere. The Tharsis clouds were prominent in 1993 at $L_s = 20.2°$, which is earlier than they historically have appeared; the distribution of Tharsis clouds was different, and the clouds were more optically thick by a factor of two at $L_s = 63.7°$ in 1995 (Figure 2). Ultraviolet images acquired using the F255W filter on March 30, 1997 as part of two different HST observing programs were used to determine cloud displacements and winds. Winds of 15 to 44 m/sec from the southeast were characteristic for a wide range of latitudes (Mischna et al 1998).

Dust storms on Mars range from small dust devils to huge, planet-encircling storms. Events of the largest scale historically have occurred during the season in which HST has not been able to observe Mars because of the solar interdict. However, considerable dust activity in the form of local and regional storms was observed during 1993, 1995, 1996, and 1997. Several storms were observed in the vicinity of the receding seasonal north polar cap during its spring retreat (James et al 1998) and over the residual north polar cap in summer (Wolff et al 1997). Dust clouds were also detected in Valles Marineris, shortly before the landing of Pathfinder (Wolff et al 1998) and in Hellas Basin (Wolff et al 1997). Optical depths of these localized dust clouds were generally between 1.0 and 1.5. The optical depths of diffuse background dust were also modeled using the HST images from 1992 to 1997 (Wolff et al 1998); between $L_s = 350°$ and $90°$ dust optical depths varied between 0.1 and 0.3, while after summer solstice the dust loading increased to between 0.3 and 0.4.

Interannual variability in the sublimation of the seasonal CO_2 cap is of interest because of the potential sensitivity of this process to other climate phenomena. HST observed four seasonal recessions of the Martian north polar cap at substantially higher resolution than is possible using ground based facilities. WFPC2 images showed that the north polar cap assumes a hexagonal form in mid spring, $L_s \sim 60°$, when the circumpolar dunes also become visible as a dark albedo ring beneath the surface frost. Differences between the four 1990s regressions were identified, but these were small compared to variations between recessions observed by ground-based astronomers over the last few decades (Cantor et al 1998). The Lambert albedos of the surface cap were determined using the F673N and F410M bandpasses. The albedos at both wavelengths tend to increase linearly during the spring and summer seasons, but the red-to-violet ratio drops significantly around summer solstice, possibly due to erosion of dust deposited in the seasonal cap during winter. The only opportunity to observe the south polar cap occurred in 1992 when Mars was between only five and six arcseconds in diameter. The WFPC1 images established that this cap regression was similar to that observed in 1977 by Viking observations; i.e., it is also near the mean of previous ground-based observations (James et al 1996b).

Multi-filter WFPC2 images of Mars fill a valuable niche between high-spatial–/low-spectral–resolution spacecraft images and low-spatial–/high-spectral–resolution ground-based observations (Bell et al 1997). The linear ramp filters on WFPC2 can be used to acquire images at specific wavelengths of interest, such as the 860 nm ferric band. The change in time allocations to units of orbits, which permits acquisition of a large number of short exposures during a single orbit, has also greatly enhanced the value of HST for surface mineralogy studies. Bell et al (1997) identified two distinct units that are separated by their red (673 nm) to violet (410 nm) ratio within the classic bright units on Mars. The signature of ferric iron absorption is found in all regions of the planet while ferrous iron absorption, measured by the depth of the band at 953 nm, is concentrated in the dark regions. A large number of images created using the F1042M filter were analyzed in order to provide accurate reflectance maps as support for the Mars Orbiter Laser Altimeter (MOLA) experiment on Mars Global Surveyor, which uses reflections from a Nd:YAG laser at 1060 nm to map Martian topography (Bell et al 1998). The analysis indicates substantial changes in the surface albedo patterns compared to the broadband albedos determined by the Viking Infrared Thermal Mapper (IRTM). As two examples among many, the boundaries of Syrtis Major and Hesperia Planum have receded by several hundred kilometers during the two decades and the dark albedo feature in the Cerberus region has essentially disappeared (see also James et al 1996a).

FOS UV spectra of Mars have complemented the WFPC observation program since 1991. Modeling these spectra, which span the 225-to-330 nm range, yields ozone abundance, cloud and dust opacities, and the albedo of surface ice deposits in the polar regions. Ozone abundance is controlled by the abundance of water vapor in the Martian atmosphere and therefore varies substantially both seasonally and geographically. Clancy et al (1996) analyzed FOS spectra from February 1995. They found low-latitude ozone abundances that are elevated at this season ($L_s = 63.5°$, roughly aphelion) by more than a factor of two relative to earth-based measurements prior to perihelion. They argue that this result supports the low-altitude saturation and condensation of water vapor mentioned previously in connection with the global cloud bands. Analysis of two FOS spectra obtained in 1996 and 1997 at $L_s = 10°$ and $61°$ showed a factor-of-two increase in low-latitude ozone from 1.8 to 3.6 μm-atm, confirming the increase as the planet nears aphelion (Clancy et al 1998). The aerosol distribution in the atmosphere determined from the FOS spectra for this northern spring season is most consistent with a globally distributed equatorial cloud band that increases in opacity from 0.04 to 0.07 in the interval between observations and a dust opacity that is less than 0.2.

Knowing the ratio of deuterium to hydrogen (D/H) in the Martian atmosphere is essential to understanding the evolution of the Martian atmosphere and current and past reservoirs of water. The ratio of HD/H_2 in the upper atmosphere of Mars has been measured recently (Krasnopolsky et al 1998) using GHRS spectra of Lyman α emissions: $HD/H_2 = 1.5 \pm 0.6 \times 10^{-4}$. Knowing the value of this ratio allows direct determination of a "fractionation factor," F, that is a measure of the efficiency of the escape of D relative to H. Previously the best value for F has been determined through a photochemical model for the Martian atmosphere to be $F = 0.32$ (Yung et al 1988). Krasnopolsky et al find a much smaller value, $F = 0.02$ from their measured D emission. This lower value of F would imply a significantly larger reservoir of water that is exchangeable with the atmosphere; but reconciling the current result with the known value of the ratio $HDO/H_2O = 1.7 \times 10^{-3}$ presents a challenge.

VESTA

Vesta's basaltic surface composition is unique among large asteroids. For this reason Vesta is suspected to be the parent body for about six percent of meteorites: those having igneous compositions. Binzel & Xu (1993) discovered several small asteroids, with Vesta-like spectra, extending from Vesta's orbit to the 3:1 resonance with Jupiter, at which material can be ejected into the inner solar system. Binzel and Xu proposed that these asteroids were formed in a large impact on Vesta. Recent HST observations have provided the strongest evidence to date supporting this hypothesis.

The first WFPC2 observations of Vesta were acquired during its 1994 opposition at about 51 km/pixel resolution. These 56 images were used to produce a geologic map of the region between -15° and 50° latitude (Binzel et al 1997) and to determine the size, shape, and spin axis of the asteroid (Thomas et al 1997b). The 1994 images were combined with a set of 78 WFPC2 images acquired during the more favorable 1996 opposition at 36 km/pixel (Thomas et al 1997a). These images extended the coverage to both poles (a color map projection is shown in Color Plate 1). The major discovery of the Vesta imaging program is a large impact crater, 430 km in diameter, near the South Pole of the asteroid. This crater has a rim raised roughly seven km above the average asteroid surface and 13 km above the floor of the crater. A central peak rises 13 km above the crater floor. The composition determined from ratios of images obtained with various filters is consistent with excavation into a high calcium-pyroxene–rich crust or olivine upper mantle. Thomas et al (1997b) calculate that roughly one percent of the mass of Vesta was ejected in the impact that formed the crater, and this amount is sufficient to account for the small Vesta-like asteroids mentioned previously. Thomas et al (1997b) also use their images to determine the best ellipsoid for Vesta with semi-axes of 289, 280, and 229 km. Using previous determinations of Vesta's mass, this leads to an average density of 3500 to 3900 kg m^{-3}.

JUPITER AND SATURN SYSTEMS

Jupiter

WFPC2 afforded the best resolution of Jupiter since the Voyager encounters until the arrival of Galileo; the available spectral range was broader than that of the Voyager cameras. Therefore a large number of HST orbits were devoted to the SL-9 campaign in 1994. Chanover et al (1996) used the 1994 observations to derive the absolute reflectivity of Jupiter as a function of latitude for several passbands between 255 and 953 nm. The 1994 data were combined with additional WFPC2 images from February 1995 for a study of the cloud features and winds in Jupiter's troposphere (Simon & Beebe 1996). Maps were generated from a combination of WF and PC images from six consecutive HST orbits (the field of view of the PC is too small to encompass Jupiter). Changes in the appearance and location of features such as the Great Red Spot and a number of white and dark ovals were determined by comparing data from the two years. Near infrared WFPC images separated by a few HST orbits were used to measure drift speeds of clouds; a zonal wind profile that agrees very closely with Voyager observations was derived from these data. The 1994 and 1995 WFPC2 data were also used to show that the 160–220 m/sec winds measured by the Galileo atmospheric probe in late1995 are not incompatible with generally lower speeds ($\sim$100 m/sec) measured by IRTF and Voyager at that latitude (Beebe et al 1996). Very small clouds located near the jet at 6°N in 1994 gave wind speeds of $\sim$150 m/sec, while larger clouds give speeds that are characteristic of larger weather systems and can therefore be considered as lower limits.

WFPC2 data from August 1994, February 1995, and October 1996 have been used to study the behaviors of the cyclonic and anticyclonic systems in conjunction with higher spatial-resolution Galileo images (Simon et al 1998). The translation rates of a series of white ovals (anticyclonic systems) centered at approximately $-33°$ longitude, that were first observed when the band first clouded over in 1939, have slowed since Voyager observations; these systems now constitute a fairly regularly spaced chain of alternating cyclonic and anticyclonic features (Figure 3). Previous events involving slowing and close approaches of these features had resulted in the ovals "repelling" each other and moving apart; in this case the configuration has been stable for several years. This type of study is a prime example of the value of the long term synoptic observations of atmospheric dynamics at good resolution that can be obtained with HST.

Emerich et al (1996) used the Echelle A grating on the GHRS to study the profiles of Lyman α emission lines in a region of enhanced Lyman α brightness on Jupiter called the Lyman α bulge. They find dramatic changes at low latitudes on time scales of a few minutes, and the structure of the emission lines is indicative of supersonic velocity parcels at high altitudes.

Saturn

Saturn was selected as an example of an extended but high signal-to-noise target for purposes of assessing the capabilities of HST when the spherical aberration problem was discovered (Westphal et al 1991). The initial set of images, acquired in August of 1990, revealed that the polar hexagon reported by Godfrey (1988) from analysis of Voyager images was still present and that it therefore represented a persistent rather than a transient phenomenon; the images also showed that the visible profile of the belts and zones had changed

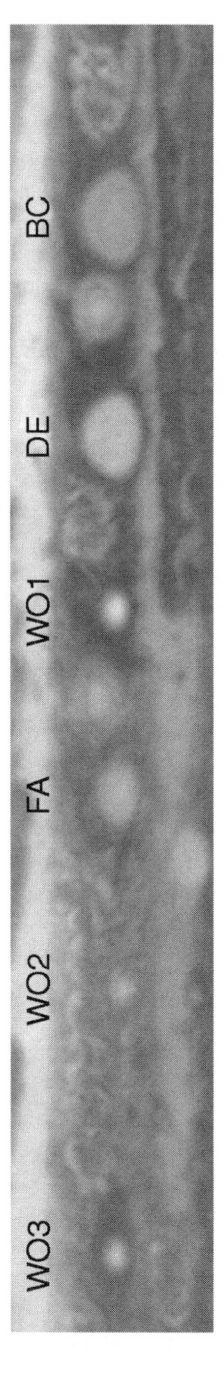

Figure 3 This close-up of the alternating cyclonic and anticyclonic features near 33° S on Jupiter was acquired by HST's Wide Field and Planetary Camera (WFPC2) in October 1996. The anticyclonic features are labeled, and the corresponding cyclonic cells can be seen between each pair of anticyclones and to the east of BC. The figure, originally published by Simon et al (1998), is reproduced here with permission of *Icarus*.

somewhat from Voyager. The work of Westphal et al was most importantly a test of the efficacy of deconvolution on the WFPC images and confirmation that good solar system science could be done with HST even with the spherical aberration handicap.

In September of 1990 a major equatorial disturbance appeared on Saturn that spread in longitude until it completely circled the planet. HST WFPC1 images of Saturn obtained in August 1990 (prior to the storm), November1990, and June 1991 (when the disturbance had mostly dissipated) were used to study this unusual event. The disturbance was similar in size, duration, and appearance to events in 1876 and 1933 (Beebe et al 1992). These three major events and two non-equatorial disturbances occurred during the summer season in the Northern Hemisphere, suggesting a seasonal dependence. Beebe et al used HST and ground-based observations to identify three major phases in the development of the storm: onset and expansion of the initial disturbance; the eastward and westward expansion of the bright cloud; and the ultimate evolution into wave-like structures that circle the globe near the equator.

A photometric analysis of the Hubble image set in four wavelength pass-bands showed that the particles constituting the disturbance differ from those in the usual Saturnian clouds (Westphal et al 1992). The storm clouds have consistently larger Minnaert coefficients (therefore stronger limb darkening) and are brighter than the normal equatorial cloud belt. Such findings could be interpreted as evidence that the materials in the two clouds were different or that the storm material had not yet had time to be processed by external interactions with particles or solar radiation. Several images taken with the F547M (roughly corresponding to Voyager imaging) and F889N filters were used to study the zonal winds near the equator of Saturn by observing the motions of cloud features (Barnet et al 1992). The low-latitude zonal winds were found to be significantly slower than those observed by Voyager (Figure 4). In addition, the wind speeds derived from the F889N observations were less than those found from the F547M filter. Winds near the top of the reflecting cloud layer are probed by the visible filter while F889N, which is located at a strong methane absorption, is sensitive to events roughly a scale height higher in the atmosphere. This suggests that the very strong zonal winds on Saturn decay with height in the equatorial region.

Karkoshka and Tomasko (1993) used WFPC1 images in seven passbands between 284 and 889 nm to study the distribution of aerosols on Saturn. Stratospheric haze optical depths are small at equatorial and mid-latitudes (<0.2) but are much larger in the polar regions, suggesting a separate production mechanism there (see related discussion in the section on auroras). Conversely, in the portion of the troposphere probed by these images (between 100 and 300 mb), optical depths due to tropospheric aerosols increase from pole to equator. The differences between the belts and zones, which are fairly subtle on Saturn,

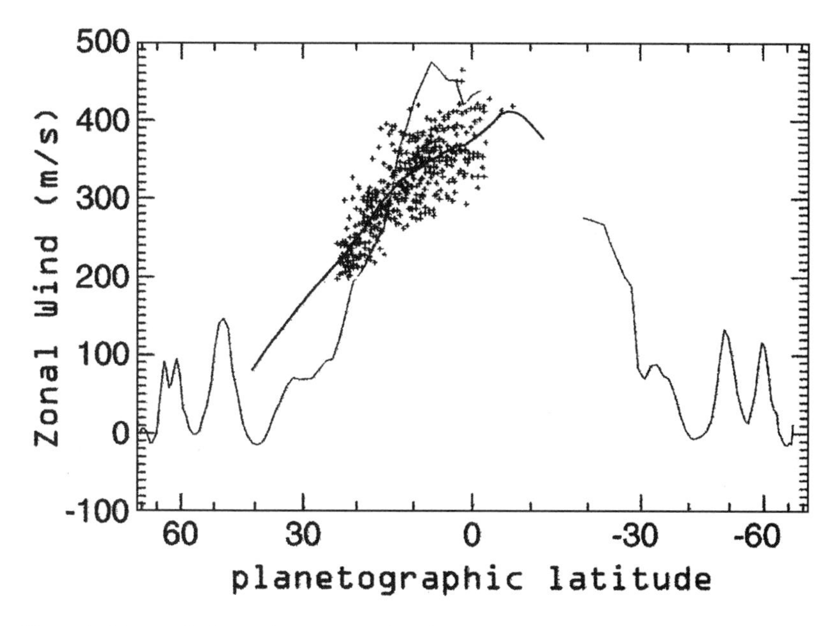

Figure 4 The zonal wind on Saturn determined from pairs of HST images in meters per second, plotted as a function of latitude. The *dark line* represents the average of the HST cloud track measurements, and the *light line* represents the Voyager measurements binned and averaged in the same manner as the HST data. The figure, originally published by Barnet et al (1992), is reproduced here with permission of *Icarus*.

are attributed to small differences in the mean size of the scatterers. Such size differences could be maintained by the strength of vertical motions with latitude.

Opportunities for observations of Saturn when the ring system appears edge on occur twice during Saturn's orbital period of 29.5 years. These ring-crossing events are important because they afford the opportunity to observe vertical structure associated with the rings and small satellites that are otherwise obscured by the glare from the bright rings. HST observations were scheduled when Earth passed through the plane of Saturn's rings on August 10, 1995 and May 22, 1995 as well as when the sun was in the plane of the rings in November 1995. The small F ring, which lies outside the major part of the Saturn ring system, was observed clearly in the HST images. Nicholson et al (1996) found that the F ring dominates the perceived 1.2-to-1.5 km edge-on thickness of the ring system. The F ring therefore masks the much thinner main rings in ring plane crossings.

Bosh & Rivkin (1996) and Nicholson et al (1996) used WFPC2 to re-acquire images of most of the known small satellites close to their predicted positions. The major discrepancy was in the case of Prometheus, which was displaced in longitude by roughly six or seven times the expected uncertainty. The fact that the displacement did not change in the time between HST observations does not support a collision of Prometheus with the F ring material. Nicholson et al suggest that an encounter with a small, co-orbital satellite could have produced the discrepancy, but such a satellite should have been detected by Voyager and HST. Bosh and Rivkin also found the moon Atlas even farther from its projected position, but this deviation can be accounted for by the uncertainties in the orbit determination by Voyager. Prometheus actively exchanges angular momentum with the rings. Future observations of Prometheus by HST and Cassini should provide valuable tests for dynamical ring calculations.

Shemansky et al (1993) first used FOS to detect neutral OH emissions from the $A^2\Sigma^+ \rightarrow X^2\Pi(0,0)$ band near the orbit of Tethys in the Saturn system; their result suggests that quenching by neutral gas derived from the icy satellites determines electron temperatures in the magnetosphere. Subsequently, Hall et al (1996) observed Saturn with the FOS during the August ring plane crossing. They detected UV emissions from a tenuous atmosphere of OH surrounding the rings. OH column densities of 0.2 to $1.2 \times 10^{13}/\text{cm}^2$ were found for the five observations. They calculate that the rings must produce between 10^{25} and 10^{29} OH molecules per second to maintain this atmosphere.

Auroras

Auroras are the visible manifestation of interaction between the atmosphere of a planet and its magnetosphere. Observations of the aurora therefore constrain planetary fields, magnetospheric physics, and atmospheres. Prior to HST, ultraviolet observations of auroras on Jupiter by International Ultraviolet Explorer (IUE) and Voyagers were made with large-aperture spectrographs that were unable to resolve the spatial distribution of the emissions. Since 1992 HST has been able to study the Jovian and Saturnian auroras in the UV with good spatial resolution. The absorption of ultraviolet solar radiation by aerosols in the upper atmosphere of Jupiter enhances the contrast of auroral features in the UV FOC and WFPC images. One of the main contributions of HST has been identification of the oval region that encloses the auroral activity. The oval is determined by the footprint on the planet of the field lines from the magnetosphere region that is the origin of the high-energy particles causing the excitations. These footprints are usually designated by the distance from Jupiter (R_J) at which the field lines cross the equatorial plane. Calculations of the reference ovals require models for the magnetic field. These models have improved as more data are assimilated (Connerney 1993), but uncertainties in the higher multipole moments result in some uncertainties in the positions of the reference ovals at the surface.

The first images of the North Polar aurora, isolating both Lyman α (Dols et al 1992) and H_2 Lyman band (Caldwell et al 1992a) contributions, were acquired using the FOC in early1992. These images confirmed the existence of enhanced UV emissions surrounding the north polar region and suggested that the auroral oval was consistent with particles from farther out in the magnetosphere than the Io torus, which had been the suspected source since Voyager. The analysis of Gerard et al (1993), using two Lyman α images from February 1992 together with two H_2 Lyman band images from June 1992, indicated that the emissions fit the footprint of the $R_J \sim 30$ field lines. A series of six Lyman band images covering a complete rotation of the north polar region of Jupiter was obtained in February 1993 (Gerard et al 1994a). The auroral emissions follow the 30 R_J oval, shifted slightly by local distortions of the field lines. A definite difference exists between longitudes east and west of System III longitude 180°. For longitudes greater than 180° the emission takes the form of a sharp arc, while for longitudes less than 180° the emissions are more diffuse. Gerard et al interpret this to be a real longitude effect rather than a time-of-day effect as originally suggested by Caldwell et al (1992a).

Gerard et al (1994b) observed an unusually intense aurora in July 1993 using the FOC. A very bright arc along the $R_J = 30$ oval was in co-rotation with the planet. The power radiated in H_2 bands during this event was 10^{12} watts compared to 0.3 terra watts in images acquired 20 hours later, more typical of the Jovian aurora. Comparison of the two sets of images suggested that the intense display overlay a more diffuse region that remained relatively static during the event. The second set of 1530 Å images as well as additional images at 1250 and 1300 Å acquired in the aftermath of the intense display were analyzed by Grodent et al (1996). They observed no bright arc after the event, but some diffuse H_2 emissions extended equatorward of that boundary.

The first observation in FUV of the footprint of Io as distinct from the auroral oval was made with FOC after the addition of COSTAR by Prange et al (1996). Subsequently, Clarke et al (1996, 1998) have used the WFPC2 with alkali metal "Woods" filters, which increase sensitivity in the UV and improve rejection of longer wavelengths, providing the most definitive studies to date of the morphology of the auroral emissions at both poles. The major features at both poles are the main auroral ovals with similar latitude and longitude structures on different days, patchy and variable emissions poleward of these main ovals, and discrete emission from the foot of the Io flux tube. Trauger et al (1998) recently imaged the aurora of Saturn with WFPC2. Ultraviolet images reveal narrow bands of emissions near both the north and south poles of Saturn. These are most prominent in the morning sector, and the pattern of emissions tends to be fixed in local time.

A recent STIS image of the North Polar aurora is shown in Figure 5. The pattern of emissions from and within the main oval co-rotates with Jupiter, reflecting the co-rotation of the inner portions (<30 R_J) of the Jovian

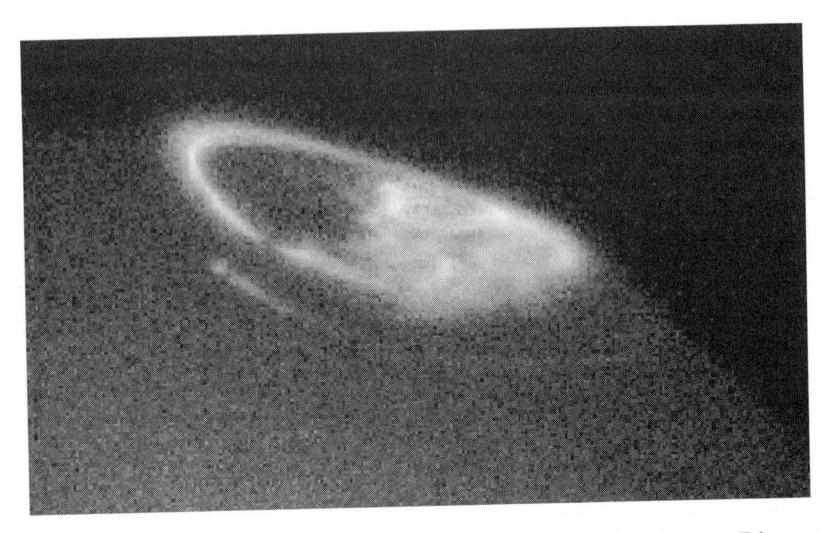

Figure 5 This image of the north polar aurora was taken in ultraviolet light by the Space Telescope Imaging Spectrograph (STIS). The resolution is sufficient to show the "curtain" of auroral light extending several hundred miles above Jupiter's limb. The comet-shaped streak just outside the auroral oval is caused by an electrical current of charged particles ejected from Io that flow along Jupiter's magnetic field lines to the planet's north and south magnetic poles. Credits: John Clarke (University of Michigan) and NASA.

magnetosphere. The Io flux tube emissions follow the satellite's motion. Because the main oval is always 4° to 6° poleward of the Io emissions, the oval corresponds to portions of the magnetosphere well beyond Io's orbit independent of a particular magnetospheric model. The $R_J = 30$ footprint in a revised model for the magnetic field that incorporates the observations of the Io footprint to constrain the field (Connerney et al 1998) agrees fairly well with the observations of the main oval. They observe an interesting symmetry between features in the auroras at the two poles. The conjugate features are associated with opposite ends of the same $R_J = 30$ field lines. Clarke et al (1998) also identified candidates for Ganymede and Europa auroral footprints, although the evidence is not conclusive. Ballester et al (1996) discussed time variations in auroral features. They report a bright event near the dawn limb that, unlike the main aurora, is co-rotating neither with Jupiter nor the Io flux tube; and an "equatorward surge" of emissions that are tightly confined to the main oval in the morning but move equatorward in the afternoon.

High-resolution UV spectra can be used to determine the temperature and pressure at the location where the auroral emissions originate. The spectral region between 1585 Å and 1620 Å is dominated by H_2 emission from the

Lyman band and is relatively unattenuated by methane absorption in the upper atmosphere (Figure 6). Modeling of the intensity distribution among individual transitions in the band leads to constraints on the excitation process and to the temperature at the level where the emissions originate. At shorter wavelengths, methane absorption becomes more important, and the degree of attenuation constrains the pressure at the emission level. The first use of the GHRS instrument to obtain high-resolution spectra of the aurora in both the Lyman and Werner bands of H_2 occurred during a period in February 1992 bracketed by the FOC images (Trafton et al 1994). Their analysis led them to identify secondary electron impact excitation as the probable source of the auroral emissions, most of which came from altitudes near the homopause. Subsequently, Clarke et al (1994b), Kim et al (1995), and Kim et al (1997) modeled GHRS spectra obtained between 1993 and 1995. Modeling emission intensities of rotational lines, they find rotational temperatures between 400 and 850 K fitting the longer wavelength emissions. The brightest regions tend to fall near the bottom of this range, suggesting that they are formed lower in the atmosphere where cooling due to hydrocarbons is most efficient. Kim et al (1997) used attenuation of the shorter wavelength emissions by methane to deduce that the auroral emissions form at methane column densities between 1×10^{16} and 7×10^{16} corresponding to pressures from a few μbar to a few tens of μbars.

Polar aerosols on Jupiter and Saturn differ substantially in properties and abundance from those in equatorial regions. The polar auroras may be responsible for the production of hydrocarbon aerosols as a result of H_2^+ production. FOC observations of Saturn (Ben Jaffel et al 1995) at 220 nm detected a dark oval encircling the north magnetic pole of the planet centered at about 79°N latitude. Nearly simultaneous observations of the H_2 auroral emissions at 153 nm and the haze at 210 nm by Gerard et al (1995) indicated that auroral emissions were centered at about 80°N and that the maximum optical depth due to haze occurred a few degrees equatorward. Moreno (1996) presented a model for the microphysical processes affecting Jovian aerosol particles in conjunction with aerosol distributions derived from WFPC2 archive images. His modeling and data support the hypothesis that aerosols in the polar region on Jupiter are formed from phenomena related to the aurora.

Io

Io, the Galilean satellite closest to Jupiter, has been the object of several studies based on HST data. HST proved itself to be a valuable instrument for monitoring surface changes on the dynamic satellite, especially with the corrected optics available since the repair mission. Though HST observing from Earth orbit can never compete with Galileo in terms of surface resolution, the opportunity for

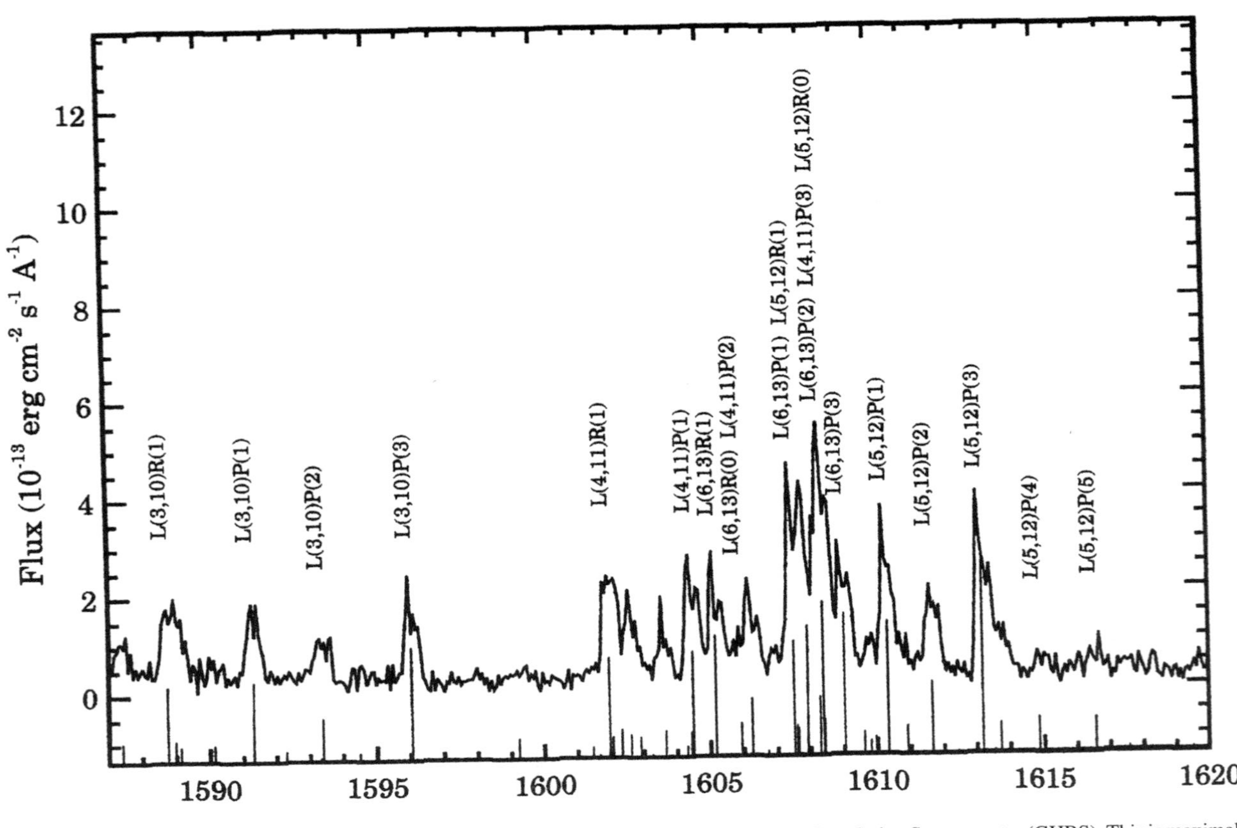

Figure 6 Ultraviolet spectrum of the northern aurora of Jupiter obtained by the Goddard High Resolution Spectrometer (GHRS). This is maximal (one diode) resolution with no smoothing. The rotation/vibrational emission lines indicated are part of the H_2 Lyman band system. The figure, originally published by Kim et al (1997), is reproduced here with permission of *Icarus*.

extended comparisons and the greater wavelength resolution for images is an advantage.

FOC observations at UV wavelengths shorter than the 0.32 μm SO_2 absorption edge confirmed the presence of SO_2 frost covering large portions of Io's surface (Sartoretti et al 1994). Sartoretti et al (1996) found significantly more surface SO_2 frost, roughly 50 to 60% on both hemispheres, than was determined from Voyager data. SO_2 frost extended to $+/-60^\circ$ latitude as opposed to $+/-30^\circ$ in the earlier data. However, because the albedos depend on grain size and photometric angles as well as on wavelength, there may be no actual conflict.

WFPC images of Io in five wavelengths obtained in March of 1994 have been used to study changes on the surface of Io that have resulted from its active volcanism (Spencer et al 1997). Three of the wavelengths overlapped the series of filters used by Voyagers I and II, permitting a direct comparison between Voyager results and the new HST observations. Lower-resolution pre-repair HST images did not show significant changes in large-scale albedo patterns since Voyager II (Sartoretti et al 1995). The Spencer et al comparisons confirmed the large-scale stability noted by Sartoretti et al, but also showed that several smaller-scale changes associated with known volcanic centers had occurred since 1979. Remarkably, additional WFPC2 images obtained in 1995 and shown here in Figure 7 revealed an albedo change more dramatic than any that had occurred during the preceding 15 years. The observed changes appear to be connected to activity of Ra Patera during the 16 months separating the two observations.

Wavelengths accessible to WFPC2 included the region from 700 to 1000 nm, which was not included in Voyager imaging. Ratio maps revealed that the 550-to-700 nm absorption edge on Io has a different spatial distribution from the 400-to-500 nm absorption edge studied by Voyager and therefore represents a different chemical species. In particular, the longer wavelength feature is strongly concentrated in the ejecta blanket of the volcano Pele and in the polar regions. Spencer et al consider several possibilities as spectral analogs for the Pele deposits and conclude that condensed S_2O that has partially thermally decomposed into S_4 and S_3 provides the best match.

A combination of FOC, FOS, and GHRS observations of Io in the UV have been used to study the distribution of the tenuous SO_2 atmosphere on the satellite. The first detection of an SO_2 atmosphere on Io was made by the Voyager IRIS experiment (Pearl et al 1979); it was subsequently detected by observations of rotational line emission by Lellouch et al (1992). The density and distribution of the SO_2 atmosphere have been studied by HST using the effects of atmospheric absorption by SO_2 at UV wavelengths shorter than 320 nm. Two possible sources could create Io's atmosphere: outgassing from active

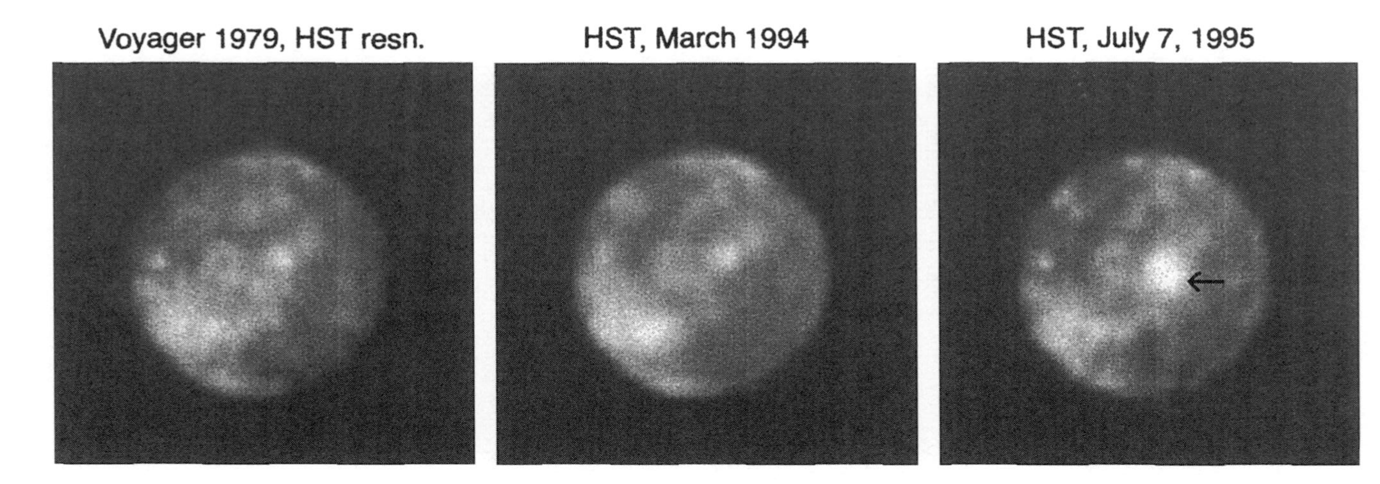

Figure 7 Three images showing the emergence of an enhanced bright spot on Io near Ra Patera. The *left* image is a Voyager image that has been degraded to HST resolution. It is compared to HST images acquired in March 1994 (*center*) and July 1995 (*right*). Little change in this area occurred between Voyager and the earlier HST observations, but substantial change took place during the months separating the two HST views. The figure, originally published by Spencer et al (1997), is reproduced here with permission of *Icarus*.

volcanoes and sublimation from the substantial deposits of SO_2 on Io's surface. In general, SO_2 frost has high albedo in the near UV but is very dark for wavelengths shorter than 300 nm. The albedos depend on grain size and on phase angle, however, complicating comparisons between various data sets. Io keeps the same face toward Jupiter at all times; the distributions of both SO_2 solid and gas appear to be different on the leading and trailing hemispheres of Io, possibly reflecting the difference in the distribution of volcanism on the two hemispheres.

Ballester et al (1994) made the first positive detection of SO_2 absorption bands in the ultraviolet in FOS spectra of the trailing hemisphere using the G190H grating. They detected emissions of neutral and ionized sulfur in addition to the SO_2 absorption band between 200 and 220 nm. These observations were disk average and therefore did not provide specific information concerning the distribution or density of the atmosphere. Ballester et al were able to fit their data with either a tenuous ($N \sim 10^{16}/cm^2$) uniformly distributed atmosphere or a denser SO_2 atmosphere, with $N \sim 3 \times 10^{17}/cm^2$, that is confined to about eight percent of the surface area.

Observations with the FOC using ultraviolet filters give resolutions of ~200 km on Io and permit the study of the relative contributions of the SO_2 atmosphere and surface albedo variations in determining the reflectance spectrum of Io. The 450 nm observations of Io's trailing hemisphere by Sartoretti et al (1994) showed that no significant changes in albedo distribution had occurred in the visible since Voyager. Additional observations at 285 nm, where SO_2 gas absorbs strongly, added constraints on the atmospheric component. Their analysis favored an optically thick ($\tau > 1$) SO_2 atmosphere covering 13 to 20% of the surface. As much as half of the atmospheric contribution was associated with Pele; in addition, the absence of thick gas over subsolar points and of any dependence on solar zenith angle suggested that the thick patches are the result of volcanism rather than sublimation. Sartoretti et al (1996) augmented the 285 nm coverage with FOC images at 232.5 and 260 nm in order to further constrain the atmospheric contributions. Images from March 1991 to July 1993 were used in their analysis. A model with only two distinct surface components is consistent with the observations if a non-linear portion of the curve of growth pertains at 285 nm where the SO_2 absorption is strongest. In their preferred solution the UV albedo of the surface frost is sufficiently low that there is essentially no vapor over the SO_2 frosted areas. This solution again favors an atmosphere predominantly due to direct volcanic emission rather than sublimation. They find that the vapor covers 11 to 15% of the surface with peak column densities $\sim 10^{18}/cm^2$.

The FOS also was used to observe the behavior of species in Io's atmosphere as the moon went into eclipse (Clarke et al 1994a). Atomic oxygen and sulfur emissions at 0.5 to 1.0 Io radii above the surface were detected between 115 and 157 nm. The emissions decrease as the moon is eclipsed, and then level off. This finding is consistent with recombination with photoelectrons, if the emissions are ionospheric, or with a rapid decrease in the atmospheric column due to SO_2 condensation, if torus plasma is the source. Residual emissions at longer wavelengths (227 to 330 nm) are entirely due to scattered light from Jupiter.

Trafton et al (1996) observed both hemispheres of Io with the GHRS in May and June 1992. Their analysis of the $C^1B_2 \rightarrow X^1A_1$ electronic transition band of SO_2 between 209.5 and 213.5 nm suggests a larger fractional coverage and a lower SO_2 density than the FOC analysis. The zero gas continuum in the UV is 20% brighter on the trailing hemisphere, opposite to the case for visual observations. Because the reflectivity of SO_2 is so low (geometric albedo = .018), this implies more SO_2 frost on the leading hemisphere. They find a lower limit for the fractional coverage of about 23% with a maximum SO_2 column density of about $10^{17}/cm^2$, roughly an order of magnitude lower than that in the FOC analysis. The evident importance of time-variable volcanic events, especially at Pele, in determining the distribution of the atmosphere is a possible explanation of the discrepancy because the GHRS observations were separated in time from the FOC images.

Atmospheres of Icy Satellites

HST-based spectroscopic observations of the icy satellites of Jupiter and Saturn have resulted in new insight regarding the nature of their atmospheres. Yung (1977) and Kumar & Hunten (1982) predicted that those satellites with water-rich surfaces would have oxygen atmospheres as a result of photolysis of water vapor and subsequent escape of hydrogen. Hall et al (1995) detected an oxygen atmosphere on Europa using the far UV grating (G140L) of the GHRS. They observed emission at wavelengths of 130.4 and 135.6 nm from transitions in neutral atomic oxygen from the $(2p)^3(3s)$ configuration to the ground state. The emission is consistent with dissociative excitation of molecular oxygen in a tenuous Europan atmosphere by electrons from Jupiter's magnetosphere, within which Europa orbits. The inferred atmospheric density, $1.5 \pm .5 \times 10^{15}$ cm^{-2}, corresponds to a surface pressure of roughly 10^{-5} μbar.

The Galilean satellites are tidally locked to Jupiter so that the same hemisphere of a moon is always facing the direction of its orbital motion. Because the orbital velocities are smaller than the co-rotation velocity, the training hemisphere receives the greater flux of charged particles trapped in Jupiter's magnetosphere. This is the origin of a number of dichotomies between the two hemispheres. Noll et al (1996) used the G190H and G270H gratings of FOS to observe the ultraviolet reflectance spectra of both the leading and trailing hemispheres of Ganymede. The large difference in albedo between the two hemispheres is due to an absorption band centered at 260 nm which is identified as the Hartley band of ozone. The density of ozone is estimated as $4.5 \times 10^{16}/cm^2$ on the trailing hemisphere. The continuous dissociation of the ozone in the absorption process requires a source for O_3 in the form of molecular oxygen.

Molecular oxygen on Ganymede's trailing face had previously been observed using ground-based spectroscopy by Spencer et al (1995). The visible absorption bands that they observed require either condensed phases of O_2 or gas at pressures orders of magnitude higher than upper limits for Ganymede. So the O_2-O_3 "atmosphere" must be contained in the icy surface, perhaps in cracks and pores within the ice. Calvin et al (1996) estimate that the equivalent of a layer between 20 and 200 μm thick is needed to explain the visible bands. The resulting O_3/O_2 ratio, about .0004, is about ten times the ratio in Earth's stratosphere. Chemical kinetics require an atmospheric density orders of magnitude larger than the limits imposed by Voyager observations, supporting the argument that the ozone and the molecular oxygen are trapped in the surface (Noll et al 1996). Noll et al (1996) note that this non-biological ozone must be considered in ozone-detection–based searches for Earth-like planets.

Recently, Calvin & Spencer (1997) obtained a pole-to-pole FOS scan and accompanying narrow-band WFPC2 images of Ganymede and used them to study the distributions of oxygen and ozone on its trailing hemisphere. Absorptions at 577 and 627 nm due to the dense-phase oxygen were observed in the FOS spectrum of Ganymede (Figure 8). The observations suggest that the dimer phase is concentrated in equatorial latitudes and is absent in the polar regions. There is no strong correlation between the oxygen abundance and albedo. This is consistent with formation of O_2 through plasma bombardment and favors defect trapping of oxygen in the surface.

Ultraviolet spectra of Europa (Noll et al 1995) and Callisto (Noll et al 1997a) show a broad absorption feature attributed to SO_2 on the trailing hemisphere of Europa and the leading hemisphere of Callisto. The spectral ratios of leading to trailing hemispheres of Callisto and trailing to leading hemispheres of Europa are similar. In the case of Europa, a likely source is bombardment of the trailing hemisphere by co-rotating plasma that has been enriched in sulfur from Io. However, the plasma is greatly reduced both in overall density and in sulfur content at Callisto, and this source would not be expected to enrich the leading hemisphere anyway. Micrometeorites or indigenous sulfur are possible sources for the SO_2 on the outermost Galilean satellite. At this time the data from the three ice-rich worlds are not sufficient to be able to identify the specific processes that result in the interesting differences between their spectra.

Recently FOS observations of the Saturnian satellites Rhea, Dione, and Iapetus have discovered the 260 nm absorption band of ozone on the first two

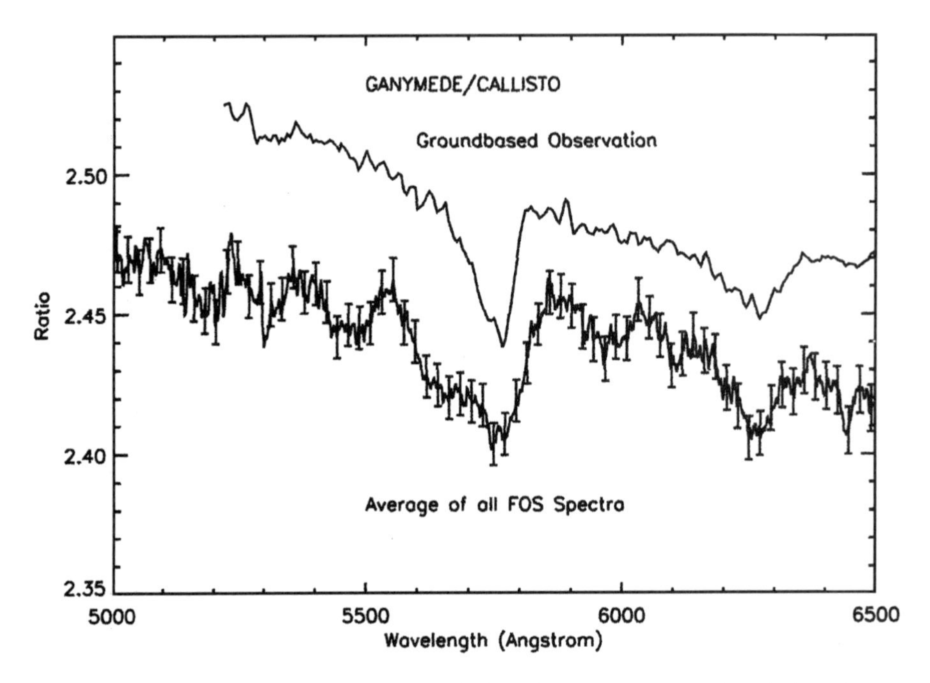

Figure 8 The 5770 and 6275 Angstrom absorption lines due to oxygen are prominent in this ration of Faint Object Spectrograph (FOS) spectra of Ganymede and Callisto. The oxygen on Ganymede that produces this absorption appears to be in a condensed phase within pores in the surface of the satellite. Similar features have been seen recently in the spectra of the Saturn satellites Rhea and Dione. The figure, originally published by Calvin & Spencer (1997), is reproduced here with permission of *Icarus*.

(Noll et al 1997b). These observations indicate that the processes that lead to surface accumulations of oxygen and to the associated production of ozone are not unique to one object or system. These observations reinforce the need to consider this process in proposals for detection of terrestrial planets that involve detection of ozone.

Titan

Despite considerable attention by Voyager I, Titan has been an enigmatic world because a significant atmospheric haze effectively prevented observation of its surface. However, Voyager did reveal a marked asymmetry in albedo between the northern and southern hemispheres of Titan, with the Southern Hemisphere brighter than the Northern Hemisphere at the Voyager imaging wavelengths (less than 550 nm). Titan was among the earliest solar system targets of HST in August of 1990 (Caldwell et al 1992). These observations showed that the

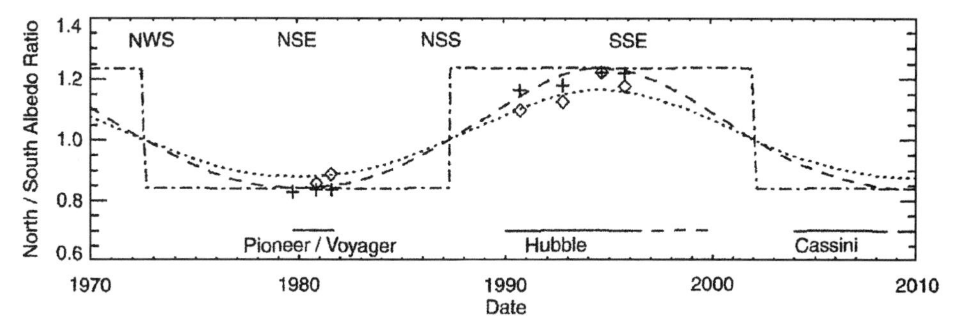

Figure 9 The time evolution in the north-south brightness asymmetry on Titan plotted using both Pioneer/Voyager observations and Hubble Space Telescope images. The *crosses* represent violet images using the F439W filter; *diamonds* are visible F547M data. The fits to the data at the two wavelengths are sinusoidal, and the *dot-dash curve* is a square wave fit. The figure, originally published by Lorenz et al (1997), is reproduced here with permission of *Icarus*.

asymmetry had reversed, with the Northern Hemisphere being brighter in 1990, in concert with expectations that the asymmetry was produced by seasonal variations and would be periodic with a 29.4-year period. An HST image at 889 nm suggested that the contrast reversed at longer wavelengths. Lorenz et al (1997) reexamined the asymmetry problem using all HST images of Titan between 260 and 1040 nm acquired through October 1995. The amplitude of the asymmetry was fairly constant during the time period of the HST observations because the variation was near its maximum (Figure 9). This is consistent with the 85–90° phase lag between this variation and the insolation first derived by Sromovsky et al (1986) and Lockwood et al (1986) from Voyager and ground-based observations. Using their much-improved spectral coverage, Lorenz et al found that the asymmetry reverses at about 700 nm, thus confirming the suggestion of Caldwell et al.

WFPC2 images of Titan with the 850 LP and 1042M filters sample windows in which the atmosphere is less opaque than at 940 and 1080 nm in the methane continuum. Using several observations obtained between October 4 and October 18, 1994, Smith et al (1996) were able to make the first albedo maps of Titan's surface at a resolution of 295 km/pixel. Images are still subject to atmospheric limb effects and therefore cannot constrain albedos in the polar regions. The major feature in the maps is a large, high-albedo area covering roughly 4000 by 2500 km^2 centered at 110°W and 10°S. There are also some smaller, bright spots and dark regions that are seen in several images, but these are smaller than the area mentioned above. The location of the large feature is consistent with ground based imaging at longer wavelengths as well as with radar images. It therefore appears to be a permanent feature, at least on the time scale of a few years. The 14-day gap between the first and last images used in the study was not sufficient to discriminate between synchronous and slightly supersynchronous rotation rates for Titan. No clouds that could be used for determining wind speeds were detected.

The UV capabilities of the FOS were used to extend the wavelength dependence of Titan's albedo to 180 nm. Titan is very dark at the shortest wavelengths with geometric albedo of 0.02 (McGrath et al 1998). This is darker in the UV than the "Titan" tholins analog produced in the laboratory, although a mixture of these and HCN polymers will suffice. The spectrum does not show absorption features that can be attributed to hydrocarbons known to be present in the atmosphere, such as acetylene; but the modeled acetylene concentrations in the upper atmosphere are consistent with Voyager observations.

Karkoschka & Lorenz (1997) used the shadow of Titan on Saturn, imaged by WFPC2, to infer the optical radius of Titan as a function of wavelength and latitude. These observations have an advantage over techniques involving reflected light because they are independent of albedos, phase functions, and gas density. The almost–edge-on rings on August 6, 1995 provided a positional reference whereby the different wavelength images could be registered. Karkoschka and Lorenz divide Titan into three distinct latitude regions: latitudes north of $-5°$ with a high limb; a region south of $-50°$ with a low limb; and a linear transition region in between. The variation in height is a strong function of wavelength ranging from nearly zero for the 336W filter to 130 km for 953N. Karkoschka and Lorenz interpret their observations in terms of latitudinal dependence in aerosol sizes ranging from 0.1 μm in the south to 0.3 μm in the north. Comparing these results to Voyager observations when the seasons on Titan were reversed, they conclude that there are two distinct haze layers, a main layer and a detached layer, now in the north, that contains the larger particles.

URANUS, NEPTUNE, AND PLUTO

Uranus

Observations of clouds on Uranus by Voyager were limited by very low contrast in the visible region of the spectrum <620 nm. Forty-three NICMOS and long wavelength WFPC2 observations of Uranus covering five orbits in 1997 revealed several discrete clouds with contrast that increased with increasing wavelength; in fact, these clouds were essentially invisible in concurrent HST images at wavelengths similar to those of Voyager (Karkoschka 1998). Most of the clouds were roughly circular with diameters estimated to be a few thousand km. Two clouds appear to be streaks or plumes similar to the clouds seen by Voyager. Rotation rates were determined for seven clouds. These were observed whenever they were on the illuminated side of the disk, suggesting that these phenomena either remained relatively unchanged during the ~100-day term of observations or regenerated during the intervals in which they could not be seen. Rotation rates for the three southern clouds agree with determinations for southern hemisphere clouds by Voyager, while those at northern latitudes, not imaged by Voyager, rotate more slowly.

Karkoschka (1997) used six HST images obtained in 1995 to study the photometry of the satellites and rings of Uranus at approximately 1° phase angle. Their paper discusses the photometry of the five major Uranian satellites (Miranda, Ariel, Umbriel, Titania, and Oberon), four minor satellites (Puck, Juliet, Portia, and Belinda), the bright ε ring, and nine rings inside the ε ring resolved into three groups. Pseudo-geometric albedos (extrapolated to 0° from HST data taken at larger phase angles) are inconsistent with most estimates extrapolated from Voyager data that in most cases have phase angles greater than 15°. The phase functions of the satellites between 1° and 90° are similar to those of asteroids. The HST data on the ε ring are used to show that the geometric albedo of the ring particles is 0.061, almost twice the Voyager value, indicating that the Uranian ring particles are not as dark as previously supposed. Finally, the HST spectral coverage makes it possible to specify the colors of the Uranian satellites. Thought to be gray after Voyager, the satellites are found to display subtle shades of blue (Miranda), yellow-orange (Ariel), red (Umbriel), and yellow (Titania and Oberon).

Neptune

Voyager 2's visit to Neptune in 1989 revealed the planet's dynamic atmosphere. The most interesting feature Voyager observed was the Great Dark Spot (GDS-89) in the southern hemisphere at -22° latitude. During eight months, Voyager watched the dark oval drift northward at 1.2° per month. Unfortunately, the size and contrast of the feature was such that it could not be observed by ground-based or pre-refurbishment HST imaging. Multiwavelength WFPC2 images of Neptune acquired in late 1994 showed that GDS-89 had disappeared but a new dark spot was discovered: GDS-94, similar in size to GDS-89 but in the northern hemisphere (Hammel et al 1995). A comprehensive study of WFPC2 images of Neptune during 1994, 1995, and 1996 confirmed the GDS-94 at $+31.4^\circ$ latitude (Hammel & Lockwood 1997); the contrast of the feature at 467 nm, 5%, is only about half that in GDS-89 at a similar wavelength. 1995 images revealed a dark spot, GDS-95, at $+31.5^\circ$ of similar size and contrast. The similarities in location and properties of GDS-94 and 95 strongly suggest that they are the same feature, but its survival at the same latitude over the eleven-month period separating the 1994 and 1995 images then implies different dynamical circumstances from the rapidly migrating 1989 storm. A complex group of clouds that were bright in methane bands, indicating their high altitude, accompanied GDS-94. Despite extending almost to the equator, this complex maintained its alignment in the face of latitude variation of zonal wind speed, suggesting these were perhaps due to a wave phenomenon associated with GDS-94. An isolated bright feature seen in 1995 just poleward of GDS-95 was similar to a feature associated with GDS-89. The rotation periods for both storms were 17.29 hr. A compilation of the periods calculated for all clouds observed by HST is in excellent agreement with Voyager and ground-based profiles of zonal wind measurements. Although stable during the Hubble observations from 1994 to 1996, the pattern of zonal bands seen by HST was different from the pattern seen by Voyager in 1989.

Triton

Faint object spectrograph (FOS) spectra of the major Neptunian satellite Triton were obtained in 1992 and 1993 in two bandpasses, 1770–2330 Å and 2220–3278 Å (Stern & Trafton 1994, Stern et al 1995). The previously established increase in geometric albedo with increasing wavelength reverses at about 2750 Å, suggesting a broad, shallow absorption feature at that wavelength due to solid materials on the surface. Additional absorption features between 2000 and 2100 Å may be due to photochemical byproducts of methane-nitrogen chemistry in the atmosphere that are deposited on the surface. Nondetection of anticipated absorptions was used to set upper bounds to the mixing ratios of oxygen species OH, CO, and NO in the atmosphere of Triton. CO, which is known to be on the surface, has a mixing ratio $[CO]/[N_2] < .015$.

Pluto and Charon

The first FOC solar system observations were of the Pluto-Charon system (Albrecht et al 1991), but they suffered from the spherical aberration problem so further high-resolution observations were postponed pending the correction of the optics. The first post-COSTAR images of Pluto were acquired using FOC in February, 1994 (Albrecht et al 1994). The radii and albedos of Pluto and Charon were determined; reduced contrast at 342 nm compared to 550 nm suggested the presence of a scattering layer of haze in the atmosphere. The best study to date of the surface of Pluto relied on four sets of FOC images of the planet at 410 and 278 nm acquired later in 1994. The image sets were centered at four sub-Earth longitudes approximately 90° apart so that they could be used to construct Mercator projection maps of Pluto at both wavelengths (Stern et al 1997). The 410 nm map is shown in Figure 10. At the maximum resolution possible with HST, the images have about 7.5 pixels across the diameter of Pluto. The resulting maps show that Pluto is dominated by bright polar regions; the equator tends to be dark except for one very bright spot at the antipodes of the sub-Charon point, which is itself one of the darkest regions. There are both similarities and dissimilarities between the direct imaging HST map and

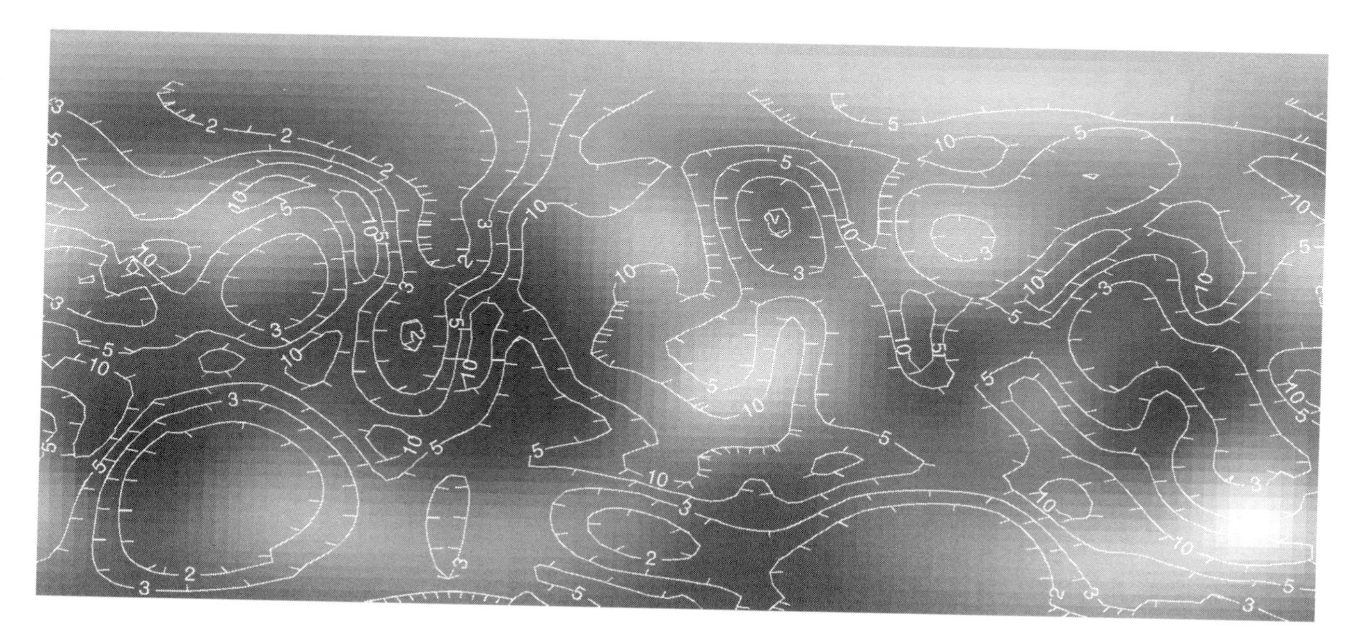

Figure 10 A rectangular projection of Faint Object Camera images of Pluto at 410 nm obtained during the summer of 1994. The superimposed contours are defined by the value of reflectance divided by its uncertainty in units of standard deviation; larger numbers correspond to higher confidence. The South Pole was not included in these images. The figure, originally published by Stern et al (1997), is reproduced here with permission of *The Astronomical Journal.*

previous maps compiled from ground based rotational and mutual event light curves.

The first use of HST images to determine the orbital parameters of the Pluto-Charon binary system relied on three days of data acquired by WFPC1 in 1991 (Null et al 1993). The same images were also used to search for additional satellites of Pluto (Stern et al 1994). No satellite was found in these images, so they derived only upper limits on their magnitudes. Two subsequent analyses of the masses and orbital elements of the Pluto-Charon system used considerably more WFPC1 images (Tholen & Buie 1997, Null & Owen 1996). The two papers agree within errors on most orbital parameters as well as on GM for the system. Of particular interest, both studies lead to a nonzero eccentricity of approximately .0076, although the Null and Owen result is almost consistent with zero within the stated uncertainty. This is significant because the Pluto-Charon system is tidally evolved. The result may suggest a large, recent impact on one of the bodies. A 1.5 σ discrepancy in orbital inclination remains, however.

Buie et al (1997) used 60 WFPC1 images of the Pluto Charon system acquired over a 15-month period to determine the individual light curves of the two bodies. The amplitude of Charon's lightcurve at 555 nm, 0.08 magnitudes, is much less than that of Pluto, 0.33 magnitudes. The variations in the previously unresolved lightcurve of the system can therefore be attributed mostly to Pluto. Pluto, with B-V $=$ 0.873, was also found to be considerably redder than Charon at B-V $=$ 0.710. Trafton & Stern (1996) used FOS to obtain UV spectra at seven rotational phases of Pluto between 2500 and 4800 Å. They were able to set an upper limit of $\tau < 0.54$ to the contribution of atmospheric hazes at 2600 Å to Pluto's reflectance. Pluto and Triton provide an interesting comparison (Stern et al 1995). In the visible, the photometric properties are substantially different; in the mid UV, Triton has about twice the albedo of Pluto, but the wavelength dependence is very similar once the overall difference is normalized out.

SUMMARY

Hubble Space Telescope has explored all of the major planets and satellites in the solar system since its launch in 1990. HST imaging has made several significant discoveries, including first-ever resolution of surface details on some of the smaller bodies. Equally important, HST has extended the time line of monitoring of planets and satellites started by earlier space missions and provided data that have given new insights into dynamic surface and atmospheric phenomena on planets and satellites. The ultraviolet capability of HST has afforded a unique opportunity to study spectacular and important phenomena such as the auroras on Jupiter and Saturn. It is clear that HST will remain of great value in supplementing spacecraft studies of some planets and satellites as well as in studying many solar system objects that have not yet been visited by robot explorers.

This article is based on observations with NASA/ESA Hubble Space Telescope obtained at the Space Telescope Science Institute, which is operated by the Association of Universities for Research in Astronomy, Inc., under NASA contract NAS5-26555. Support for this work was provided by NASA through Grant # GO-7792 from the Space Telescope Science Institute.

Literature Cited

Albrecht R, Barbier C, Adorf HM, Corrain G, Gemmo A, et al. 1994. High resolution imaging of the Pluto-Charon system with the Faint Object Camera of the Hubble Space Telescope. *Ap. J.* 435:L75–L78

Albrecht R, Barbieri C, Blades JC, Boksenberg A, Crane P, et al. 1991. First results from the Faint Object Camera: High resolution imaging of the Pluto-Charon system. *Ap. J.* 374:L65–L67

Ballester GE, McGrath MA, Strobel DF, Zhu X, Feldman PD, Moos HW. 1994. Detection of the SO_2 atmosphere on Io with the Hubble Space Telescope. *Icarus* 111:2–17

Ballester GE, Clarke JT, Trauger JT, et al. 1996. Time-resolved observations of Jupiter's far-ultraviolet aurora. *Science* 274:409–412

Barnet CD, Westphal JA, Beebe RF, Huber LF. 1992. Hubble Space Telescope observations of the 1990 equatorial disturbance on Saturn: Zonal winds and central meridian albedos. *Icarus* 100:499–511

Beebe RF. 1991. A review of planetary opportunities and observations with the Hubble Space Telescope. In *The First Year of HST Observations*, eds. AL Kinney, JC Blades, pp. 147–60. STScI: Baltimore, MD 297 pp.

Beebe RF, Barnet C, Sada PV, Murrell AS. 1992. The onset and growth of the 1990 equatorial disturbance on Saturn. *Icarus* 95:163–72

Beebe RF, Simon AA, Huber LF. 1996. Comparison of Galileo Probe and Earth-based translation rates of Jupiter's equatorial clouds. *Science* 272:841

Bell JF III, Wolff MJ, James PB, Clancy RT, Lee SW, Martin LJ. 1997. Mars surface mineralogy from Hubble Space Telescope imaging during 1994–1995: Observations, calibration, and initial results. *J. Geophys. Res.* 102:9109–23

Bell JF III, Wolff MJ, Daley TC, Crisp D, James PB, Lee SW, Trauger JT, Evans RW. 1999. Near infrared imaging of Mars from HST: Surface reflectance, photometric properties, and implications for MOLA data. *Icarus* (in press)

Ben Jaffel L, Leers V, Sandel BR. 1995. Dark auroral oval on Saturn discovered in Hubble Space Telescope ultraviolet images. *Science* 269:951–52

Binzel RP, Gaffey MJ, Thomas PC, Zellner BH, Storrs AD, Wells EN. 1997. Geologic mapping of Vesta from the 1994 Hubble Space Telescope images. *Icarus* 128:95–103

Binzel RP, Xu S. 1993. Chips off of Asteroid 4 Vesta: Evidence for the parent body of basaltic achondrite meteorites. *Science* 260:186–91

Bosh AS, Rivkin AS. 1996. Observations of Saturn's inner satellites during the May 1995 ring-plane crossing. *Science* 272:518–21

Buie MW, Tholen DJ, Wasserman LH. 1997. Separate lightcurves of Pluto and Caron. *Icarus* 125:233–44

Caldwell J, Turgeon B, Hua XM. 1992a. Hubble Space Telescope imaging of the north polar aurora on Jupiter. *Science* 257:1512–15

Caldwell JJ, Cunningham CC, Anthony D, White HP, Groth EJ, et al. 1992b. Titan: Evidence for seasonal change—a comparison of HST and Voyager images. *Icarus* 96:1–9

Calvin WM, Johnson RE, Spenser JR. 1996. O_2 on Ganymede: Spectral characteristics and plasma formation mechanisms. *Geophys. Res. Lett.* 23:673–76

Calvin WM, Spencer JR. 1997. Latitudinal distribution of O_2 on Ganymede: Observations with the Hubble Space Telescope. *Icarus* 130:505–16

Cantor BA, Wolff MJ, James PB, Higgs E. 1999. Recession of Martian north polar cap: 1990–1997. Hubble Space Telescope observations. *Icarus* (in press)

Chanover NJ, Kuehn DM, Banfield D, Momary

T, Beebe RF, et al. 1996. Absolute reflectivity spectra of Jupiter: 0.25–3.5 micrometers. *Icarus* 121:351–360
Clancy RT, Grossman AW, Wolff MJ, James PB, Rudy DJ, et al. 1996a. Water vapor saturation at low altitudes around Mars aphelion: A key to Mars climate? *Icarus* 122:36–62
Clancy RT, Wolff MJ, James PB. 1999. Minimal aerosol loading and global increases in atmospheric ozone during the 1996–97 Martian northern spring season. *Icarus* (in press)
Clancy RT, Wolff MJ, James PB, Smith E, Billiwala YN, et al. 1996b. Mars ozone measurements near the 1995 aphelion: Hubble Space Telescope ultraviolet spectroscopy with the Faint Object Spectrograph. *J. Geophys. Res.* 101:12, 777–83
Clarke JT, Ajello J, Luhmann J, Schneider N, Kanik I . 1994a. Hubble Space Telescope UV spectral observations of Io passing into eclipse. *J. Geophys. Res.* 99:8387–402
Clarke JT, Ballester GE, Trauger JT, et al. 1996. Far-ultraviolet imaging of Jupiter's aurora and the Io "footprint." *Science* 274:404–08
Clarke JT, Ballester GE, Trauger JT, Ajello J, Pryor W, et al. 1998. HST imaging of Jupiter's UV aurora during the Galileo orbiter mission. *J. Geophys. Res.* 103:20, 217–20, 236
Clarke JT, Ben Jaffel L, Vidal-Madjar A, Gladstone GR, Waite JH, et al. 1994b. Hubble Space Telescope Goddard High Resolution Spectrograph H2 rotational spectra of Jupiter's aurora. *Ap. J.* 430:L73–L76
Connerney JEP. 1993. Magnetic fields of the outer planets. *J. Geophys. Res.* 98:18, 659–79
Dols V, Gerard JC, Paresce F, Prange R, Vidal-Madjar A. 1992. UV imaging of the Jovian aurora with HST. *Geophys. Res. Lett.* 19:1803–06
Elliot JL, Bosh AS, Cooke ML, Bless RC, Nelson MJ, et al. 1993. An occultation by Saturn's rings on 1991 October 2–3 observed with the Hubble Space Telescope. *Ast. J.* 106:2544–72
Emerich C, Ben Jaffel L, Clarke JT, Prange R, Gladstone GR, et al. 1996. Evidence for supersonic turbulence in the upper atmosphere of Jupiter. *Science* 273:1085–87
Gerard JC, Dols V, Grodent D, Waite JH, Gladstone GR, Prange R. 1995. Simultaneous observations of the Saturnian aurora and polar haze with the HST/FOC. *Geophys. Res. Lett.* 22:2685–88
Gerard JC, Dols V, Paresce F, Prange R. 1993. Morphology and time variation of the Jovian far UV aurora: Hubble Space Telescope observations. *J. Geophys. Res.* 98:18793–801
Gerard JC, Dols V, Prange R, Paresce F. 1994a. The morphology of the north Jovian ultraviolet aurora observed with the Hubble Space Telescope. *Planet. Space Sci.* 42:904–17
Gerard JC, Grodent D, Prange R, Waite JH, Galdstone GR, et al. 1994b. A remarkable auroral event in Jupiter observed in the ultraviolet with the Hubble Space Telescope. *Science* 266:1675–78
Godfrey DA. 1988. A hexagonal feature around Saturn's north pole. *Icarus* 76:335–56
Grodent D, Dols V, Gerard JC, Rego D. 1996. The equatorial boundary of the ultraviolet Jovian north aurora observed with multispectral Hubble Space Telescope images. *J. Geophys. Res.* 101:2163–68
Hall DT, Feldman PD, Holberg JB, McGrath MA. 1996. Fluorescent hydroxyl emissions from Saturn's ring atmosphere. *Science* 272:516–18
Hall DT, Strobel DF, Feldman PD, McGrath MA, Weaver HA. 1995. Detection of an oxygen atmosphere on Jupiter's moon Europa. *Nature* 373:677–79
Hammel HB, Lockwood GW. 1997. Atmospheric structure of Neptune in 1994, 1995, and 1996: HST imaging at multiple wavelengths. *Icarus* 129:466–81
Hammel HB, Lockwood GW, Mills JR, Barnet CD. 1995. Hubble Space Telescope imaging of Neptune's cloud structure in 1994. *Science* 268:1740–42
James PB, Clancy RT, Lee SW, Kahn R, Zurek R, et al. 1991. Observations of Mars using Hubble Space Telescope Observatory. In *The First Year of HST Observations*. eds. AL Kinney and JC Blades, p. 161–78. Baltimore, MD: STScI 297 pp.
James PB, Bell JF III, Clancy RT, Lee SW, Martin LJ, Wolff MJ. 1996a. Global imaging of Mars by Hubble Space Telescope during the 1995 opposition. *J. Geophys. Res.* 101:18, 883–91
James PB, Clancy RT, Lee SW, Martin LJ, Bell JF III. 1996b. Seasonal recession of Martian south polar cap: 1992 HST observations. *Icarus* 123:87–100
James PB, Clancy RT, Lee SW, Martin LJ, Singer RB, et al. 1994. Monitoring Mars with the Hubble Space Telescope: 1990–1991 observations. *Icarus* 109:79–101
James PB, Hollingsworth J, Lee SW, Wolff MJ. 1999. North polar dust storms in early spring on Mars. *Icarus* (in press)
Karkoschka E. 1997. Rings and satellites of Uranus: Colorful and not so dark. *Icarus* 125:348–63
Karkoschka E. 1998. Clouds of high contrast on Uranus. *Science* 280:570–72
Karkoschka E, Lorenz RD. 1997. Latitudinal variation of aerosol sizes inferred from Titan's shadow. *Icarus* 125:369–79
Karkoschka E, Tomasko MG. 1993. Saturn's upper atmosphere hazes observed by the

Hubble Space Telescope. *Icarus* 106:428–41
Kim YH, Caldwell JJ, Fox JL. 1995. High resolution ultraviolet spectroscopy of Jupiter's aurora with the Hubble Space Telescope. *Ap. J.* 447:906–14
Kim YH, Fox JL, Caldwell JJ. 1997. Temperatures and altitudes of Jupiter's ultraviolet aurora inferred from GHRS observations with the Hubble Space Telescope. *Icarus* 128:189–201
Krasnopolsky VA, Mumma MJ, Gladstone GR. 1998. Detection of atomic deuterium in the upper atmosphere of Mars. *Science* 280:1576–80
Kumar S, Hunten DM. 1982. The atmospheres of Io and other satellites. In *Satellites of Jupiter*, ed. D Morrison, p. 782–806. Tucson, AZ: University of Arizona Press. 972 pp.
Lellouch E, Belton M, de Pater I, Paubert G, Gulkis S, Enccrenaz T. 1992. The structure, stability, and global distribution of Io's atmosphere. *Icarus* 98:271–95
Lockwood GW, Thompson DT, Sromovsky LA. 1986. Photometry of Titan: Evidence supporting the seasonal contrast model of albedo variations. *Bull. Am. Astron. Soc.* 18:809
Lorenz RD, Smith PH, Lemmon MT, Karkoschka E, Lockwood GW, Caldwell J. 1997. Titan's north-south asymmetry from HST and Voyager imaging: Comparison with models and ground based photometry. *Icarus* 127:173–89
McGrath MA, Courtin R, Smith TE, Feldman PD, Srobel DF. 1998. The ultraviolet albedo of Titan. *Icarus* 131:382–92
Mischna M, Bell JF III, James PB, Crisp D. 1998. Synoptic measurements of Martian winds using the Hubble Space Telescope. *Geophys. Res. Lett.* 25:611–14
Moreno F. 1996. The structure of the stratospheric aerosol layer in the equatorial and south polar regions of Jupiter. *Icarus* 124: 632–44
Morrissey PF, Feldman PD, Clarke JT, Wolven BC, Strobel DF, et al. 1997. Simultaneous spectroscopy and imaging of the Jovian aurora with the Hopkins Ultraviolet Telescope and the Hubble Space Telescope. *Ap. J.* 476:918–23
Na CY, Esposito LW. 1995. UV observations of Venus with HST. *BAAS* 27:17
Nicholson PD, Showalter MR, Dones L, French RG, Larson SM, et al. 1996. Observations of Saturn's ring-plane crossings in August and November 1995. *Science* 272:509–15
Noll KS, Johnson RE, Lane AL, Domingue DL, Weaver HA. 1996. Detection of ozone on Ganymede. *Science* 273:341–43
Noll KS, Johnson RE, McGrath MA, Caldwell JJ. 1997a. Detection of SO_2 on Callisto with the Hubble Space Telescope. *Geophys. Res. Lett.* 24:1139–42
Noll KS, Roush TL, Cruikshank DP, Johnson RE, Pendleton YJ. 1997b. Detection of ozone on Saturn's satellites Rhea and Dione. *Nature* 388:45–48
Noll KS, Weaver HA, Gonnella AM. 1995. The albedo spectrum of Europa from 2200 Å to 3300 Å. *J. Geophys. Res.* 100:19, 057–59
Null GW, Owen WM, Synot SP. 1993. Masses and densities of Pluto and Charon. *Astron. J.* 105:2319–35
Null GW, Owen WM. 1996. Charon/Pluto mass ratio obtained with HST CCD observations in 1991 and 1993. *Astron. J.* 111:1368–81
Pearl JC, Hanel R, Kunde V, Maguire W, Fox K, et al. 1979. Identification of gaseous SO_2 and new upper limits for other gasses on Io. *Nature* 280:755–57
Prange R, Rego D, Southwood D, Zarka P, Miller S, Ip W. 1996. Rapid energy dissipation and variability of the Io-Jupiter electrodynamic circuit. *Nature* 379:323–25
Sartoretti P, Belton MJS, McGrath MA. 1996. SO_2 distributions on Io. *Icarus* 122:273–87
Sartoretti P, McGrath MA, McEwen AS, Spencer JR. 1995. Post-Voyager brightness variations on Io. *J. Geophys. Res.* 100:7523–30
Sartoretti P, McGrath MA, Paresce F. 1994. Disk-resolved imaging of Io with the Hubble Space Telescope. *Icarus* 108:272–84
Shemansky DE, Matheson P, Hall DT, Hu HY, Tripp TM. 1993. Detection of the hydroxyl radical in the Saturn magnetosphere. *Nature* 363:329–31
Simon AA, Beebe RF. 1996. Jovian tropospheric features–wind field, morphology, and motion of long-lived systems. *Icarus* 121:319–30
Simon AA, Beebe RF, Gierasch PJ, Vasavada AR, Belton MJS, Galileo Imaging Team. 1998. Global context of the Galileo-E6 observations of Jupiter's white ovals. *Icarus* 135:220–29
Smith PH, Lemmon MT, Lorenz RD, Sromovsky LA, Caldwell JJ, Allison MD. 1996. Titan's surface, revealed by HST imaging. *Icarus* 119:336–49
Smith PH, Tomasko MG, Weaver HA. 1992. Titan: Evidence for seasonal change–a comparison of HST and Voyager images. *Icarus* 96:1–9
Spencer JR, Calvin WM, Person MJ. 1995. Charge coupled device spectra of the Galilean satellites: molecular oxygen on Ganymede. *J. Geophys. Res.* 100:19049–56
Spencer JR, McEwen AS, McGrath MA, Sartoretti P, Nash DB, et al. 1997. Volcanic resurfacing of Io: Post-repair HST imaging. *Icarus* 127:221–37

Sromovsky LA, Lockwood GW, Thompson DT. 1986. Titan's albedo variation: Relative influence of solar UV and seasonal contrast mechanisms. *Bull. Am. Astron. Soc.* 18:809
Stern SA, Buie MW, Trafton LM. 1997. HST high resolution images and maps of Pluto. *Astron. J.* 113:827–43
Stern SA, Parker JW, Duncan MJ, Snowdall JC, Levinson HF. 1994. Dynamical and observational constraints on satellites in the inner Pluto-Charon system. *Icarus* 108:234–42
Stern SA, Trafton LM. 1994. The ultraviolet spectrum of Triton from HST: First results. *Icarus* 112:533–36
Stern SA, Trafton LM, Flynn B. 1995. Rotationally resolved studies of the mid-ultraviolet spectrum of Triton. II. HST surface and atmospheric results. *Astronomical J.* 109:2855–68
Tholen DJ, Buie MW. 1997. The orbit of Charon. I. New Hubble Space Telescope observations. *Icarus* 125:245–60
Thomas PC, Binzel RP, Gaffey J, Storrs AD, Wells EN, Zellner BH. 1997a. Impact excavation on Asteroid 4 Vesta: Hubble Space Telescope results. *Science* 277:1492–94
Thomas PC, Binzel RP, Gaffey MJ, Zellner BH, Storrs AD, Wells E. 1997b. Vesta: Spin pole, size and shape from HST images. *Icarus* 128:88–94
Trafton LM, Caldwell JJ, Barnet C, Cunningham CC. 1996. The gaseous sulfur dioxide abundance over Io's leading and trailing hemispheres: HST spectra of Io's $C^1B_2 \rightarrow X^1A_1$ band of SO_2 near 2100 Å. 1996. *Ap. J.* 456:384–92
Trafton LM, Gerard JC, Munhoven G, Waite JH. 1994. High resolution spectra of Jupiter's northern auroral ultraviolet emission with the Hubble Space Telescope. *Ap. J.* 421:816–27
Trafton LM, Stern SA. 1996. Rotationally resolved spectra studies of Pluto from 2500 to 4800 Å obtained with HST. *Astron. J.* 112:1212–24
Trauger JT, Clarke JT, Ballester GE, et al. 1998. Saturn's hydrogen aurora: Wide field and planetary camera 2 imaging from the Hubble Space Telescope. *J. Geophys. Res.* 103:20, 237–20, 244
Westphal JA, Baum WA, Ingersoll AP, Barnet CD, De Jong EM, Danielson GE, Caldwell J. 1992. Hubble Space Telescope observations of the 1990 equatorial disturbance on Saturn: Images, albedos, and limb darkening. *Icarus* 100:485–98
Westphal JA, Baum WA, Lauer TR, Danielson GE, Currie DG, et al. 1991. Hubble Space Telescope wide field/Planetary Camera images of Saturn. *Ap. J.* 369:L51–L53
Wolff MJ, Lee SW, Clancy RT, Martin LJ, Bell JF III, James PB. 1997. 1995 Observations of Martian dust storms using the Hubble Space Telescope. *J. Geophys. Res.* 102:1679–92
Wolff MJ, James PB, Bell JF III, Clancy RT, Lee SW. 1999. Hubble Space Telescope observations of the Martian aphelion cloud belt prior to the Pathfinder mission: seasonal and interannual variations. *J. Geophys. Res.* (in press)
Yung YL, McElroy MB. 1977. Stability of an oxygen atmosphere on Ganymede. *Icarus* 30:97–103
Yung YL, Wen JS, Pinto JP, Allen M, Pierce KK, Paulson S. 1988. HDO in the Martian atmosphere: Implications for the abundance of crustal water. *Icarus* 76:146–59

Annu. Rev. Earth Planet. Sci. 1999. 27:149–82

THE DEGLACIATION OF THE NORTHERN HEMISPHERE: A Global Perspective

Richard B. Alley[1] *and Peter U. Clark*[2]
[1]Earth System Science Center and Department of Geosciences, The Pennsylvania State University, University Park, Pennsylvania 16802; e-mail: ralley@essc.psu.edu
[2]Department of Geosciences, Oregon State University, Corvallis, Oregon 97331; e-mail: clarkp@ucs.orst.edu

KEY WORDS: paleoceanography, paleoclimatology, ice ages, Heinrich events, Dansgaard-Oeschger events

ABSTRACT

Orbitally induced increase in northern summer insolation after growth of a large ice sheet triggered deglaciation and associated global warming. Ice-albedo, sea-level, and greenhouse-gas feedbacks, together with tropical warming from weakening winds in response to polar amplification of warming, caused regional-to-global (near-) synchronization of deglaciation. Effects were larger at orbital rather than millennial frequencies because ice sheets and carbon dioxide vary slowly. Ice-sheet–linked changes in freshwater delivery to the North Atlantic, and possibly free oscillations in the climate system, forced millennial climate oscillations associated with changes in North Atlantic deep water (NADW) flow. The North Atlantic typically operates in one of three modes: modern, glacial, and Heinrich. Deglaciation occurred from a glacial-mode ocean that, in comparison to modern, had shallower depth of penetration of NADW formed further south, causing strong northern cooling and the widespread cold, dry, and windy conditions associated with the glacial maximum and the cold phases of the millennial Dansgaard-Oeschger oscillations. The glacial mode was punctuated by meltwater-forced Heinrich conditions that caused only small additional cooling at high northern latitudes, but greatly reduced the formation of NADW and triggered an oceanic "seesaw" that warmed some high-latitude southern regions centered in the South Atlantic.

0084-6597/99/0515-0149$08.00

Introduction

Complex processes and feedbacks in the coupled Earth system involving atmosphere, ocean, ice, land surface, and biota caused northern-hemisphere and global deglaciation. Long-term geological changes produced conditions favorable for the recent ice ages. Over the last million years, warming and cooling have occurred with strong periodicities ranging from about 1500 years to about 100,000 years. The slower variations appear to have been driven by changes in the distribution of sunlight on Earth associated with features of the Earth's orbit (Hays et al 1976, Lorius et al 1990, Imbrie et al 1992, 1993). The faster variations are plausibly related to changes in ice sheets and in the deep circulation of the ocean. The deglaciation must be understood as a superposition of these and other variations.

We briefly summarize the orbital hypothesis and its implications for nearly synchronous global climate change. We then review the characteristics and possible mechanisms of the faster variations in terrestrial and marine environments. Changes in freshwater fluxes to the oceans during the last deglaciation provided links between these environments. The hemispheric phasing of responses to this forcing provides further clues to the mechanisms involved, but raises additional questions about near-synchrony at orbital frequencies. Identification of atmospheric and shallow-ocean mechanisms synchronizing the hemispheres leads to an overview of deglaciation.

Milankovitch Control

Most relevant research shows that features of Earth's orbit cause changes in the seasonality of incoming solar radiation that control glacial cycles (the Milankovitch or Milankovitch-Croll model). Some of the forcing is out of phase between the hemispheres. Climate has cooled and ice has grown globally when middle and high northern latitudes had short, cool summers, with global warming and ice shrinkage linked to long, hot northern summers (Imbrie et al 1992, 1993). A complicating factor—that large seasonality variations with periodicity of 19,000, 23,000, and 41,000 years caused small climate changes, but small variations with 100,000-year periodicity caused large climate changes—is likely explained by the effects of the ice sheets themselves, particularly the tendency of large ice sheets that have depressed the Earth's crust and warmed their beds to change more rapidly than smaller ice sheets (reviewed by Imbrie et al 1992, 1993; also see Tarasov & Peltier 1997).

Global changes have not been entirely synchronous, and the detailed phasing provides critical information on mechanisms (Imbrie et al 1992, 1993), but the near-synchrony is striking. For example, Genthon et al (1987) and Lorius et al (1990) attempted to explain the stable isotopic record of ice from the Vostok ice core, a proxy for temperature at high southern latitude, as a linear superposition of possible controlling variables (northern insolation, southern insolation, atmospheric carbon dioxide concentration, etc.) with weighting determined by multiple regression. Uncertainties were introduced by difficulties in dating and choice of members for the set of possible controlling variables. Nonetheless, through a range of models it proved necessary to include a northern control as well as a southern one to explain southern temperature. Most of the variance of the Vostok isotopic record is well explained by changes in greenhouse-gas concentrations and global ice volume, which in turn are largely linked to northern insolation and ice-sheet processes. Strong covariation of CO_2 and ice volume prevented clear identification of their relative importance.

Global cooling and warming have been caused by little net change of incoming solar radiation, so large positive feedback must have been active. Sea-level changes, ice-albedo feedbacks, and changes in greenhouse gases likely were important. Below we discuss these and other mechanisms, and how they may have contributed to the global near-synchrony of climate changes at orbital frequencies. However, explaining southern temperatures requires consideration of southern as well as northern influences (Genthon et al 1987).

Most available data support this broad picture, with a possible exception: At one site in Nevada, groundwater isotopes shifted to heavier values before summertime insolation began rising at the end of the previous glaciation (Winograd et al 1992). This exception may reflect non-temperature effects in the records (Grootes 1993) or other complications. The orbital features also might have caused variations in the extraterrestrial dust encountered by Earth (Muller & McDonald 1997), but available sedimentary evidence does not show large changes in dust accretion (Marcantonio et al 1996). The Milankovitch model remains the best description of the ice-age cycles over tens to hundreds of millennia. A deglaciation—a major, rapid reduction in ice volume—has occurred each time northern-summer insolation rose significantly after a big ice sheet had grown (Raymo 1997). However, the picture is less clear for more rapid climate changes, which we review next.

Heinrich-Bond and Dansgaard-Oeschger Cycles

Deglaciation has not been monotonic. Early European work documented returns to cold conditions, such as the Younger Dryas (YD) interval. Longer records show that such millennial variability has been quite common, especially during orbitally driven coolings and warmings (Figure 1). Prominent events have had an approximate spacing of 1500 years (Dansgaard-Oeschger oscillations) (Bond & Lotti 1995, Bond et al 1997, Grootes & Stuiver 1997, Mayewski et al 1997), and of a few thousand years (Heinrich-Bond cycles) (Heinrich 1988, Bond et al 1993). Whether these are true periodicities phase-locked to the calendar, or whether they are simply general bands in which variability has occurred

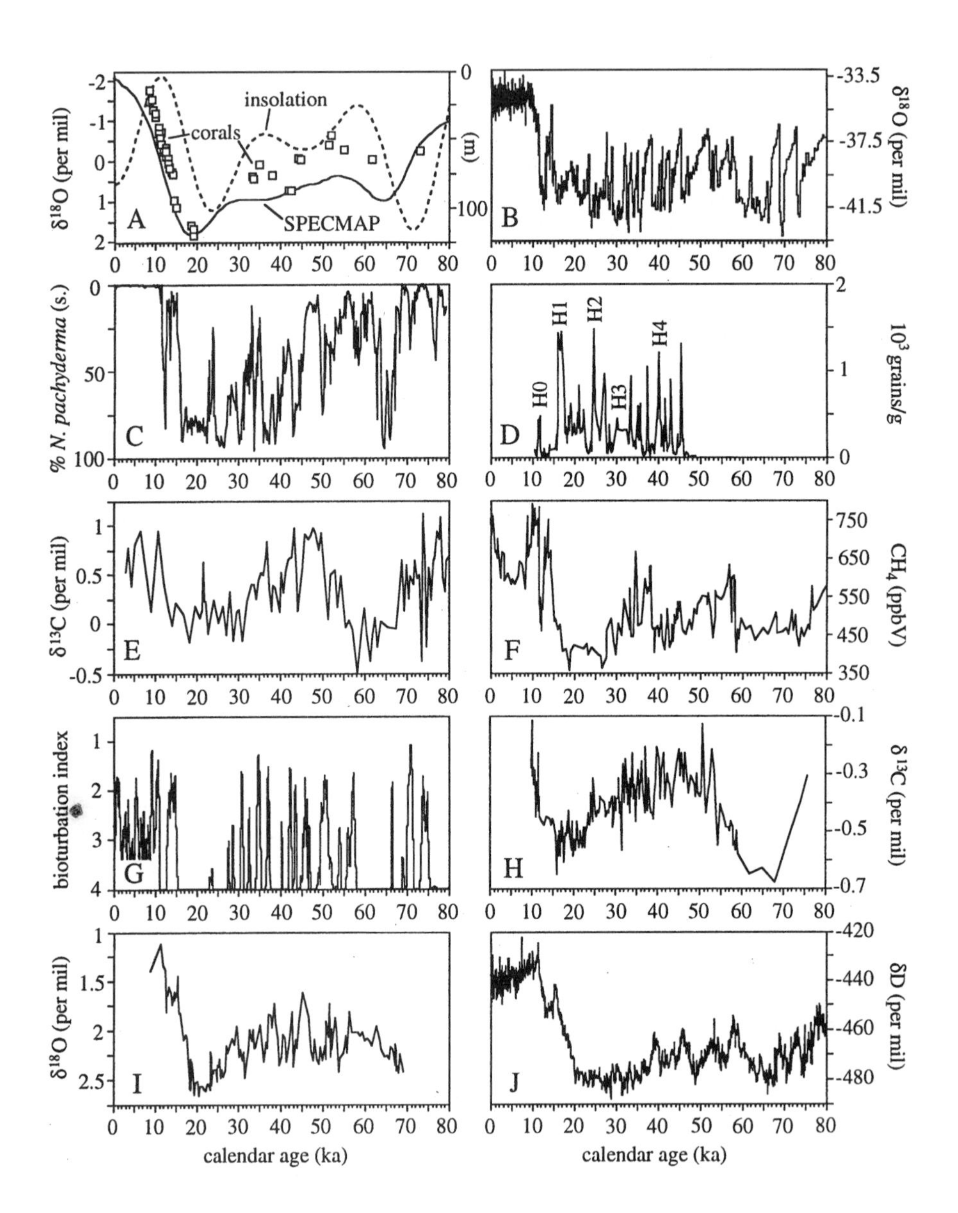

δ18O (per mil)
(m)
insolation
corals
SPECMAP
A
B
δ18O (per mil)
% N. pachyderma (s.)
C
D
H0
H1
H2
H3
H4
10³ grains/g
δ13C (per mil)
E
F
CH4 (ppbV)
bioturbation index
G
H
δ13C (per mil)
δ18O (per mil)
I
J
δD (per mil)
calendar age (ka)
calendar age (ka)

(Heinrich 1988, Mayewski et al 1997), is a question yet unanswered. Evidence on meltwater routing (discussed below) favors variability in a general band, as does observed variability in spacing (e.g. from 700 to 2200 years for the Holocene events of Bond et al 1997).

The coolings of the Dansgaard-Oeschger (D-O) cycles have been marked by abrupt terminations, and often by abrupt onsets (Figure 1*b*). For example, the termination of the YD—a prominent event that may also have been a Heinrich event (H event)—about 11.5 thousand calendar years before present (cal kyear) in central Greenland involved 5 to 10°C warming, accumulation-rate doubling, threefold (fine-grained dust and sea salt) to sevenfold (coarse-grained dust, calcium) drop in windblown materials, and other changes, with most of the change occurring in only a few years (Alley et al 1993, 1995; Mayewski et al 1997; Taylor et al 1997; Severinghaus et al 1998). Methane began a large (40 percent) rise at the end of the YD (Figure 1*f*), 0 to 30 years (<1 sample) after the warming in Greenland (Severinghaus et al 1998). Because the low-latitude hydrologic cycle was probably involved in the methane rise (Chappellaz et al 1993), one can argue for a widespread to globally significant and synchronous event.

Large changes are recorded at about the same time from widespread regions across much of the northern hemisphere and into the southern (e.g. Peteet 1993, Denton & Hendy 1994). In general, YD conditions were cold, dry, or windy compared to modern across much of the Earth. The drop in windiness

←

Figure 1 Climate records covering the last 80,000 calendar years. (*a*) Insolation (June, 60°N) (vertical scale, not shown, ranges from 450 to 515 W/m^2), SPECMAP stack of the global ice volume signal ($\delta^{18}O$ in per mil) (Imbrie et al 1984), and U/Th ages on raised corals (elevations corrected for tectonic uplift) from Barbados (Bard et al 1993) and New Guinea (Edwards et al 1993; Chappell et al 1996) showing changes in sea level. (*b*) GISP2 $\delta^{18}O$ record (Grootes et al 1993). (*c*) Changes in the percentage of *Neogloboquadrina pachyderma* (s.) from North Atlantic core VM23-081 (Bond et al 1993). (*d*) Concentrations of lithic grains (in numbers of grains $>150\ \mu m$ per gram of sediment) from North Atlantic core VM23-081. Data from Bond & Lotti (1995). Radiocarbon ages calibrated to calendar ages using relation of Bard et al (1997). Given uncertainties in calibrating radiocarbon ages on marine shell material to calendar years, we do not expect a perfect correlation of timescales. However, recent additional radiocarbon calibration in the interval from 10 to 30 cal kyear (Kitagawa & van der Plicht 1998) strongly supports the calibration from paired U-Th and ^{14}C ages on corals (Bard et al 1993). (*e*) $\delta^{13}C$ data for *Cibicidoides wuellerstorfi* from equatorial Atlantic core EW9209-1JPC (5°N) (Curry & Oppo 1997). (*f*) GISP2 record of variations in atmospheric methane (Brook et al 1996). (*g*) Changes in bioturbation index from ODP Site 893 (Santa Barbara basin) (1 = laminated sediments, 4 = massive sediments) (Behl & Kennett 1996). (*h*) $\delta^{13}C$ record (*Cibicidoides wuellerstorfi*) from northeastern Pacific core W8709A-13PC (Lund & Mix 1998). (*i*) $\delta^{18}O$ data for *Neogloboquadrina pachyderma* from Southern Ocean core RC11-83 (~42°S) (Charles et al 1996). (*j*) Vostok, Antarctica, δD record (data from Jouzel et al 1987, age model from Bender et al 1994).

documented in the Cariaco Basin off Venezuela, at the same time within the dating uncertainties of about a century, occurred in less than a decade (Hughen et al 1996).

Growing evidence indicates that this YD pattern occurred for most or all of the D-O events (e.g. Chappellaz et al 1993). The geographic pattern of the cold event about 8.2 cal kyear is quite similar to that of the YD (e.g. Alley et al 1997). Over the available record, virtually all of the D-O coolings are recorded in sediments from the Santa Barbara Basin, off the coast of California (Figure 1*g*), as times of enhanced bioturbation related to enhanced oxygenation of these shallow waters (475 m modern sill depth; Behl & Kennett 1996). The cold phases of the D-O oscillations in the North Atlantic were marked by increased iceberg rafting of debris into a cold, fresh surface ocean (Figures 1*c*, 1*d*), and by reduced formation of North Atlantic deep water (NADW) (e.g. Lehman & Keigwin 1992b, Oppo & Lehman 1995, Bond et al 1993, Bond & Lotti 1995) (Figure 1*e*).

Heinrich (H) events were identified as prominent layers of ice-rafted debris in North Atlantic sediment cores (Heinrich 1988) (Figure 1*d*). Six events were originally identified over the most recent 100,000 years (H1 to H6 with increasing age); another is possible at the YD (H0; Andrews et al 1994). Most studies have focused on the more recent ones (H1 to H3 or H4). Of these, H3 (Gwiazda et al 1996a) and possibly H6 may have been somewhat anomalous.

The H layers in the Atlantic thicken towards Hudson Strait, from typically less than 1 cm in the eastern Atlantic to more than 50 cm near the strait (Andrews & Tedesco 1992, Grousset et al 1993, Andrews et al 1994). With the probable exceptions of H3 and perhaps H6, the thicker parts of these layers were deposited with greatly enhanced rates of sedimentation compared to times between H events (by perhaps an order of magnitude or more; Bond et al 1992, McManus et al 1998). Ice-rafted sediment from widespread sources around the north Atlantic is present in the thin edges of the H layers far away from Hudson Strait (Bond & Lotti 1995, Revel et al 1996) and in H3, but the thick regions of H1, H2, and H4 are dominated by material with composition indicating a Hudson Strait/Hudson Bay source region, based on data such as Pb isotopes in feldspars (Gwiazda et al 1996a), Ar-Ar dates of amphiboles (Gwiazda et al 1996b), Sm-Nd isotope systematics (Grousset et al 1993), and carbonate lithologies (Bond et al 1992).

The mass flux of ice-rafted debris in H layers thus appears to have been almost entirely dominated by Hudson Bay-area sources, although increased ice rafting did occur from other sources. The ice-rafted-debris signal from multiple ice sheets may be explained by enhanced survival of icebergs in a cold ocean, by advance of small ice sheets related to cooling, or by other processes (McCabe & Clark 1998), but the much larger sediment flux with Hudson Bay affinity seems to require major changes in the Laurentide ice sheet in Hudson Bay/Strait (Alley & MacAyeal 1994).

The H events occurred with increased turbidite activity in the Northeast Atlantic Mid-Ocean Channel (NAMOC) (Stoner et al 1996, Hesse & Khodabakhsh 1998), which likely was caused by ice-sheet advance onto the outer continental shelf to trigger failures (Mulder & Moran 1995). Although circumstantial, this evidence further strengthens the argument for advance of Laurentide grounded ice during the H events.

The H events are correlated with especially cold times in the North Atlantic. The Heinrich-Bond (H-B) variability has been proposed to consist of cooling over a few millennia achieved by successively colder millennial D-O oscillations, followed by an H event during the cold phase of a D-O oscillation and then rapid warming (Bond et al 1993). The available data seem to indicate that the cooling of the ocean and reduction in deepwater formation preceded H events significantly (by centuries to 1 to 2 millennia? see Bond et al 1993, Zahn et al 1997), and that resumption of deepwater formation may have lagged the termination of enhanced ice-rafted-debris sedimentation (Zahn et al 1997).

MacAyeal (1993) proposed that the H events were surges from the Laurentide ice sheet in the Hudson Bay region through Hudson Strait, in response to a thermal oscillation. Thickening of ice over a frozen bed insulates that bed from the cold surface, and the reduction in snow accumulation associated with thickening leads to reduced vertical ice flow and so less cooling of the bed by the downward-moving, cold surface ice. The bed then thaws, allowing rapid ice motion that could have supplied debris-laden (Alley & MacAyeal 1994) bergs to the North Atlantic. Thinning from rapid ice motion brings cold surface ice near the bed, causing freezing and a cessation of rapid motion. The ice flow is then unable to evacuate the snow accumulation, causing the ice to thicken until thawing occurs again. A periodic oscillation results—quasi-periodic if boundary conditions are allowed to vary.

Subsequent work has demonstrated that it is very difficult to explain the H layers based on penetration of surface forcing through the Laurentide ice sheet to affect the bed (e.g. Oerlemans 1993). An ice-shelf mechanism has been invoked, involving debris "packaging" by basal freeze-on driven by shelf thickness gradients, but such a mechanism would probably have produced a different mix of sediment sources than observed (Hulbe 1997). More sophisticated modeling confirms the possibility of a thermally oscillating ice sheet in Hudson Bay (Marshall & Clarke 1997), although with somewhat smaller ice velocities and fluxes than in the MacAyeal (1993) model. Available evidence, although not conclusive, continues to be most consistent with the model that H events involved surging of the Laurentide ice sheet through Hudson Strait.

The phasing of cooling preceding H events is of considerable interest (Bond et al 1993, Zahn et al 1997). If the D-O and H-B oscillations were decoupled, it would be unlikely for the H events to have always occurred during the cold phases of the faster oscillation. This observation raises the possibility that the ice-sheet surges were triggered by D-O cooling or some process associated with that cooling. H events may have occurred during the first cold phase of the D-O oscillation following thawing of the ice sheet in Hudson Bay. A trigger might have involved marine instability (Weertman 1974), thermal state of the marginal regions (Alley et al 1996), or some ice-shelf process. Marginal regions with short response times must have been involved to have allowed triggering in centuries to a millennium or two, as observed.

Large oscillations of the southern margin of the Laurentide ice sheet occurred at about the time of, and possibly related to, H-B oscillations (Clark 1994, Mooers & Lehr 1997), probably because of the cooling. The H events are associated with the cold phase of the D-O oscillation, and so with cold and fresh surface-water conditions (Bond et al 1992, Maslin et al 1995, Cortijo et al 1997), reduced deepwater formation in the North Atlantic (Keigwin & Lehman 1994, Vidal et al 1997, Zahn et al 1997), and large climate changes in and beyond the immediate North Atlantic basin (Figure 1) (e.g. Grimm et al 1993; Clark & Bartlein 1995; Lowell et al 1995; Porter & An 1995; Benson et al 1996, 1998; Charles et al 1996; McIntyre & Molfino 1996; Phillips et al 1996; Little et al 1997; Curry & Oppo 1997; Lund & Mix 1998; Schulz et al 1998). Because of the role of NADW in H-B and D-O variability, we next review North Atlantic changes.

North Atlantic Surface-Water Hydrology and Deep Circulation

Milankovitch, H-B, and D-O variability are all linked to changes in oceanic circulation centered on the North Atlantic. Relevant paleoceanographic data suggest that the ocean has exhibited three fundamental modes of behavior during the deglaciation (Figure 2) (cf. Sarnthein et al 1994, Stocker 1998): modern (vigorous NADW formation in high-latitude seas that warms Northern Europe and Greenland greatly), glacial (less vigorous but active NADW formation, but not sinking as deeply, and occurring at lower latitudes with less warming of Greenland and Northern Europe; this may be the mode for the cold phases of non-H D-O oscillations), and Heinrich (greatly reduced NADW formation). (Sarnthein et al called this third mode "meltwater," but we chose a different name because many meltwater events apparently did not trigger it.) The modern (labeled as such in Figure 2) and glacial states (in Figure 2, time slice 21.2 cal kyear) can be compared to the two-pump and one-pump modes of Imbrie et al (1992, 1993), with the Heinrich mode (in Figure 2, time slice 15.8 cal kyear)

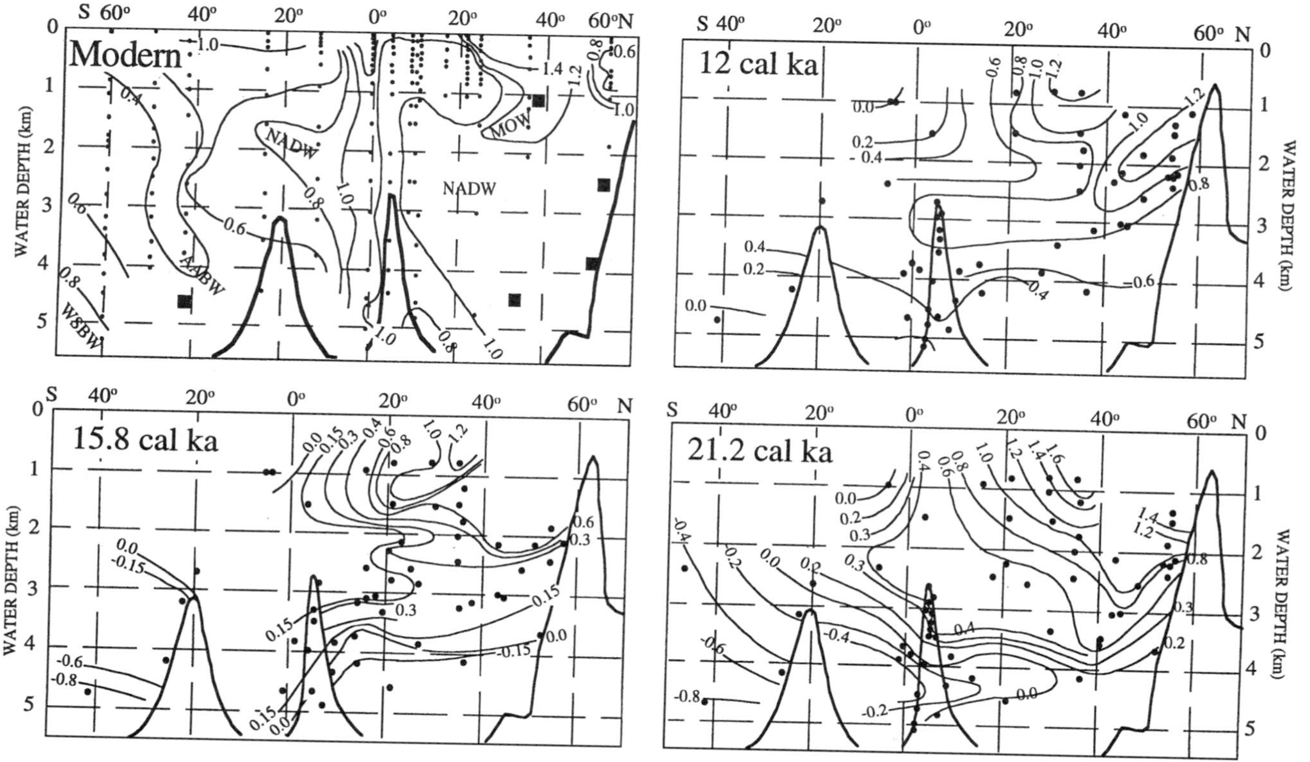

Figure 2 Distribution of $\delta^{13}C$ in the eastern Atlantic basin for four time slices: modern, 12 cal kyear, 15.8 cal kyear (Heinrich mode), and 21.2 cal kyear (glacial mode) (from Sarnthein et al 1994). *Black squares* indicate the locations of core sites with high-resolution records shown in Figure 4. Modern nutrient-depleted Mediterranean outflow water (MOW) is characterized by the highest $\delta^{13}C$ values measured on the total dissolved CO_2, North Atlantic deep water (NADW) is characterized by intermediate $\delta^{13}C$ values, and nutrient-enriched Antarctic bottom water (AABW) is characterized by the lowest $\delta^{13}C$ values. None of the $\delta^{13}C$ records are corrected for the glacial-interglacial difference.

operating at time scales shorter than those considered by Imbrie et al. Gradations, intermediates, or other modes may exist, but large jumps over years to decades have occurred between modes. As discussed below, the atmospheric effects probably have been largest in transition from modern to glacial, and the deep-ocean effects probably have been largest in transition from glacial to Heinrich, with implications for their transmission to other regions.

High-resolution geochemical records from the North Atlantic and Nordic Seas identify significant geographic and temporal variability in surface-water hydrology—sea-surface temperatures (SSTs) and sea-surface salinities (SSS)—and deep-ocean circulation during the last deglaciation, which can be linked to changes in sea-surface and air temperatures. Changes in $\delta^{18}O$ (Figure 3) represent episodic injection of icebergs and meltwater (Jones & Keigwin 1988, Lehman et al 1991, Keigwin et al 1991), varying rates of advection associated with deepwater formation (Duplessy et al 1992, Fairbanks et al 1992), and

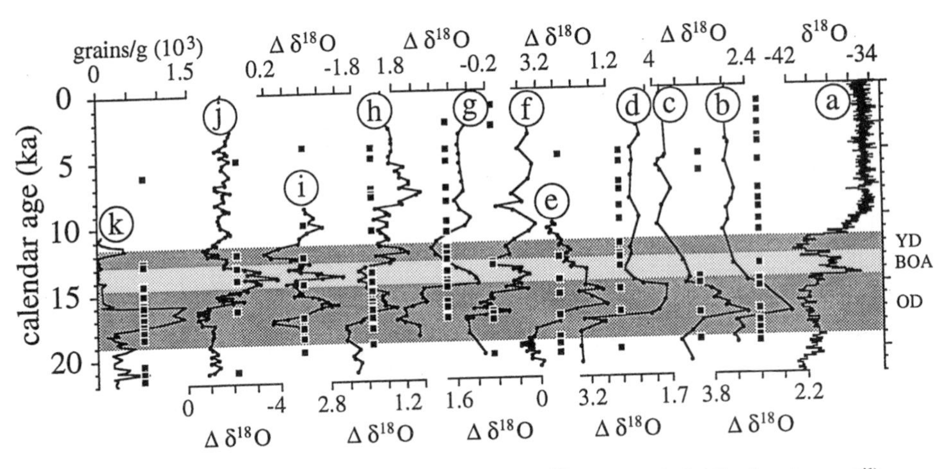

Figure 3 High-resolution North Atlantic planktonic $\Delta\delta^{18}O$ records (*b–j*) (all values in per mil) compared to (*a*) the GISP2 $\delta^{18}O$ record (in per mil) (Grootes et al 1993) and (*k*) the lithic-grain record from North Atlantic core VM23-081 (Bond & Lotti 1995). We have subtracted the global ice volume component (derived from the Barbados sea-level record, Fairbanks 1989) from the $\delta^{18}O$ records in order to illustrate anomalies ($\Delta\delta^{18}O$) that reflect some combination of changes in sea-surface salinity and temperature. *Gray-scale horizontal bars* indicate the Oldest Dryas (OD), Bølling-Allerød (BOA), and Younger Dryas (YD) intervals. *Small black squares* show location of AMS radiocarbon ages in each core. Radiocarbon timescales were converted to calendar years using the relation of Bard et al (1997). Core records are: (*b*) PS21295-4 (Jones & Keigwin 1988), (*c*) HM94-34 (Koç & Jansen 1994), (*d*) V28-14 (Lehman et al 1991), (*e*) ODP 609 (Bond et al 1993), (*f*) CHN82-20 (Keigwin & Lehman 1994), (*g*) SU81-18 (Bard et al 1987), (*h*) stacked $\delta^{18}O$ records from continental slope of Nova Scotia (Keigwin & Jones 1995), (*i*) KNR31 GPC5 (Keigwin et al., 1991), (*j*) EN32 PC-6 (Keigwin et al 1991).

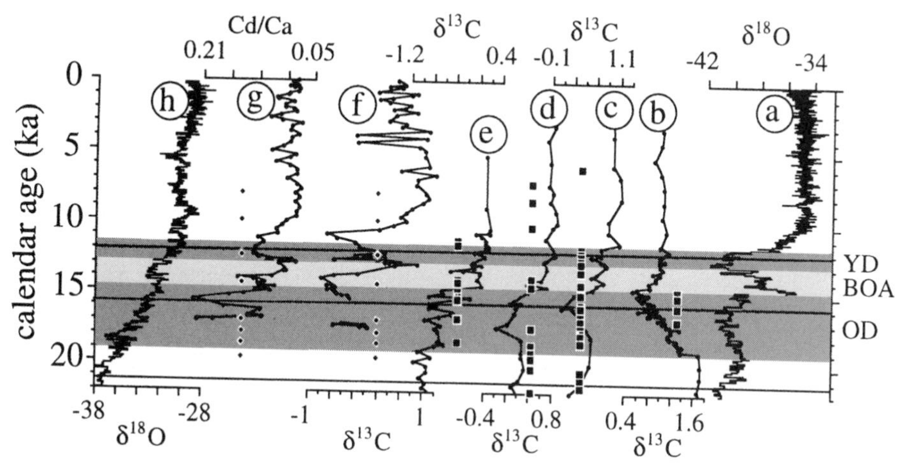

Figure 4 High-resolution nutrient records ($\delta^{13}C$ in per mil and Cd/Ca in μmole/mole) (*b–j*) spanning the last deglaciation, compared to the (*a*) GISP2 (Grootes et al 1993) and (*h*) Byrd (Sowers & Bender 1995) $\delta^{18}O$ records (in per mil). *Gray-scale horizontal bars* indicate the Oldest Dryas (OD), Bølling-Allerød (BOA), and Younger Dryas (YD) intervals. *Small black squares and diamonds* show location of AMS radiocarbon ages in each core. The *three solid horizontal lines* correspond to the time slices shown in Figure 2. Age model for core EN120-GGC1 (*f, g*) is based on graphic correlation to nearby core KNR31 GPC5 (Keigwin et al 1991), with AMS ages shown as *black diamonds* being from GPC5. Radiocarbon timescales were converted to calendar years using the relation of Bard et al (1997). The nutrient records are arranged from shallowest (1099 m) to deepest (4450 m) sites (see also Figure 2), and are thus sampling temporal variability across a depth transect of intermediate to deep waters of the North Atlantic. Core records are: (*b*) S075-26KL (1099 m) (Zahn et al 1997), (*c*) V23-081 (2393 m) (Jansen & Veum 1990), (*d*) 17045 (3663 m) (Sarnthein et al 1994), (*e*) RC11-83 (4718 m, South Atlantic) (Charles & Fairbanks 1992), (*f, g*) EN120-GGC1 (4450 m) (Keigwin et al 1991).

changes in precipitation minus evaporation (Duplessy et al 1992), with meltwater and precipitation having light isotopes. Geochemical tracers of deep-ocean nutrient distributions, primarily $\delta^{13}C$ but also Cd/Ca, monitor past variations in deepwater circulation (Figures 2, 4) (Boyle & Keigwin 1982, 1987; Curry & Lohmann 1983). Surface-water productivity strips from the water isotopically light carbon, and nutrients including phosphate and Cd (which follows phosphate). The carbon and nutrients accumulate in deep water over time through dissolution of sinking organic matter. Newly formed deep waters thus have low Cd and high $\delta^{13}C$. Over time, Cd increases and $\delta^{13}C$ decreases.

We first compare glacial maximum to modern conditions, and then consider further changes during abrupt events. For the last glacial maximum, the $\delta^{13}C$ records show that a northward-flowing nutrient-rich water mass replaced the nutrient-poor lower NADW that now fills the deep northern and tropical Atlantic

(Figures 2, 4) (Sarnthein et al 1994, Beveridge et al 1995). A strong gradient at about 3500 m separated this nutrient-rich water mass from nutrient-poor "embryonic" NADW (Sarnthein et al 1994), which ventilated the intermediate and upper deep glacial North Atlantic and can be traced as far as 10°S (Figure 2). Finally, the most positive $\delta^{13}C$ values of this interval (>1.2 per mil) identify Mediterranean Outflow Water (MOW) at shallow water depths (<1500 m) between 20° and 40°N. This water mass persisted throughout the last deglaciation (Sarnthein et al 1994) (Figure 2).

Sea-surface salinity and temperature estimates from $\delta^{18}O$ records (Figure 3) suggest that some glacial-maximum convection occurred in the Nordic Seas, where seasonally ice-free waters existed (Weinelt et al 1996), but that most North Atlantic intermediate water (NAIW) and deep water (NADW) formed by open-ocean convection somewhere in the subpolar North Atlantic, probably south of Iceland and in the northwest (Duplessy et al 1992, Labeyrie et al 1992, Oppo & Lehman 1993, Sarnthein et al 1995, Weinelt et al 1996). Ventilation rates of NADW (or NADW+NAIW) were similar to today's rates, although higher fluxes characterized intermediate waters and lower fluxes occurred at depth (Oppo & Lehman 1993, Boyle 1995). Recent $^{231}Pa/^{230}Th$ data support the notion of a glacial NADW forming at near-modern rates and feeding waters of the Southern Ocean (Yu et al 1996).

Little change occurred in the $\delta^{13}C$ values of the eastern Atlantic basin between the last glacial maximum ~18 ^{14}C kyear (thousand radiocarbon years before present) (21 cal kyear) and 15.5 ^{14}C kyear (18.3 cal kyear), suggesting that glacial NADW continued to form in similar fashion throughout this interval, although the site(s) of convection may have shifted slightly (Sarnthein et al 1994). Geochemical data suggest a major change in North Atlantic surface-water hydrology and intermediate and deep-ocean circulation starting at 15.5–16 ^{14}C kyear (18.3–19 cal kyear), which is the start of the cooling often identified with the Oldest Dryas and leading to H1 and the "Heinrich" state of ocean circulation. North Atlantic and Nordic Sea $\delta^{18}O$ records indicate a widespread low-salinity event during the Oldest Dryas (Figure 3). Estimates of North Atlantic SST and SSS indicate that surface-water densities were too low to form intermediate or deep water during this interval (Sarnthein et al 1995, Maslin et al 1995, Labeyrie et al 1995).

Within the Nordic Seas (Figures 3*b*, *c*, *d*), this distinct event, which Sarnthein et al (1995) identify as the largest excursion of the last 60 kyear, is thought to represent the first major retreat of marine margins of the Barents and Fennoscandian ice sheets (Jones & Keigwin 1988, Lehman et al 1991). Discharge of icebergs from the Laurentide, Icelandic, and Greenland Ice Sheet during H1 (Figure 3*k*) (Bond & Lotti 1995), which accompanied this event, likely accounts for the $\delta^{18}O$ minima seen in cores extending across the subpolar North Atlantic

(Figures 3*e*, 3*f*, 3*h*) and possibly as far south as the Bermuda Rise (Figure 3*i*). Cold, low-salinity surface waters are also registered off the Portugal coast (Figure 3*g*) (Bard et al 1987, Duplessy et al 1992).

Oldest Dryas $\delta^{13}C$ and Cd/Ca reached the most extreme values characteristic of nutrient enrichment of any time during the last deglaciation (Figure 4). This nutrient distribution at 13.5 ^{14}C kyear (15.8 cal kyear) clearly signals: (*1*) a shutdown of NADW formation, (*2*) reduced NAIW formation, and (*3*) southward flow of NAIW, which did not extend beyond the equator (Sarnthein et al 1994) (Figure 2).

At the onset of the Bølling-Allerød warm interval, $\delta^{18}O$ values increased throughout the Nordic Seas (Figures 3*b*, 3*c*, 3*d*) and the NE North Atlantic (Figure 3*e*), and showed no significant variability through the remainder of the deglaciation although some high-resolution records in the North Sea suggest meltwater influences associated with drainage of the Baltic Lake late in the deglaciation (Lehman & Keigwin 1992b, Boden et al 1997). Elsewhere in the North Atlantic, SST records (Bond & Lotti 1995, Sarnthein et al 1995) indicate that sea-surface warming was primarily responsible for the $\delta^{18}O$ minimum recorded during the Bølling-Allerød interval. In conjunction with increased warm-water planktonic foraminifera, the $\delta^{18}O$ maximum recorded off Nova Scotia indicates increased salinity (Figure 3*h*) (Keigwin & Jones 1995), whereas the large decrease in $\delta^{18}O$ values ($\sim$4 per mil) from the Gulf of Mexico reflects increased melting from the Laurentide Ice Sheet (Figure 3*j*) (Clark et al 1996). Nutrient levels during this time changed to values comparable to interglacial values (Figure 4), suggesting the establishment of a deep-ocean circulation in the North Atlantic similar to modern circulation, with only small-scale variability.

Only the records off Nova Scotia (Keigwin & Jones 1995, de Vernal et al 1996) (Figure 3*h*) suggest a significant $\delta^{18}O$ anomaly during the YD, which is consistent with the known routing of Lake Agassiz waters from the Mississippi to the St. Lawrence River at this time (Teller 1990). This routing event is also well expressed in the Gulf of Mexico $\delta^{18}O$ record (Broecker et al 1988, Clark et al 1996). Because only cores monitoring the deepest waters in the western North Atlantic basin (Boyle & Keigwin 1987, Keigwin et al 1991) show a significant change in nutrients during the YD (Figures 4*f*, 4*g*), some controversy has existed regarding the extent of change in deep-ocean circulation at this time (Jansen & Veum 1990, Veum et al 1992, Lehman & Keigwin 1992a, 1992b, Boyle 1995). The balance of evidence suggests a western deep-water response, with no signals or smaller signals in shallower and eastern deep waters (Figures 4*b*, 4*c*, 4*d*, Figure 2: 12 cal kyear), either because the change was restricted to western deep waters or because low sedimentation rate and bioturbation have combined to remove the signal in other records (Lehman & Keigwin 1992a, Boyle 1995).

For records older than H1, high-resolution coverage is less common, and orbital variations complicate interpretation. Nonetheless, other H events seem to have had a similar effect on the ocean as H1 had (especially H2, H4, and H5; H3 and perhaps H6 seem "smaller" or otherwise anomalous). The record of Zahn et al (1997) is long enough to capture H4 and shows great reduction in NAIW at H2 and H4 as well as at H1. Sarnthein et al (1994) have records across H2, and interpreted NADW weakening at about the age of H2 (Figure 21 in Sarnthein et al 1994). At a site sampling modern NADW, Oppo & Lehman (1995) found a strong carbon-isotopic signature of reduced NADW formation at H5, as well as for younger H events and cold phases of D-O oscillations. Maslin et al (1995) reconstructed sea-surface conditions in the northeast Atlantic from just preceding H4 to the present. They found that H1-H4 were times of cold and especially low salinity, with more effect on surface-water salinity than for other D-O stadials. Maslin et al also found evidence for a three-step pattern of warm, almost-modern conditions just after an H event, then glacial conditions, and then Heinrich conditions.

Rasmussen et al (1996) reconstructed deep-water outflow from the Nordic seas in a core covering H1-H5. They found strong outflow correlated with warm times, transitional intervals between warm and cold conditions, and little or no outflow during each of the cold D-O stadials. Interestingly, one of their key indicators shows a much stronger response to H1-H5 than to other D-O stadials (last panel of Figure 5 in Rasmussen et al 1996).

Strong oscillations can be produced in models of the ocean thermohaline circulation (e.g. Broecker et al 1990, Birchfield & Broecker 1990, Sakai & Peltier 1995). Deep-water formation in the north Atlantic is linked to cooling of initially high-salinity waters. Freshwater dilution of these waters can reduce their salinity sufficiently to prevent deep-water formation, reducing the oceanic heat flux to high northern latitudes and triggering other climatic changes. An oscillator can be modeled in which deep-water formation leads to ice-sheet melting, increased freshwater supply, and deep-water reduction or shutdown, followed by ice-sheet growth, ocean-salinity increase, and resumption of deep-water formation. "Preconditioning" of the North Atlantic with freshwater may cause deep-water formation to become "inherently unstable" (Tziperman 1997; also see Sakai & Peltier 1995), such that it may rapidly strengthen, collapse, or oscillate.

Of course, the ability to model such a signal does not mean that the cause has been identified. Perhaps it is related to solar variability or other processes, although large solar changes are unlikely, based on cosmogenic isotopes in ice cores (e.g. Finkel & Nishiizumi 1997, Lal et al 1997). Similarity between the D-O frequency and the characteristic timescale for deep-ocean circulation supports a role for the thermohaline circulation. We next review evidence strengthening this relation.

Rates and Routings of Meltwater Flux to the North Atlantic

Broecker et al (1988, 1989) first suggested that the well-dated change in routing of meltwater from the Mississippi to the St. Lawrence River 11 ^{14}C kyear, followed by rerouting of these waters down the Mississippi at 10 ^{14}C kyear, was responsible for a reduction in NADW formation and the YD cooling. In light of the Barbados sea-level record (Figure 5*b*) (Fairbanks 1989), and considering that meltwater-routing could not explain each of the earlier D-O events, Broecker et al (1990) proposed the salt oscillator (see above), and relegated the meltwater routing hypothesis to a secondary forcing that triggered the YD through its effect on an already sluggish NADW weakened by the preceding meltwater pulse-1A (mwp-IA).

Clark et al (1996) argued that evidence identifying the origin of and response to mwp-IA was elusive. Given that much smaller meltwater events have clearly left their signal in North Atlantic records (Figure 3), it is somewhat surprising that an event the size of mwp-IA (order of 0.5 Sv $= 0.5 \times 10^6$ m^3 s over 300–500 years) should go largely unnoticed. Lehman et al (1993) suggested that advection and downwelling of low-salinity surface waters in an actively running "conveyor belt" may result in the loss of a planktonic foraminiferal δ^{18}O signal. However, most ocean models predict that thermohaline circulation would decrease rapidly and dramatically in response to a freshwater forcing the magnitude of mwp-IA (Maier-Raimer & Mikolajewicz 1989, Stocker et al 1992, Stocker & Wright 1996, Rahmstorf 1995, Mikolajewicz et al 1997, Manabe & Stouffer 1997), and thus not be able to advect the signal from North Atlantic surface waters. The timescale of modeled response (order of 100 years) is much shorter than the thousand-year interval between mwp-IA and the YD, whereas the magnitude and duration of the modeled response are much larger than the Intra-Bølling Cold Period (IBCP), which occurred at the same time as mwp-IA (Figure 5) (Bard et al 1996, Clark et al 1996).

Thus we are left with an apparently very large forcing (mwp-IA) coincident with a minor response (IBCP) versus a much smaller forcing (meltwater routing to the St. Lawrence) coincident with a large response (YD). In view of these issues, we consider the origin and impact of mwp-IA to remain an unresolved but critical question in understanding the dynamics of the last deglaciation.

If the meltwater-routing event through the St. Lawrence River 11-10 ^{14}C kyear was responsible for the YD cooling, other meltwater-routing events may have similarly affected North Atlantic climate. To evaluate this hypothesis, we compare the meltwater-routing history of the southern margin of the Laurentide Ice Sheet to high-resolution climate records of the last deglaciation (Figure 5). Ocean models predict a greater deep-water response to Laurentide meltwater delivered to the North Atlantic from eastern outlets (Hudson River,

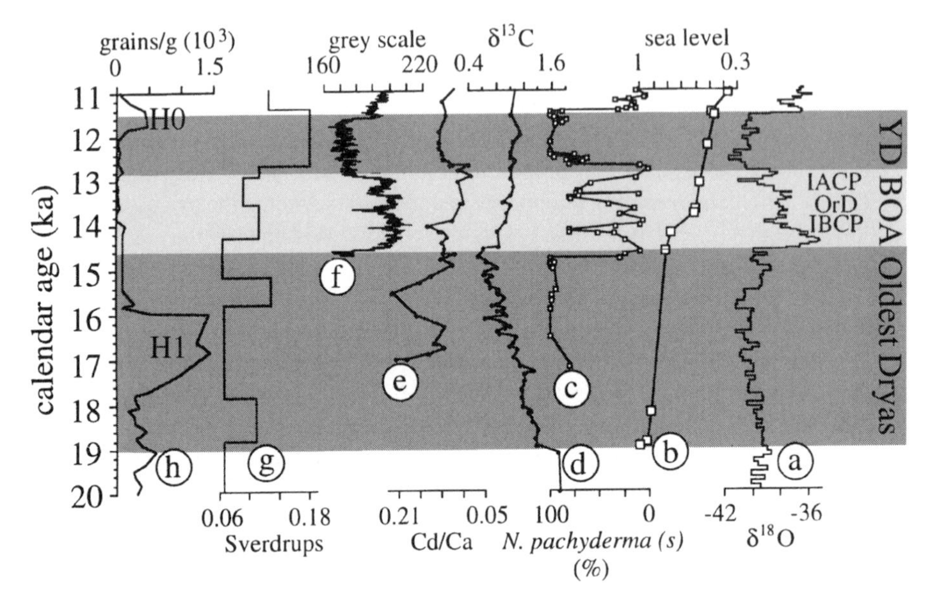

Figure 5 Climate records of the last deglaciation (*a–f*) compared to records of freshwater flux to the North Atlantic (*g, h*). (*a*) GISP2 $\delta^{18}O$ record (in per mil) (Grootes et al 1993), (*b*) Barbados sea-level record (normalized) (Fairbanks 1989), (*c*) percentage of *Neogloboquadrina pachyderma* (s.) from North Atlantic core Troll 8903 (Haflidason et al 1995), (*d*) $\delta^{13}C$ record from core S075-26KL (Zahn et al 1997), (*e*) Cd/Ca record from core EN120-GGC1 (Keigwin et al 1991), (*f*) Cariaco basin grey-scale record (Hughen et al 1996), (*g*), freshwater fluxes routed through the Hudson and St. Lawrence rivers (Teller et al 1998), and (*h*) concentrations of lithic grains from core V23-081 (Bond & Lotti 1995). *Gray-scale horizontal bars* indicate the Oldest Dryas (OD), Bølling-Allerød (BOA), and Younger Dryas (YD) intervals. We constructed the age model for Troll 8903 by linear interpolation between AMS radiocarbon ages, after subtracting 800 years. This correction gave the best age agreement with subtropical (Bard et al 1987) and tropical (Hughen et al 1996) radiocarbon dated records (with 400-year reservoir correction) for age of onset of the Bølling and Younger Dryas intervals, suggesting an 800-year reservoir correction for high-latitude North Atlantic sites may apply for the Bølling-Allerød interval as well as the Younger Dryas (Bard et al 1994). We have calibrated radiocarbon-dated records using the relation in Bard et al (1997). We used the radiocarbon-dated Cariaco Basin grey-scale record, which on a calendar year timescale is in phase with the GISP2 record (Hughen et al. 1996), to directly compare the radiocarbon ages of land and marine records with the calendar ages of the three centennial-scale climate events that occurred during the Bølling-Allerod interval (the intra-Bølling cold period (IBCP), the Older Dryas (OrD), and the intra-Allerød cold period (IACP) in the GISP2 record.

St. Lawrence River) than from the southern outlet (Mississippi River) (Maier-Raimer & Mikolajewicz 1989, Manabe & Stouffer 1997), and that preconditioning of North Atlantic surface waters by meltwater delivered through the southern outlet may be important in increasing sensitivity to meltwater subsequently routed through the eastern outlets (Fanning & Weaver 1997). To test the meltwater-routing hypothesis, identifying the timing of the opening of these outlets and the meltwater flux through them is required.

Teller (1990) first quantified these amounts for the interval surrounding the YD (12-9 ^{14}C kyear). Teller et al (1998) revised these calculations and extended them over the entire deglaciation. In general, they found (Figure 5) that when retreat of the Laurentide ice margin during interstadials uncovered an eastern outlet (the Hudson and, subsequently, St. Lawrence Rivers), the eastward freshwater flux increased, and North Atlantic climate cooled. When readvance of the ice margin closed off the eastern outlet, the eastward freshwater flux decreased and climate warmed.

The Oldest Dryas interval is a composite of successive freshwater forcings, beginning with eastward routing, followed by icebergs from H1 and then followed by eastward routing (Figure 5). While the onset of the IACP at ~11.5 ^{14}C kyear is associated with little change in magnitude of freshwater supply to the North Atlantic, it is associated with a shift in routing from the Hudson River to the St. Lawrence River. Because the St. Lawrence River routing would deliver freshwater directly to the Labrador Sea, this routing likely had a greater effect on North Atlantic salinities than routing through the Hudson River, in the same fashion that ocean models distinguish a greater response to St. Lawrence versus Mississippi routing (Manabe & Stouffer 1997).

The excellent agreement between the timing of meltwater routing or iceberg discharge events and climate change in the North Atlantic during the last deglaciation (Figure 5) supports the notion that freshwater forcing of NAIW/NADW played a key role in driving centennial- to millennial-scale climate change during the last deglaciation. The interplay between routing events and North Atlantic climate change suggests a possible feedback mechanism explaining the rapid and abrupt ice-marginal oscillations along the southern Laurentide margin superimposed on general margin retreat of the last deglaciation (Clark 1994); in addition, a similar oscillatory system may have existed at other times when the southern margin was at these latitudes.

However, we do not suggest that a routing mechanism was responsible for earlier D-O events, particularly during marine isotope stage 3. We believe instead that the significance of the meltwater-routing hypothesis is in identifying the sensitivity of the thermohaline circulation to freshwater forcing under a range of conditions. It is possible that the ocean oscillates freely at 1500 years or somewhat longer (e.g. Bond et al 1997), and that as the system approaches the threshold for a spontaneous mode switch, progressively smaller forcing is capable of triggering a mode switch (Rahmstorf 1995). If this is so, then in a variable system all mode switches will be triggered, but timing as well as magnitude of forcing will be significant in effectiveness. Such variable effectiveness of forcing may have played some role in apparent mismatches between forcing and response magnitudes during the deglaciation, discussed above.

N-S Phasing

As noted above, climate changes for much of the Earth seem to have been similar to those in the North Atlantic region, with a cold North Atlantic correlated to generally cold, dry, or windy conditions elsewhere. Whereas this correlation appears to be global at Milankovitch frequencies, it is widespread but not global at D-O frequencies, and has some out-of-phase regions at H-B frequencies.

To further evaluate this, we compare high-resolution marine and ice-core records from the Southern Hemisphere spanning the last deglaciation (Figures 6, 7). More and better-dated records are needed from some regions; we summarize many of the best records here.

These records suggest two different responses, which Bender (1998) refers to as a "northern" and "southern" response. The northern response shows the same signature of abrupt climate changes during the last deglaciation as seen in Greenland ice-core records, whereas in the southern response warming following the last glacial maximum is interrupted by the Antarctic Cold Reversal (ACR) between ~13 to 15 cal kyear (Figure 6) (Jouzel et al 1995, Sowers & Bender 1995). Alternately, one might view the southern response as one of anomalously rapid warming at the time of H1, with the ACR a return to "normal" deglaciation.

Well-dated time series of the northern response are represented by records of sea-surface temperatures (SSTs) in the Indian Ocean east of southern Africa (Figures 6, 7) (Labracherie et al 1989, Pichon et al 1992, Bard et al 1997). In particular, these records show abrupt warming at the onset of the Bølling-Allerød period, and a small cooling during the YD. The $\delta^{18}O$ signal in core MD84-527 (Figure 6*e*) (Labracherie et al 1989) suggests some combination of an abrupt increase in temperature and decrease in salinity at the onset of the Bølling, and a reversal during the YD, whereas summer SSTs estimated from transfer functions (Figure 6*d*) (Pichon et al 1992) suggest more gradual warming, beginning with the onset of the Bølling and culminating with abrupt warming at the end of the YD. Records of monsoonal variations in the Arabian Sea show enhanced (weaker) monsoons during Greenland interstadials (stadials) (Sirocko et al 1996, Schulz et al 1998), suggesting a broad region directly east of the African continent with the characteristic structure of the northern response.

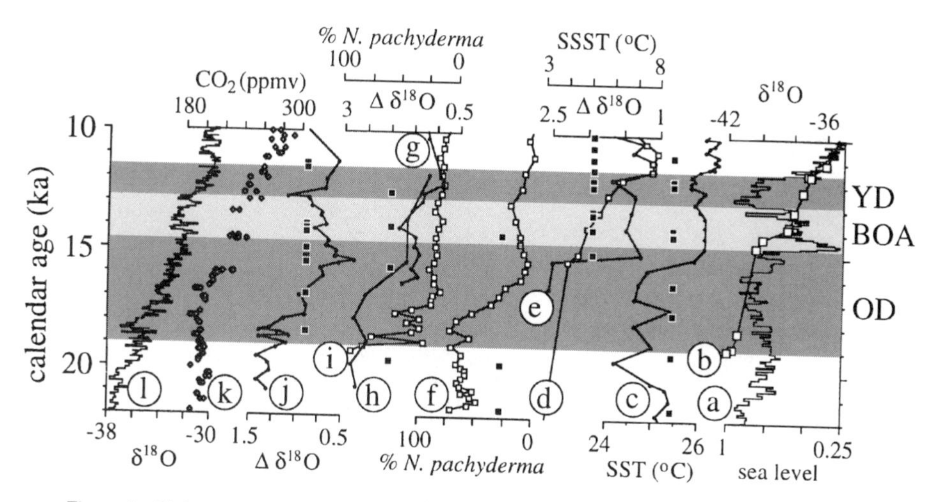

Figure 6 High-resolution records from the southern hemisphere (*c–l*) compared to (*a*) the GISP2 $\delta^{18}O$ record (in per mil) (Grootes et al 1993) and (*b*) the Barbados sea-level record (normalized) (Fairbanks 1989) (see Figure 7 for location of records). We have subtracted the global ice volume component (derived from the Barbados sea-level record; Fairbanks 1989) from the planktonic $\delta^{18}O$ records in order to illustrate anomalies ($\Delta\delta^{18}O$) which reflect some combination of changes in sea-surface salinity and temperature. *Gray-scale horizontal bars* indicate the Oldest Dryas (OD), Bølling-Allerød (BOA), and Younger Dryas (YD) intervals. *Small black squares* show location of AMS radiocarbon ages in each core. The records are: (*c*) alkenone-derived sea-surface temperature (SST) estimates from MD 79257 (Bard et al 1997), (*d*) summer sea-surface temperature (SSST) estimates and (*e*) $\Delta\delta^{18}O$ record (in per mil) from core MD 84 527 (Labracherie et al 1989; Pichon et al 1992), (*f*) percentage of *Neogloboquadrina pachyderma* (s.) from core GeoB 1711 (Little et al 1997), (*g*) $\Delta\delta^{18}O$ record (in per mil) measured on *Globigerina bulloides*, (*h*) $\Delta\delta^{18}O$ record (in per mil) measured on *Neogloboquadrina pachyderma* (s.), and (*i*) percentage of *Neogloboquadrina pachyderma* (s.) from core MD88-770 (Labeyrie et al 1996), (*j*) $\Delta\delta^{18}O$ record (in per mil) from core RC11-83 (Charles et al 1996), (*k*) atmospheric CO_2 record from the Byrd core (Sowers & Bender 1995), and (*l*) $\delta^{18}O$ record (in per mil) from the Bryd core (Sowers & Bender 1995).

Other records (not shown) with a northern response include those of mountain-glacier advances in South America and New Zealand (Figure 7) during the Oldest Dryas (H1) and the YD, respectively (Denton & Hendy 1994, Lowell et al 1995). Finally, the ice-core record from Taylor Dome, Antarctica also shows the northern response (Mayewski et al 1996, Bender 1998, Steig et al 1998).

Records showing the southern response include other Antarctic ice cores (e.g. Jouzel et al 1995). Using the $\delta^{18}O$ of atmospheric O_2 for correlation, Sowers & Bender (1995) placed the Byrd ice-core chronology on a common timescale with the GISP II chronology (Figure 6). Using methane, Blunier et al (1997) similarly correlated the Vostok ice-core chronology to the GRIP record. These two Antarctic ice-core records suggest that significant deglacial warming

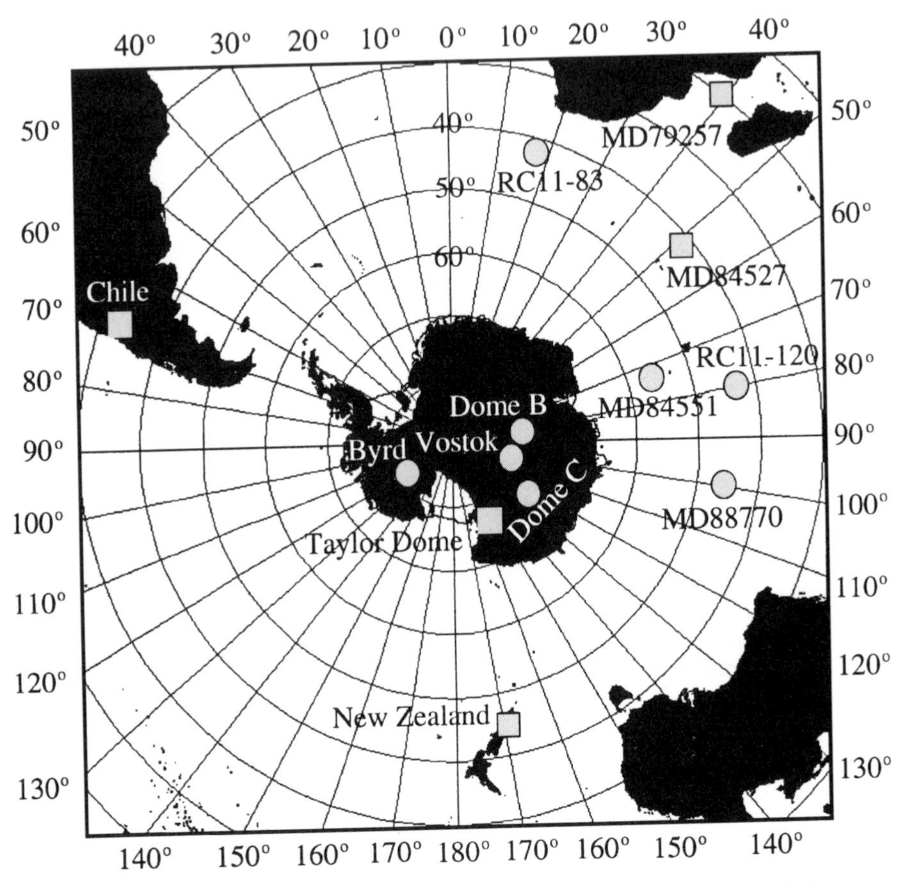

Figure 7 Map of the southern hemisphere showing distribution of sites with evidence of either a northern response (*squares*) or a southern response (*circles*) during the last deglaciation (see text for further explanation).

began between 19 and 20 cal kyear, or near the onset of Oldest Dryas cooling seen in Greenland records (Figure 6). Note that the isotopically coldest time in Greenland was 23 to 24 cal kyear (Grootes & Stuiver 1997). A consistent though not unique interpretation is that northern warming led southern warming and was nearly synchronous with northern summer insolation, but was interrupted by meltwater events (McCabe & Clark 1998). Byrd and Vostok warming continued until ~14.2 to 14.7 cal kyear, or near the onset of Bølling warming, when the ACR began. The ACR continued until ~13.6 to 13.8 cal kyear, coincident with the onset of the YD, when Antarctic temperatures warmed again, reaching their interglacial maximum by ~11.5 cal kyear.

This southern response is also present in several well-dated marine records (Figures 6, 7). Two SST records from the South Atlantic (GeoB1711 at 23°S, Figure 6*f*; RC11-83 at 41°S, Figures 6*j*, 7) and two records from the Indian Ocean (MD88770 at 45°S, Figures 6*g*, 6*h*, 6*i*, 7; MD84551 at 55°S, Figure 7) show onset of deglacial warming at the same time as for Byrd and Vostok. The *Neogloboquadrina pachyderma* record from core MD88-770 in the central Indian Ocean (Figure 6*i*) (Labeyrie et al 1996) suggests an abrupt change at the onset of deglacial warming. Insofar as the core is now situated near the Subtropical Front, the abrupt decrease in *Neogloboquadrina pachyderma* may reflect passage of this front. The SST records also suggest cooling during the ACR, followed by rapid warming (Figure 6).

The spatial variability in the distribution of the response of the Southern Hemisphere during the last deglaciation (Figure 7) is significant in identifying a mechanism or mechanisms that transmit North Atlantic climate change across the equator to produce a northern response in some regions versus other regions in which a southern response is registered. Many more data sets are needed, but it appears that the southern response is focused on the South Atlantic and the central Indian Ocean.

Imbrie et al (1992, 1993) proposed that an early response of Southern Ocean SSTs which led global ice volume changes (i.e. a southern response) resulted from an increase in the rate of formation of NADW. Because NADW is warmer than Antarctic bottom water (AABW), upwelling of NADW in the Southern Ocean followed by formation of AABW may provide heat to the Antarctic. However, drawing upon the out-of-phase climate behavior between the Byrd and GISP II records during the Bølling-Allerød-YD interval (Figure 6), Broecker (1998) proposed that a reduction (increase) in formation of NADW would be balanced by increased (decreased) convection somewhere in the Southern Ocean (Stocker & Wright 1996, Schiller et al 1997), resulting in cooling (warming) in the North Atlantic and corresponding warming (cooling) in the Southern Hemisphere. In addition, NADW formation causes water warmed in the South Atlantic to release that heat in the high northern latitudes. Reduction in NADW formation then would tend to leave that heat in the South Atlantic, warming it (Crowley 1992). Based on radiocarbon data and at least consistent with available dating of climate records, Broecker (1998) suggested that the deepwater-formation contribution to southern warming during northern cooling lags the northern cooling on the century scale.

Data in Figure 6 suggest that a southern-northern seesaw also operated during the Oldest Dryas, when significant warming in some regions of the Southern Hemisphere coincided with severely reduced rates of NADW formation (see above). Longer time series similarly show out-of-phase interhemispheric behavior between rates of NADW formation and Vostok ice isotopes and South

Atlantic SSTs (Bender et al 1994, Charles et al 1996, Little et al 1997). In long records, the southern response is especially prominent for H events and less so for non-H D-O oscillations (Bender et al 1994, Blunier et al 1998).

Similar behavior is exhibited in the North Pacific. The Santa Barbara Basin has a clear northern signal (Figure 1*g*), in which oxygenated conditions related to increased intermediate-water circulation, cooler surface conditions or stronger winds occurred during the cold phases of Atlantic D-O oscillations, with little or no time lag (Behl & Kennett 1996). However, deep Northeast Pacific waters seem to record an "anti-northern" or southern signal, in which increased ventilation is related to reduced North Atlantic ventilation (Figure 1*h*) (Lund & Mix 1998). This deep Pacific response seems to have lagged the North Atlantic changes by a few centuries for the YD (older dating is not sufficiently accurate to assess such lags), and seems to have recorded H events but missed D-O oscillations (Lund & Mix 1998). This is consistent with oceanic transmission of the southern climate-change signal from the southern ocean (Lund & Mix 1998).

The presence of a northern response in some regions of the Southern Hemisphere (Figure 7) shows that the influence of Broecker's "bipolar seesaw" mechanism is not hemispherically symmetric. Instead, the northern response suggests that the influence of changes in NADW is rapidly transmitted across the equator in some fashion to be registered directly in some regions of the Southern Hemisphere.

Manabe & Stouffer (1997) and Tziperman (1997) found from model studies that large areas of the Southern Hemisphere cooled in response to a reduced rate of NADW formation. Southern Hemisphere sites with a northern response (Figure 7) generally occur in the areas of negative temperature anomalies corresponding to "NADW off" (Manabe & Stouffer 1997).

In addition, changes in North Atlantic SSTs corresponding to changes in NADW may be transmitted across the equator as a northern signal through their influence on the African and Asian monsoon, thus explaining the data from east of Africa (Street-Perrott & Perrott 1990, Hostetler et al 1999) (Figures 6, 7). This mechanism probably is part of a more general class of processes that act to synchronize the hemispheres at orbital as well as millennial timescales, as discussed next.

N-S Orbital Synchronization

An emerging picture, which still requires more confirmation, is that the entire Earth experiences a northern response at orbital frequencies (complicated though it is by local insolation and other effects). For larger and longer-lasting millennial events (especially H events), some regions at high southern latitudes centered on the South Atlantic and into the deep Pacific experience an anti-northern or southern response transmitted through the ocean over centuries, while the rest of the Earth has a northern response transmitted through the atmosphere with little time delay. For smaller or shorter-lived D-O oscillations, the North Atlantic and regions somewhat beyond experience a northern response, but the high southern latitudes appear to have little or no response. The out-of-phase behavior between north and south is explainable through the oceanic seesaw (Broecker 1998), but then how are the hemispheres synchronized at orbital timescales and D-O timescales?

Citing Manabe & Broccoli (1985), Broecker & Denton (1989) argued that the direct atmospheric transmission of cooling at high northern latitudes to sites at high southern latitudes is small or zero. However, using a general circulation model with mixed-layer ocean lacking thermohaline circulation, Pollard & Thompson (1997) recently showed that setting appropriately low CO_2 levels, large ice sheets, and orbital parameters for glacial-maximum conditions produces significant global cooling, including cooling at high southern latitudes. Pollard & Thompson (1997) argued further that the results of Manabe & Broccoli (1985) are consistent with these newer simulations in showing that glacial-maximum boundary conditions produce strong and more or less hemispherically symmetrical cooling, if allowance is made for the extra cooling achieved by the elevation and radiative effects of the large ice sheets that grew primarily in the north.

However, in model-data comparison for glacial-maximum conditions (Table 1 in Pollard & Thompson 1997) at 18 sites spanning all latitudes (using the accurate borehole temperature calibrated results for central Greenland), Pollard & Thompson (1997) obtained modeled temperature changes less than observed for 16 of the 18 sites, and new results from Vostok, Antarctica now show that modeled temperatures changed less than observed for this 17th of the 18 sites (Salamatin et al 1998). The difference for all 18 sites as tabulated averages 2.5°C, with no latitudinal trend.

This difference can be interpreted in several ways. Perhaps the GENESIS model is not sufficiently sensitive to changes in CO_2 and ice albedo (the success of the model in many tests argues against this). Perhaps the CO_2 level chosen was not sufficiently low: as discussed by Pollard & Thompson (1997), they chose a slightly high late–glacial-maximum CO_2 level to allow comparison to recent runs with recent CO_2 because of disequilibrium between recent climates and CO_2 levels. However, we consider it more likely that there is an additional change between glacial maximum and modern, not directly related to greenhouse gases or ice sheets, that caused glacial-maximum cooling to be larger than calculated by Pollard & Thompson.

Some possibilities include changes in clouds, vegetation, and dust (again, we do not consider changes in NADW production as a likely explanation because response to the NADW changes associated with D-O events has had little effect in much of Antarctica, and response to the further changes of H events has warmed portions of Antarctica). The strong dissimilarity between dust and temperature time series from various sites including Vostok, Antarctica (Lorius et al 1990) casts doubt on this mechanism, as does the difficulty of obtaining such a large change based on aerosols (see discussion in Pollard & Thompson 1997). The models calculate clouds. Although one cannot rule out the possibility that some significant cloud feedback has been missed, we follow Pollard and Thompson (1997) in assuming that this is not the case. In simulations using the GENESIS general circulation model and fixed sea-surface temperatures, Crowley & Baum (1997) found that vegetation feedbacks can have large regional impacts but essentially no globally averaged effect (biome-wide effects to a few degrees, continent-wide effects to a few tenths of a degree but globally averaged effects of only 0.15°).

The most likely feedback may be the oceanographic effects of the stronger winds associated with cold times (e.g. Ganopolski et al 1998, Bush & Philander 1998). Stronger winds increase mixing in the ocean, cooling the warmest regions. Because of the exponential increase of evaporation rate and equilibrium vapor pressure with temperature, mixing will tend to reduce the global atmospheric water-vapor loading, which in turn will tend to cool the planet (Broecker 1997). Ganopolski et al (1998) found that in glacial-maximum reconstructions using an intermediate-complexity ocean-atmosphere model, a switch from specified modern heat flux in a slab ocean to full ocean dynamics caused tropical cooling, strong northern cooling, but slight high-latitude southern warming. Their model also shows that tropical cooling is related to both the enhanced meridional temperature gradient and to enhanced Ekman circulation caused by stronger northern hemisphere trade winds, as well as water-vapor feedbacks. Bush & Philander (1998), in a fully coupled ocean-atmosphere GCM, similarly simulated strong ice-age tropical cooling in response to stronger winds causing stronger upwelling, mixing of cooler extratropical waters into the tropics, and associated reduction in atmospheric water vapor.

Numerous authors have presented evidence for stronger winds, and stronger tropical upwelling, during glacial times. For example, Patrick & Thunell (1997) compared oxygen-isotopic changes in surface-dwelling, thermocline-dwelling, and subthermocline-dwelling foraminifera from the eastern and western Pacific. They found glacial-maximum cooling larger than interpreted by CLIMAP reconstructions, linked to a shoaling of the thermocline in the eastern Pacific. They interpreted this as indicating the effect of increased wind-driven upwelling associated with stronger trade winds.

Pisias & Mix (1997) computed radiolarian-based empirical-orthogonal-function transfer functions for cores from the eastern equatorial Pacific. They found greater glacial-maximum cooling than in CLIMAP, again plausibly linked to increased wind strength.

Andreasen & Revelo (1997) used factor-analyzed core-top foraminiferal abundances compared to modern hydrography to develop transfer functions that they applied to glacial-maximum samples from the equatorial Pacific. For glacial maximum, they reconstructed a steeper east-west thermocline slope than modern, suggesting stronger zonal wind stress (Walker circulation), and an equatorward compression of climate zones. They emphasized that this enhanced wind stress is unlikely to have caused major temperature changes, such as cooling of more than 3°C in the western equatorial Pacific. However, the imbalance remaining in the Pollard & Thompson (1997) simulations is $<3°C$. Lyle (1988) documented enhanced organic carbon concentrations in glacial-age sediments in the equatorial Atlantic and Pacific. Of the several possible explanations, this is at least plausibly linked to enhanced productivity and consistent with enhanced wind-driven upwelling in glacial times.

Mix et al (1986) documented that the most significant mode of glacial-interglacial climate change in the tropical Atlantic Ocean was centered on the eastern boundary currents and tropical upwelling areas. In this hemispherically symmetrical mode of variability, cold times are taken as probably indicating stronger or more zonal trade winds.

Many Pacific records of wind-blown dust show a spectral peak at about 31 kyear (e.g. Rea 1994), which may be related to but is not a direct result of orbital forcing. The presence of a 31 kyear peak in Vostok, Antarctica isotopic (temperature) records (Yiou et al 1991) is at least consistent with an important role for tropical, wind-related processes in affecting southern temperatures. The presence of this peak also shows that the system is more complex than any simple story we tell.

Much evidence thus supports the hypothesis that stronger, and possibly more zonal, glacial-maximum trade winds enhanced vertical and/or lateral mixing, reducing tropical sea-surface temperatures. This may have been achieved by more persistent La Niña-type conditions or in other ways. However it was achieved, the dominant effect of tropical conditions on global climate would have allowed even a small tropical change to affect widespread regions. It is likely that water-vapor feedbacks (Broecker 1997, Bush & Philander 1998) were particularly important.

Such a mechanism would act at orbital, H-B, and D-O timescales. However, some of the positive feedbacks (notably ice-sheet area and atmospheric CO_2 concentration) change more slowly than the D-O timescale but change greatly on the orbital timescale. Thus the D-O climate signal should be smaller in amplitude or geographical coverage than the orbital climate signal, as observed.

Differences in H-B and D-O, then, are likely related to oceanic feedbacks, as discussed next.

Synthesis

Mid-latitude continental ice sheets have grown and shrunk largely in response to changes in the seasonal distribution of solar heating linked to features of the Earth's orbit, as modulated through the dynamics of those ice sheets. Global cooling has been linked to ice growth during cool, short summers at high northern latitudes, with a much smaller role for solar-radiation anomalies elsewhere, suggesting important controls for synchronizing the hemispheres.

The presence of mid-latitude continents in the north (but not the south) is probably critical. Large land-ice fluctuations driving changes in ice albedo, sea level—hence possibly CO_2—and other factors can occur only in the north. Controls on CO_2 are especially problematic (Broecker 1995), but global ice volume (hence sea-level fall) and CO_2 show a strong inverse correlation, and many plausible mechanisms exist by which changes in sea level, temperature, windiness, and other factors could influence CO_2 (Broecker 1995).

Ice-albedo and CO_2 feedbacks cause changes in atmospheric and shallow oceanic circulation that further amplify the climate changes. Extratropical cooling increases the equator-to-pole temperature gradient and thus the wind strength. Stronger winds increase tropical upwelling and other mixing processes, cooling surface waters, reducing tropical water vapor, and thus causing additional cooling in and beyond the tropics. Cold northern conditions tend to weaken African and Asian monsoons, reducing tropical water vapor and also affecting cross-equatorial exchange. These processes have contributed to synchronizing the hemispheres.

The most recent deglaciation has been punctuated by millennial-scale warming and cooling events. Changes were large and rapid around the North Atlantic, and probably smaller and slower elsewhere. The changes may represent some combination of response to a free oscillation or oscillations in the climate system, and forced oscillations linked to changes in the mid-latitude ice sheets changing water and iceberg drainage from their margins. Most or all of the deglacial events are traceable to changes in the ice sheets and their meltwater routing, but this is not (yet) known to be true for some of the events during the growth of the mid-latitude ice sheets. The roughly consistent spacing of the events suggests a preferred timescale in any case, possibly linked to oceanic processes.

The abrupt climate changes are linked to reduction or elimination of North Atlantic deep water formation. Three suggested modes of operation (Sarnthein et al 1994, Stocker 1998) are modern, with vigorous high-latitude formation of North Atlantic deep water through release of heat to the atmosphere that warms Greenland and Northern Europe; glacial, with vigorous formation of North Atlantic intermediate water and shallower North Atlantic deep water at lower latitude that does not provide much heat to Greenland and Northern Europe; and Heinrich, with greatly reduced sinking in the North Atlantic that leads to more rapid formation of Antarctic bottom waters and warming of some southern regions centered on the South Atlantic. The modern state applies to warm times, and cold times may reach the glacial or Heinrich modes. More data are needed, but the Heinrich mode appears to have caused, or been caused by, the larger, longer-lasting cold events of the North Atlantic and especially the Heinrich-event surges of the Laurentide ice sheet.

The modern-to-glacial switch greatly cools the high-latitude North Atlantic. Effects are transmitted through the atmosphere to the North Pacific (Behl and Kennett 1996, Mikolajewicz et al 1997), the monsoon belt of Africa and Asia (Street-Perrott & Perrott 1990), and other regions. Prominent methane changes occur for glacial-interglacial and D-O changes, related to high-latitude and monsoon-belt effects (Chappellaz et al 1993). Glacial-maximum and D-O cooling are both associated with the modern-to-glacial switch in ocean circulation. D-O events are too short to allow the large changes in CO_2 (Stauffer et al 1998), ice sheets and sea level that produce large global climate changes on the slower orbital timescale. The North Atlantic circulation spends most of its time during interglacials in the modern (hot) mode, and most of its time during glacials in the glacial mode. D-O oscillations represent jumps between these modes during transitional times.

The additional freshwater supply of H events forces jumps to the Heinrich mode, which have a smaller effect on the North Atlantic atmosphere than do D-O oscillations because the heat transport to the high-latitude North Atlantic cannot be reduced much more, but have a larger effect on the ocean circulation because they shut down NADW formation and reduce NAIW formation, creating a "demand" for deepwater formation elsewhere. These oceanic changes during Heinrich coolings cause a temperature seesaw in which most of the Earth follows the north Atlantic but with antiphase, probably slightly lagged behavior centered on the South Atlantic and likely involving other regions (Lund & Mix 1998). CO_2, which is primarily controlled by oceanic processes, changes during H events (Stauffer et al 1998).

The most direct tests of this synthesis would be provided by additional well-dated, high-resolution climate-change records, especially from the Antarctic continent, the Pacific sector of the southern ocean, tropical oceans and the North Pacific. Improved interpretation of paleoclimatic proxies, and improved modeling, would also be highly beneficial.

ACKNOWLEDGMENTS

We thank Michael Bender, Lloyd Keigwin, Jerry McManus, Alan Mix, and Thomas Stocker for helpful discussions, and Edouard Bard, Thomas Blunier, Gerard Bond, Ed Brook, Chris Charles, Haflidi Haflidason, Konrad Hughen, Lloyd Keigwin, Laurent Labeyrie, Scott Lehman, Alan Mix, Anne de Vernal, Todd Sowers, and Rainer Zahn for generously sharing data. NSF grants to Alley and Clark provided support for this paper.

Literature Cited

Alley RB, Anandakrishnan S, Cuffey KM. 1996. Subglacial sediment transport and ice stream behavior. *Antarct. J. US.* 31:81–82

Alley RB, Finkel RC, Nishiizumi K, Anandakrishnan S, Shuman CA, et al. 1995. Changes in continental and sea-salt atmospheric loadings in central Greenland during the most recent deglaciation. *J. Glaciol.* 41:503–14

Alley RB, MacAyeal DR. 1994. Ice-rafted debris associated with binge/purge oscillations of the Laurentide Ice Sheet. *Paleoceanography* 9:503–11

Alley RB, Mayewski PA, Sowers T, Stuiver M, Taylor KC, Clark PU. 1997. Holocene climatic instability: A prominent, widespread event 8200 years ago. *Geology* 25:483–86

Alley RB, Meese DA, Shuman CA, Gow AJ, Taylor KC, et al. 1993. Abrupt increase in snow accumulation at the end of the Younger Dryas event. *Nature* 362:527–29

Andreasen DJ, Ravelo AC. 1997. Tropical Pacific Ocean thermocline depth reconstructions for the last glacial maximum. *Paleoceanography* 12:395–413

Andrews JT, Erlenkeuser H, Tedesco K, Aksu AE, Jull AJT. 1994. Late Quaternary (Stage 2 and 3) meltwater and Heinrich events, northwest Labrador Sea. *Quat. Res.* 41:26–34

Andrews JT, Tedesco K. 1992. Detrital carbonate-rich sediments, northwestern Labrador Sea: implications for ice-sheet dynamics and iceberg rafting (Heinrich) events in the North Atlantic. *Geology* 20:1087–90

Bard E, Arnold M, Fairbanks RG, Hamelin B. 1993. ^{230}Th-^{234}U and ^{14}C ages obtained by mass spectrometry on corals. *Radiocarbon* 35:191–99

Bard E, Arnold M, Mangerud J, Paterne M, Labeyrie L, et al. 1994. The North Atlantic atmosphere-sea surface ^{14}C gradient during the Younger Dryas climatic event. *Earth Planet. Sci. Lett.* 126:275–87

Bard E, Arnold M, Maurice P, Duprat J, Moyes J, Duplessy JC. 1987. Retreat velocity of the North Atlantic polar front during the last deglaciation determined by ^{14}C accelerator mass spectrometry. *Nature* 328:791–94

Bard E, Hamelin B, Arnold M, Montaggioni L, Cabioch G, et al. 1996. Deglacial sea-level record from Tahiti corals and the timing of global meltwater discharge. *Nature* 382:241–44

Bard E, Rostek F, Sonzogni C. 1997. Interhemispheric synchrony of the last deglaciation inferred from alkenone palaeothermometry. *Nature* 385:707–10

Behl RJ, Kennett JP. 1996. Brief interstadial events in the Santa Barbara basin, NE Pacific, during the past 60 kyr. *Nature* 379:243–46

Bender ML. 1998. Interhemispheric phasing of millennial-duration climate events during the last 100 ka. *AGU Chapman Conf. Mechanisms of Millennial-Scale Global Climate Change. Abstracts with Program*, p. 10

Bender M, Sowers T, Dickson M-L, Orchardo J, Grootes P, et al. 1994. Climate correlations between Greenland and Antarctica during the past 100,000 years. *Science* 372:663–66

Benson LV, Burdett JW, Kashgarian M, Lund SP, Phillips FM, Rye RO. 1996. Climatic and hydrologic oscillations in the Owens Lake basin and adjacent Sierra Nevada, California. *Science* 274:746–49

Blunier T, Chappellaz J, Schwander J, Dällenbach A, Stauffer B, Stocker TF, Raynaud D, Jouzel J, Clausen HB, Hammer CU, Johnsen SJ. 1998. Asynchrony of Antarctic and Greenland climate change during the last glacial period. *Nature* 394:739–43

Benson LV, Lund SP, Burdett JW, Kashgarian

M, Rose TP, et al. 1998. Correlation of Late-Pleistocene lake-level oscillations in Mono Lake, California, with North Atlantic climate events. *Quat. Res.* 49:1–10

Beveridge NAS, Elderfield H, Shackleton NS. 1995. Deep thermohaline circulation in the low-latitude Atlantic during the last glacial. *Paleoceanography* 10:643–60

Birchfield GE, Broecker WS. 1990. A salt oscillator in the glacial Atlantic? 2. A "scale analysis" model. *Paleoceanography* 5:835–43

Blunier T, Schwander J, Stauffer B, Stocker T, Dallenbach A, et al. 1997. Timing of the Antarctic cold reversal and the atmospheric CO_2 increase with respect to the Younger Dryas event. *Geophys. Res. Lett.* 24:2683–86

Boden P, Fairbanks RG, Wright JD, Burckle LH. 1997. High-resolution stable isotope records from southwest Sweden: The drainage of the Baltic ice lake and Younger Dryas ice margin oscillations. *Paleoceanography* 12:39–49

Bond G, Broecker W, Johnsen S, McManus J, Labeyrie L, et al. 1993. Correlations between climate records from North Atlantic sediments and Greenland ice. *Nature* 365:143–47

Bond G, Heinrich H, Broecker W, Labeyrie L, McManus J, et al. 1992. Evidence for massive discharges of icebergs into the North Atlantic ocean during the last glacial period. *Nature* 360:245–49

Bond GC, Lotti R. 1995. Iceberg discharges into the North Atlantic on millennial time scales during the last deglaciation. *Science* 267:1005–10

Bond G, Showers W, Cheseby M, Lotti R, Almasi P, et al. 1997. A pervasive millennial-scale cycle in North Atlantic Holocene and glacial climates. *Science* 278:1257–66

Boyle E. 1995. Last-glacial-maximum North Atlantic deep water: on, off or somewhere in-between? *Philos. Trans. R. Soc. London Ser. B* 348:243–53

Boyle EA, Keigwin L. 1987. North Atlantic thermohaline circulation during the past 20,000 years linked to high-latitude surface temperature. *Nature* 330:35–40

Boyle EA, Keigwin LD. 1982. Deep circulation of the North Atlantic over the last 200,000 years: geochemical evidence. *Science* 218:784–87

Broecker WS. 1995. *The Glacial World According to Wally*. Palisades, NY: Lamont-Doherty Earth Obs. 318 pp.

Broecker WS. 1997. Mountain glaciers: recorders of atmospheric water vapor content? *Glob. Biogeochem. Cycles* 11:589–97

Broecker WS. 1998. Paleocean circulation during the last deglaciation: A bipolar seesaw? *Paleoceanography* 13:119–21

Broecker WS, Andree M, Wolfli W, Oeschger H, Bonani G, et al. 1988. The chronology of the last deglaciation: Implications to the cause of the Younger Dryas event. *Paleoceanography* 3:1–19

Broecker WS, Bond G, Klas M. 1990. A salt oscillator in the glacial Atlantic? 1. The concept. *Paleoceanography* 5:469–77

Broecker WS, Denton GH. 1989. The role of ocean-atmosphere reorganization in glacial cycles. *Geochim. Cosmochim. Acta* 53:2465–501

Broecker WS, Kennett JP, Flower BP, Teller JT, Trumbore S, et al. 1989. Routing of meltwater from the Laurentide ice sheet during the Younger Dryas cold episode. *Nature* 341:318–21

Brook EJ, Sowers T, Orchardo J. 1996. Rapid variations in atmospheric methane concentration during the past 110,000 years. *Science* 273:1087–1091

Bush ABG, Philander SGH. 1998. The role of ocean-atmosphere interactions in tropical cooling during the last glacial maximum. *Science* 279:1341–44

Chappell J, Omura A, Esat T, McCulloch M, Pandolfi J, et al. 1996. Reconciliation of late Quaternary sea levels derived from coral terraces at Huon Peninsula with deep sea oxygen isotope records. *Earth Planet. Sci. Lett.* 141:227–36

Chappellaz J, Blunier T, Raynaud D, Barnola JM, Schwander J, Stauffer B. 1993. Synchronous changes in atmospheric CH_4 and Greenland climate between 40 and 8 kyr b.p. *Nature* 366:443–45

Charles CD, Fairbanks RG. 1992. Evidence from Southern Ocean sediments for the effect of North Atlantic deep-water flux on climate. *Nature* 355:416–18

Charles CD, Lynch-Stieglitz J, Ninnemann US, Fairbanks RG. 1996. Climate connections between the hemispheres revealed by deep sea sediment core/ice core correlations. *Earth Planet. Sci. Lett.* 142:19–27

Clark PU. 1994. Unstable behavior of the Laurentide ice sheet over deforming sediment and its implications for climate change. *Quat. Res.* 41:19–25

Clark PU, Alley RB, Keigwin LD, Licciardi JM, Johnsen SJ, Wang HX. 1996. Origin of the first global meltwater pulse following the last glacial maximum. *Paleoceanography* 11:563–77

Clark PU, Bartlein PJ. 1995. Correlation of late Pleistocene glaciation in the western United States with North Atlantic Heinrich events. *Geology* 23:483–86

Cortijo E, Labeyrie L, Vidal L, Vautravers M, Chapman M, et al. 1997. Changes in sea surface hydrology associated with Heinrich event 4 in the North Atlantic Ocean between 40° and 60°N. *Earth Planet. Sci. Lett.* 146:29–45

Crowley TJ. 1992. North Atlantic deep water cools the southern hemisphere. *Paleoceanography* 7:489–97

Crowley TJ, Baum SK. 1997. Effect of vegetation on an ice-age climate model simulation. *J. Geophys. Res.* 102:16463–80

Curry WB, Lohmann GP. 1983. Reduced advection into Atlantic Ocean deep eastern basins during last glaciation maximum. *Nature* 306:577–80

Curry WB, Oppo DW. 1997. Synchronous, high-frequency oscillations in tropical sea surface temperatures and North Atlantic deep water production during the last glacial cycle. *Paleoceanography* 12:1–14

Denton GH, Hendy CH. 1994. Documentation of an advance of New Zealand's Franz Josef Glacier at the onset of Younger Dryas time. *Science* 264:1434–37

de Vernal A, Hillaire-Marcel C, Bilodeau G. 1996. Reduced meltwater outflow from the Laurentide ice margin during the Younger Dryas. *Nature* 381:774–77

Duplessy JC, Labeyrie L, Arnold M, Paterne M, Duprat J, van Weering DCE. 1992. Changes in surface salinity of the North Atlantic ocean during the last deglaciation. *Nature* 358:485–87

Edwards RL, Beck JW, Burr GS, Donahue DJ, Chappell J, et al. 1993. A large drop in atmospheric $^{14}C/^{12}C$ and reduced melting in the Younger Dryas, documented with ^{230}Th ages of corals. *Science* 260:962–68

Fanning AF, Weaver AJ. 1997. Temporal-geographical meltwater influences on the North Atlantic conveyor: Implications for the Younger Dryas. *Paleoceanography* 12:307–20

Fairbanks RG. 1989. A 17,000-year glacio-eustatic sea level record: influence of glacial melting rates on the Younger Dryas event and deep-ocean circulation. *Nature* 342:637–42

Fairbanks RG, Charles CD, Wright JD. 1992. Origin of global meltwater pulses. *Radiocarbon (After Four Decades)*, p. 473–500

Finkel RC, Nishiizumi K. 1997. Beryllium 10 concentrations in the Greenland Ice Sheet Project 2 ice core from 3-40 ka. *J. Geophys. Res.* 102:26699–706

Ganopolski A, Rahmstorf S, Petouknov V, Claussen M. 1998. Simulation of modern and glacial climates with a coupled global model of intermediate complexity. *Nature* 319:351–56

Genthon C, Barnola JM, Raynaud D, Lorius C, Jouzel J, et al. 1987. Vostok ice core: Climatic response to CO_2 and orbital forcing changes over the last climatic cycle. *Nature* 329:414–18

Grimm EC, Jacobson GL Jr, Watts WA, Hansen BCS, Maasch KA. 1993. A 50,000-year record of climate oscillations from Florida and its temporal correlation with the Heinrich events. *Science* 261:198–200

Grootes PM. 1993. Interpreting continental oxygen isotope records. In *Climate Change in Continental Isotopic Records. Am. Geophys. Union Geophys. Monogr.* 78:37–46

Grootes PM, Stuiver M. 1997. Oxygen 18/16 variability in Greenland snow and ice with 10^{-3}- to 10^5-year time resolution. *J. Geophys. Res.* 102:26455–70

Grootes PM, Stuiver M, White JWC, Johnsen SJ, Jouzel J. 1993. Comparison of oxygen isotope records from the GISP2 and GRIP Greenland ice cores. *Nature* 366:552–54

Grousset FE, Labeyrie L, Sinko JA, Cremer M, Bond G, et al. 1993. Patterns of ice-rafted detritus in the glacial North Atlantic (40-55°N). *Paleoceanography* 8:175–92

Gwiazda RH, Hemming SR, Broecker WS. 1996a. Provenance of icebergs during Heinrich event 3 and the contrast to their sources during other Heinrich episodes. *Paleoceanography* 11:371–78

Gwiazda RH, Hemming SR, Broecker WS, Onsttot T, Mueller C. 1996b. Evidence from $^{40}Ar/^{39}Ar$ ages for a Churchill province source of ice-rafted amphiboles in Heinrich layer 2. *J. Glaciol.* 42:440–46

Haflidason H, Sejrup HP, Kristensen DK, Johnsen S. 1995. Coupled response of the late glacial climatic shifts of northwest Europe reflected in Greenland ice cores: Evidence from the northern North Sea. *Geology* 23:1059–62

Hays HD, Imbrie J, Shackleton NJ. 1976. Variations in the earth's orbit: pacemaker of the ice ages. *Science* 194:1121–32

Heinrich H. 1988. Origin and consequences of cyclic ice rafting in the northeast Atlantic Ocean during the past 130,000 years. *Quat. Res.* 29:143–52

Hesse R, Khodabakhsh S. 1998. Depositional facies of late Pleistocene Heinrich events in the Labrador Sea. *Geology* 26:103–6

Hostetler SW, Clark PU, Bartlein PJ, Mix AC, Pisias NG. 1999. Atmospheric transmission of North Atlantic Heinrich events. *J. Geophys. Res.* (in press)

Hughen KA, Overpeck JT, Peterson LC, Trumbore S. 1996. Rapid climate changes in the tropical Atlantic region during the last deglaciation. *Nature* 380:51–54

Hulbe CL. 1997. An ice shelf mechanism for

Heinrich layer production. *Paleoceanography* 12:711–17
Imbrie J, Berger A, Boyle EA, Clemens SC, Duffy A, et al. 1993. On the structure and origin of major glaciation cycles: 2. The 100,000-year cycle. *Paleoceanography* 8: 699–735
Imbrie J, Boyle EA, Clemens SC, Duffy A, Howard WR, et al. 1992. On the structure and origin of major glaciation cycles 1. Linear responses to Milankovitch forcing. *Paleoceanography* 7:701–38
Imbrie J, Hays JD, Martinson DG, McIntyre A, Mix AC, et al. 1984. The orbital theory of Pleistocene climate: Support from a revised chronology of the marine $\delta^{18}O$ record. In *Milankovitch and Climate, Part I*, ed. A Berger J Imbrie, J Hays, G Kukla, B Saltzman, p. 269–305. Norwell, MA: D Riedel. 895 pp.
Jansen E, Veum T. 1990. Evidence for two-step deglaciation and its impact on North Atlantic deep-water circulation. *Nature* 343:612–16
Jones GA, Keigwin LD. 1988. Evidence from Fram Strait (78°N) for early deglaciation. *Nature* 336:57–59
Jouzel J, Lorius C, Petit JR, Genthon C, Barkov NI, et al. 1987. Vostok ice core: A continuous isotope temperature record over the last climatic cycle (160,000 years). *Nature* 329:403–8
Jouzel J, Vaikmae R, Petit JR, Martin M, Duclos Y, et al. 1995. The two-step shape and timing of the last deglaciation in Antarctica. *Clim. Dyn.* 11:151–61
Keigwin LD, Jones GA. 1995. The marine record of deglaciation from the continental margin off Nova Scotia. *Paleoceanography* 10:973–85
Keigwin LD, Jones GA, Lehman SJ, Boyle E. 1991. Deglacial meltwater discharge, North Atlantic deep circulation, and abrupt climate change. *J. Geophys. Res.* 96:16811–26
Keigwin LD, Lehman SJ. 1994. Deep circulation change linked to Heinrich event 1 and Younger Dryas in a middepth North Atlantic core. *Paleoceanography* 9:185–94
Kitagawa H, van der Plicht J. 1998. Atmospheric radiocarbon calibrations to 45,000 yr B.P.: Late glacial fluctuations of cosmogenic isotope production. *Science* 279:1187–90
Koç N, Jansen E. 1994. Response of the high-latitude northern hemisphere to orbital climate forcing: Evidence from the Nordic seas. *Geology* 22:523–26
Labeyrie L, Labracherie M, Gorfti N, Pichon JJ, Vautravers M, et al. 1996. Hydrographic changes of the Southern ocean (southeast Indian sector) over the last 230 kyr. *Paleoceanography* 11:57–76
Labeyrie LD, Duplessy JC, Duprat J, Juillet-Leclerc A, Moyes J, et al. 1992. Changes in the vertical structure of the North Atlantic ocean between glacial and modern times. *Quat. Sci. Rev.* 11:401–13
Labeyrie L, Vidal L, Cortijo E, Paterne M, Arnold M, et al. 1995. Surface and deep hydrology of the Northern Atlantic ocean during the past 150,000 years. *Philos. Trans. R. Soc. London Ser. B* 348:255–64
Labracherie M, Labeyrie LD, Duprat J, Bard E, Arnold M, et al. 1989. The last deglaciation in the southern ocean. *Paleoceanography* 4:629–38
Lal D, Jull AJT, Burr GS, Donahue DJ. 1997. Measurements of in situ ^{14}C concentrations in Greenland Ice Sheet Project 2 ice covering a 17-kyr time span: Implications to ice flow dynamics. *J. Geophys. Res.* 102:26505–10
Lehman SJ, Jones GA, Keigwin LD, Andersen ES, Butenko G, Ostmo SR. 1991. Initiation of Fennoscandian ice-sheet retreat during the last deglaciation. *Nature* 349:513–16
Lehman SJ, Keigwin LD. 1992a. Deep circulation revisited. *Nature* 358:197–98
Lehman SJ, Keigwin LD. 1992b. Sudden changes in the North Atlantic circulation during the last deglaciation. *Nature* 356:757–62
Lehman SJ, Wright DG, Stocker TF. 1993. Transport of freshwater into the deep ocean by the conveyor. *NATO ASI Ser. I* 12:187–209
Little MG, Schneider RR, Kroon D, Price B, Summerhayes CP, Segl M. 1997. Trade wind forcing of upwelling seasonality, and Heinrich events as a response to sub-Milankovitch climate variability. *Paleoceanography* 12:568–76
Lorius C, Jouzel J, Raynaud D, Hansen J, Le Treut H. 1990. The ice-core record: climate sensitivity and future greenhouse warming. *Nature* 347:139–45
Lowell TV, Heusser CJ, Andersen BG, Moreno PI, Hauser A, et al. 1995. Interhemispheric correlation of Late Pleistocene glacial events. *Science* 269:1541–49
Lund DC, Mix AC. 1998. Millennial-scale deep water oscillations: reflections of the North Atlantic in the deep Pacific from 10 to 60 ka. *Paleoceanography* 13:1–19
Lyle M. 1988. Climatically forced organic carbon burial in equatorial Atlantic and Pacific Oceans. *Nature* 335:529–32
MacAyeal DR. 1993. A low-order model of growth/purge oscillations of the Laurentide Ice Sheet. *Paleoceanography* 8:767–73
Maier-Raimer E, Mikolajewicz U. 1989. Experiments with an OGCM on the cause of the Younger Dryas. In *Oceanography, 1988*, ed. A Ayala-Castanares, W Wooster, A Yanez-Arancibia, p. 87–100. Mexico City: Univ. Nac. Auton. Mex. Press

Manabe S, Broccoli AJ. 1985. The influence of continental ice sheets on the climate of an ice age. *J. Geophys. Res.* 90:2167–90

Manabe S, Stouffer RJ. 1997. Coupled ocean-atmosphere model response to freshwater input: Comparison to Younger Dryas event. *Paleoceanography* 12:321–36

Marcantonio F, Anderson RF, Stute M, Kumar N, Schlosser P, Mix A. 1996. Extraterrestrial ^{3}He as a tracer of marine sediment transport and accumulation. *Nature* 383:705–7

Marshall SJ, Clarke GKC. 1997. A continuum mixture model of ice stream thermomechanics in the Laurentide Ice Sheet. 2. Application to the Hudson Strait ice stream. *J. Geophys. Res.* 102:20615–37

Maslin MA, Shackleton NJ, Pflaumann U. 1995. Surface water temperature, salinity, and density changes in the northeast Atlantic during the last 45,000 years: Heinrich events, deep water formation, and climatic rebounds. *Paleoceanography* 10:527–44

Mayewski PA, Meeker LD, Twickler MS, Whitlow S, Yang Q, et al. 1997. Major features and forcing of high-latitude northern hemisphere atmospheric circulation using a 110,000-year-long glaciochemical series. *J. Geophys. Res.* 102:26345–66

Mayewski PA, Twickler MS, Whitlow SI, Meeker LD, Yang Q, et al. 1996. Climate change during the last deglaciation in Antarctica. *Science* 272:1636–38

McCabe AM, Clark PU. 1998. Ice-sheet variability around the North Atlantic Ocean during the last deglaciation. *Nature* 392:373–77

McIntyre A, Molfino, B. 1996. Forcing of Atlantic equatorial and subpolar millennial cycles by precession. *Science* 274:1867–70

McManus JF, Anderson RF, Broecker WS, Fleisher MQ, Higgins SM. 1998. Radiometrically determined sedimentary fluxes in the sub-polar North Atlantic during the last 140,000 years. *Earth Planet. Sci. Lett.* 155:29–43

Mikolajewicz U, Crowley TJ, Schiller A, Voss R. 1997. Modeling teleconnections between the North Atlantic and North Pacific during the Younger Dryas. *Nature* 387:384–7

Mix AC, Ruddiman WF, McIntyre A. 1986. Late Quaternary paleoceanography of the tropical Atlantic, 1: Spatial variability of annual mean sea-surface temperatures, 0-20,000 years bp. *Paleoceanography* 1:43–66

Mooers HD, Lehr JD. 1997. Terrestrial record of Laurentide Ice Sheet reorganization during Heinrich events. *Geology* 25:987–90

Mulder T, Moran K. 1995. Relationship among submarine instabilities, sea level variations, and the presence of an ice sheet on the continental shelf, An example from the Verrill Canyon Area, Scotian Shelf. *Paleoceanography* 10:137–54

Muller RA, MacDonald GJ. 1997. Glacial cycles and astronomical forcing. *Science* 277: 215–18

Oerlemans J. 1993. Evaluating the role of climate cooling in iceberg production and the Heinrich events. *Nature* 364:783–86

Oppo DW, Lehman SJ. 1993. Mid-depth circulation of the subpolar North Atlantic during the last glacial maximum. *Science* 259:1148–52

Oppo DW, Lehman SJ. 1995. Suborbital timescale variability of North Atlantic Deep Water formation during the last 200,000 years. *Paleoceanography* 12:191–205

Patrick A, Thunell RC. 1997. Tropical Pacific sea surface temperatures and upper water column thermal structure during the last glacial maximum. *Paleoceanography* 12:649–57

Peteet DM, ed. 1993. Global Younger Dryas, a special issue of *Quat. Sci. Rev.* 12:277–355 (Suppl.)

Phillips FM, Zreda MG, Benson LV, Plummer MA, Elmore D, Sharma P. 1996. Chronology for fluctuations in Late Pleistocene Sierra Nevada glaciers and lakes. *Science* 274:749–51

Pichon JJ, Labeyrie LD, Bareille G, Labracherie M, Duprat J, Jouzel J. 1992. Surface water temperature changes in the high latitudes of the southern hemisphere over the last glacial-interglacial cycle. *Paleoceanography* 7:289–318

Pisias NG, Mix AC. 1997. Spatial and temporal oceanographic variability of the eastern equatorial Pacific during the late Pleistocene: evidence from radiolaria microfossils. *Paleoceanography* 12:381–93

Pollard D, Thompson SL. 1997. Climate and ice-sheet mass balance at the last glacial maximum from the GENESIS Version 2 global climate model. *Quat. Sci. Rev.* 16:841–64

Porter SC, An Z. 1995. Correlation between climate events in the North Atlantic and China during the last glaciation. *Nature* 375:305–8

Rahmstorf S. 1995. Bifurcations of the Atlantic thermohaline circulation in response to changes in the hydrological cycle. *Nature* 378:145–49

Rasmussen TL, Thomsen E, van Weering TCE, Labeyrie L. 1996. Rapid changes in surface and deep water conditions at the Faeroe Margin during the last 58,000 years. *Paleoceanography* 11:757–72

Raymo ME. 1997. The timing of major climate terminations. *Paleoceanography* 12:577–85

Rea DK. 1994. The paleoclimatic record provided by eolian deposition in the deep sea: the geologic history of wind. *Rev. Geophys.* 32:159–95

Revel M, Sinko JA, Grousset FE, Biscaye PE. 1996. Sr and Nd isotopes as tracers of North Atlantic lithic particles: paleoclimatic implications. *Paleoceanography* 11:95–113

Sakai K, Peltier WR. 1995. A simple model of the Atlantic thermohaline circulation: internal and forced variability with paleoclimatological implications. *J. Geophys. Res.* 100:13,455–79

Salamatin AN, Lipenkov VY, Barkov NI, Jouzel J, Petit JR, Raynaud D. 1998. Ice core age dating and paleothermometer calibration based on isotope and temperature profiles from deep boreholes at Vostok Station (East Antarctica). *J. Geophys. Res.* 103:8963–77

Sarnthein M, Jansen E, Weinelt M, Arnold M, Duplessy JC, et al. 1995. Variations in Atlantic surface ocean paleoceanography, 50°-80°N: Time-slice records of the last 30,000 years. *Paleoceanography* 10:1063–94

Sarnthein M, Winn K, Jung SJA, Duplessy JC, Labeyrie L, et al. 1994. Changes in east Atlantic deepwater circulation over the last 30,000 years: Eight time slice reconstructions. *Paleoceanography* 9:209–67

Schiller A, Mikolajewicz U, Voss R. 1997. The stability of the North Atlantic thermohaline circulation in a coupled ocean-atmosphere general circulation model. *Clim. Dyn.* 13:325–47

Schulz H, von Rad U, Erlenkeuser H. 1998. Correlation between Arabian Sea and Greenland climate oscillations of the past 110,000 years. *Nature* 393:54–57

Severinghaus JP, Sowers T, Brook EJ, Alley RB, Bender ML. 1998. Timing of abrupt climate change at the end of the Younger Dryas interval from thermally fractionated gases in polar ice. *Nature* 391:141–46

Sirocko F, Garbe-Schönberg D, McIntyre A, Molfino B. 1996. Teleconnections between the subtropical monsoons and high-latitude climates during the last deglaciation. *Science* 272:526–29

Sowers T, Bender M. 1995. Climate records covering the last deglaciation. *Science* 269: 210–14

Stauffer B, Blunier T, Dällenback A, Indermühle A, Schwander J et al. 1998. Atmospheric CO_2 concentration and millennial-scale climate change during the last glacial period. *Nature* 392:59–62

Steig EJ, Brook EJ, White JWC, Sucher CM, Bender ML, Lehman SJ, Morse DL, Waddington ED, Clow GD, 1998. Synchronous climate changes in Antarctica and the North Atlantic. *Science* 282:92–95

Stocker TF. 1998. Is there a unique mechanism for abrupt changes in the climate system? *AGU Chapman Conf. Mechanisms of Millennial-Scale Global Climate Change. Abstracts with Program*, p. 30

Stocker TF, Wright DG. 1996. Rapid changes in ocean circulation and atmospheric radiocarbon. *Paleoceanography* 11:773–95

Stocker TF, Wright DG, Broecker WS. 1992. The Influence of high-latitude surface forcing on the global thermohaline circulation. *Paleoceanography* 7:529–41

Stoner JS, Channell JET, Hillaire-Marcel C. 1996. The magnetic signature of rapidly deposited detrital layers from the deep Labrador Sea: relationship to North Atlantic Heinrich layers. *Paleoceanography* 11:309–26

Street-Perrott FA, Perrott RA. 1990. Abrupt climate fluctuations in the tropics: the influence of Atlantic Ocean circulation. *Nature* 358:607–12

Tarasov L, Peltier WR. 1997. A high-resolution model of the 100 ka ice-age cycle. *Ann. Glaciol.* 25:58–65

Taylor KC, Mayewski PA, Alley RB, Brook EJ, Gow AJ, et al. 1997. The Holocene/Younger Dryas transition recorded at Summit, Greenland. *Science* 278:825–7

Teller JT. 1990. Meltwater and precipitation runoff to the North Atlantic, Arctic, and Gulf of Mexico from the Laurentide ice sheet and adjacent regions during the Younger Dryas. *Paleoceanography* 5:897–905

Teller JT, Licciardi J, Clark PU. 1998. North American meltwater routing to the North Atlantic during the last deglaciation. AGU Chapman Conference, *Mechanisms of Millennial-Scale Global Climate Change. Abstracts with Program*. p. 26

Tziperman E. 1997. Inherently unstable climate behavior due to weak thermohaline ocean circulation. *Nature* 386:592–95

Veum T, Jansen E, Arnold M, Beyer I, Duplessy JC. 1992. Water mass exchange between the North Atlantic and the Norwegian Sea during the past 28,000 years. *Nature* 356:783–5

Vidal L, Labeyrie L, Cortijo E, Arnold M, Duplessy JC, et al. 1997. Evidence for changes in the North Atlantic Deep Water linked to meltwater surges during the Heinrich events. *Earth Planet. Sci. Lett.* 146:13–27

Weertman J. 1974. Stability of the junction of an ice sheet and ice shelf. *J. Glaciol.* 5:145–58

Weinelt M, Sarnthein M, Pflaumann U, Schulz H, Jung S, Erlenkeuser H. 1996. Ice-free Nordic Seas during the last glacial maximum? Potential sites of deepwater formation. *Paleoclimates* 1:283–309

Winograd IJ, Coplen TB, Landwehr JM, Riggs AC, Ludwig KR, et al. 1992. Continuous 500,000-year climate record from vein calcite

Annu. Rev. Earth Planet. Sci. 1999. 27:183–229

K-Ar AND ^{40}Ar/^{39}Ar GEOCHRONOLOGY OF WEATHERING PROCESSES

P. M. Vasconcelos
University of Queensland, Department of Earth Sciences, Brisbane, Queensland 4072, Australia; e-mail: paulo@sol.earthsciences.uq.edu.au

KEY WORDS: alunite, jarosite, Mn-oxides, paleoclimates, landscape evolution

ABSTRACT
Recent developments in the application of K-Ar and ^{40}Ar/^{39}Ar dating of continental weathering process demonstrate the method's suitability for dating minerals present in weathering profiles. Alunite-group sulfates and hollandite-group manganese oxides, which often precipitate through weathering reactions, were first analyzed by the K-Ar method 30 years ago. Recently these minerals were shown to be suitable to ^{40}Ar/^{39}Ar geochronology, despite their fine-grained habits. The bulk nature of the K-Ar technique and the complex mineral assemblages in weathering profiles restrict K-Ar dating of weathering processes. The single-crystal approach possible with the ^{40}Ar/^{39}Ar method allows the study of weathering profiles where alunite- and hollandite-group minerals occur as minor phases. Step-heating analysis possible with the ^{40}Ar/^{39}Ar method provides information about the Ar and K retention histories, the presence of hypogene contaminants, and possible ^{39}Ar recoil during sample irradiation. Fully automated, modern ^{40}Ar/^{39}Ar systems enable analysis of several samples, providing a comprehensive weathering database. These results are useful in the study of continental paleoclimates and the geochemical, geomorphological, and tectonic histories of an area.

WHY DATE WEATHERING PROCESSES?

Weathering geochronology, as treated in this review, consists of the application of radiogenic-isotope dating techniques to the determination of the precise and accurate timing of precipitation of supergene minerals in soils and weathering profiles on continental landscapes. Application of these geochronological

0084-6597/99/0515-0183$08.00

techniques permits dating of low-temperature chemical reactions resulting from rock-water interaction and is responsible for the formation of weathering profiles. The ultimate objective of weathering geochronology is to determine the timing, rate, and mechanisms of propagation of weathering fronts through the profile and to unravel the climatic and geochemical history implicit in the evolution of continental regolith covers.

Continental weathering encompasses inorganic and biologically mediated chemical and mechanical processes. The ultimate effect of weathering reactions is the transfer of elements from the continental lithosphere to the oceans. However, the product of weathering processes may remain metastably as thick covers on continents as long as the weathering blanket resists chemical and physical erosion. The formation and preservation of weathering profiles reflect the balance between chemical and mechanical weathering promoting the formation of a regolith blanket and mechanical and chemical erosion acting to transport the products of weathering into the drainage system and ultimately into the oceans.

The intensity and rate of interactions among the lithosphere, the hydrosphere, the atmosphere, and the biosphere control the global cycle of elements. They also affect the buildup and the residence time of nutrients and greenhouse gases in global reservoirs (Berner et al 1983, Berner 1994). Climate-weathering feedback mechanisms and our ability to predict anthropogenic influences on future climate depend on accurate quantification of chemical weathering rates (Brady & Caroll 1994). Much of our present knowledge about the variation in the intensity of rates of weathering processes in the geological past is derived from variations in ocean water composition and the sedimentary record preserved in the oceanic environment (Robert & Chamley 1987). The transfer of elements and sediments from continents to oceans identifies increases in erosion (chemical and mechanical), which may or may not be simultaneous with intervals of enhanced mechanical and chemical weathering. Geochronology of weathering profiles provides the database necessary for the correlation between events of global significance (volcanism, tectonism, sea-level changes, chemical changes in oceanic water composition, extinction events, and changes in physical oceanographic patterns) and the weathering record preserved on continents. This database is necessary if cause-and-effect relationships are to be established between global events and continental weathering processes.

In addition, weathering processes and the rate of dispersion of elements in weathering profiles through geological time are natural analogues to the dispersion of toxic elements in polluted environments. The reliability of predictive models simulating mechanisms and rates of dispersion of pollutants in nature can be tested by comparison with mechanisms and rates measured through the study of the geochronology of natural weathering processes. From an economic viewpoint, many deposits of mineral commodities important to modern life (bauxites, lateritic Ni deposits, supergene-enriched Fe, Mn, Co, Au, and Cu deposits, for example) owe their existence and economic viability to weathering. The discovery of such deposits often requires exploration in deeply weathered terrains. Geochronology of weathering processes permits the evaluation of deeply weathered profiles in distinct geomorphological provinces as potential hosts for supergene mineralization.

THE CONTINENTAL WEATHERING RECORD

Relatively thick, chemically and mineralogically stratified weathering profiles may develop when the balance between weathering and erosion favors the former. An idealized complete profile developed on average continental crust (granodioritic composition) is illustrated in Figure 1. A profile composed of all or most of the horizons shown in Figure 1 is defined as a lateritic profile (Nahon 1986). Some of these horizons may be absent, and the stratigraphy of lateritic profiles may vary based on host-rock composition. Studies in stratigraphy of

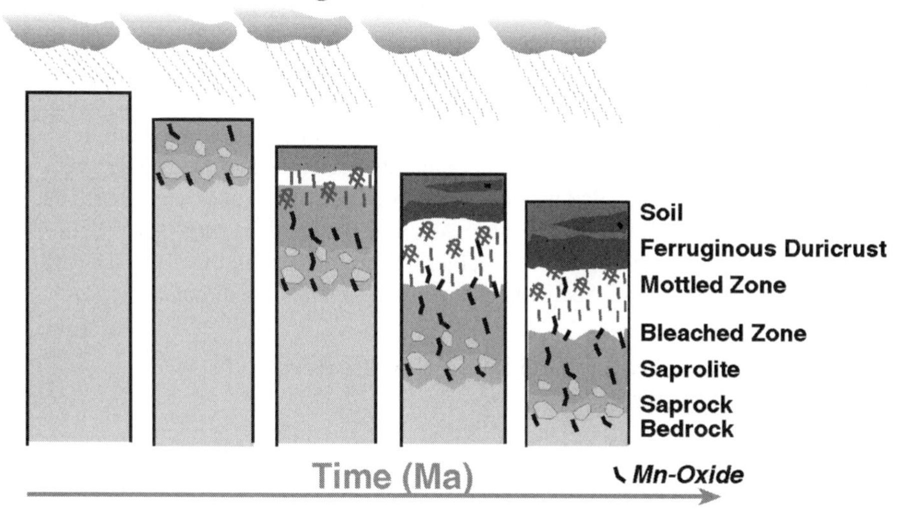

Figure 1 Depiction of the evolution of lateritic weathering profiles with time. The presence or absence of all the stratigraphic horizons depicted depends on the composition of the bedrock and on the erosion history of the profile. These weathering profiles are commonly 50 to 100 m deep in some regions of South America, Australia, and Africa. Their complex chemical stratigraphic evolution and their longevity suggest that they may host evidence of past climatic conditions. The application of geochronological methods to unraveling the history of mineral precipitation in these deep weathering profiles may help to retrieve information related to the climate at the time each component of the profile evolved.

weathering profiles show that some profiles in continental landscapes record a complex history, with several periods of weathering, truncation, partial burial, and reweathering (Finkl 1984, Tardy & Roquin 1992). Stratigraphic evidence also suggests that these superimposed conditions may have operated for tens of millions to hundreds of millions of years, and the resulting profiles may record processes developed under extremely variable climatic conditions (Tardy & Roquin 1992).

Weathering fronts migrate downward from the bedrock-atmosphere interface, creating inverted age profiles. Ideally, older weathering assemblages occur at the top, whereas more recently precipitated supergene minerals are present at the bottom of the profile. However, preferential migration of oxygenated surface waters along permeable horizons or fracture planes may cause the formation of early precipitated, older weathering minerals at lower levels in the profile. This leads to a great complexity in the time-stratigraphic record preserved because zones representative of different weathering stages may occur at the same horizon (Figure 2). In addition, superimposed weathering events complicate the age pattern owing to recurrent dissolution and reprecipitation of supergene minerals.

Finally, weathering profiles are inherently open systems. The continuous flow of mass and energy through the profiles may suggest that these systems are in a continuous state of flux, constantly re-equilibrating with newly imposed environmental conditions. To provide information about paleoenvironments, profiles must contain minerals precipitated under conditions different from those existing at present. Field and petrographic observations and thermodynamic-kinetic models indicate that weathering profiles may exist for extended periods of time under metastable equilibrium. Absolute dating of the products of chemical reactions preserved in these weathering profiles can, in this case, provide information about past geochemical and environmental conditions.

K-Ar AND $^{40}Ar/^{39}Ar$ METHODS: APPLICATION TO WEATHERING

Several qualitative, semiquantitative, and quantitative methods have been used to determine the age of weathered material. In this review I will present recent advances in the application of K-Ar and $^{40}Ar/^{39}Ar$ dating of supergene minerals, particularly K-bearing Mn-oxides and sulfates, to the study of weathering processes.

The application of the K-Ar and $^{40}Ar/^{39}Ar$ methods to the study of continental weathering processes had been, until relatively recently, generally neglected by the geochronological community. This lack of interest may have been derived from the complexity of the mineralogical assemblages and paragenetic

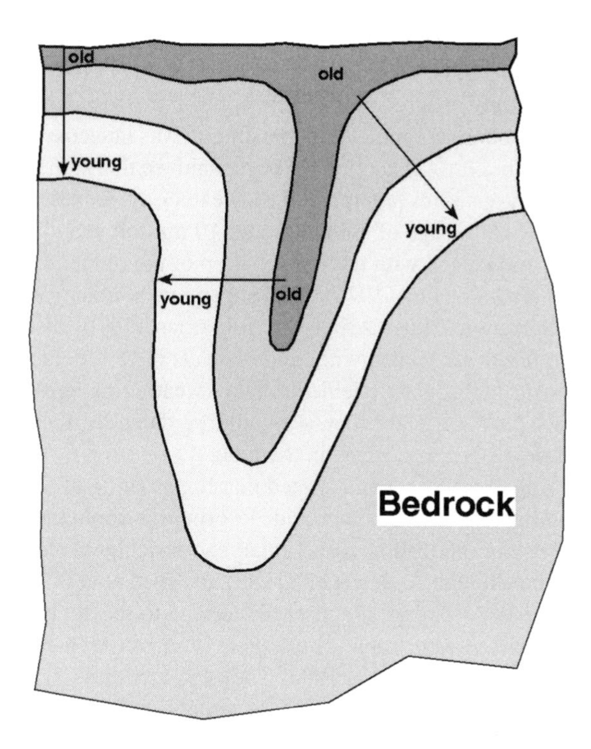

Figure 2 The penetration of oxygenated, organic acid-rich weathering solutions along preferential fluid migration paths. Penetration may allow the weathering front to propagate more rapidly in more permeable zones than in less permeable adjacent zones. This heterogeneous fluid flow through the profile creates a complex distribution of ages of supergene minerals.

sequences present in weathering profiles. Alternatively, this lack of interest may have arisen from the commonly held belief that weathered rocks are open systems, constantly re-equilibrating with percolating meteoric solutions, which would preclude the quantitative preservation of parent and daughter isotope systems used in geochronology. Despite this paucity in the application of K-Ar and $^{40}Ar/^{39}Ar$ to date weathering processes, researchers recognized long ago that some minerals often present in weathering profiles were suitable to K-Ar dating. Alunite-group minerals such as alunite—$KAL_3(SO_4)_2{\cdot}(OH)_6$— and jarosite—$KFe_3(SO_4)_2{\cdot}(OH)_6$)—phases, which precipitate through weathering reactions in sulfur-rich systems but also occur in the hydrothermal alteration zones of some ore deposits, were first analyzed by the K-Ar method 30 years ago. Shanin et al (1968) and Webb & McDougall (1968) showed that K-Ar dating of alunite provided reliable information about the hydrothermal alteration of ore deposits. Subsequent studies by Chuchrov et al (1969) provided an extensive database of K-Ar results for alunites obtained from sulphate deposits in the Soviet Union. Silberman & Ashley (1970) conducted a well-constrained study, where stratigraphic control provided limits to the alunite ages that were obtained, indicating that alunite was indeed a suitable phase for K-Ar dating. Chukhrov et al (1969) also recognized that some of the alunite samples they analyzed were precipitated by weathering processes. They concluded that "the absolute age of alunite (about 10 million years) indicates that this process was associated with the rejuvenation of the old land surface in the Neogene." Gustafson & Hunt (1975) dated supergene alunite from weathering profiles at El Salvador, Chile. Ashley & Silberman (1976) also recognized that supergene alunite associated with gold deposits in Goldfield, Nevada, was suitable for geochronology and yielded reliable weathering ages. Despite the early recognition that surficial processes could be dated by K-Ar analysis of alunite, very little progress occurred in the field.

Alpers & Brimhall (1988) demonstrated that K-Ar dating of alunite precipitated by weathering reactions could provide information about the environmental and climatic conditions leading to the supergene enrichment of porphyry copper systems in the Chilean Andes. Bird (1988) and Bird et al (1990) expanded the application of K-Ar dating of supergene alunite to the study of the Cenozoic climatic history of Australia. Subsequent studies (Arehardt et al 1992, Arehardt & O'Neil 1992, Rye et al 1993, Cook 1994, Sillitoe & McKee 1996, Itaya et al 1996, Virtue 1996) confirmed the suitability of K-Ar dating of supergene alunite-jarosite minerals in weathering geochronology. However, the bulk nature of the K-Ar technique (0.1 to 1 g of pure sample is needed for age determination) and the complex and intimately intergrown mineral assemblages present in most weathering profiles (supergene phases often occur closely associated with unweathered primary minerals) restrict the application of K-Ar dating to the study of few geological environments (e.g. weathered ore bodies, salt lake deposits). Vasconcelos (1992) and Vasconcelos et al (1994b) showed that single crystals of alunite and jarosite could be dated by the laser-heating $^{40}Ar/^{39}Ar$ technique. This single-crystal approach allows the study of weathering profiles where alunite and jarosite occur as minor phases. It also helps to obviate contamination problems. In addition, step-heating analysis possible with the $^{40}Ar/^{39}Ar$ method provides information about the Ar and K retention history of the mineral analyzed and permits detection of the presence of excess/inherited Ar. The full automation of modern $^{40}Ar/^{39}Ar$ systems enables the analyses of several samples from each site or weathering profile, providing a comprehensive weathering database.

The use of K-Ar and $^{40}Ar/^{39}Ar$ dating of supergene K-bearing Mn-oxides followed a similar history. In the first reported application of Mn-oxides in K-Ar geochronology, Chukhrov et al (1966) recognized that the minerals analyzed were precipitated by weathering processes. After this pioneering investigation, several subsequent studies (Yashvili & Gukasyan 1974, Varentsov & Golovin 1987, Pracejus 1989, Segev et al 1991, Vasconcelos et al 1992, Lippolt & Hautmann 1995) confirmed that K-Ar dating of hollandite group Mn-oxides provided reliable age information. However, most of these studies focused primarily on hypogene processes. Although primarily interested in dating the formation of marine manganese deposits, Varentsov & Golovin (1987) and Pracejus (1989) recognized that K-Ar dating of Mn-oxides provided reliable ages of weathering reactions that affected these deposits when exposed on continental landmasses. Vasconcelos et al (1992, 1994a) confirmed that reliable weathering ages were obtained from K-Ar dating of supergene hollandite group Mn-oxides. In addition, they demonstrated the suitability of these minerals to $^{40}Ar/^{39}Ar$ geochronology. A comprehensive study of lateritic weathering profiles in the Amazon showed that the weathering ages obtained from K-Ar and $^{40}Ar/^{39}Ar$ analyses of supergene Mn-oxides provided a wealth of information about the climatic history of continental landscapes. Subsequently, several researchers (Grasselly et al 1994, Segev et al 1995, Dammer et al 1996, Ruffet et al 1996, Vasconcelos 1998) have used a combination of K-Ar and $^{40}Ar/^{39}Ar$ dating of supergene Mn-oxides to unravel the weathering history of continental environments.

As is often the case when new results are available for old problems, the knowledge gained from the study of weathering geochronology has created as much controversy as enlightenment. In the following sections, I will address some of the technical problems associated with K-Ar and $^{40}Ar/^{39}Ar$ dating of supergene minerals, I will discuss the suitability of several supergene sulfates and Mn-oxides to K-Ar and $^{40}Ar/^{39}Ar$ geochronology, and I will discuss some of the more important problems potentially resolvable through the methodology and will address some of the current divergences in the interpretation of weathering geochronology results.

WHERE TO FIND DATABLE PHASES?

Supergene alunite and jarosite are present in weathering profiles encompassing a large range of geological environments. Alunite and jarosite are common in the oxidation zones of ore deposits (Blanchard 1968, Anderson 1981, Bladh 1982, Scott 1987, Alpers & Brimhall 1988, Scott 1990), acid mine drainages (Nordstrom 1982, Alpers & Brimhall 1989), acid soils (van Breeman 1988), weathering coal deposits, paleosols (Myer & Reis 1985), silcretes (Myer & Reis 1985), salt lake sediments (Alpers et al 1992, Long et al 1992), cave deposits (Polyak et al 1998), and weathering river or marine terraces or deltaic deposits (Ivarson et al 1992).

Most studies on weathering geochronology have concentrated on investigations of weathered ore bodies, where access to the sampling sites and suitable minerals is generally not a problem. The common occurrence of supergene jarosite and alunite partially intergrown with primary or other supergene minerals requires selection, in the field, of samples where relatively pure alunite-jarosite precipitates in veins or cavities (Alpers & Brimhall 1988, Bird et al 1990, Vasconcelos et al 1994b, Sillitoe & McKee 1996). These samples are often relatively coarsely crystalline and easy to separate from possible contaminants. In opal deposits and some weathering profiles in Australia (Bird 1988, Bird et al 1990), alunite also occurs as lenticular concretions and nodules, easily separated from the hosting assemblage.

As suggested by Itaya et al (1996), "the most significant step in obtaining pure alunite separates is the collection in the field of the best possible specimens." However, selection criteria targeting better crystallized and pure supergene phases risk biasing the sample. By targeting a specific sample type in an area where many generations or paragenesis of supergene minerals occur, one may risk selectively analyzing one generation to the detriment of others. Only through a systematic approach and through the analyses of many samples from a single location can this issue be properly addressed. Unfortunately, no comprehensive studies of supergene jarosite or alunite from a single site (targeting samples where these minerals occur in different microenvironments and paragenesis) have been carried out to date.

Mn-oxides are also very common in weathering profiles. Manganese is the second most abundant transition element in the Earth's crust and is most stable in oxygenated surface waters as Mn^{4+}-oxides. Soils, lateritic profiles, lake and marine sediments, river terraces, rock varnish, cave deposits, some silcretes, and zones of groundwater upwelling and seepage contain one or more Mn-oxide species. Whether these minerals are appropriate for geochronology or occur in large enough concentrations to be successfully identified and separated depends on the nature of the parent material, the complexity of the weathering profile, and the accessibility to the entire profile.

Sampling Mn-oxides requires identification, in the field, of the most suitable phases for geochronology. In general, Mn-oxides form botryoidal growth bands, precipitating as concretions, nodules, fracture plane infills, or coating mineral surfaces or pseudomorphically replacing primary Mn-carbonates and silicates. Phases suitable for geochronology are indurated. Porous and friable Mn-oxides are not suitable for K-Ar or $^{40}Ar/^{39}Ar$ dating. These samples generally disaggregate during sample preparation. Manganese dendrites and other fine coatings are only suitable for geochronology if the substrate on which they occur is devoid of potassium. Relatively thick dendrites precipitated on a fracture plane in limestone were successfully dated by the $^{40}Ar/^{39}Ar$ method (PM Vasconcelos, unpublished results).

The ideal source of Mn-oxides for geochronological purposes is weathered manganese ore deposits. Large concentrations of Mn-oxides in these systems and the large Mn/Fe abundance in these environments ensure that each generation of Mn-oxide precipitated during the weathering history of the deposit is preserved. Since one of the objectives of weathering geochronology is to identify each and all generations of supergene minerals formed in a profile, these deposits are the most suitable to preservation of a complete history of continental weathering in a region.

WHAT IS THE METHOD DATING?

In weathering studies the ultimate goal of K/Ar and $^{40}Ar/^{39}Ar$ dating is to determine precisely and accurately the timing of chemical reactions. The dated chemical reactions depend on the environmental factors controlling the precipitation and preservation of the mineral phase to be analyzed.

Alunite-Jarosite Systems

The most important environmental parameters controlling the distribution of alunite-jarosite in nature are pH, Eh, (SO_4^{2-}), (K^+), (Fe^{3+}), (Al^{3+}) (where the symbol ( ) denotes activity of the aqueous species), H_2O fugacity, and temperature. In alunite-jarosite systems the chemical reactions dated by the K-Ar or the $^{40}Ar/^{39}Ar$ methods may be the simple precipitation of jarosite or alunite from aqueous solution (Equations 1 and 4) or more complex mineral transformations as illustrated in Equations 2, 3, 5, 6, and 7:

Direct precipitation of jarosite from solution:

$$K^+_{(aq)} + 3\,Fe^{3+}_{(aq)} + 2\,SO^{2-}_{4(aq)} + 6\,OH^-_{(aq)} \Leftrightarrow KFe_3(SO_4)_2(OH)_{6(Jrs)} \tag{1}$$

K^+ and SO_4^{2-} metasomatism of iron oxides:

$$3\,Fe(OH)_{3(s)} + K^+_{(aq)} + 2\,SO^{2-}_{4(aq)} + 3\,H^+_{(aq)} \Leftrightarrow KFe_3(SO_4)_2(OH)_{6(Jrs)} + 3\,OH^-_{(aq)} \tag{2}$$

Oxidation and K^+-metasomatism of pyrite:

$$12\,FeS_{2(s)} + 4\,K^+_{(aq)} + 30\,H_2O_{(l)} + 45\,O_{2(aq)} \Leftrightarrow 4\,KFe_3(SO_4)_2(OH)_{6(Jrs)} + 16\,SO^{2-}_{4(aq)} + 36\,H^+_{(aq)} \tag{3}$$

Direct precipitation of alunite from solution:

$$K^+_{(aq)} + 3\,Al^{3+}_{(aq)} + 2\,SO^{2-}_{4(aq)} + 6\,OH^-_{(aq)} \Leftrightarrow KAl_3(SO_4)_2(OH)_{6(Aln)} \tag{4}$$

K^+ and SO_4^{2-} metasomatism of kaolinite:

$$3\,Al_2Si_2O_5(OH)_{4(Kln)} + 2\,K^+_{(aq)} + 4\,SO^{2-}_{4(aq)} + 9\,H_2O_{(l)} + 6\,H^+_{(aq)} \Leftrightarrow 2\,KAl_3(SO_4)_2(OH)_{6(Aln)} + 6\,H_4SiO_{4(aq)} \tag{5}$$

SO_4^{2-} metasomatism of K-feldspar:

$$3\,KAlSi_3O_{8(Kfs)} + 2\,SO_{4(aq)}^{2-} + 18\,H_2O_{(l)} + 6\,H_{(aq)}^{+}$$
$$\Leftrightarrow KAl_3(SO_4)_2(OH)_{6(Aln)} + 9\,H_4SiO_{4(aq)} + 2\,K_{(aq)}^{+} \qquad (6)$$

SO_4^{2-} metasomatism of muscovite:

$$KAl_3Si_3O_{10}(OH)_{2(Ms)} + 2\,SO_{4(aq)}^{2-} + 6\,H_2O_{(l)} + 4\,H_{(aq)}^{+}$$
$$\Leftrightarrow KAl_3(SO_4)_2(OH)_{6(Aln)} + 3\,H_4SiO_{4(aq)} \qquad (7)$$

The interpretation of geochronological results in weathering studies requires the identification of the dated reaction. For this purpose the dated mineral sample must be characterized petrographically, and the reaction must be identified from paragenetic relationships.

Mn-Oxide Systems

During weathering processes, manganese is mobile as Mn^{2+} (Hem 1963). The cation Mn^{2+} predominates for most of the range of conditions characteristic of natural-water systems (Hem 1963). At pH > 10.5, the complex $MnOH^+$ becomes the predominant form, and in solutions with high concentrations of HCO_3^- or $SO_4^=$ the complexes $MnHCO_3^+$ and $MnSO_{4(aq)}$ may be important (Hem 1989). Organic acids also control manganese solubility in the surficial environment (Crerar et al 1972; Stone & Morgan 1984a, 1984b; Stone 1987). Mn^{3+} species may occur in strongly acid or organic-rich solutions. However, the tendency for the Mn^{3+} species to disproportionate indicates that it does not play a major role in the solution chemistry of manganese under surface conditions.

The chemical oxidation of Mn^{2+} and Mn^{3+} and the disproportionation of Mn^{3+} in aerated surface water lead to the precipitation of Mn^{4+} oxides. These oxides further catalyze the oxidation process. Thus, Mn-oxide precipitation tends to occur on previously deposited oxides, forming accretionary growth bands. Supergene Mn-oxides also catalyze the precipitation of other cations from solution (Ba, Ni, Co, Cu, K, Pb, etc.), leading to the formation of complex supergene phases (Burns & Burns 1979). Hollandite, coronadite, cryptomelane, birnessite, romanéchite, todorokite, and vernadite are some of the complex Mn-oxides commonly found in soils and weathering profiles (Burns & Burns 1979).

Similar to jarosite-alunite systems, geochronology of supergene Mn-oxides may determine the direct precipitation of these oxides from aqueous solutions (Equation 8), or it may determine the timing of more complex metasomatic reactions in the weathering environment (Equation 9, 10, and 11):

Direct precipitation of cryptomelane from solution:

$$K^+_{(aq)} + 8\,Mn^{2+}_{(aq)} + 8\,H_2O_{(l)} + 4\,O_{2(aq)} + OH^-_{(aq)}$$
$$\Leftrightarrow KMn_8O_{16}{\cdot}(OH)_{(cml)} + 16\,H^+_{(aq)} \qquad (8)$$

K^+ metasomatism of pyrolusite:

$$8\,MnO_{2(pyr)} + K^+_{(aq)} + OH^-_{(aq)} \Leftrightarrow KMn_8O_{16}{\cdot}(OH)_{(cpt)} \qquad (9)$$

Oxidative dissolution and K^+ metasomatism of rhodochrosite:

$$8\,MnCO_{3(Rds)} + K^+_{(aq)} + 4\,O_{2(aq)} + 8\,H_2O_{(l)} + OH^-_{(aq)}$$
$$\Leftrightarrow KMn_8O_{16}{\cdot}(OH)_{(cml)} + 8\,H_2CO_{3(aq)}. \qquad (10)$$

Oxidative dissolution and K^+ metasomatism of Mn-silicates (tephroite, rhodonite, etc.):

$$8\,MnSiO_{3(Rdn)} + K^+_{(aq)} + 4\,O_{2(aq)} + 16\,H_2O_{(l)} + OH^-_{(aq)}$$
$$\Leftrightarrow KMn_8O_{16}{\cdot}(OH)_{(cml)} + 8\,H_4SiO_{4(aq)} \qquad (11)$$

When multiple precipitation-dissolution-reprecipitation reactions occur, geochronology of supergene Mn-oxides may date the influx of solutions, promoting the dissolution of previously precipitated Mn-oxides, the transport of aqueous Mn^{2+} elsewhere in the system, and the reprecipitation of new generations of Mn-oxides in suitable sites. Because the most significant aqueous manganese species in the surficial environment is Mn^{2+}, the partial dissolution of previously precipitated Mn-oxides implies the influx of acid and/or reducing solutions:

Reductive dissolution by acidification:

$$KMn_8O_{16}{\cdot}(OH)_{(cml)} + 17\,H^+_{(aq)} \Leftrightarrow K^+_{(aq)} + 8\,Mn^{2+}_{(aq)} + 9\,H_2O_{(l)} + 4\,O_{2(g)} \qquad (12)$$

Reductive dissolution by organic ligands:

$$KMn_8O_{16}{\cdot}(OH)_{(cml)} + 8\,CHO^-_{2(aq)} + 25\,H^+_{(aq)}$$
$$\Leftrightarrow K^+_{(aq)} + 8\,Mn^{2+}_{(aq)} + 8\,CO_{2(g)} + 17\,H_2O_{(1)} \qquad (13)$$

Reductive dissolution by soluble Fe^{2+}:

$$KMn_8O_{16}{\cdot}(OH)_{(cml)} + 16\,Fe^{2+}_{(aq)} + 16\,H_2O_{(l)}$$
$$\Leftrightarrow 16\,FeOOOH_{(s)} + K^+_{(aq)} + 8\,Mn^{2+}_{(aq)} + 15\,H^+_{(aq)} \qquad (14)$$

The aqueous Mn^{2+} species generated by reactions 12 through 14 must be reprecipitated nearby within the weathering profile if these reactions are to be identified and dated. The kinetics of the oxidation of Mn^{2+} aqueous species and the precipitation of Mn^{4+}-oxides is generally slow in neutral to slightly acidic oxygenated weathering solutions (Crerar et al 1972). In alkaline conditions ($pH > 8$) these reactions proceed more rapidly (Hem 1989). The reactions are also catalyzed by bacteria (Mandernack et al 1995a,1995b) and by mineral surfaces in the natural environment (Davies & Morgan 1989). The identification of the exact reaction controlling the precipitation of datable Mn-oxides is desirable if the ages obtained are to be used to interpret paleochemical and paleoclimatic conditions.

SAMPLE PREPARATION

One of the major concerns in the geochronology of weathering profiles is the difficulty in obtaining pure fractions of the supergene phase of interest completely devoid of hypogene contaminants. This problem was recognized by Chukhrov et al (1966), who applied a two-stage analytical process to identify and subtract the contribution of hypogene contaminants to the date obtained for the weathered assemblage. This approach permits a true calculated weathering age to be obtained. Variations of this same approach have been applied more recently (Bird et al 1990, Grassely et al 1994, Dammer et al 1996) to subtract the effects of hypogene contamination on K-Ar dates obtained for supergene alunite and Mn-oxides.

The bulk samples needed in K-Ar analysis limit the number of weathering profiles and minerals suitable for geochronology. Large samples also enhance the probability of contamination. The fine-scale resolution possible with the $^{40}Ar/^{39}Ar$ method permits the selection of visually pure small (100 μm to 2 mm grains, depending on age and K content) samples. The purity of these samples can be further tested through scanning electron microscopy before isotope analyses. A further advantage of the $^{40}Ar/^{39}Ar$ method is that any contaminant not identified during sample preparation will be detected during incremental heating study of the sample. The K-Ar method, however, cannot detect the presence of contaminants.

Gentle crushing, combined with sieving and heavy liquid separation, is often enough to provide a pure alunite or jarosite separate. Itaya et al (1996) have also used chemical partial digestion to rid slightly contaminated samples from clays, Fe-oxides and hydroxides, native sulphur, silicates, and sulfides. Treatment with HF to rid samples of hypogene silicate contaminants will partially dissolve alunite and jarosite, affecting the accuracy of the isotopic analyses; it should be avoided.

Most Mn-oxides present in weathering profiles contain rhythmic growth bands, which, if dated separately, will provide information on the rate of Mn-oxide precipitation. In addition, selective analysis of accretionary growth bands provides an independent geological constraint on the veracity of the results (Segev et al 1991, Vasconcelos 1992). If a stratigraphically younger sample yields an older date, the suitability of the mineral to K-Ar or $^{40}Ar/^{39}Ar$ dating becomes questionable. To maintain proper microstratigraphic control on the analyzed samples, microsampling (with microdrill or diamond saw) is necessary. For samples where growth bands are not visible the most appropriate procedure consists of crushing, cleaning in distilled water in ultrasonic bath, and hand picking pure clusters in the binocular microscope. Acid treatment should be avoided because it will partially dissolve manganese oxide minerals.

K-Ar AND $^{40}Ar/^{39}Ar$ METHODOLOGY

The K-Ar and the $^{40}Ar/^{39}Ar$ methods are well-established geochronological techniques. Faure (1986) provides succinct descriptions of these methods. Extensive discussion on technical aspects of the K-Ar method is available from Dalrymple & Lamphere (1969); a detailed description of the $^{40}Ar/^{39}Ar$ technique is available from McDougall & Harrison (1988). In the study of hypogene processes (igneous, metamorphic, or hydrothermal), K-Ar and $^{40}Ar/^{39}Ar$ dating is generally used to determine the timing of cooling of a rock or mineral. When the system's temperature falls below a threshold value (the closure temperature), Ar is quantitatively retained. In contrast, K-Ar and $^{40}Ar/^{39}Ar$ dating of weathering processes determines the age of precipitation of a K-rich mineral by low-temperature chemical reactions. In weathering studies, closure temperature is not a relevant concept because the minerals under investigation were precipitated and remained at surface conditions. However, the activation energy of the K- and Ar-bearing site, which ultimately controls the ease of exchange of K and Ar with the external environment, plays a very important role in determining the suitability of a supergene phase to geochronology.

The basic requirements for the application of the K-Ar method in weathering geochronology are:

1. The mineral has remained closed to gains or losses of radiogenic $^{40}Ar^*$ soon after its formation.

2. The mineral has remained closed to gains or losses of K soon after its formation.

3. No excess Ar was incorporated into the mineral.

4. Correction can be made for atmospheric ^{40}Ar adsorbed onto or included in the mineral structure at the time of precipitation.

5. The isotopic composition of K in the mineral is known and has changed only due to radioactive decay of ^{40}Ar.

6. The decay constants of K are known and have been constant through time.

The application of the $^{40}Ar/^{39}Ar$ technique to weathering studies is also subject to requirements 1, 2, 4, 5, and 6. Although the incorporation of excess Ar may pose problems in the $^{40}Ar/^{39}Ar$ technique, the application of isotope correlation diagrams enables the identification of and correction for the excess Ar component. An additional requirement in the application of the $^{40}Ar/^{39}Ar$ technique is that the mineral be stable and not subject to the losses of $^{40}Ar^*$ by heating, or ^{39}Ar by recoil, during neutron irradiation (Harrison 1983, McDougall & Harrison 1988).

During neutron irradiation the ^{39}Ar generated from ^{39}K may recoil from its original site. Estimated recoil distances in minerals range from 0 to 0.5 μm (McDougall & Harrison 1988), depending on the density of the mineral structure. The effect of recoil on the age spectra of large crystals ($>$100 μm) is volumetrically insignificant, but it becomes a problem in dating fine-grained minerals (Huneke & Smith 1976). If crystal domains are $<$5 μm, as is often the case in supergene Mn-oxides, the volume subject to ^{39}Ar recoil may be significant (Huneke & Smith 1976), and it may potentially prevent accurate dating of certain types of finely crystalline supergene phases by the $^{40}Ar/^{39}Ar$ method. Vasconcelos et al (1994a, 1995) showed that some Mn-oxides were subject to recoil, but that recoil losses were relatively small in the analyzed phases and did not prevent successful dating of Mn-oxides. Ruffet et al (1996) identified the same amount of recoil as previously identified by Vasconcelos et al (1994a, 1995) and also concluded that recoil did not prevent successful dating of Mn-oxides.

A brief comparison between the K-Ar and $^{40}Ar/^{39}Ar$ dating techniques, as applied to weathering studies, is summarized in Table 1.

In the K-Ar method a date (time, *t*) is given by:

$$t = (1/\lambda_c)\ \ln\left(1 + (\lambda/\lambda_c)(^{40}\mathrm{Ar}/^{40}\mathrm{K}\right), \tag{15}$$

where λ is the total decay constant for ^{40}K (5.543×10^{-10} year^{-1}); λ_c is the decay constant of ^{40}K to ^{40}Ar (5.81×10^{-9} year^{-1}); $^{40}Ar^*$ is the number of moles of radiogenic ^{40}Ar calculated from the total amount of ^{40}Ar measured by mass spectrometric techniques and corrected for atmospheric ^{40}Ar; ^{40}K is the number of moles of ^{40}K calculated from the total number of moles of K measured by

Table 1 Comparison between the K-Ar and $^{40}Ar/^{39}Ar$ methods

K-Ar	Parameter	$^{40}Ar/^{39}Ar$
>~1 Ma	Age range	>250,000 ka
0.1–1.0 g	Sample size	0.5 μm to 1.0 g
AAS or flame photometry, on separate aliquot	K determination by	Mass spectrometric analyses of ^{39}Ar generated by neutron irradiation of ^{39}K
Isotope dilution mass spectrometry	Ar determination by	Mass spectrometry
~1–10 × 10^{-11} mol	Ar detection limit	~1 × 10^{-16} mol
0.5–5%	Analytical precision	0.05–1%
No	Able to identify excess Ar	Yes
No	Able to identify Ar loss	Yes
No	Able to identify mixed phases	Yes
No	Able to determine thermal history	Yes
0.5–1 day	Duration of one analyses	15–30 min
Partial, data reduction only	Automation	Full automation
No	Crystallographic information	Yes

atomic absorption spectroscopy or flame photometry. The numerical value obtained for t is a date. A date is simply a numerical result from the application of a geochronological technique. No quality filter is applied to that result, and the date may represent an accurate geological time (an age) or may be an artifact resulting from the study of an unsuitable sample, from contamination, from analytical error, etc. In the definition given above, a date may or may not have age significance. An age, however, is a numerical result from the application of a geochronological technique, which, given the uncertainties about decay constants for radioactive elements and the analytical uncertainties inherent to the method used, is as good a result as can be obtained. In addition, the analyzed sample is representative of its geological environment, it is suitable to geochronology by the method applied, and it is devoid of contaminants. An age, therefore, represents the best estimate of when a geological phenomenon took place.

In the $^{40}Ar/^{39}Ar$ technique, time is obtained from:

$$t = (1/\lambda)\ln[(^{40}Ar^*/^{39}Ar)\cdot J + 1], \tag{16}$$

where λ is the total decay constant for ^{40}K (5.543×10^{-10} year^{-1}); $^{40}Ar^*$ is the number of moles of radiogenic ^{40}Ar calculated from the total amount of ^{40}Ar measured by mass spectrometric techniques and corrected for atmospheric ^{40}Ar and nucleogenic interferences; ^{39}Ar is the total amount of ^{39}Ar generated by the reaction $^{39}K(n,p)^{39}Ar$, also measured by mass spectrometry and corrected for nucleogenic interferences; and J is a dimensionless irradiation parameter empirically measured through the analyses of neutron fluency monitors of known ages. In this approach the atmospheric $^{40}Ar/^{36}Ar$ value is assumed to be 295.5 (McDougall & Harrison 1988).

Two distinct analytical procedures are possible with the $^{40}Ar/^{39}Ar$ technique: total fusion or incremental heating (step heating) analyses. In the absence of recoil problems a total fusion $^{40}Ar/^{39}Ar$ result is equivalent to a K-Ar date. Step-heating analysis, however, yields several dates (apparent ages) for a single sample, each date corresponding to one analyzed step. Step-heating results are generally illustrated in two different ways: through step-heating spectra or by $^{39}Ar/^{40}Ar$ versus $^{36}Ar/^{40}Ar$ isotope correlation diagrams. Three age estimates are obtained from the $^{40}Ar/^{39}Ar$ technique: the plateau age, the integrated age, and the isochron age. In addition, several total fusion analyses for grains from a single sample or the results of step-heating analyses for one grain can be combined in ideogram plots, which yield "most probable" age estimates for these results (Deino and Potts 1990). Plateau, integrated, isochron, and ideogram ages are simply numerical results from the application of the $^{40}Ar/^{39}Ar$ method. These results may or may not represent a true geological age.

Step-Heating Spectra, Plateau Ages, and Integrated Ages

K-Ar and total fusion $^{40}Ar/^{36}Ar$ analyses provide information only on the total K and the total ^{40}Ar contained in a sample. Several assumptions (absence of significant inherited Ar component, absence of Ar or K losses by exchange or recoil, absence of contaminants) are implicit if t is interpreted as an age.

Step-heating analyses of a single grain or crystal are the most powerful application of the $^{40}Ar/^{39}Ar$ method. The results of step-heating analyses are generally plotted as the cumulative percent of ^{39}Ar released (x axis) against apparent age (y axis) for each step (Figure 3*a*). Degassing temperature increases toward the right-hand side of Figure 3*a*. Apparent ages are calculated for each step using Equation 16. An integrated age is the apparent age calculated from the total gas yielded by the sample. In the absence of ^{39}Ar recoil, the integrated age should be the same as a K-Ar date for the sample. If recoil, $^{40}Ar^{*}$ loss, or contamination are absent, the integrated age should correspond to the plateau age

→

Figure 3 (*a*) Step-heating spectra for four distinct cryptomelane grains, illustrating the reproducibility of the results obtained for each of the grains analyzed. Reproducibility of results in the analyses of supergene Mn-oxides is essential to test the suitability of different mineral structures to geochronology. (*b*) Isotope correlation diagram obtained for the same four grains illustrated in Figure 3*a* yielding a well-defined isochron with atmospheric intercept and a slope that corresponds to the ages calculated by the plateaus in the step-heating spectra. (*c*) Ideogram illustrating the well-defined peak for the highly radiogenic steps (age = 8.6 Ma).

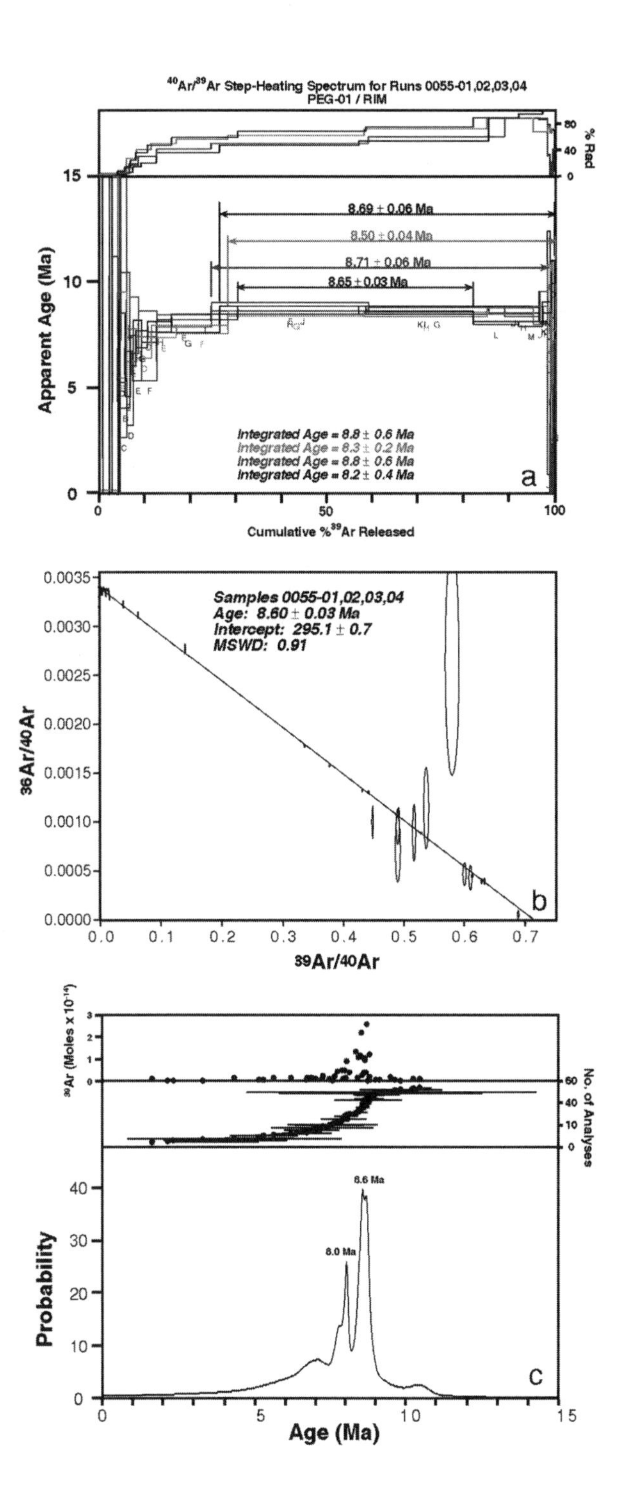

$^{40}Ar/^{39}Ar$ Step-Heating Spectrum for Runs 0055-01,02,03,04
PEG-01 / RIM
8.69 ± 0.06 Ma
8.50 ± 0.04 Ma
8.71 ± 0.06 Ma
8.65 ± 0.03 Ma
Integrated Age = 8.8 ± 0.6 Ma
Integrated Age = 8.3 ± 0.2 Ma
Integrated Age = 8.8 ± 0.6 Ma
Integrated Age = 8.2 ± 0.4 Ma
Apparent Age (Ma)
% Rad
Cumulative $\%^{39}Ar$ Released
a
Samples 0055-01,02,03,04
Age: 8.60 ± 0.03 Ma
Intercept: 295.1 ± 0.7
MSWD: 0.91
$^{36}Ar/^{40}Ar$
$^{39}Ar/^{40}Ar$
b
^{39}Ar (Moles x 10^{-14})
No. of Analyses
8.6 Ma
8.0 Ma
Probability
Age (Ma)
c

for the sample. Step-heating spectra provide information on the history of the Ar (and K) retentivity for the sample, the physical properties of the sites where these elements occur, the presence or absence of contaminants, and possible ^{39}Ar recoil losses. Figure 4 illustrates several end-member step-heating spectra obtained by the analysis of cryptomelane samples.

Figure 4*a* represents the ideal spectrum in which recoil is absent, the sample is closed to Ar and K, the $^{40}Ar^*$ is hosted in stable tunnel sites that release their occupants only during mineral breakdown at high temperature, and the sample does not host any contaminants. Steps A through H yield negligible amounts of ^{39}Ar and large amounts of atmospheric ^{40}Ar (adsorbed onto the grain surface or trapped in intercrystalline sites) (Figure 5). The large ratio between atmospheric and radiogenic Ar accounts for the large errors in the initial measurements. Steps I through M represent mixtures of atmospheric ^{40}Ar released from intercrystalline sites and $^{40}Ar^*$ released from tunnel sites. The amount of released radiogenic gas progressively increases with temperature, and the apparent age can be more precisely determined. At steps N and O, the tunnel sites have released most of their $^{40}Ar^*$ and ^{39}Ar contents; any remaining released $^{40}Ar^*$ and ^{39}Ar are derived from small amounts of gas still trapped in the collapsed Mn-oxide structure. At this stage the sample is undergoing phase transformations from Mn_3O_4 to MnO (Vasconcelos et al 1994c), releasing large volumes of oxygen into the vacuum system and thus increasing the measurement uncertainties.

Figure 4*b* also represents a well-defined spectrum. Recoil is still absent, most of the $^{40}Ar^*$ is hosted in stable tunnel sites that release their occupants only during mineral breakdown at high temperature, and the sample does not host any contaminants. However, this spectrum shows evidence for the presence of small $^{40}Ar^*$ losses from less retentive, poorly crystalline sites. Steps A, B, and C yield negligible amounts of ^{39}Ar and large amounts of atmospheric ^{40}Ar hosted in the intercrystalline sites. The amount of atmospheric gases released decreases with temperature, whereas the $^{40}Ar^*$ progressively increases (steps C through G). The apparent ages obtained represent mixtures of atmospheric and radiogenic gases. However, the radiogenic gases in these steps are derived from poorly crystallized sites that have undergone some $^{40}Ar^*$ loss since the time of mineral precipitation. The apparent ages obtained are younger than the

→

Figure 4 The most common end-member spectra obtained from the analyses of supergene Mn-oxides. These end-member spectra provide information on the Ar and K retentivity of the phase analyzed and on the presence or absence of recoil, contaminants, or mixed phases. A detailed description of each spectrum is presented in the text. Incremental heating techniques are the only analytical procedures that will provide a quality filter for the dates obtained.

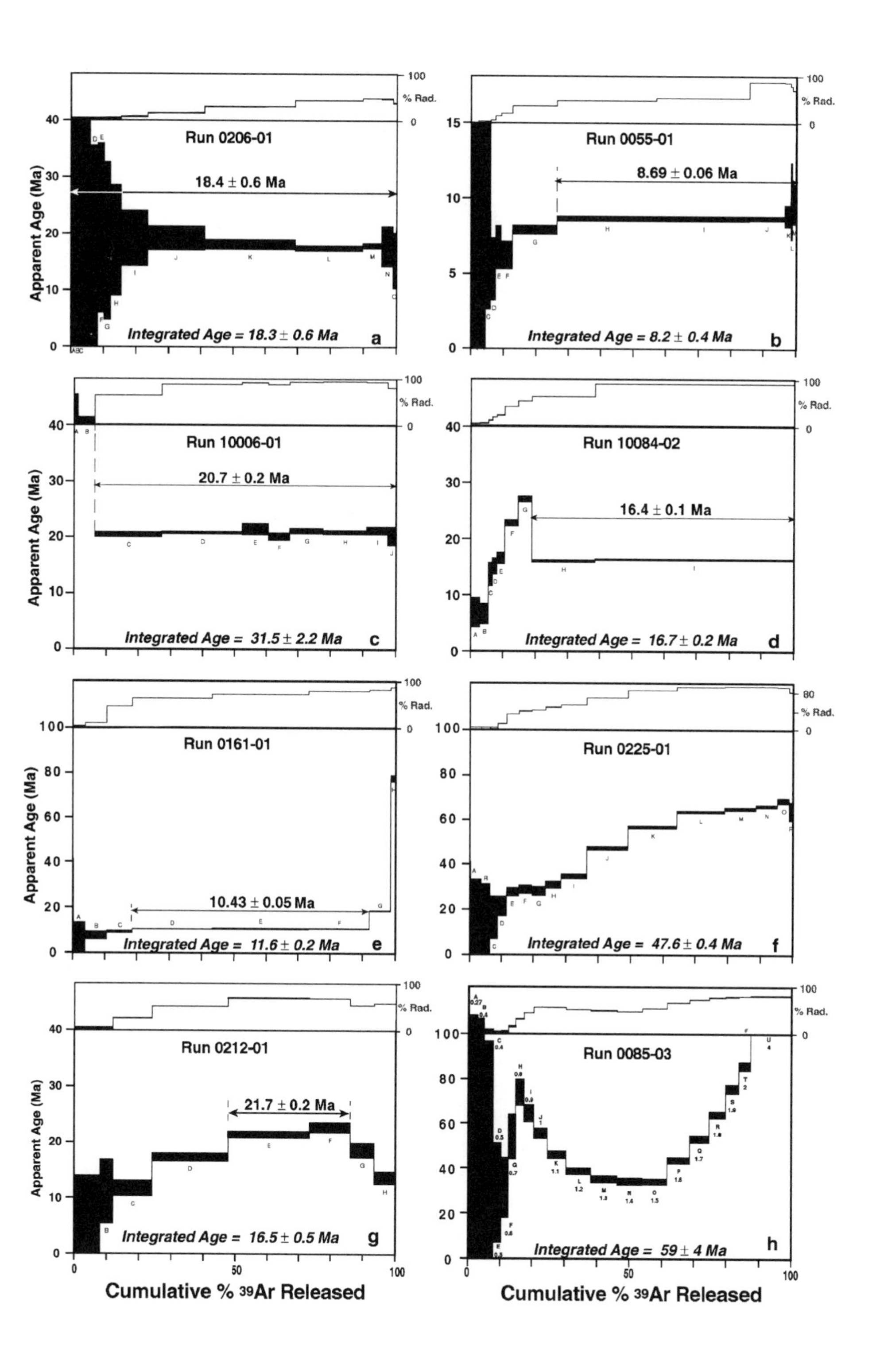
Run 0206-01
18.4 ± 0.6 Ma
Integrated Age = 18.3 ± 0.6 Ma
a
Run 0055-01
8.69 ± 0.06 Ma
Integrated Age = 8.2 ± 0.4 Ma
b
Run 10006-01
20.7 ± 0.2 Ma
Integrated Age = 31.5 ± 2.2 Ma
c
Run 10084-02
16.4 ± 0.1 Ma
Integrated Age = 16.7 ± 0.2 Ma
d
Run 0161-01
10.43 ± 0.05 Ma
Integrated Age = 11.6 ± 0.2 Ma
e
Run 0225-01
Integrated Age = 47.6 ± 0.4 Ma
f
Run 0212-01
21.7 ± 0.2 Ma
Integrated Age = 16.5 ± 0.5 Ma
g
Run 0085-03
Integrated Age = 59 ± 4 Ma
h
Apparent Age (Ma)
% Rad.
Cumulative % 39Ar Released

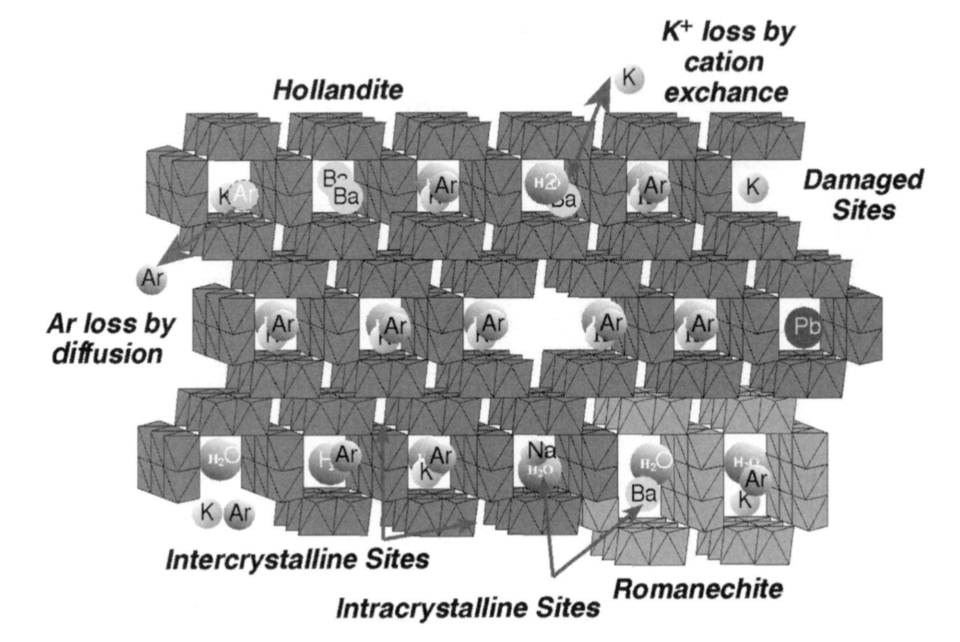

Figure 5 Depiction of the complexity of natural supergene manganese oxides, which are often composed of intergrown phases (hollandite and romanechite structures, in this case). These natural manganese oxides contain large surface areas and intercrystalline sites that may host significant volumes of atmospheric Ar. This ultimately imposes a lower limit on the resolution of the dating technique. Currently, this lower age limit is approximately 200,000 ka.

true precipitation age for the mineral, as shown below:

$$t = (1/\lambda)\ \ln\left[\left(\left({}^{40}\mathrm{Ar}^{*} - {}^{40}\mathrm{Ar}^{*}_{\mathrm{LOST}}\right)/{}^{39}\mathrm{Ar}\right)\cdot J + 1\right]. \tag{17}$$

Steps H, I, and J represent the temperature range at which the well-crystallized, retentive tunnel sites break down and release $^{40}Ar^*$. The plateau age represented by these steps is the best estimate of the true precipitation age for the sample. Steps K, L, and M represent small amounts of gas still trapped in the collapsed Mn_3O_4 or MnO structure at high temperature. The small amount of radiogenic gas and the release of oxygen during phase transformations occurring at this temperature again account for the large uncertainties. As shown in this spectrum, in cases in which a small fraction of the mineral sites have lost $^{40}Ar^*$, the integrated age will be slightly younger than the plateau age. If the open sites are abundant and the losses are volumetrically significant, the integrated ages will be significantly younger than the plateau ages.

If the amount of ^{39}Ar lost by recoil is extreme, a reliable age cannot be obtained by the $^{40}Ar/^{39}Ar$ method unless special irradiation and analytical procedures are followed (Smith et al 1993). However, if only a small fraction of the ^{39}Ar is lost by recoil, $^{40}Ar/^{39}Ar$ step-heating analyses will permit identification and quantification of ^{39}Ar recoil losses. When ^{39}Ar recoil occurs, there are three possible fates for the recoiled ^{39}Ar: it may be completely lost to the surrounding atmosphere; it may be loosely lodged in poorly retentive intercrystalline or damaged sites; or it may be reimplanted into other sites in the mineral structure. When a significant amount of ^{39}Ar recoil occurs, it may be clearly identified by the discrepancies between K-Ar and $^{40}Ar/^{39}Ar$ dates for aliquots of the same sample or by discrepancies in plateau and integrated ages for samples analyzed by the $^{40}Ar/^{39}Ar$ step-heating method.

Figure 4*c* illustrates the spectrum obtained for a sample that has lost the recoiled ^{39}Ar to the surrounding atmosphere during irradiation or bake-out procedure. The initial steps (A and B) contain mostly atmospheric ^{40}Ar. Small amounts of radiogenic $^{40}Ar^*$ released during these steps are derived from sites that have lost ^{39}Ar by recoil. The apparent age obtained is significantly older than the true sample age because the ^{40}K content is underestimated:

$$t = (1/\lambda)\ \ln\left[\left({}^{40}\mathrm{Ar}^*/\left({}^{39}\mathrm{Ar} - {}^{39}\mathrm{Ar}_{\mathrm{RECOILED}}\right)\right) \cdot J + 1\right]. \tag{18}$$

If only a small fraction of the sample has lost ^{39}Ar by recoil, progressively higher temperatures will release $^{40}Ar^*$ and ^{39}Ar from retentive tunnel sites; an age plateau is obtained. This plateau represents the best estimate of the true sample precipitation age. In the example above, the integrated age is older than the plateau age.

The spectrum in Figure 4*c* could be alternatively explained by the presence of an excess ^{40}Ar component adsorbed to the mineral surface or trapped in intercrystalline sites. Although possible, this scenario is less likely due to the large amount of excess Ar necessary to be detected in reservoirs overwhelmed by atmospheric ^{40}Ar. In addition, any excess Ar component should remain or fractionate into the aqueous solution during mineral precipitation because of the large Bunsen coefficient for Ar (Murray & Riley 1970).

A second type of spectrum can be obtained from samples subject to ^{39}Ar recoil. Figure 4*d* illustrates the spectrum for a sample in which the recoiled ^{39}Ar has been reimplanted or trapped within less retentive sites in the mineral. The initial steps (A through E), although mostly atmospheric Ar, contain radiogenic $^{40}Ar^*$ and $^{39}Ar_{RECOILED}$ easily released from their implantation or trapping sites. An apparent age younger than the true age for the sample is obtained because of the extra amount of ^{39}Ar released at these steps:

$$t = (1/\lambda)\ \ln\left[\left({}^{40}\mathrm{Ar}^*/\left({}^{39}\mathrm{Ar} + {}^{39}\mathrm{Ar}_{\mathrm{RECOILED}}\right)\right) \cdot J + 1\right]. \tag{19}$$

At progressively higher temperatures (F, G), $^{40}Ar^*$ is released from sites that lost ^{39}Ar by recoil, and apparent ages older than the true age for the sample will be obtained, as shown in Equation 18. If, at higher temperatures (H, I), retentive sites not subject to recoil losses and hosting significant amounts of $^{40}Ar^*$ and ^{39}Ar are sampled, a plateau age representing the true precipitation age for the mineral is obtained. In this case, the anomalously young and anomalously old steps may cancel each other, and the integrated age may be close to the value obtained for the plateau age (Figure 4*d*).

The spectrum in Figure 4*e* illustrates the effect of a hypogene contaminant that occurs intergrown with supergene cryptomelane. This spectrum shows similar evolution to the spectrum in Figure 4*b*. However, as higher temperatures are reached and the hypogene contaminant yields $^{40}Ar^*$, the ages obtained are much older than the age of the supergene phase, and a climbing spectrum is obtained. In the ideal scenario illustrated in Figure 4*e*, the supergene and the hypogene phases yield their Ar contents at discrete temperatures, and the age of the supergene mineral is retrievable. The hypogene contaminant (Archean magnetite) in this case hosts a very small amount of $^{40}Ar^*$. However, if old high-K (i.e. Archean or Proterozoic micas) hypogene contaminants occur intergrown with the supergene phase and if this hypogene contaminant yields $^{40}Ar^*$ throughout the heating schedule, a reliable precipitation age for the supergene phase cannot be obtained. When hypogene contaminants are present, the integrated ages obtained for the sample are older than the true precipitation ages.

The spectrum illustrated in Figure 4*f* represents a sample that has lost Ar and/or K throughout its history. The initial steps (A and B) also represent the atmospheric component. All the subsequent steps represent mixtures of gases derived from sites that have lost Ar, K, or both after mineral precipitation. Sometimes such spectra will reach a plateau; in other cases, a plateau is never reached. This type of spectrum provides no information about the age of the sample. However, if this sample were analyzed by the K-Ar method or the total fusion $^{40}Ar/^{36}Ar$ method, it would misleadingly yield a relatively well constrained date equivalent to the integrated age illustrated in Figure 4*f*.

Figure 4*g* illustrates a spectrum displaying ascending apparent ages at low temperatures (similar to Figure 4*b* or 4*f*); the spectrum reaches a plateau (steps E and F), and the apparent ages decline at high temperatures. This spectrum may represent a mixture of two or more distinct minerals, which are intimately intergrown. The younger phase is better crystallized and more retentive of $^{40}Ar^*$ than the older phase. An alternative explanation is that this spectrum may represent the implantation of recoiled ^{39}Ar into crystallographic sites that are more retentive than the sites from where ^{39}Ar was derived.

The spectra illustrated in Figures 4*a* through 4*g* are simply end-members. Complex spectra, displaying various combinations of the spectra in Figures 4*a* through 4*g*, are possible. For example, the complex spectrum illustrated in Figure 4*h* may be the combined result of open system behavior (Figure 4*f*), recoil (Figure 4*d*), and contamination (Figure 4*e*). This type of spectrum does not yield reliable age information. However, K-Ar and total fusion $^{40}Ar/^{39}Ar$ analysis of this sample would also yield a well-constrained date, which could be wrongly interpreted as the mineral precipitation age.

Isochron ($^{39}Ar/^{40}Ar$ Versus $^{36}Ar/^{40}Ar$ Correlation Diagrams) and Isochron Ages

In age calculations using Equation 16, the $^{40}Ar/^{36}Ar$ ratio is assumed to be 295.5. This assumption may be incorrect when excess ^{40}Ar or ^{36}Ar is incorporated into the mineral during the time of precipitation. An alternative way of determining an age for a sample by the $^{40}Ar/^{39}Ar$ method is to plot $^{39}Ar/^{40}Ar$ versus $^{36}Ar/^{40}Ar$ ratios (Faure 1986, McDougall & Harrison 1988). When a sample hosts argon in different crystallographic sites, with varying ratios of atmospheric, nucleogenic, and radiogenic components, the results plotted as a $^{39}Ar/^{40}Ar$ versus $^{36}Ar/^{40}Ar$ isotope correlation diagram will yield straight lines given by the relationship:

$$(^{40}\mathrm{Ar}/^{39}\mathrm{Ar})_m = (^{40}\mathrm{Ar}/^{36}\mathrm{Ar})_c + (^{40}\mathrm{Ar}^*/^{39}\mathrm{Ar})_k \cdot (^{39}\mathrm{Ar}/^{36}\mathrm{Ar})_m, \qquad (20)$$

where m are measured ratios, c are ratios of Ar gas contamination (excess ^{40}Ar or ^{36}Ar), and k identifies the argon isotopes produced by potassium in the samples (Faure 1986). The slope of these lines is the ratio $(^{40}Ar^*/^{39}Ar)_k$, which can be used in Equation 16 to calculate the age for the sample. In this approach, contaminant Ar components can be identified by the discrepancy between the $^{40}Ar/^{36}Ar$ intercept and the atmospheric value 295.5 (Figure 3*b*). When excess ^{40}Ar is present, the initial ratio obtained is significantly larger than 295.5. If excess ^{36}Ar is present, this ratio is less than 295.5.

Under ideal circumstances, plateau ages and isochron ages should yield exactly the same results. Figure 3*a* illustrates four step-heating spectra for the rim of a botryoidal Mn-oxide sample analyzed at the University of Queensland Argon Geochronology in Earth Sciences Laboratory (UQ-AGES). Previous analysis of this same sample at the Berkeley Geochronology Center (BGC) yielded a plateau age of 8.6 ± 0.1 Ma and an isochron age of 8.6 ± 0.2 Ma (2σ) (Vasconcelos 1992, Vasconcelos et al 1994a, 1995). The current results (Figure 3*a*) show remarkable reproducibility both between the four grains analyzed and between the analysis of these four grains at UQ-AGES and previous analyses at the BGC. These results illustrate the ideal case of Mn-oxide samples that yield consistent, reproducible plateau (Figure 3*a*) and isochron ages (Figure 3*b*) and are free of contamination by excess ^{40}Ar or ^{36}Ar.

Ideograms

Ideograms are age-probability diagrams (Figure 3*c*). They are based on the assumption that errors in a date have a Gaussian distribution. To construct an age probability plot, the sum of the values for the individual Gaussian curves for each date is plotted (Deino & Potts 1990). One of the main disadvantages of using ideograms to illustrate the probability of age distributions is the fact that ideograms do not take into consideration the amount of gas released during each step. As a result, an age-step or a sequence of age-steps, which yield(s) insignificantly small amounts of gas but whose age(s) is(are) analytically very precise, may misleadingly plot as a very high-probability age estimate for the sample. A useful criterion for differentiating between true high-probability peaks and artifacts of the ideogram method is to compare the ideogram peaks with plateau ages (Figure 3*c*). If the most probable peaks correspond to well-defined plateau ages, the ideogram method reliably identifies the most probable precipitation age for the sample. If the most probable peaks have no plateau age counterparts, these peaks are analytical artifacts and should be disregarded (Figure 3*c*).

MINERALOGICAL CONSIDERATIONS

Alunite-Jarosite

Alunite and jarosite have the general formula $AB_3(SO_4)_2(OH)_6$. The trivalent cation in the B site (Al^{3+} in alunite, Fe^{3+} in jarosite) occurs in sixfold coordination with two O^{3+} ions from the $SO_4^=$ tetrahedra and four OH^- groups (Figure 6) (Brophy et al 1962). K^+ and other monovalent cations (i.e. Na^+, H_3O^+, NH_4^+, and Ag^+) occur in the large twelvefold coordination A site surrounded by six O^- ions from the $SO_4^=$ tetrahedra and six OH^- ions (Figure 6) (Brophy et al 1962).

The site occupied by K^+ in alunite-jarosite minerals can easily accommodate the large radiogenic ^{40}Ar atom (atomic radius = 1.96 Å) generated by the radioactive decay of ^{40}K. One of the requirements for the successful analyses of jarosite-alunite minerals by the K-Ar and $^{40}Ar/^{39}Ar$ methods is the quantitative retention of K^+ and radiogenic $^{40}Ar^*$ in the mineral structures. Experimental studies (Stoffregen & Cygan 1990, Stoffregen et al 1994) suggest that the loss of K^+ from the A site is controlled by the rate of alkali exchange between alunite and water and that this exchange takes place by dissolution-reprecipitation. The activation energy for the alunite-jarosite exchange reaction is $156.5 \pm 6.3\ KJ \cdot mol^{-1}$, comparable to the values obtained for alkali-feldspar water Na-K exchange.

Modeling the experimental results with an empirical rate equation, Stoffregen et al (1994) suggested that temperature and grain size play a significant role in the rate of exchange between A-site cations in alunite and weathering solution. High Cl^- ion concentration in solution favors the dissolution of alunite by increasing the stability of the alkali aqueous complex. Based on these

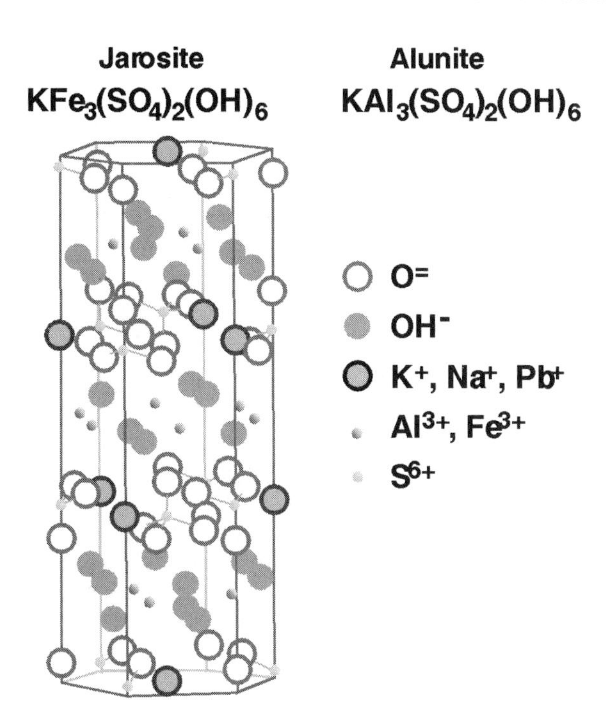

Figure 6 llustration of the alunite-jarosite structure showing the large *A* sites capable of hosting the large argon atom. Jarosite-alunite structures are not as complex as the supergene manganese oxide phases and are more amenable to K-Ar analysis (after Brophy et al 1962).

experimental results, Stoffregen et al (1994) conclude that coarse crystalline hydrothermal alunite is suitable to K-Ar dating but that fine-grained supergene alunites may experience limited alkali exchange, resulting in alteration of their K-Ar ages. However, resistance furnace results on the Ar release systematics of hypogene and supergene alunite (Itaya et al 1996) indicate that hypogene and supergene alunite have similar retentivity for Ar. The results indicate that the activation energies for Ar release from the A site is the same for coarse- and fine-grained alunite, suggesting that fine-grained alunite may yield reliable precipitation ages.

Geological evidence for the stability of alunite in the weathering environment and for its suitability to K-Ar dating is provided by Sillitoe & McKee (1996). Analyses of six samples representing a 200 m vertical section through a weathering profile yielded seven ages of 19.0 ± 0.7, 18.1 ± 0.7, 17.6 ± 0.6, 16.8 ± 1.2, 16.5 ± 0.5, 16.3 ± 0.6, and 15.2 ± 0.5 Ma. They interpret the narrow age grouping obtained and the progression from greater age at the top to smaller age at the bottom of the profile as confirmation of the suitability of alunite for K-Ar dating, concluding that neither Ar loss by diffusion nor potassium exchange with younger supergene solutions poses a problem to alunite K-Ar dating.

The low values for K_{sp} for alunite ($10^{-80.95}$) and jarosite ($10^{-93.21}$, $10^{-89.28}$, and $10^{-75.39}$ for jarosite, natrojarosite, and hydronium jarosite, respectively; Alpers et al 1989) imply that, once precipitated, these minerals should be relatively stable in the presence of meteoric water in the surficial environment. However, alunite and jarosite may be unstable when exposed to saline groundwater. Consequently, the presence of jarosite-alunite in weathering profiles is interpreted to record a transition toward arid conditions, when a drop in the water table would allow the preservation of these minerals in the unsaturated zone (Alpers & Brimhall 1988, Bird et al 1990).

No experimental studies on the alkali exchange and Ar release properties for jarosite are available. Also, no comprehensive analyses of jarosite crystals from a single profile have been carried out. Judgments on the suitability of jarosite for K-Ar and $^{40}Ar/^{39}Ar$ analyses are based solely on the geological consistency of geochronological results obtained for a few samples analyzed from a few localities (Stokowski & Krueger 1981; Vasconcelos et al 1994a,1994b; Cook 1994).

Mn-Oxides

The suitability of Mn-oxides to geochronology by the K-Ar and $^{40}Ar/^{39}Ar$ methods arises from the fact that some Mn-oxide minerals host K and other cations in well-defined crystallographic sites. These sites are relatively retentive of K and radiogenic $^{40}Ar^*$, depending on the Mn-oxide structure under investigation. Figures 7 and 8 and Table 2 illustrate the most common K-bearing Mn-oxides possibly suitable for K-Ar and $^{40}Ar/^{39}Ar$ geochronology.

Table 2 K-bearing Mn-oxides commonly found in continental weathering profiles (Burns & Burns 1979)

Mineral	Approximate formula	Group	Structure
Hollandite	$(Ba,K)_{1\text{-}2}Mn_8O_{16}.xH_2O$	Hollandite	2 × 2 tunnel
Cryptomelane	$(K,Ba)_{1\text{-}2}Mn_8O_{16}.xH_2O$	Hollandite	2 × 2 tunnel
Coronadite	$(Pb,Ba,K)_{1\text{-}2}Mn_8O_{16}.xH_2O$	Hollandite	2 × 2 tunnel
Romanechite	$(Ba,K,Mn^{2+},Co)_2Mn_5O_{10}.xH_2O$	Romanechite	2 × 3 tunnel
Todorokite	$(Ca,Na,K)(Mg,Mn^{2+})Mn_5O_{12}.xH_2O$	Todorokite	3 × 3 tunnel
Vernadite	$MnO_2.nH_2O.m(R_2O, RO, R_2O_3)$ R = Na, K, Ca, Co, Fe, Mn	Disordered birnessite?	Layer
Birnessite	$(Na,Ca,K)(Mg,Mn)Mn_6O_{14}.5H_2O$	Birnessite	7 Å layer

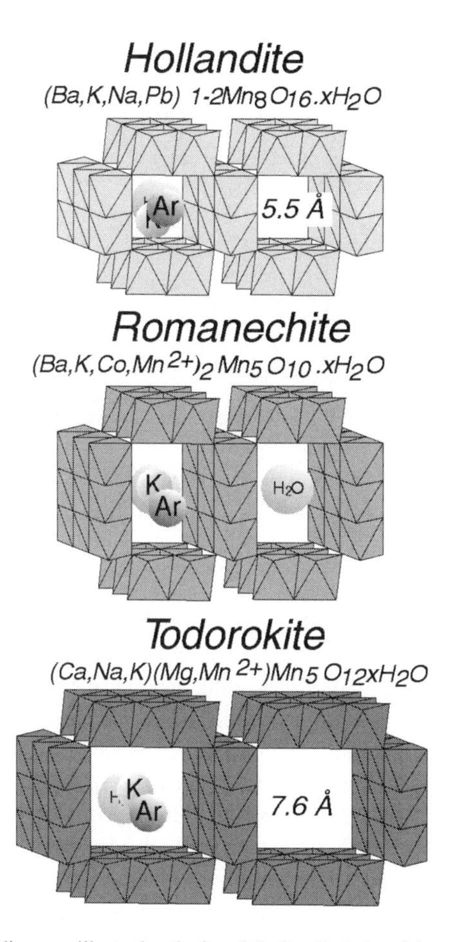

Figure 7 Schematic diagrams illustrating the 2 × 2 (hollandite), 2 × 3 (romanechite), and 3 × 3 (todorokite) tunnel structures of supergene manganese oxides. The cation exchange capacity and the thermal stability of the tunnel structures vary with the size of the tunnels. The larger tunnels in todorokite host more water and are more amenable to cation exchange than the smaller 2 × 2 hollandite structure. These properties eventually determine the suitability of manganese oxides for K-Ar and $^{40}Ar/^{39}Ar$ geochronology. The physical properties and the Ar retentivity of hollandite structures have been relatively well tested. Such tests are not yet available for the 2 × 3 and 3 × 3 structures.

Birnessite

$(Na,Ca,K,Mg)_{0.3\text{-}0.6}Mn_{1.4\text{-}1.9}O_4.1.4\text{-}1.7\ H_2O$

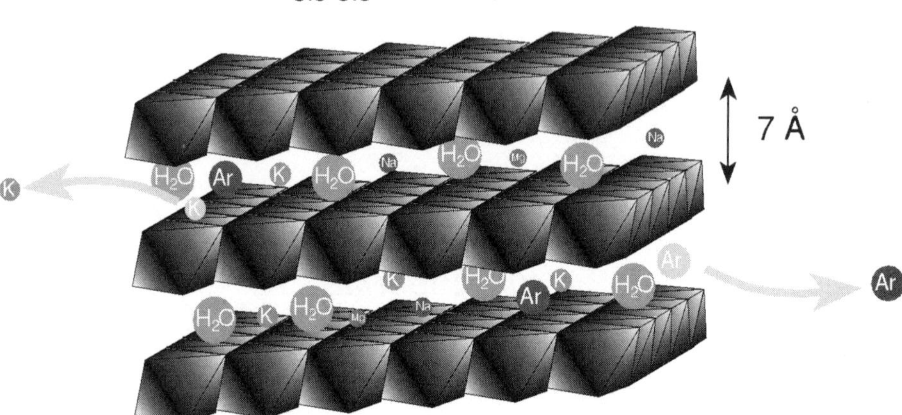

Figure 8 Layered phyllomanganate structure of birnessite. This structure also hosts considerable amounts of K. The common occurrence of birnessite in soil profiles (Golden et al 1986) makes these phases ideally suited to dating of soil processes. However, the large cation exchange capacity (Golden et al 1986) of birnessite and the tendency of this mineral to collapse in vacuum may suggest that birnessite is not suitable to geochronology by the K-Ar and the $^{40}Ar/^{39}Ar$ methods.

Except for birnessite and vernadite, which display layered structures (Figure 8), all other K-bearing Mn-oxide minerals listed above display tunnel structures (Figure 7). Tunnel structure minerals vary considerably in physical properties, depending on the size of the tunnels, the number of vacancies in these sites, and the nature of the tunnel cation. Natural Mn-oxides vary considerably in composition. Figure 9 illustrates the range in Mn, O, K, Ba, and Pb contents of approximately 90 Mn-oxide samples (>2800 spot analyses) from several weathering profiles in Australia. Petrographic inspection and X-ray diffraction (XRD) analyses for these samples indicate the presence of hollandite-group minerals with some pyrolusite and lithiophorite. Most of the analyzed samples are consistent with a hollandite stoichiometry. Comparisons between the electron microprobe results and $^{40}Ar/^{39}Ar$ analyses for each sample indicate that the phases most suitable for geochronology contain between 1 and 5.5 wt% K in their structure. Extremely Ba-rich samples (hollandite field, Figure 9) generally contain only 0 to 0.5 wt% K and are unsuitable or only marginally suitable for geochronology because of the very high ratios of atmospheric to radiogenic ^{40}Ar. The minimum K content necessary for reliable analyses of Cenozoic Mn-oxides has not been precisely determined, but this limit probably lies in the 0.2 to 0.4 wt% range.

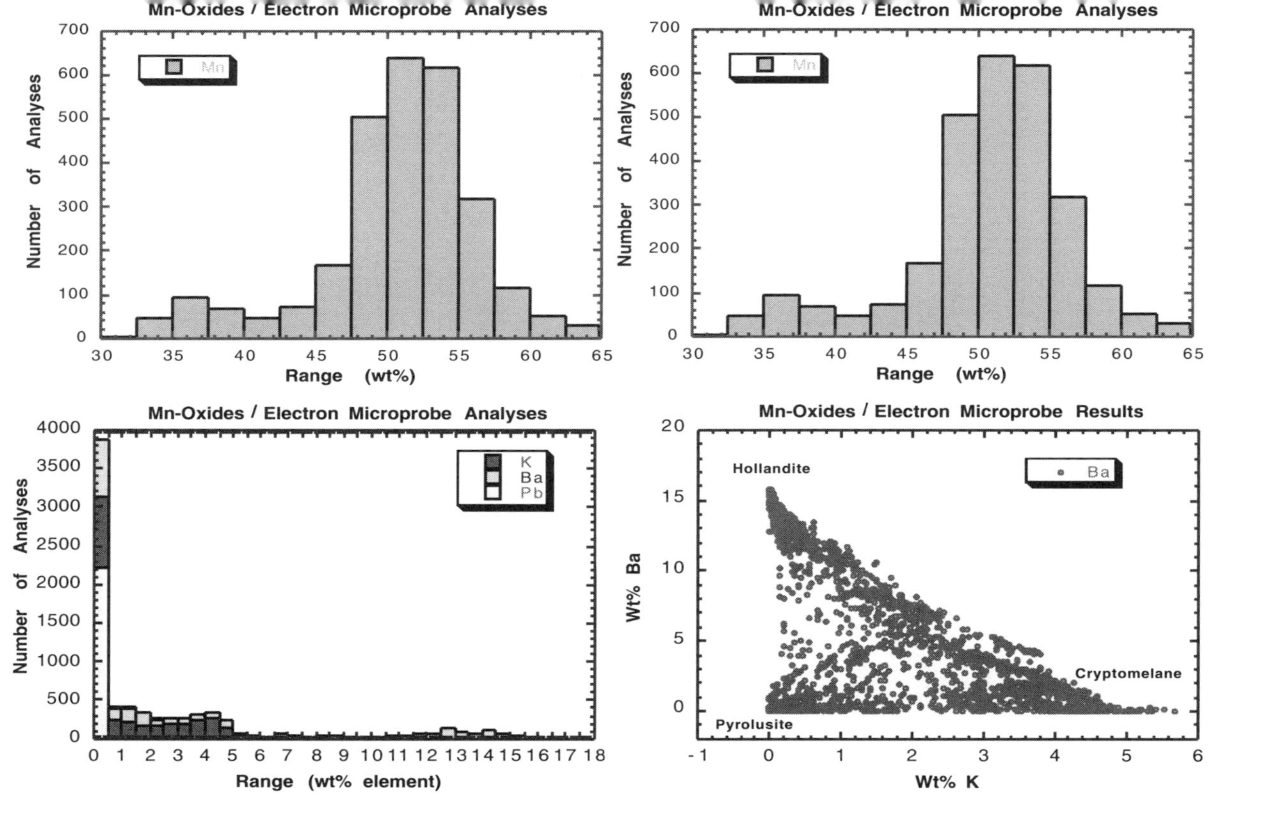

Figure 9 Natural manganese oxides with a very large range of compositions. Electron microprobe analyses of more than 2800 spots in approximately 90 samples of supergene manganese oxides from weathering profiles in Australia illustrate this variation in composition. These samples were collected for geochronology, and the hollandite mineral was preferentially targeted. The stoichiometry obtained from the electron microprobe analyses is consistent with a hollandite stoichiometry. The analyzed samples show a large range of Ba and K contents. The only other significant tunnel site element in the analyzed samples is Pb.

HOLLANDITE The hollandite group ($A_{1\text{-}2}Mn_8O_{16}.xH_2O$) (Figure 8), one of the most common groups of supergene Mn-oxides, includes the minerals hollandite, cryptomelane, coronadite, and manjiroite (Ba, K, Pb, and Na end members, respectively). In hollandite group minerals, the *A* site is filled by H_2O, Ba^{2+}, K^+, Pb^+, Na^+, Cu^+, Sr^{2+}, and Rb^{2+}. The *B* site may host Mn^{3+}, Mn^{3+}, Fe^{3+}, Al^{3+}, Si^{4+}, Co^{2+}, and Mg^{2+} (Burns & Burns 1979). The large channels (tunnel sites) formed by double or triple chains of edge-sharing Mn,O octahedra are partially filled by H_2O and *A* cations in eightfold coordination with oxygen atoms from the double chains (Byström & Byström 1950). These tunnel sites are only partially filled (Burns & Burns 1979). The large spaces within the tunnel site can accommodate the large Ar (1.96 Å radius) atoms derived from the radioactive decay of K and may also host atmospheric Ar either as a trapped air component or dissolved in tunnel H_2O.

Large variations in degrees of crystallinity, a wide range of grain sizes, and significant variations in the ionic radius of cations and amount of vacancies in the tunnel site contribute to the variation in the Ar retentivity of hollandite group minerals. Ba^{2+}, K^+, Pb^+, and other large cations occupying the *A* site give stability to the hollandite structure (Burns & Burns 1979, Miura 1986, Tsuji et al 1993); smaller cations in the *A* sites will make the structure more susceptible to cation exchange (Burns & Burns 1979) and thermal transformations (Vasconcelos et al 1994c).

The suitability of cryptomelane and hollandite to K-Ar and $^{40}Ar/^{39}Ar$ geochronology has been substantially tested in the past few years (Vasconcelos et al 1992, 1994a, 1994b, 1995; Grassely et al 1994; Segev et al 1995; Lippolt & Hautmann 1995; Dammer et al 1996; Ruffet et al 1996). Segev et al (1991) have also dated coronadite samples.

CRYPTOMELANE Cryptomelane was the first Mn-oxide mineral to be analyzed by the K/Ar (Chukhrov et al 1966) and $^{40}Ar/^{39}Ar$ methods (Vasconcelos 1992). The suitability of this mineral to geochronology arises from its high structural K content (up to 5.5 wt%) and from the high retentivity of K and $^{40}Ar^*$ in the tunnel site (Vasconcelos et al 1995). Cryptomelane occurs as a hydrothermal mineral or a supergene phase, which demonstrates its wide temperature stability range. In situ heating studies of cryptomelane crystals in the transmission electron microscope (TEM) indicate that the pure K end member is stable to 650°C, when cryptomelane crystals undergo structural transformation into hausmannite and manganosite (Vasconcelos et al 1994c). These results are compatible with the thermal behavior of other hollandite group Mn-oxides determined from thermogravimetric analyses (Bish & Post 1989).

Comparison among the thermal behavior of cryptomelane in the TEM (Vasconcelos et al 1994c), the results of temperature-controlled resistance furnace (RF) Ar analyses (Vasconcelos et al 1995), and the thermogravimetric analyses for the same samples (Figure 10) indicate that the release of K and Ar from the tunnel sites occurs primarily by the collapse of the tunnel structure and is not the result of volume diffusion of these elements along the tunnel sites. Attempts to calculate the activation energy for the release of ^{40}Ar from the tunnel sites by plotting the results of vacuum experiments on Arrhenius diagrams (Grassely et al 1994) are not valid, since the mechanism of Ar release is not volume diffusion but mineral breakdown (Musset 1969, Giletti 1974).

Despite evidence showing that K and Ar are relatively retentive in the cryptomelane structure, electrodialysis experiments indicate that natural and synthetic cryptomelane display significant cation exchange in deionized water (Sreenivas & Roy 1961). These experiments may suggest that cryptomelane easily exchanges tunnel K^+ with weathering solutions and that geochronological results may have no geological significance. Given the large range of composition and crystallinity of natural cryptomelane, it is premature to conclude that all cryptomelane samples will demonstrate the same cation exchange properties with weathering solutions.

Geological evidence for the suitability of cryptomelane for K-Ar and ^{40}Ar/^{39}Ar geochronology arises from the analyses of botryoidal growth bands. Textural and geochemical considerations indicate that these bands grow by accretionary precipitation of Mn-oxides outward from a core. This growth mechanism implies that the outer bands must be younger than the inner bands. The ^{40}Ar/^{39}Ar analyses of inner and outer bands from several Mn-oxide samples in different studies, conducted at different laboratories, confirm the age relationships implicit in the accretionary growth mechanism. In these studies, outer bands are consistently younger than inner bands, confirming the geological consistency of the results. Estimated growth rates calculated from age differences between core and rim range from $6.4 \times 10^{-3} \pm 1.2 \times 10^{-3}$ mm ka^{-1} (Vasconcelos et al 1992) to $0.37–1.5 \times 10^{-3}$ mm ka^{-1} (Dammer et al 1996). These growth rates are similar to values estimated for growth rates of ocean floor manganese nodules (0 to 10×10^{-3} mm ka^{-1}) (Heye 1979), suggesting an analogous mechanism (adsorption-surface precipitation) controlling the formation of nodular Mn-oxides in nature.

CORONADITE The similarity in the properties of hollandite group minerals (Bish & Post 1989) suggests that coronadite should be suitable to geochronology. The only K-Ar and ^{40}Ar/^{39}Ar studies of coronadite samples were performed by bulk methods, which provide no information on the behavior of the mineral analyzed (Segev et al 1991). At the moment, the suitability of coronadite to Ar geochronology is undetermined.

ROMANÉCHITE The structure of romanéchite is similar to that of hollandite, except that the tunnels along the *b* axes are composed of 2 × 3 Mn,O octahedra chains. Ba (plus Na^+, K^+, Sr^{2+}, etc.) and H_2O occupy the tunnel sites. Rietveld refinement of the romanéchite structure (Turner & Post 1988) suggests ordering of Ba and H_2O in the tunnel sites. Thermogravimetric and water analyses also indicate that the water in the tunnel site is relatively stable and quantitatively retained up to 500°C, suggesting a high activation energy for the displacement of occupants of the tunnel site (Turner & Post 1988). These results are similar to results obtained for hollandite structures but differ significantly from results obtained for todorokite (Bish & Post 1989).

Relatively high K contents and the relative stability of tunnel water suggest that romanéchite structures may be retentive of K and Ar in the natural environment and may be suitable for geochronology. Despite evidence suggesting structural stability for romanéchite, heating to 300°C induces minor structural changes; coronadite, on the other hand, does not display any structural changes below 600°C. This structural change in romanéchite may suggest that this phase will release Ar from the tunnel sites at lower temperatures than 2 × 2 Mn-oxide structures. No systematic studies on the cation exchange capacity, the activation energy for the tunnel site occupants, or the Ar degassing behavior of romanéchite are available. The larger 2 × 3 sites and the susceptibility to structural transformations at relatively low temperature raise questions about the stability of romanéchite and its suitability to geochronology by the K-Ar and $^{40}Ar/^{39}Ar$ methods. Consequently, K-Ar results obtained from romanéchite analyses (Dammer et al 1996) are suspect until the Ar and K retention properties and the suitability of this mineral structure to geochronology are demonstrated.

TODOROKITE Todorokite is a complex 3 × 3 tunnel structure Mn-oxide mineral that has the general formula $(Ca,Na,K,Ba,Sr,)_{0.3\text{-}0.7}(Mn,Mg,Al)_6O_{12}$.3.2-4.5 H_2O (Post & Bish 1988). The monovalent and divalent cations and the water molecules occupying the tunnel sites display various degrees of ordering. The cations occupying the large tunnel sites formed by the triple chains of edge-sharing Mn,O octahedra are octahedrally coordinated by water molecules also occupying tunnel positions. In addition to water molecules occupying well-defined crystallographic positions, the tunnel sites also host a disordered water component. Bish & Post (1989) showed that the 3 × 3 todorokite structure hosts much more water (10 to 12 wt%) in the tunnel sites than similar 2 × 2 and 2 × 3 tunnels. This water is lost from the tunnel sites upon heating from 200° to 500°C, with significant weight losses. Thermogravimetric analyses suggest that the water in the 3 × 3 tunnel structures is much more mobile than the water in smaller tunnels. Cation exchange experiments indicate that the

Table 3 K-Ar analyses of todorokite sample UCB-547

Sample	Grain size (μm)	K (wt%)	Method	Weight (g)	% ^{40}Ar*	Age (Ma)	$\pm 1\delta$ (Ma)
BGC							
6247	250–350	0.363	AAS	.50436	5.7	26.2	3.4
6248	150–250	0.363	AAS	.50256	10.5	15.6	1.6
USGS							
6247	250–350	0.37	FP	.53980	1.9	7.6	1.8
6248	150–250	0.37	FP	.43050	0.9	3.7	2.1

tunnel cations in todorokite are more easily displaced than cations occupying equivalent, but smaller, tunnel sites in the 2 $\times$ 2 hollandite and the 2 $\times$ 3 romanéchite structures (Shen et al 1993). Bish & Post (1989), however, concluded that tunnel structure Mn-oxides, including the 3 $\times$ 3 tunnel in todorokite structures, do not have the same cation exchange capacity of zeolites because the tunnels in the Mn-oxides are not linked infinitely in the tunnel direction.

In a pilot study, a massive todorokite sample from Hokkaido, Japan (University of California, Berkeley Collection #547), was analyzed by K-Ar and ^{40}Ar/^{39}Ar laser-heating methods at the BGC. Aliquots of the same sample were subsequently analyzed at the geochronology laboratory at the United States Geological Survey (USGS) in Menlo Park, California. The results are presented in Table 3. As shown in Table 3, grains crushed to different sizes yield substantially different results, with the coarser-grained samples yielding dates twice as old as the fine-grained samples. In addition, the samples analyzed at the USGS and BGC show disparate K-Ar dates, despite the consistency in the K contents. A possible interpretation for these disparate results is that the todorokite samples analyzed do not retain Ar quantitatively in their tunnel sites and that Ar loss may be promoted by crushing or by the bake-out procedure preceding mass spectrometry. This hypothesis is consistent with the fact that younger ages were obtained for the fine-grained samples in both laboratories. The discrepancy between the two laboratories may be attributed to different Ar losses in the preheating process. Similar comparison for cryptomelane samples yielded indistinguishable results between the two laboratories.

One todorokite grain was also analyzed by the step-heating method at the BGC. The analyses yielded a relatively well-defined plateau age consistent with one of the K-Ar results obtained in the USGS laboratory. Subsequently, three grains from sample 6247 and four grains from sample 6248 were analyzed by the laser step-heating method at UQ-AGES. The results are substantially younger than the K-Ar results in Table 3 and plateau age obtained at the BGC, suggesting that todorokite may not be suitable to K-Ar or $^{40}Ar/^{39}Ar$ geochronology, despite the fact that it yields a date when analyzed by these methods.

Using the K-Ar method, Dammer et al (1996) dated Ba-todorokite samples from weathering profiles in Groote Eylandt, Australia. From their data, it is impossible to correlate geochronology results with the identity of the mineral analyzed. Dates obtained by the K-Ar analyses of todorokite crystals may not have any age significance, and results obtained from the analyses of these minerals should be clearly identified when presenting geochronology results.

BIRNESSITE Birnessite has the general formula $(Na,Ca,K,Mg)_{0.3\text{-}0.6}Mn_{1.4\text{-}1.9}O_4.1.4\text{-}1.7\,H_2O$. Electron diffraction patterns and Rietveld refinement show that birnessite displays a layer structure and that the interlayer water molecules and cations are probably ordered (Post & Veblen 1990). Layer structure Mn-oxides are often called phyllomanganates, in analogy to phyllosilicates (Giovannoli 1985) with which they share many properties. Estimates of the distance between the Mn-octahedra sheets in the layer structure in birnessite range from 7.1 to 7.2 Å (Golden et al 1986, Post & Veblen 1990).

Despite the occurrence of some relatively K-rich birnessite minerals in nature, the suitability of these minerals to geochronology is uncertain. Post & Veblen (1990) have shown that the Na-birnessite structure collapses from 7 to 5 Å in the vacuum of the electron microscope, probably due to loss of the interlayer water. This structural instability may prevent the successful analyses of birnessite by K-Ar and $^{40}Ar/^{39}Ar$ geochronology. In addition, Golden et al (1986) demonstrated that birnessite structures display large cation exchange capacity. However, they conclude that only Li-, Na-, Mg-, and Ca-birnessites show cation exchange properties. In their experiments, K-birnessite did not show appreciable cation exchange properties. Before birnessite K-Ar and $^{40}Ar/^{39}Ar$ results can be used in the interpretation of geological processes, systematic work demonstrating the suitability of this mineral to K-Ar and $^{40}Ar/^{39}Ar$ geochronology must be carried out.

VERNADITE (δ-MNO_2) Vernadite is a poorly crystalline oxide regarded as a structurally disordered birnessite (Burns & Burns 1979, McKenzie 1989). The same reservations regarding the use of birnessite in geochronology apply to vernadite.

INTERPRETATION OF RESULTS

Geochronology of Soils and Weathering Profiles

K-Ar and $^{40}Ar/^{39}Ar$ dating of some supergene minerals may provide reliable and reproducible geochronological information about chemical processes occurring in the weathering environment. Sulfates (alunite and jarosite) and some manganese oxides (cryptomelane, hollandite, and perhaps coronadite) are suitable for geochronology by the methods described above. The relative simplicity of the mineral structures of supergene alunite and jarosite allows these phases to be dated equally well by the K-Ar and the $^{40}Ar/^{39}Ar$ techniques provided that large enough quantities of pure mineral separates can be obtained.

However, the complex nature of supergene Mn-oxides and the large number of mineral structures encompassed by the Mn-oxide group demand care in the application of radiogenic isotope techniques to date these phases. It is imperative that the phase analyzed by isotopic methods be properly characterized by XRD and electron microprobe analysis to ensure proper mineral identification. In addition, the numerous possibilities for deviations from a true age (contamination, recoil, open system behavior, mixed generations of the same phase, mixed phases, etc.) require that the applied geochronological technique be suitable for determining, in addition to a date, the quality of the result obtained.

Although two-step K-Ar analysis of manganese oxides improves the reliability of the geochronological technique by subtracting contributions from contaminants, it does not provide information on whether the analyzed sample is composed of a single generation or phase or of multiple Mn-oxide generations or phases. Also, it does not provide a test for ^{40}Ar loss after mineral precipitation. Only the $^{40}Ar/^{39}Ar$ incremental-heating technique may provide the necessary checks and balances to ensure that the result obtained from the analyses of Mn-oxides is a reliable age. Some of the surficial-process problems potentially solvable by the application of geochronology of supergene minerals will be properly addressed only if a database of reliable analytical results is available. Most of the weathering geochronology studies carried out until now (Chukhrov et al 1966, 1969; Gustafson & Hunt 1975; Ashley & Silberman 1976; Varentsov & Golovin 1987; Alpers & Brimhall 1988; Pracejus 1989; Bird et al 1990; Arehardt et al 1992; Arehardt & O'Neil 1992; Cook 1994; Grasselly et al 1994; Segev et al 1995; Dammer et al 1996; Sillitoe & McKee 1996; Virtue 1996) have relied on K/Ar results. Given the uncertainties intrinsic to the method, particularly when applied to Mn-oxide dating, it would be advisable that future studies focus on the application of the $^{40}Ar/^{39}Ar$ incremental-heating methodology.

Longevity and Evolution of Lateritic Profiles

One of the main difficulties in determining the age of deep lateritic weathering profiles is that these profiles are not the product of a single weathering event in a well-defined time in the past, but are the combined result of long-lasting superimposed weathering processes active during extended periods of time.

In addition, even in areas subject to extremely low erosion rates, erosion still occurs (Bierman & Turner 1995). If weathering fronts migrate downward and the oldest weathered material is located at the top horizons, it is probable that the oldest, initially precipitated supergene minerals have been removed by erosion. In this scenario, weathering geochronology can provide only a minimum age for the profile.

However, in extremely Fe-rich weathering profiles, particularly in the ferruginous duricrusts associated with lateritic profiles, Mn-oxides may be absent from the Fe-rich zones due to the reduction of Mn^{4+}-oxides into soluble Mn^{2+} aqueous species by locally abundant Fe^{2+}. Manganese is also susceptible to reduction and dissolution by organic ligands in soils. In lateritic systems, manganese is leached from the soils and ferruginous duricrusts and is redeposited in the bleached zone or saprolite horizon where iron is less abundant (Figure 1). If this process was active early during the history of the lateritic profile, Mn-oxides precipitated at lower levels in the profile may actually reflect the oxidation-reduction couple between iron and manganese developed at incipient stages of weathering. The Mn-oxides precipitated at greater depths are relatively protected from erosion and may actually reflect the initial age of formation of the profile.

K-Ar and $^{40}Ar/^{39}Ar$ dating of supergene hollandite group minerals from weathering profiles in Brazil (Figure 11) (Vasconcelos et al 1994a, Ruffet et al 1996) and Australia (Dammer et al 1996, Vasconcelos 1998) confirms that some of these lateritic profiles are older than 65 to 70 Ma and have been continuously exposed to weathering solutions throughout their history. Some geomorphologists have proposed that some of the lateritic plateaus and land surfaces in relatively tectonically stable southern hemisphere cratons are the product of weathering and landscape evolution processes active when these land masses were still assembled as supercontinents (King 1949, 1953). Weathering geochronology results may prove their hypothesis correct. The oldest ages obtained for supergene Mn-oxides in Brazil (65 to 70 Ma) (Vasconcelos et al 1994a, Ruffet et al 1996) and Australia (also 65 to 70 Ma) (Vasconcelos 1996, Fleming 1997, Vasconcelos 1998) attest to the longevity of these weathering profiles.

The distribution of lateritic weathering profiles has been used as a paleoclimatic indicator (Frakes 1979, Nahon 1986, Crowley & North 1991). The presence of lateritic weathering profiles is generally interpreted as indicative of warm temperatures with seasonal rainfall (Frakes 1979, Crowley & North 1991). More recently, Bird & Chivas (1993) have suggested that deep weathering may take place under cool to cold humid climates. Precise and accurate geochronological information about the intensity of weathering reactions occurring within the profiles throughout their history may be necessary

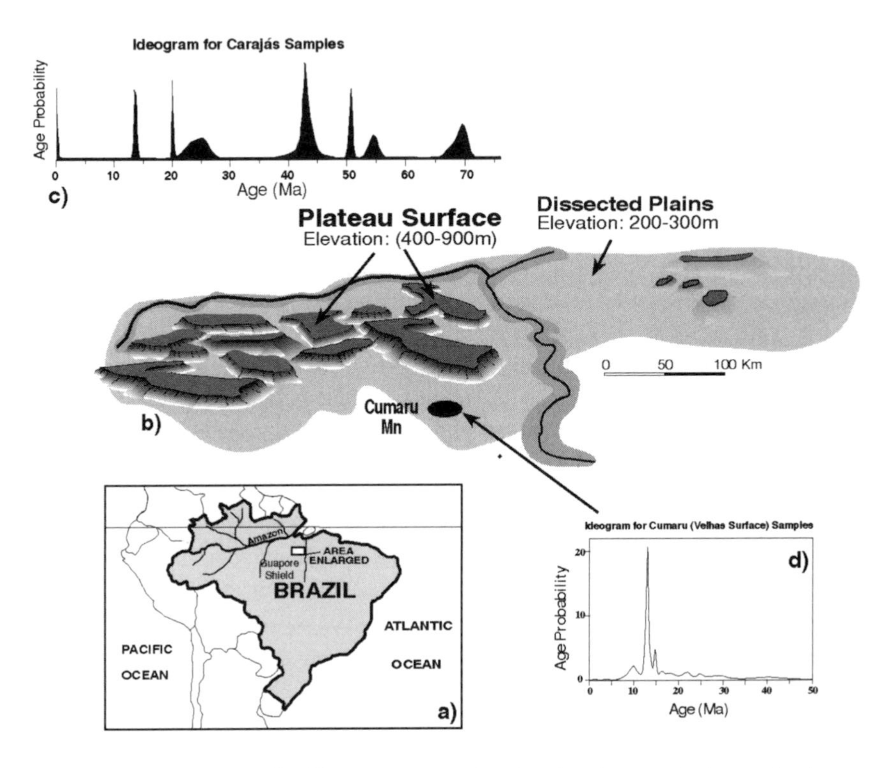

Figure 11 Distribution of $^{40}Ar/^{39}Ar$ ages obtained for Mn-oxides from weathering profiles from the Carajas region, Amazon, Brazil (Vasconcelos et al 1994a). The results confirm that some of these weathering profiles are extremely long-lived, having been continuously exposed to surface conditions at least since the end of the Mesozoic. Subsequent analyses of Mn-oxides from the dissected landscape surrounding the Carajas plateaus (Cumaru Mn occurrence) yielded Miocene ages, possibly suggesting that the landscape was dissected before the Miocene and the newly exposed bedrocks were weathered during the Miocene. Systematic studies of weathering ages in different parts of the landscape can provide useful constraints on geomorphological evolution for a region.

for determining the exact climatic conditions under which these profiles develop. This information will make lateritic profiles more reliable indicators of paleoclimates.

Weathering and Denudation Rates

Weathering geochronology provides constraints on maximum rates of denudation for continental landscapes. Assuming that the average thickness for deep weathering profiles is 100 m and that the whole profile was formed at once, if a mineral present at the surface today was formed at the bottom of the profile and yields an age of 50 Ma, the maximum denudation rate for the area is constrained at 2 m Ma^{-1}. Therefore, the mere presence of ancient, deep weathering profiles in a region constrains the maximum denudation possible for an area. The longevity of some of the weathering profiles in the Carajás region, Amazon (Figure 11) (Vasconcelos 1992, Vasconcelos et al 1994a, Ruffet et al 1996), and the Mount Isa region, Australia (Vasconcelos 1998), indicates that some parts of the landscape have undergone slow denudation rates during the past 70 Ma.

Apatite fission track thermochronology in these regions, however, suggests significant rates of denudation for the same time interval. Harman et al (1999) have interpreted the results of modeling thermal history from apatite fission track length as indicating 3 to 4 km of denudation in the Guapore Shield, in the Amazon region of Brazil, during the past 80 Ma. Spikings et al (1997) have proposed 1.2–2.0 km of denudation for the Mount Isa region, in Queensland, for the last 100 Ma. It is difficult to reconcile weathering geochronology results with the estimated denudation rates obtained from fission track studies. If the weathering geochronology results for the Carajás region, substantiated by two independent studies, are correct, the denudation rates obtained from modeling apatite fission track data are unreasonably large. Combined studies, in the same profiles, of fission track thermochronology and weathering geochronology will be necessary to resolve these discrepancies.

Continuous Versus Episodic Weathering

Weathering and soil formation processes are ultimately controlled by five major factors: the nature of the bedrock, climate, topography, biological activity, and time (Jenny 1941). In addition, chemical weathering is essentially a liquid water-rock interaction process. Limited chemical weathering may occur by atmosphere-rock interaction, by ice-rock interaction, or through the mediation of organisms without the direct involvement of liquid water. However, in the absence of liquid water, reactive species cannot be easily introduced to the weathering interface, solutes cannot be effectively removed from the reacting interface, and microorganisms may have limited mobility and shortened life cycles. Consequently, weathering is favored by high water-to-rock ratios.

Chemical weathering is also strongly dependent on temperature. Chemical kinetics predicts that reaction rates increase by a factor of two for a 10°C increase in ambient temperature. Recently, experimental results have shown that silicate weathering rates, in the presence of organic acids, may increase by up to 150 times for an increase in temperature from 25° to 60°C (Brady & Carroll 1994). Laboratory (Drever 1994) and field studies (Cawley et al 1969) also suggest that rates of chemical weathering increase in the presence of vegetation and soluble organic species.

The considerations provided suggest that the evolution of weathering profiles should reflect variations in the amount, temperature, and composition of the weathering solutions percolating through the profile. Since the amount of rainfall, the ambient temperature, and the abundance and type of vegetation vary with climate, weathering rates should also vary with climate. In the absence of mechanical erosion, the volume of newly formed minerals in weathering profiles should be large during favorable climates and relatively small during unfavorable conditions.

Vasconcelos (1992) and Vasconcelos et al (1994a) interpreted the results of K-Ar and $^{40}Ar/^{30}Ar$ analyses of supergene hollandite group minerals from weathering profiles in the Carajás region (Figure 11), Amazon, as an indication that weathering profiles do not evolve continuously through time but are formed in an episodic mode. The episodic evolution of weathering profiles reflects the influence of external factors, global climates, according to Vasconcelos et al (1994a). This interpretation was based on the clustering of ages obtained from the K-Ar (47 independent analyses) and $^{40}Ar/^{39}Ar$ (51 grains analyzed) analyses of 40 samples of supergene cryptomelane and hollandite from five weathering profiles in the Carajás region, Amazon.

In one of these profiles, Vasconcelos et al (1994a) identified three major periods of weathering (65 to 69, 51 to 56, and 40 to 43 Ma). Ruffet et al (1996) analyzed 13 Mn-oxide samples from the same profile and obtained ages that correspond to the 65-to-69 and the 51-to-56 Ma episodes identified by Vasconcelos et al (1994a). However, Ruffet et al (1996) did not obtain any result in the 40-to-43 Ma age range but had several samples yield ages with a mean value of 47.4 ± 0.1 Ma. Based on the new data they concluded that the record preserved in the Azul weathering profiles is continuous, at least from 70 to about 40 Ma, and that the evolution of weathering profiles is independent of external factors, such as global climate.

Episodic versus continuous evolution of weathering profiles is a semantic problem. As an open system continuously exposed to the atmosphere and the biosphere, weathering profiles will continuously witness the influx of weathering solutions. However, weathering reactions are chemical reactions; they are subject to the same kinetic and thermodynamic constraints as any chemical process. In simplistic form, the law of mass action predicts that the increase in the activity of a reactant will drive the reaction forward. All the chemical reactions studied by weathering geochronology (reactions 1 through 14) have water as an explicit or implicit reactant. This implies that an increase in the availability of water in the system should drive these reactions forward and should promote the formation of supergene minerals.

Volumetrically, more supergene minerals will form during favorable, wet conditions. Consequently, the weathering profile will host a larger volume of

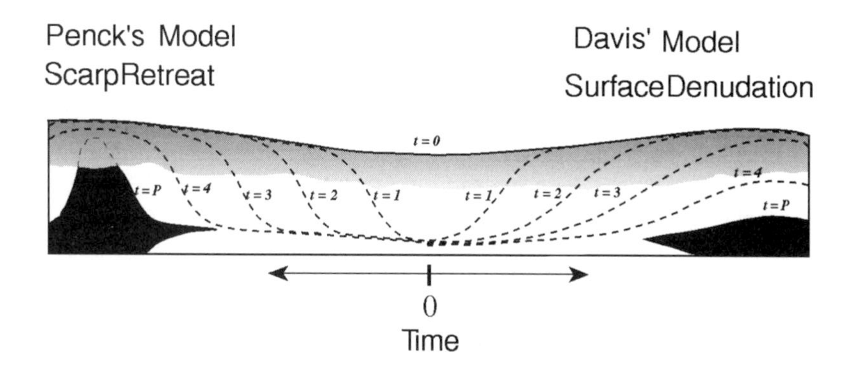

Figure 12 Possible end-member landscape evolution models and their bearing on the distribution of weathering ages in a region. The presence of old weathering ages in some weathering profiles in Brazil and Australia suggests that these regions have evolved primarily by scarp retreat mechanisms.

minerals formed during favorable conditions. Random samples from this profile should produce a database biased toward the times when more supergene minerals precipitated. The resolution of the difference in interpretation given to the geochronology results obtained for the Carajás profiles by Vasconcelos et al (1994c) and Ruffet et al (1996) will have to wait until a larger, statistically meaningful database is available.

Landscape Evolution

Two competing end-member landscape evolution models, scarp retreat and surface denudation (Figure 12), can be tested through the study of weathering geochronology. Figure 12 illustrates the effects of each of these landscape evolution models on the dissection of a land surface covered by a deep weathering profile. In the Davisian model, surface denudation will progressively remove the products of weathering. If the landscape in the areas studied by weathering geochronology (the Carajás region—Vasconcelos et al 1994a, Ruffet et al 1996; and the Mount Isa region—Vasconcelos 1998) had evolved in a Davisian fashion, no relic of the ancient weathering profiles that once blanketed the region would be preserved. However, if the landscape followed a scarp retreat evolution, remnants of the ancient weathering profile should be preserved. The geochronological results obtained by the analyses of supergene Mn-oxides in the Amazon and in the Mount Isa region suggest an evolution by scarp retreat.

This interpretation can be further tested through the application of weathering geochronology to the determination of the time at which the dissected plains became exposed (Figures 11 and 12). If the scarp retreat model is correct, weathered material at dissected parts of the landscape should yield younger ages than the ages obtained from the adjacent plateaus that preserve the ancient weathering profiles. One such study in the Amazon, based on laser-heating $^{40}Ar/^{30}Ar$ analyses of cryptomelane samples from the Cumaru profile (Figure 11), shows that the weathering age obtained for the dissected part of the landscape is much smaller than the ages obtained from the deeply weathered plateaus. In addition, the age obtained for the dissected part of the landscape corresponds to one of the age episodes previously identified for the plateau (Figure 11), as should be expected given that the plateau has been continuously exposed and should record all weathering events in the region. A comprehensive study (>110 cryptomelane and hollandite samples analyzed by the $^{40}Ar/^{30}Ar$ incremental heating method) of weathering profiles in the Mount Isa region, Queensland, also indicates that the dissected terrains host younger supergene minerals than the adjacent plateaus hosting ancient weathering profiles (Vasconcelos 1998).

Global Correlations

As stated in the introduction, the ultimate goal of weathering geochronology is to determine the timing, rate, and mechanisms of propagation of weathering fronts through the weathering profile and to unravel the climatic and geochemical history implicit in the evolution of continental regolith covers. The weathering geochronology database can be used to correlate variations in the intensity and patterns of global events such as volcanism, tectonism, sea-level changes, chemical changes in oceanic water composition, extinction events, changes in physical oceanographic patterns, and the weathering record preserved on continents. This correlation can be used to determine cause-and-effect relationships between global events and continental weathering processes.

At this early stage the available weathering geochronology record is inadequate to provide a meaningful, statistically valid correlation between the intensity of formation of weathering profiles on continents and timing of global events. This inadequacy arises from the few results available and the lack of an appropriate filter to screen meaningful results from artifacts resulting from the analysis of inappropriate samples. However, Figure 13 illustrates an ideogram for all the K-Ar and $^{40}Ar/^{30}Ar$ dates available in the literature on the geochronology of weathering profiles.

Two features are salient in the ideogram: the Mn-oxide record preserves information about more ancient weathering processes, while the alunite-jarosite record is strongly skewed toward the younger Miocene and post-Miocene ages. The other important feature is the indication that Miocene weathering ages are also strongly represented in the Mn-oxide record. Whether this clustered age distribution reflects sampling bias or bias in choices of study sites or whether it represents an increase in the weathering rates in the Miocene remains to be resolved.

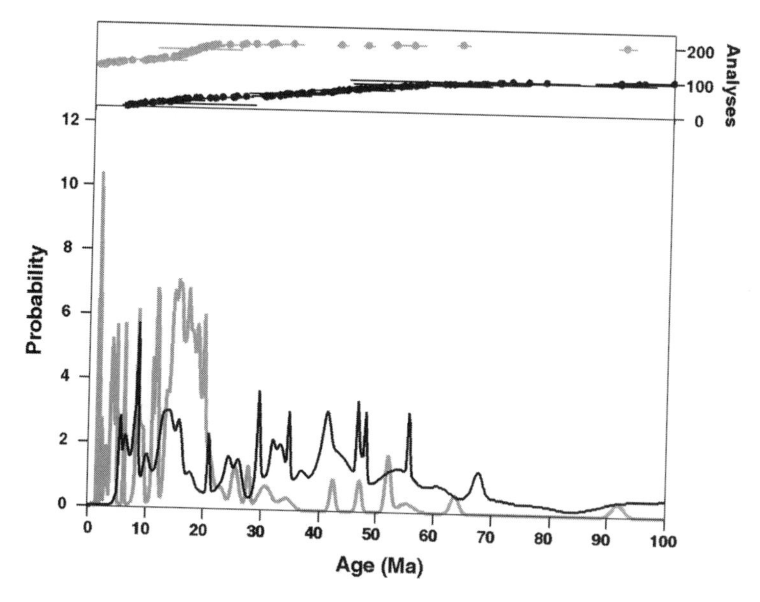

Figure 13 The distribution of weathering ages obtained from currently published results (Chukhrov et al 1966, 1969; Gustafson & Hunt 1975; Ashley & Silberman 1976; Varentsov & Golovin 1987; Alpers & Brimhall 1988; Pracejus 1989; Bird et al 1990; Arehardt et al 1992; Arehardt & O'Neil 1992; Cook 1994; Grasselly et al 1994; Segev et al 1995; Dammer et al 1996; Sillitoe & McKee 1996), indicating that Mn-oxide ages represent weathering processes active throughout the Cenozoic and extending into the Mesozoic. Alunite-jarosite age distribution is strongly skewed toward the past 20 Ma, possibly reflecting the sulfates' greater susceptibility to recrystallization when exposed to moist conditions, as suggested by Bird et al (1990).

The widespread and careful application of geochronology of supergene minerals to the study of the weathering processes will provide, in the near future, the database necessary for researchers to correlate continental surficial processes with other processes of global significance.

Literature Cited

Alpers CN, Brimhall GH. 1988. Middle Miocene climatic change in the Atacama Desert, northern Chile: evidence from supergene mineralization at La Escondida. *Geol. Soc. Am. Bull.* 100:1640–56

Alpers CN, Nordstrom DK, Ball JW. 1989. Solubility of jarosite solid solutions precipitated from acid mine waters, Iron Mountain, California, USA. *Sci. Géol. Bull.* 42:281–98

Alpers CN, Rye RO, Nordstrom DK, White LD, King BS. 1992. Chemical, crystallographic, and isotopic properties of alunite and jarosite from acid hypersaline Australian lakes. *Chem. Geol.* 96:203–26

Anderson JA. 1981. Characteristics of leached capping and techniques of appraisal. In *Advances in Geology of the Porphyry Copper Deposits, Southwestern North America*, ed. SR Titley, p. 245–87. Tucson: Univ. Ariz. Press

Arehart GB, Kesler SE, O'Neil J, Foland KA. 1992. Evidence for the supergene origin of alunite in sediment-hosted micron gold deposits, Nevada. *Econ. Geol.* 87:263–70

Arehart GB, O'Neil JR. 1992. Western U.S. continental climate record since 30 Ma as recorded in alunite; comparison with the marine record. *Geol. Soc. Am. Abstr. Programs* 24:268 (Abstr.)

Ashley RP, Silberman ML. 1976. Direct dating of mineralization at Goldfield, Nevada, by potassium-argon and fission-track methods. *Econ. Geol.* 71:904–24

Berner RA. 1994. Geocarb II: a revised model of atmospheric CO_2 over Phanerozoic time. *Am. J. Sci.* 294:56–91

Berner RA, Lasaga AC, Garrels RM. 1983. The carbonate-silicate geochemical cycle and its effect on atmospheric carbon dioxide over the past 100 million years. *Am. J. Sci.* 283:641–83

Bestland EA, Retallack GJ, Swisher CC III. 1997. Stepwise climate change recorded in Eocene-Oligocene paleosol sequences from central Oregon. *J. Geol.* 105:153–72

Bierman P, Turner J. 1995. ^{10}Be and ^{26}Al evidence for exceptionally low rates of Australian bedrock erosion and the likely existence of pre-Pleistocene landscapes. *Quat. Res.* 44:378–82

Bird MI. 1988. An isotopic study of the Australian regolith. PhD thesis. Canberra: Aust. Natl. Univ. 226 pp.

Bird MI, Chivas AR, McDougall I. 1990. An isotopic study of surficial alunite in Australia. 2. Potassium-argon geochronology. *Chem. Geol.* 80:133–45

Bish DL, Post JE. 1989. Thermal behavior of complex, tunnel-structure manganese oxides. *Am. Mineral.* 74:177–86

Bladh KW. 1982. The formation of goethite, jarosite, and alunite during the weathering of sulfide-bearing felsic rocks. *Econ. Geol.* 77:176–84

Blanchard R. 1968. Interpretation of leached outcrops. *Bull. Nev. Bur. Mines 66.* 196 pp.

Brady PV, Carroll SA. 1994. Direct effects of CO_2 and temperature on silicate weathering: possible implications for climate control. *Geochim. Cosmochim. Acta* 58:1853–56

Brophy GP, Scott ES, Snellgrove RA. 1962. Sulfate studies. II. Solid solution between alunite and jarosite. *Am. Mineral.* 47:112–26

Burns RG, Burns VM. 1979. Manganese oxides. In *Marine Minerals*, ed. PH Ribbe, p. 1–46, Washington, DC: Mineral. Soc. Am.

Byström A, Byström AM. 1950. The crystal structure of hollandite, the related manganese oxide minerals, and a-MnO_2. *Acta Crystallogr.* 3:146–54

Cawely JL, Burruss RC, Holland HD. 1969. Chemical weathering in central Iceland: an analog of pre-Silurian weathering. *Science* 165:391–92

Chukhrov FV, Shanin LL, Yermilova LP. 1966. Feasibility of absolute-age determination for potassium-carrying manganese minerals. *Int. Geol. Rev.* 8:278–80

Chukhrov FV, Yermilova LP, Shanin LL. 1969. Age of alunite from certain deposits. *Trans. Acad. Sci. USSR Dokl. Earth Sci. Sect.* 185:49–51

Cook SS III. 1994. The geological history of supergene enrichment in the porphyry copper deposits of southwestern North America. PhD thesis. Tucson: Univ. Ariz. 163 pp.

Crerar DA, Cormick RK, Barnes BL. 1972. Organic controls on the sedimentary geochemistry of manganese. *Acta Mineral. Petrogr. Szeged.* 20:217–26

Crowley TJ, North GR. 1991. *Paleoclimatology*, London/New York: Oxford Univ. Press. 349 pp.

Dalrymple GB, Lanphere MA. 1969. *Potassium-Argon Dating*. New York: Freeman. 258 pp.

Dammer D, Chivas AR, McDougall I. 1996. Isotopic dating of supergene manganese oxides from the Groote Eylandt Deposit, Northern Territory, Australia. *Econ. Geol.* 91:386–401

Davies SHR, Morgan JJ. 1989. Manganese (II) oxidation kinetics on metal oxide surfaces. *J. Colloid Interface Sci.* 129:63–77

Deino AL, Potts R. 1990. Single-crystal $^{40}Ar/^{39}Ar$ dating of the Olorgesailie Formation, southern Kenya Rift. *J. Geophys. Res.* 95:8453–70

Drever JI. 1994. The effect of land plants on weathering rates of silicate minerals. *Geochim. Cosmochim. Acta* 58:2325–32

Faure G. 1986. *Principles of Isotope Geology*. New York: John Wiley & Sons Inc. 589 pp.

Finkl CW Jr. 1984. Chronology of weathered material and soil age determination in pedostratigraphic sequences. *Chem. Geol.* 44:311–35

Frakes LA. 1979. *Climates Throughout Geologic Time*. Amsterdam/Lausanne/New York: Elsevier

Gilleti BJ. 1974. Diffusion related to geochronology. In *Geochemical Transport and Kinetics*, ed. AW Hofman, Carnegie Inst. Washington Publ. 634:61–76

Giovanoli R. 1985. Layer structures and tunnel structures in manganates. *Chem. Erde* 44:227–44

Golden DC, Dixon JB, Chen CC. 1986. Ion exchange, thermal transformations, and oxidizing properties of birnessite. *Clays Clay Miner.* 34:511–20

Grasselly G, Balogh K, Toth M, Polgari M. 1994. K/Ar age of manganese oxide ores of Urkut, Hungary: Ar retention in K-bearing Mn minerals. *Geol. Carpathica* 45:365–73

Guftafson LB, Hunt JP. 1975. The porphyry copper deposit at El Salvador, Chile. *Econ. Geol.* 70:857–912

Harman et al. 1999. In press

Harrison TM. 1983. Some observations on the interpretation of ^{40}Ar-^{39}Ar age spectra. *Isot. Geosci.* 1:319–38

Hem JD. 1963. Chemical equilibria and rates of manganese oxidation. *US Geol. Surv. Water Supply Pap. 1667-A*

Hem JD. 1989. Study and interpretation of the chemical characteristics of natural water. *US Geol. Surv. Water Supply Pap. 2254*

Hemley JJ, Hosteler PB, Gude AJ, Mountjoy WT. 1960. Some stability relations of alunite. *Econ. Geol.* 64:599–612

Heye D. 1979. Ionium dating of manganese nodules; does diffusion of ionium occur within manganese nodules? In *La Genese des Nodules de Manganese*. Colloq. Int. Cent. Natl. Rec. Sci. 289:93–99

Huneke JC, Smith SP. 1976. The realities of recoil: ^{39}Ar recoil out of small grains and anomalous age patterns in ^{40}Ar/^{39}Ar dating. *Geochim. Cosmochim. Acta* 7:1987–2008 (Suppl.)

Itaya T, Arribas A Jr, Okada T. 1996. Argon release systematics of hypogene and supergene alunite based on progressive heating experiments from 100 to 1000°C. *Geochim. Cosmochim. Acta* 60:4525–35

Ivarson KC, Ross GJ, Miles NM. 1982. Microbiological transformations of iron and sulfur and their applications to acid sulfate soils and tidal marshes. In *Acid Sulfate Weathering; Proceedings of Symposium*. 4:57–75. Madison, Wis.: Soil Sci. Soc. Am.

Jenny H. 1941. *Factors of Soil Formation*. New York: McGraw-Hill. 281 pp.

King LC. 1949. On the ages of African landsurfaces. *Q. J. Geol. Soc. London,* 104:439–53

King LC. 1953. Canons of landscape evolution. *Geol. Soc. Am. Bull.* 64:721–52

Lippolt HJ, Hautmann S. 1995. ^{40}Ar/^{39}Ar ages of Precambrian manganese ore minerals from Sweden, India and Morocco. *Miner. Deposita* 30:246–56

Long DT, Fegan NE, McKee JD, Lyons WB, Hines ME, et al. 1992. Formation of alunite, jarosite and hydrous iron oxides in a hypersaline system; Lake Tyrrell, Victoria, Australia. *Chem. Geol.* 96:183–202

Mandernack KW, Fogel ML, Tebo BM, Usui A. 1995a. Oxygen isotope analyses of chemically and microbially produced manganese oxides and manganates. *Geochim. Cosmochim. Acta* 59:4409–25

Mandernack KW, Post J, Tebo BM. 1995b. Manganese mineral formation by bacterial spores of the marine Bacillus, strain SG-1: evidence for the direct oxidation of Mn(II) to Mn(IV). *Geochim. Cosmochim. Acta* 59:4393–408

McDougall I, Harrison TM. 1988. *Geochronology and Thermochronology by the ^{40}Ar/^{39}Ar Method*. London/New York: Oxford Univ. Press. 212 pp.

McKenzie RM. 1989. Manganese oxides and hydroxides. In *Minerals in Soil Environments*, ed. JB Dixon, SB Weed, pp. 439–65, Madison, Wis.: Soil Sci. Soc. Am., 2nd ed.

Miura H. 1986. The crystal structure of hollandite. *Mineral. J.* 13:119–29

Murray CN, Riley JP. 1970. The solubility of gases in distilled water and sea water. III. Argon. *Deep-Sea Res.* 17:203–9

Musset AE. 1969. Diffusion measurements and the potassium-argon method of dating. *Geophys. J. R. Astron. Soc.* 18:257–303

Myer R, Reis RRB. 1985. Paleosols and alunite silcretes in continental Cenozoic of western Portugal. *J. Sediment. Petrol.* 55:76–85

Nahon DB. 1986. Evolution of iron crusts in tropical landscapes. In *Rate of Chemical Weathering of Rocks and Minerals*, ed. SM Cohen, DP Dethier, 9:169–91, New York: Academic

Nordstrom DK. 1982. Aqueous pyrite oxidation and the consequent formation of secondary iron minerals. In *Acid Sulfate Weathering*, ed. JA Kittrick, DS Fanning, LR Hossner, *Soil Sci. Soc. Am. Spec. Publ.* 10:37–55

Polyak VJ, McIntosh WC, Guven N, Provencio, P. 1998. Age and origin of Carlsbad cavern and related caves from Ar-40/Ar-39 of alunite. *Science* 279:1919–22

Post JE, Bish DL. 1988. Rietveld refinement of the todorokite structure. *Am. Mineral.* 73:861–69

Post JE, Veblen DR. 1990. Crystal structure and determinations of synthetic sodium, magnesium, and potassium birnessite using TEM and the Rietveld method. *Am. Mineral.* 75:477–89

Pracejus B. 1989. *Nature and Formation of Supergene Manganese Deposits on Groote Eylandt, N.T., Australia*. Adelaide, Australia: Univ. of Adelaide. 231 pp.

Robert C, Chamley H. 1987. Cenozoic evolution of continental humidity and paleoenvironment, deduced from the kaolinite content

of oceanic sediments. *Palaeogeogr. Palaeoclimatol. Palaeoecol.* 60:171–87
Ruffet G, Innocent C, Michard A, Feraud G, Beauvais A, et al. 1996. A geochronological $^{40}Ar/^{39}Ar$ and $^{87}Rb/^{86}Sr$ study of K-Mn oxides from the weathering sequence of Azul, Brazil. *Geochim. Cosmochim. Acta* 60:2219–32
Rye RO, Bethke PM, Lanphere MA, Steven TA. 1993. Age and stable isotope systematics of supergene alunite and jarosite from the Creed mining district, Colorado: implication for supergene processes and Neogene geomorphic evolution and climate of the Southern Rocky Mountains. *Geol. Soc. Am. Abstr. Programs* 25:274 (Abstr.)
Scott KM. 1987. Solid solution in, and classification of, gossan-derived members of the alunite-jarosite family, northwest Queensland, Australia. *Am. Mineral.* 72:178–87
Scott KM. 1990. Origin of alunite- and jarosite-group minerals in the Mt. Leyshon epithermal gold deposit, northeast Queensland, Australia. *Am. Mineral.* 75:1176–81
Segev A, Halicz L, Lang B, Steinitz G. 1991. K-Ar dating of manganese minerals from the Eisenbach region, Black Forest, southwest Germany. *Schweiz. Mineral. Petrogr. Mitt.* 71:101–14
Segev A, Halicz L, Steinitz G, Lang B. 1995. Post-depositional processes on a buried Cambrian sequence in southern Israel, north Arabian Massif: evidence from new K-Ar dating of Mn-nodules. *Geol. Mag.* 132:375–85
Shanin LL, Ivanov IB, Shipulin FK. 1968. The possible use of alunite in K-Ar geochronometry. *Geokhimiya* 1:109–11
Shen YF, Zerger RP, DeGuzman RN, Suib SL, McCurdy L, et al. 1993. Manganese oxide octahedral molecular sieves: preparation, characterization, and applications. *Science* 260:511–25
Silberman ML, Ashley RP. 1970. Age of ore deposition at Goldfield, Nevada, from potassium-argon dating of alunite. *Econ. Geol.* 65:352–54
Sillitoe RH, McKee EH. 1996. Age of supergene oxidation and enrichment in the Chilean porphyry copper province. *Econ. Geol.* 91: 164–79
Smith PE, Evensen NM, York D. 1993. First successful ^{40}Ar-^{39}Ar dating of glauconies: argon recoil in single grains of cryptocrystalline material. *Geology* 21:41–44
Spikings RA, Foster DA, Kohn BP. 1997. Phanerozoic denudation history of the Mount Isa Inlier, northern Australia; response of a Proterozoic mobile belt to intraplate tectonics. *Int. Geol. Rev.* 39:107–24
Sreenivas BL, Roy R. 1961. Observations on cation exchange in some manganese minerals by electrodialysis. *Econ. Geol.* 56:198–203
Stoffregen RE, Cygan GL. 1990. An experimental study of Na-K exchange between alunite and aqueous sulfate solutions. *Am. Mineral.* 75:209–20
Stoffregen RE, Rye RO, Wasserman MD. 1994. Experimental studies of alunite: II. Rates of alunite-water alkali and isotope exchange. *Geochim. Cosmochim. Acta* 58:917–29
Stokowski SJ, Krueger HW. 1981 Pliocene potassium-argon dates of jarosite formed during Mississippi River downcutting. *Geol. Soc. Am. Abstr. Programs* 13:318 (Abstr.)
Stone AT. 1987. Reductive dissolution of manganese (III/IV) oxides by substituted phenols. *Environ. Sci. Technol.* 21:979–88
Stone AT, Morgan JJ. 1984a. Reduction and dissolution of manganese (III) and manganese (IV) by organics. 1. Reaction with hydroquinone. *Environ. Sci. Technol.* 18:450–56
Stone AT, Morgan JJ. 1984b. Reduction and dissolution of manganese (III) and manganese (IV) oxides by organics: 2. Survey of the reactivity of organics. *Environ. Sci. Technol.* 18:617–24
Tardy Y, Roquin C. 1992. Geochemistry and evolution of lateritic landscapes. In *Weathering Soils and Paleosols*, ed. IP Martini, W Chesworth, p. 407–43. Amsterdam/Lausanne/New York: Elsevier. 618 pp.
Tsuji M, Komarneni S, Abe M. 1993. Ion exchange selectivity for alakali metal ions on a crytomelane-type hydrous manganese dioxide. *Solvent Extr. Ion. Exch.* 11:143–58
Turner S, Post JE. 1988. Refinement of the substructure and superstructure of romanechite. *Am. Mineral.* 73:1155–61
van Breeman N. 1988. Redox processes of iron and sulphur involved in the formation of acid sulfate soils. In *Iron in Soils and Clay Minerals*, ed. JW Stucki, BA Goodman, U Schwertmann, p. 825–841. Brussels: NATO
Varentsov IM, Golovin DI. 1987. Manganese beds of Groote Eylandt, northern Australia: K/Ar dating of cryptomelane minerals and genetic aspects. *Dokl. Akad. Nauk USSR* 294:203–7
Vasconcelos PM. 1992. Timing and rates of evolution of hydrochemical systems in semi-arid and humid environments by application of ^{40}K-^{40}Ar and laser-heating $^{40}Ar/^{39}Ar$ dating of K-bearing weathering product minerals. PhD thesis. Berkeley: Univ. Calif. 242 pp.
Vasconcelos PM. 1996. Geochronological evidence for the preservation of Cretaceous weathering profiles in northwestern Queensland. In Mesozoic Geology of the Eastern Australia Plate Conference, *Geol. Soc. Aust. Extended Abstr.* 43:543–544 (Abstr.)

Vasconcelos PM. 1998. Geochronology of weathering in the Mount Isa and Charters Towers regions, northern Queensland. *Restricted Rep. 68R/E&M Rep. 452R*, Perth, Australia: CRC LEME

Vasconcelos PM, Becker TA, Renne PR, Brimhall GH. 1992. Age and duration of weathering ^{40}K-^{40}Ar and $^{40}Ar/^{39}Ar$ analysis of K-Mn Oxides. *Science* 258:451–55

Vasconcelos PM, Becker TA, Renne PR, Brimhall GH. 1994a. Direct dating of weathering phenomena by K-Ar and $^{40}Ar/^{39}Ar$ analysis of supergene K-Mn oxides. *Geochim. Cosmochim. Acta* 58:1635–65

Vasconcelos PM, Brimhall GH, Becker T, Renne PR. 1994b. $^{40}Ar/^{39}Ar$ analysis of supergene jarosite and alunite: implications to the paleo weathering history of western US and West Africa. *Geochim. Cosmochim. Acta* 58:401–20

Vasconcelos PM, Renne PR, Becker TA, Wenk HR. 1995. Mechanisms and kinetics of atmospheric, radiogenic, and nucleogenic argon release from cryptomelane during $^{40}Ar/^{39}Ar$ analysis. *Geochim. Cosmochim. Acta* 59:2057–70

Vasconcelos PM, Wenk HR, Echer C. 1994c. In situ study of the thermal behavior of cryptomelane by high voltage and analytical electron microscopy. *Am. Mineral.* 79:80–90

Virtue TL. 1996. Geology, mineralogy, and genesis of supergene enrichment at the Cananea porphyry copper deposit, Sonora, Mexico. Master's thesis. Univ. Tex. El Paso. 261 pp.

Webb AW, McDougall I. 1968. The geochronology of the igneous rocks of eastern Queensland. *J. Geol. Soc. Aust.* 15:313–43

Yashvili LP, Gukasyan RK. 1974. Use of cryptomelane for potassium-argon dating of manganese ore of the Sevkar-Sarigyukh Deposit, Armenia. *Trans. Acad. Sci. USSR Dokl., Earth Sci. Sect.* 212:49–51

Annu. Rev. Earth Planet. Sci. 1999. 27:231–85

THERMOHALINE CIRCULATION: High-Latitude Phenomena and the Difference Between the Pacific and Atlantic

A. J. Weaver, C. M. Bitz, A. F. Fanning, and M. M. Holland
School of Earth & Ocean Sciences, University of Victoria, P.O. Box 3055, Victoria, British Columbia, Canada V8W 3P6; e-mail: weaver@uvic.ca

KEY WORDS: thermohaline circulation, climate variability, ocean circulation, Arctic Ocean

ABSTRACT

Deepwater formation, the process whereby surface water is actively converted into deep water through heat and freshwater exchange at the air-sea interface, is known to occur in the North Atlantic but not in the North Pacific. As such, the thermohaline circulation is fundamentally different in these two regions. In this review we provide a description of this circulation and outline a number of reasons as to why deep water is formed in the North Atlantic but not in the North Pacific. Special emphasis is given to the role of interactions with the Arctic Ocean. We extend our analysis to discuss the observational evidence and current theories for decadal-interdecadal climate variability in each region, with particular focus on the role of the ocean. Differences between the North Atlantic and North Pacific are highlighted.

INTRODUCTION

The discovery by Ellis (1751), captain of the ship *Earl of Halifax*, that the deep water at 25′13″N, 25′12″W was cold appears almost trivial to us today. In fact, the realization that deep water is cold even in equatorial regions has profound implications. The analysis of water mass properties has traced the cold deep water back to its polar origin. Since the surface-intensified, wind-driven gyres are not generally thought capable of driving the deep meridional flows

0084-6597/99/0515-0231$08.00

required for such excursions, the quest for the deep circulation leads directly to the thermohaline circulation.

In his remarkable essay, Thompson, Count Rumford (1798) (Figure 1), was the first to envision a meridional overturning driven by cooling at high latitudes and warming at the equator. Thompson (1798) pointed out that surface waters that were cooled at high latitudes must sink and spread out equatorward and noted the surface and deep ocean temperature observations of Ellis (1751) and Kirwan (1787) as "incontrovertible proof of the existence of currents of cold water at the bottom of the sea." He further argued that the Gulf Stream in the North Atlantic was the surface manifestation of this thermally driven meridional overturning.

In this same essay, Thompson, Count Rumford (1798), discussed the concept of thermal stability of the oceans and noted the moderating effect the oceans had on climate. He wrote:

> The vast extent of the ocean, and its great depth, but still more numerous its currents, and the power of water to absorb a vast quantity of Heat, render it peculiarly well adapted to serve as an equalizer of Heat.

He further wrote:

> There is, however, a circumstance by which these rapid advances of winter are in some measure moderated. The earth, but more especially the *water*, having imbibed a vast quantity of Heat during the long summer days, while they receive the influence of the sun's vivifying beams; this Heat, being given off to the cold air which rushes in from polar regions, serves to warm it and soften it, and consequently to diminish the impetuosity of its motion, and take off the keenness of its blast.

These remarkable and colorful insights into the large-scale interaction between the oceans and the atmosphere, the formation of deep waters, and the moderating effect that the oceans have on climate have largely stood unchanged for two centuries.

THE THERMOHALINE CIRCULATION IN THE NORTH ATLANTIC AND PACIFIC OCEANS

The ocean is well known to have a moderating effect on climate through several mechanisms. It is the buffer that moderates daily, seasonal, and even interannual temperature fluctuations. Comparison of the maritime climate of Victoria, British Columbia (48°25′N, 123°22′W, Figure 2, with average temperatures of 4°C in January and 16°C in July), with the continental climate of Winnipeg, Manitoba (49°54′N, 97°14′W, with average temperature of −18°C in January and 20°C in July), shows the moderating effect of the ocean. The ocean also acts as a large-scale conveyor that transports heat from low to high latitudes,

Figure 1 Portrait of Sir Benjamin Thompson, Count Rumford, taken in 1798 at the age of 45 (taken from Ellis 1868).

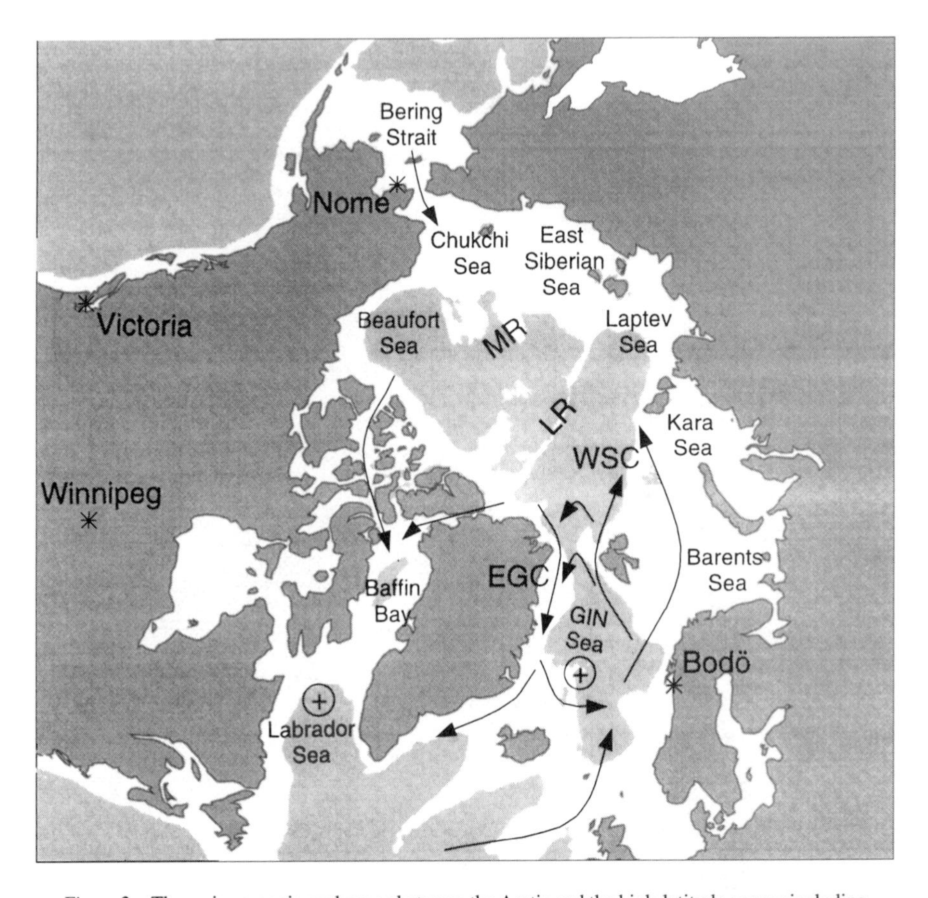

Figure 2 The major oceanic exchanges between the Arctic and the high-latitude oceans including the East Greenland Current (EGC) and the West Spitsbergen Current (WSC). Depths greater than 1000 m are shaded, and the Mendeleyev (MR) and Lomonosov (LR) ridges, as well as the general sites of deepwater formation (*circled pluses*), are also illustrated. Also shown are the locations of Victoria, British Columbia; Winnipeg, Manitoba; Bodö, Norway; Nome, Alaska.

reducing latitudinal gradients of temperature. Much of the oceanic heat transport is thought to be associated with the thermohaline circulation (that part of the ocean's circulation that is driven by fluxes of heat and fresh water through the ocean's surface). In the North Atlantic, intense heat loss to the overlying atmosphere causes deep water to be formed in the Greenland, Iceland, and Norwegian (GIN) Seas (Figure 2). These sinking regions are fed by warm, saline waters brought by the thermohaline circulation from lower latitudes. No such deep sinking exists in the Pacific. Again, if one compares the climates of

Bodö, Norway (67°17′N, 14°25′E, Figure 2, with average January temperature of −2°C and average July temperature of 14°C), to that of Nome, Alaska (64°30′N, 165°26′W, with average January temperature of −15°C and average July temperature of 10°C) (both of which are at similar latitudes and on the western flanks of continental land masses), one directly sees the impact of the poleward heat transport of the thermohaline circulation. The ocean can also regulate climate through its ability to store both anthropogenic and natural greenhouse gases.

The now classical picture of the overturning in a meridional plane driven by the poleward decrease in the net surface radiation is shown in Figure 3, taken from Wyrtki (1961). Warm, light water flows poleward at the surface and loses its buoyancy by intense cooling at high latitudes. The newly formed deep water spreads equatorward and diffuses slowly upward through the thermocline, at a rate of only a few meters per year (Bretherton 1982). With the additional constraint of geostrophy (the balance between horizontal pressure gradient and Coriolis forces), such a meridional flow must be supported by either an east-west pressure gradient, for which lateral boundaries are required, or friction, as in the circumpolar ocean—e.g. the analytic model of Wyrtki (1961).

The approach initiated by Stommel (1957) and pursued by Stommel & Arons (1960) was to isolate the deep ocean by a level surface at some depth upon which the vertical velocity was everywhere specified. The flow was supposed

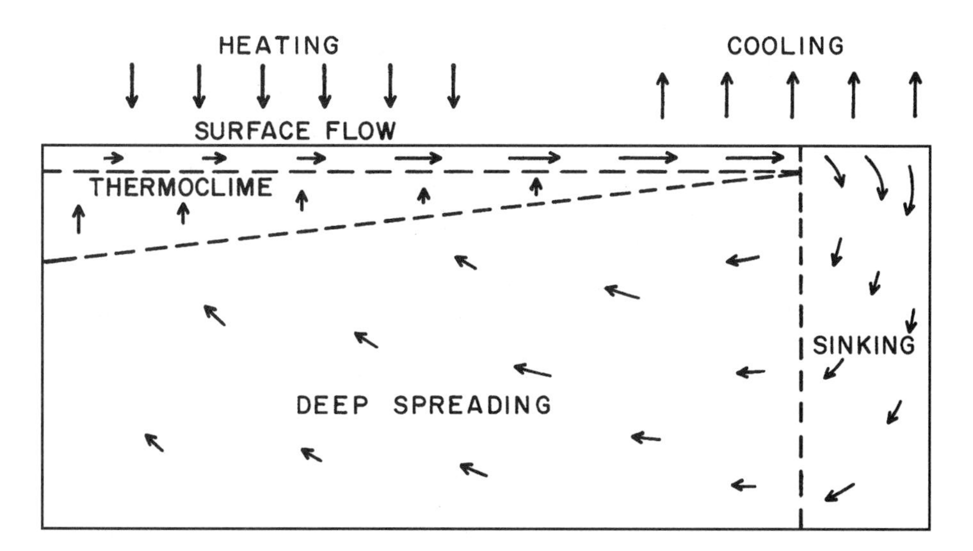

Figure 3 The now classical model of a purely thermohaline circulation driven by heating at low latitudes and cooling at high latitudes (redrawn from Wyrtki 1961).

independent of depth in the deep layer and so was dictated by the divergence/convergence needed to supply the vertical mass transport across the level of no motion. As the true distribution of the vertical velocity in the world ocean is poorly known (although Gordon 1986, Broecker 1991, and Schmitz & McCartney 1993 refer to the circumpolar ocean as the primary site for the upwelling), the assumption of uniform upwelling was made, except for specified high-latitude deep sinking regions. In the interior the geostrophic balance was expressed through the Sverdrup relation:

$$\beta v = f w_z \tag{1}$$

where v is the northward velocity and f is the Coriolis parameter with meridional gradient β. In a flat-bottomed ocean of depth H, with the vertical velocity $w = 0$ at a depth $z = -H$, the vertical integration of Equation 1 over the deep layer shows that the vertically integrated deep flow must be poleward. This can be reconciled with the above description since continuity requires that a narrow, rapid, equatorward western boundary current develop to feed the interior flow. In a zonal average the equatorward boundary current dominates the poleward interior flow, and the two-dimensional overturning cell described by Wyrtki (1961) is recreated.

Figure 4 is a sketch of an idealized deep ocean circulation as obtained using the Stommel-Arons approach. In this example, two sources of deep sinking are prescribed: one in the northern North Atlantic and one in the Southern Ocean, in approximate agreement with the observed deep circulation today. The three main oceans are represented by flat-bottomed basins bounded by parallels and meridians. The geostrophic flow in the interior is then poleward in both hemispheres, but cross-equatorial transport is possible in the western boundary currents if the strength of the deepwater source in a hemisphere is not precisely balanced by the uniform upwelling over the rest of the hemisphere.

Kawase (1987) refined the Stommel-Arons model by parameterizing (rather than prescribing) the cross-interfacial flow as proportional to the interfacial displacement. In his baroclinic (inverted reduced gravity) model, the vertical velocity then acts as a damping term in the continuity equation. When the damping is strong, the westward radiation of Rossby waves into the interior is impeded, and the deep flow is confined to boundary layers along the coasts and the equator. For weak damping, the normal Stommel-Arons circulation is duplicated.

To obtain a first idea of the importance of wind in the deep ocean, Wyrtki (1961) considered the distribution of Ekman pumping and Ekman suction at the base of the surface layer in the Southern Hemisphere. He imagined four ocean layers outcropping at successively higher latitudes and bounded by the limits of the zones of wind-driven convergence and divergence. The four layers were compared to the observed stratification of the ocean, with thermocline water

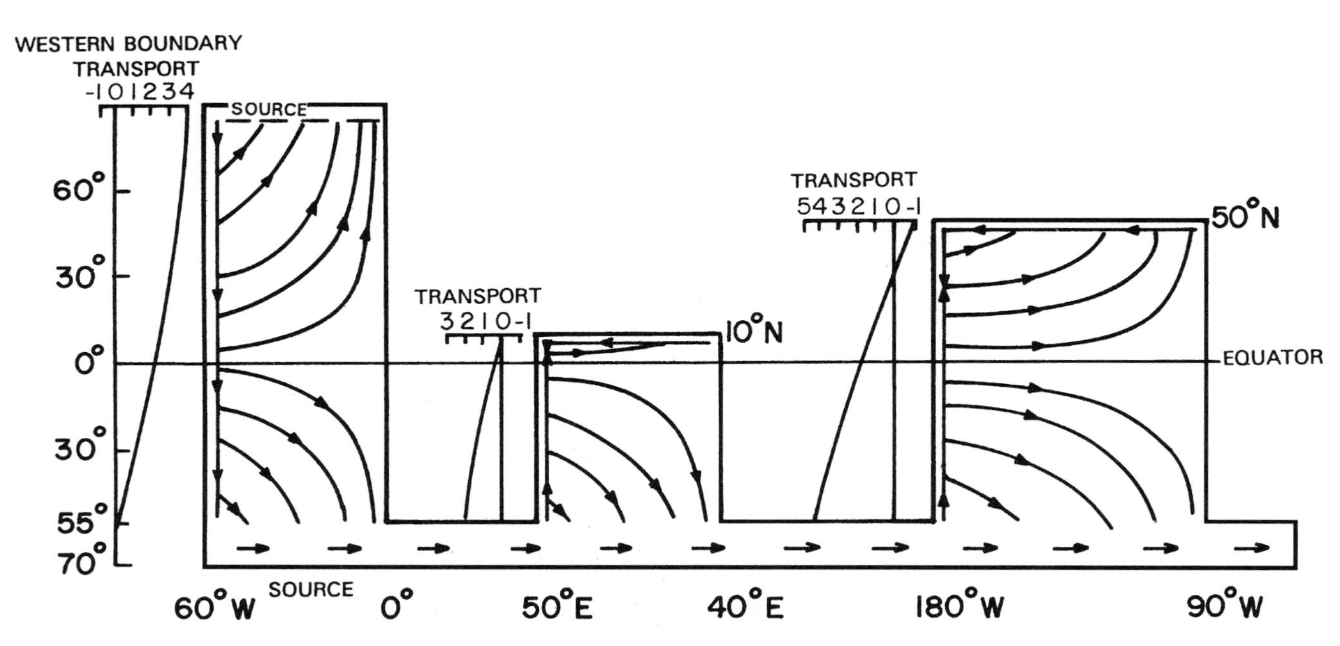

Figure 4 The Stommel-Arons model of the deep ocean thermohaline circulation with two point sources and uniform upwelling (redrawn from Warren 1981, after Kuo & Veronis 1973).

lying above intermediate water, deep water, and bottom water. In his analytic model the wind-induced meridional transport was calculated to be nonzero even in the lowest layer, although much weaker than in the overlying layers.

A similar partitioning of the sub–mixed-layer ocean into four layers was used by Yin et al (1992) in a numerical model, with the goal of studying the competition between the surface-driven Ekman pumping and the upwelling caused by the deep thermohaline circulation. In their experiments, outcropping was not permitted, but one or more point sources of deep sinking could be prescribed. The density structure was again specified, and cross-isopycnal flow was parameterized in the manner of Kawase (1987). They found that in the absence of deepwater formation the circulation of the deep ocean was several orders of magnitude weaker than that of the upper ocean. The introduction of a deepwater source at high latitudes in the Northern Hemisphere, channeling water from the first model layer into the third, created a pattern of surface convergence and divergence at depth that was superimposed on the wind-driven circulation. This resulted in the enhancement of the poleward western boundary current in the thermocline layer and the appearance of an equatorward undercurrent in the deep layer, in close analogy to the interaction between deepwater formation in the northern North Atlantic and the Gulf Stream (except that the thermodynamic processes causing the deep convection were not modeled). In Yin et al's calculations, the speed of the flow in the deep layer was increased by a factor of about 10 through the imposition of a northern deepwater source of magnitude 20 Sv (1 Sv $= 10^6$ $m^3\,s^{-1}$). When this was supplemented by a bottom-water source in the opposite hemisphere, a northward-flowing western boundary current appeared in the lowest layer, resulting in the type of layered circulation that Stommel & Arons' (1960) original theory was unable to describe. This second source had the additional effect of increasing the cross-interfacial velocities in the model so that the deep water from the north penetrated less far south before detraining completely.

In the Stommel-Arons approach, the meridional gradient of temperature at the surface and the associated secondary circulation (eastward surface flow as demanded by geostrophy, leading to an eastern sea level rise and subsequently setting up an east-west pressure gradient which drives the poleward flow) is presented as less fundamental to the thermohaline circulation than the formation of deep water at high latitudes (corresponding to a sink of mass at the surface and a source at depth). The interior flow is determined by vorticity conservation, and any poleward heat transport is more closely related to the requirement that the circulation be closed by a fast poleward (frictional) western boundary current than to the meridional temperature gradient itself.

Two distinct forms of deepwater formation are recognized: near-continent and open ocean (Killworth 1979). The former involves one of two simple processes: intense evaporation or more typically brine rejection above a continental shelf produces dense water, which sinks down and along the slope under the combined forces of gravity, friction, and the Coriolis force (Killworth 1983); alternatively, potentially supercooled water may be formed at the base of a thick ice shelf during freezing or melting, and this dense water may in turn flow downslope (Foldvik & Gammelsrøld 1988, Grumbine 1991). By contrast, open ocean convection is observed in areas remote from land and is characterized by a large-scale cyclonic mean circulation, which causes a doming of the isopycnals and weakens the static stability over an area tens of kilometers wide. The convection itself is short-lived, is restricted to a narrow circumference, and is typically driven through intense surface cooling. These narrow convective chimneys are almost purely vertical and usually do not entrain large volumes of surrounding water as in near-continent convection (Killworth 1979, 1983).

Both types of deep convection have been observed in the Mediterranean (Killworth 1983, Leaman & Schott 1991), but the dense salty outflow that exits into the Atlantic gains buoyancy by mixing with lighter environmental water and spreads out at middepth rather than sinking to the bottom (Warren 1981, see also Armi & Farmer 1988). Similar saline outflows have been detected issuing at middepth from the Red Sea and Persian Gulf, but each amounts to less than 0.5 Sv in transport (Siedler 1969, Warren 1981). Deep convection occurs at a number of locations around Antarctica, but the dense bottom water is susceptible to being trapped by topographic sills (as in the Bransfield Strait) or by local circulation patterns, not excluding the Antarctic Circumpolar Current (ACC) (Killworth 1983).

Deepwater formation in the Ross Sea has recently been estimated as 13 Sv (Jacobs et al 1985). While Ross Sea deepwater may contribute significantly to Antarctic Bottom Water (AABW) production, it is thought that the major source of AABW (approximately 80 percent) is in the Weddell Sea (Foldvik & Gammelsrøld 1988).

In summary, two main bottom-water masses exist whose trajectories can be traced by their temperature and salinity characteristics throughout the rest of the world ocean. These are (*a*) the component of AABW, which is largely produced in the Weddell Sea before mixing with Circumpolar Deep Water in the ACC and then flowing into the major ocean basins; and (*b*) North Atlantic Deep Water (NADW), which lies above the AABW at all latitudes except north of 40°N in the North Atlantic (Mantyla & Reid 1983). The North Pacific, although a source of North Pacific Intermediate Water (NPIW), is not a source of bottom water for reasons to be discussed later.

North Atlantic

The general features of the circulation in the North Atlantic have been well known for many years, although refinements are continually forthcoming. It is not the purpose of this review to give an exhaustive description of the North Atlantic Ocean circulation, and the reader is referred to the classic works of Ivers (1975), Worthington (1976), and Krauss (1986) for more historical details regarding the refinement of our understanding. More recent reviews are given by Schmitz & McCartney (1993) and Schmitz (1995). In this section we focus primarily on recent developments in our understanding of the thermohaline circulation of the North Atlantic.

The Greenland Sea has long been known to be a major deep-water formation site (Killworth 1979, Warren 1981), and in more recent years, deep convection has actually been observed (Rudels et al 1989, Rhein 1991, Schott et al 1993). Warm, very saline near-surface water from the Atlantic circulates counterclockwise from the Norwegian Sea into the Greenland Sea (Aagaard et al 1985b), where intense winter cooling can induce convection, causing the wintertime properties of the water in the cyclonic gyre to become practically homogeneous in the vertical.

The northern seas are separated from the North Atlantic by the Greenland-Scotland ridge, with a maximum depth of only about 600 to 800 m. Hence the renewal of the deep North Atlantic is actually fed by an overflow of intermediate-depth water from the GIN Seas (Aagaard et al 1985b). About 1 Sv of northern source water passes between Iceland and the Faroe Islands (Meincke 1983) and about 1.7 Sv passes between the Faroe Islands and Scotland (Borenäs & Lundberg 1988, Saunders 1990) (see Figure 5). As these overflow waters flow southwestward in a deep western boundary current, they entrain surrounding waters, yielding 3.2 Sv of transport southeast of Iceland (Saunders 1996). Although it is unclear what happens to the rest of the waters (most likely there is 0.8 Sv of recirculation in the deep Iceland Basin, south of Iceland), a strong constraint on the amount of Iceland-Faroes-Scotland ridge overflow water entering the western Atlantic basin, via transport as a deep boundary current along the Mid-Atlantic Ridge, is provided by recent measurements of 2.4 Sv through the Charlie Gibbs Fracture Zone (Saunders 1994).

In addition, a nearly equal volume (2.9 Sv, Ross 1984) of slightly colder northern source water passes over the shallow sill in the Denmark Strait, rapidly entraining surrounding water (Price & Baringer 1994) to yield about 5.1 (5.2) Sv 160 (320) km downstream from the sill (Dickson & Brown 1994). At 480 km downstream from the sill, Dickson et al (1990) and Dickson & Brown (1994) find 10.7 Sv of deep transport. Dickson & Brown (1994), in reference to McCartney (1992), provide compelling arguments suggesting that the difference between the observed transport (10.7 Sv) and the overflow transport plus Charlie Gibbs Fracture Zone transport (5.1 Sv + 2.4 Sv = 7.5 Sv) is largely caused by entrainment of recirculating cold, relatively fresh Labrador Sea water (see below). Still farther downstream, off the tip of Greenland, Clarke (1984) found 13.3 Sv of deep transport, increased from upstream through

Table 1 Thermohaline transport (in Sv) summary from Schmitz & McCartney (1993)

Water type	32°S	24°N
Thermocline water	8	13
Intermediate water (AAIW)	5	—
Deep water (NADW)	−17	−18
Bottom water (AABW)	4	5

additional recirculating components and water mass entrainment. The deep western boundary undercurrent is thought to be about 200 to 300 km wide and to transport about 13 to 14 Sv of newly formed NADW (Warren 1981, McCartney & Talley 1984, Schmitz & McCartney 1993, Schmitz 1995) southward and eventually to encounter northward flowing AABW. Despite the high salinity of the NADW (further enhanced by mixing with Mediterranean water at midlatitudes), at such great pressures the colder AABW has higher density and passes below the NADW.

Schmitz & McCartney (1993) (see also Lee et al 1996) put forth a consistent picture of the thermohaline circulation (as summarized in Table 1) based on the 32°S transport estimates of Rintoul (1991) and the 24°N estimates of Schmitz & Richardson (1991) and Schmitz et al (1992). These estimates are consistent with those of Hall & Bryden (1982), who estimated 18 Sv, and Roemmich & Wunsch (1985), who estimated 17 Sv of total overturning at 24°N. That is, about 13 Sv of cold, fresh Antarctic Intermediate Water (AAIW) formed in the sub-Antarctic South Pacific and the Scotia Sea flows into the South Atlantic, about 8 Sv of which is converted into thermocline water through mixing by the time it reaches 32°S (Rintoul 1991). Further mixing converts the remaining 5 Sv of AAIW into thermocline water as it approaches the equator (Schmitz & McCartney 1993). At 24°N there is 13 Sv of thermocline water that flows northward into the high northern North Atlantic, where it is converted into NADW. AABW, originating in the Weddell Sea, and to a lesser extent the Ross Sea, flows into the South and North Atlantic, where it is modified into and exported as NADW into the Southern Ocean.

The northward flowing AABW and southward flowing NADW in the western Atlantic are apparently well described by the Stommel-Arons theory. However, the AABW stream deviates from the theory by crossing over to the eastern side of the basin upon reaching the equator, to lie up against the western slope of the Mid-Atlantic Ridge at 8 to 16°N. Warren (1981) offers a topographic explanation for this, whereas Kawase (1987) suggests buoyancy damping of Rossby waves. In either case, by the middle latitudes this flow has diffused laterally enough that it is no longer distinguishable in the form of a boundary current (Warren 1981).

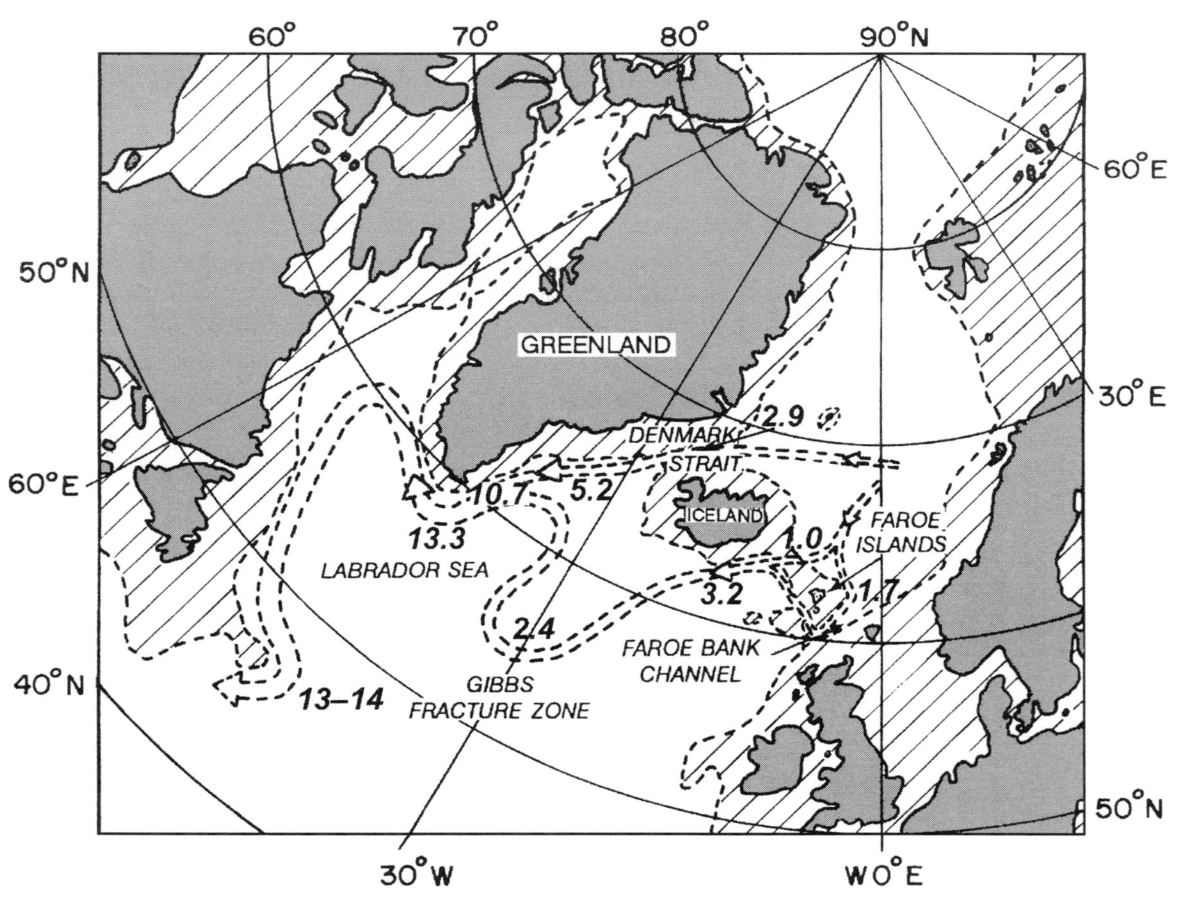

Figure 5 The deep circulation in the northern North Atlantic (redrawn from Warren 1981). The transports (in Sv) at various locations (discussed in the text) are added to the schematic (adapted from Dickson et al 1990 and Dickson & Brown 1994). *Hatched areas* show the continental shelf region.

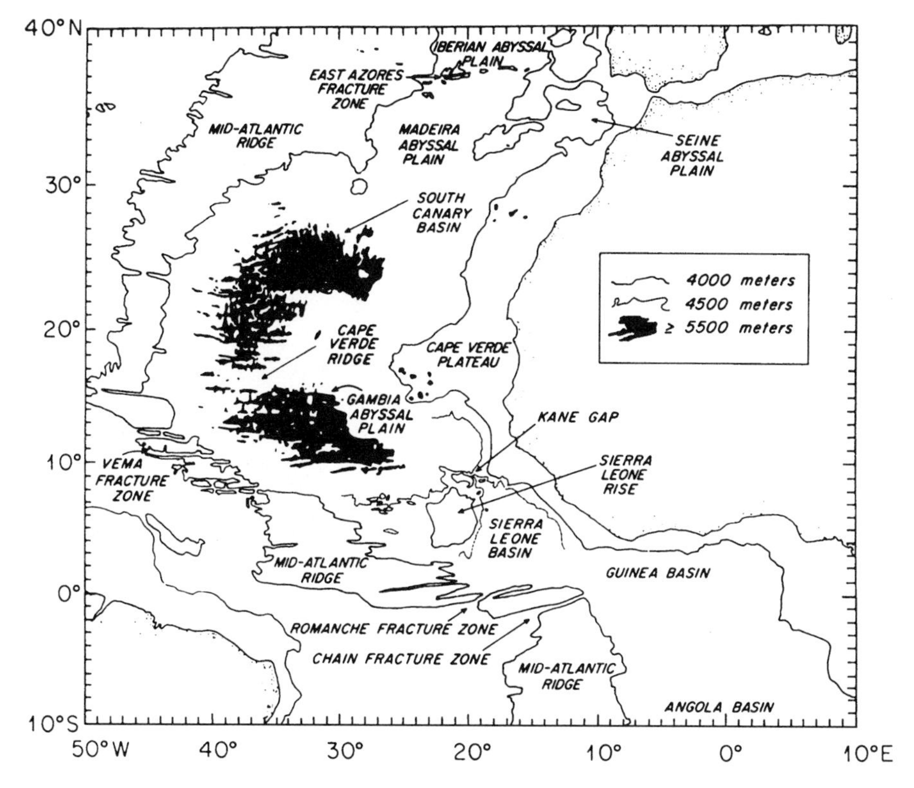

Figure 6 The major topographic features of the eastern North Atlantic (taken from McCartney et al 1991).

The Vema Fracture Zone near 10°N (see Figure 6) is the principal outlet for the deep flow of AABW into the eastern basins of the North Atlantic (Mantyla & Reid 1983, McCartney et al 1991), although farther south, the Romanche Fracture Zone (Figure 6) allows AABW to flow into the deep equatorial and southeastern Atlantic basins (Mantyla & Reid 1983, McCartney et al 1991). In the eastern North Atlantic, the deep water, which has passed through the Vema Fracture Zone, flows both northward and eastward as deep boundary currents trapped against the major topographic features (Figure 6) of the area (McCartney et al 1991).

North Pacific

The deepest connection to the Antarctic in the Pacific sector is toward Drake Passage in the east. The water supplying the central part of the ocean is therefore the lighter water from south of the Campbell Plateau (Figure 7). The

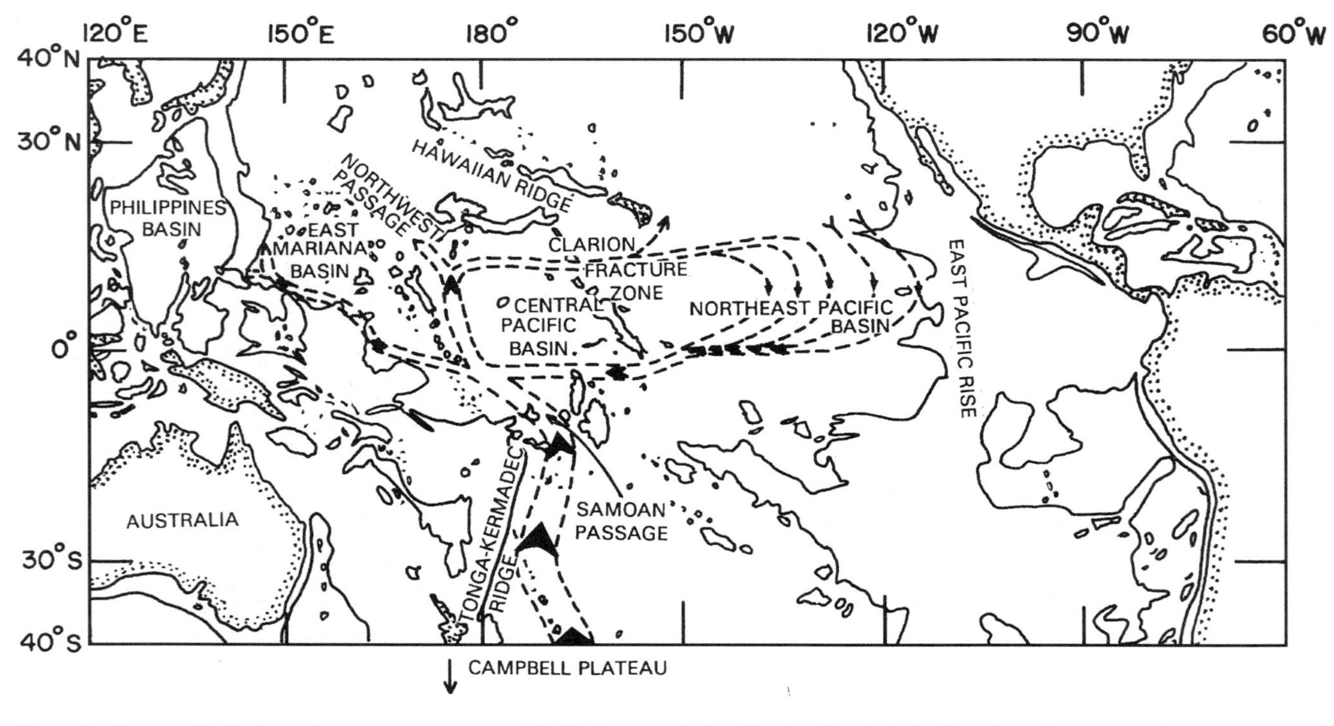

Figure 7 Topographic features of the deep near-equatorial Pacific Ocean as depicted by the 4-km isobath. The schematic representation of the spread of deep Antarctic/circumpolar water is also shown as hypothesized by Johnson (1990) and Johnson & Toole (1993) (redrawn from Johnson 1990 and Johnson & Toole 1993).

effective western boundary for the deep South Pacific is New Zealand and the Tonga-Kermadec Ridge, where a northward boundary current transporting about 19 Sv has been reported (Warren 1973). Above it, there appears to be a weak southward flow whose source is thought to be the North Pacific. Figure 7, taken from Johnson (1990) and Johnson & Toole (1993), is a schematic diagram of the transport of the Antarctic/circumpolar water through the deep basins of the near-equatorial Pacific. About 12 Sv flows northward at 12°S (Taft et al 1991) with about 6 Sv entering the Samoan Passage (Taft et al 1991). The outflow from the Samoan Passage, together with transport of unknown magnitude through gaps in the Robbie Ridge (slightly to the west of the Samoan Passage) and westward recirculated water, feeds deep western boundary currents in the East Mariana and central Pacific Basins, with estimated transports of 5.8 Sv and 8.1 Sv, respectively (Johnson 1990, Johnson & Toole 1993). A relatively weak and broad southward return flow of about 4.7 Sv has also been observed off the western side of the East Pacific Rise in the northeast Pacific Basin (Johnson & Toole 1993).

For many years there was little evidence of deep western boundary currents in the North Pacific (see Johnson 1990 and Hallock & Teague 1996 for reviews), although deep waters clearly had the signature of AABW. Off the Aleutian Islands, however, deep boundary currents were detected (Warren & Owens 1988). Warren (1981) argued that the North Pacific topography was too rough and complex, whereas Joyce et al (1986) argued that the deep flow was too slow for direct observation of deep western boundary currents. Nevertheless, recent observations by Hallock & Teague (1996) (Figure 8) provide compelling evidence for their existence off the east coast of Japan, albeit with a high degree of variability and weak nature. Figure 8, taken from Hallock & Teague (1996), shows a schematic representation of abyssal circulation as proposed by Warren & Owens (1988) and extended by Hallock & Teague (1996). Hallock & Teague (1996) further document the existence of a weak southward flowing deep western boundary current along the slope and inshore of the Japan Trench with a generally northward flow over the trench below the Kuroshio Current (see Figure 8). Their comprehensive review of earlier observations is consistent with the schematic, as are the recent deep geostrophic circulation estimates of Reid (1997).

Although there are no sources of deep water in the North Pacific, intermediate waters, characterized by a salinity minimum in the subtropical gyre at depths of 300 to 800 m within a narrow density range of 26.7 to 26.9σ_θ (Talley 1993), are locally produced.[1] This water mass is mainly constrained to

[1]σ_θ is a measure of the potential density of seawater (ρ_θ): $\sigma_\theta = \rho_\theta - 1000$. Potential density is computed by using potential temperature in the equation of state for seawater. The potential temperature of water is the temperature that water would have if it were adiabatically (no external sources or sinks of heat) brought to the surface of the ocean.

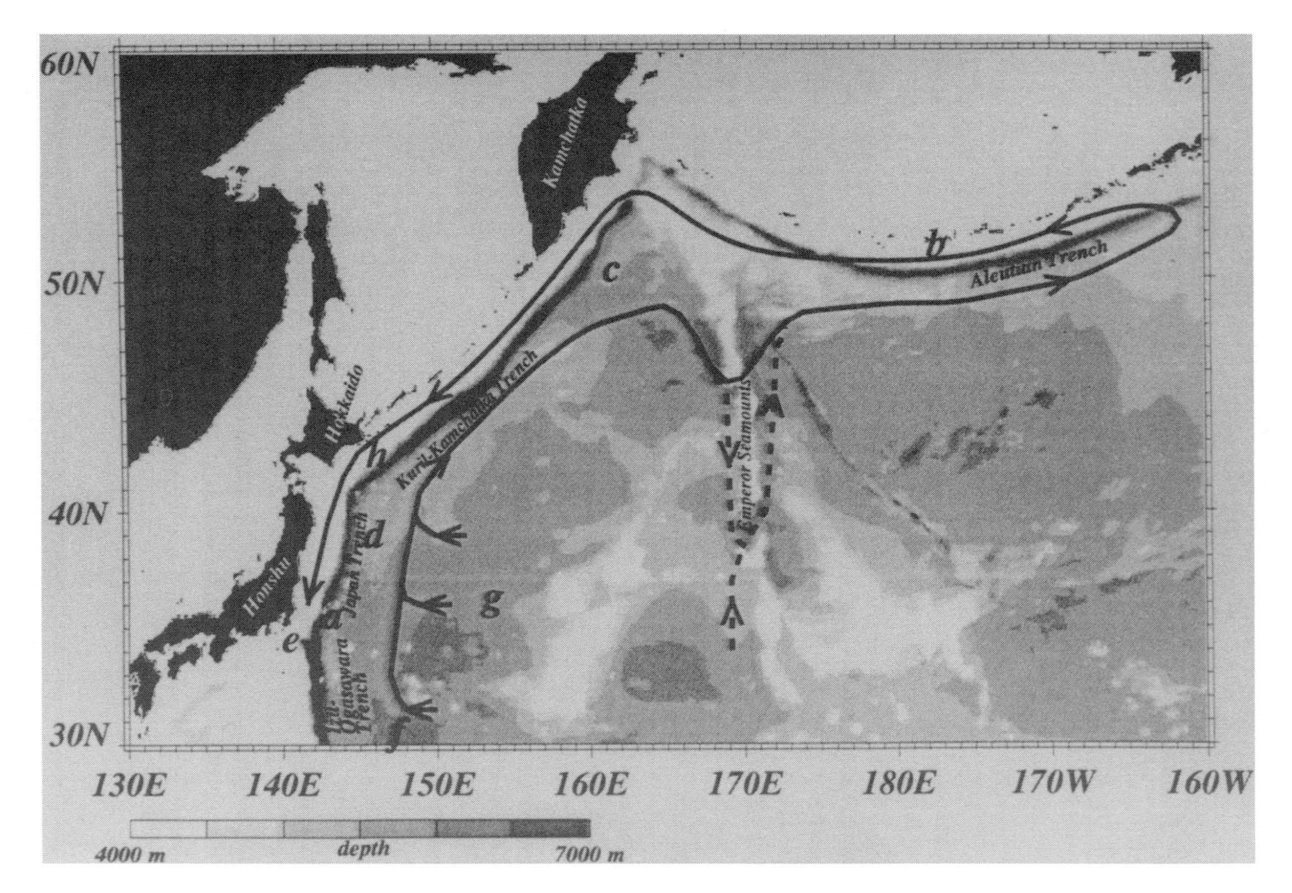

Figure 8 Diagram of the deep circulation in the North Pacific (reproduced from Hallock & Teague 1996) as suggested by Warren & Owens (1988) and modified by Hallock & Teague (1996). The letters indicate references where supporting measurements for this reconstruction are presented (see references in Hallock & Teague 1996).

the subtropical gyre, unlike the Labrador Sea intermediate water mass which is largely confined to the subpolar gyre of the North Atlantic (Talley & McCartney 1982).

Over the past few years, a consistent picture of the formation of North Pacific intermediate water (NPIW) has become apparent. It appears that the original source water for NPIW is sinking in the Sea of Okhotsk (Talley 1991, Yasuda et al 1996, Yasuda 1997, Watanabe & Wakatsuchi 1998). Yasuda (1997) suggests that about 1 Sv of Sea of Okhotsk water (within the 26.6- to 27-σ_θ range) outflows and mixes with warmer and more saline western subarctic gyre water to form Oyashio water. Near the coast, a branch of the Oyashio Current transports newly formed Oyashio water at a rate of about 3 to 7 Sv (Yasuda 1997) past the southern Kuril Islands and Hokkaido before turning eastward in the Kuroshio extension. The distinct signature of NPIW is then formed through mixing in the Kuroshio-Oyashio mixed-water region where the relatively cold, fresh Oyashio waters are overrun by warmer, more saline Kuroshio waters (Hasunuma 1978, Talley 1993, Talley et al 1995, Yasuda et al 1996, Yasuda 1997).

Why Deep Water Forms in the Atlantic but Not in the Pacific

One of the most distinctive characteristics of the present-day global thermohaline circulation is the well-known asymmetry between the Atlantic and Pacific Oceans, with active deep-water formation ongoing in the northern North Atlantic and the oldest water of the world ocean in the deep North Pacific. A central, yet unanswered question concerning the thermohaline circulation is why deep water currently forms in the northern North Atlantic but not in the North Pacific. It is still an open question whether the absence of North Pacific Deep Water (NPDW) is a fundamental aspect of global circulation or a coincidental feature of the present state of the climate. Based on the paleoclimatic record, Keigwin (1987) suggested that even at the Last Glacial Maximum (LGM) (21 KYR), when NADW formation was reduced and shallower than today, NPDW production did not occur. Shackleton et al (1988) also found that during glacial times, deep Pacific water had a ventilation age about 500 years greater than that today, consistent with the results of Keigwin (1987). They further suggested that deep-ocean ventilation rapidly increased around the time of deglaciation, to yield a ventilation age about 500 years less than today. Other paleoclimatic reconstructions for the LGM have suggested the possibility, albeit with large uncertainty, of increased NPIW formation at the LGM (Curry et al 1988, Duplessy et al 1988). Although Berger (1987) suggests that NPDW may have formed during glacial times, he also concedes that increased AABW could also explain his findings. Similarly, Dean et al (1989) provide evidence to support the seasonal formation of a North Pacific deepwater mass during the last glacial-interglacial transition, but they too concede that their results could be explained by a decreased flux of organic matter during this transition. Ventilation of the deep North Atlantic, in contrast, is thought to have reoccurred throughout the Quaternary, although the intensity may have fluctuated significantly through the glacial-interglacial cycles (Broecker & Denton 1989, Boyle 1990, Oppo & Fairbanks 1990).

It is apparent therefore that proxy indicators have by no means drawn a consistent case for the formation of NPDW during the last glaciation or the transition from it to the present Holocene. Two coupled atmosphere-ocean modeling studies available for the LGM (Ganopolski et al 1998, Weaver et al 1998) are also inconsistent in this regard. In their zonally averaged ocean model coupled to an idealized atmospheric model, Ganopolski et al (1998) find increased NPIW formation at the LGM. On the other hand, using a more realistic ocean general circulation model (GCM) coupled to a simpler atmospheric energy-moisture balance model, Weaver et al (1998) find no change in NPIW rates. Neither of these studies supports the hypothesis that NPDW occurred during the LGM.

Warren (1983) singles out the more stable stratification of the North Pacific (where surface waters are on average 32.8 psu and deep waters are 34.6 to 34.7 psu compared with 34.9 psu and 34.9 to 35.0 psu in the North Atlantic) as the explanation for the present-day occurrence of deepwater formation in the North Atlantic but not in the North Pacific. This follows because even if North Pacific waters are cooled to their freezing point (-1.8°C), they are so fresh that they can sink to a depth of only a few hundred meters (Broecker 1991). Warren (1983) identifies a number of causes for this. For example, there is nearly twice as much evaporation over the North Atlantic as over the North Pacific: 103 cm/yr versus 55 cm/yr averaged over the North Atlantic and Pacific, respectively (Baumgartner & Reichel 1975). The water introduced into the northern North Atlantic from lower latitudes is more saline than its counterpart in the Pacific, and the residence time of this water in the region of net precipitation at high latitudes is shorter. However, as Warren (1983) concedes, none of these factors is truly independent of the already existing thermohaline circulation in the North Atlantic. The salinities of the surface and bottom water masses are similar because the one is being actively converted into the other; the higher evaporation is related to higher sea surface temperatures (SSTs), which are due in part to the greater northward advection of warm subtropical water by the Gulf Stream, but the Gulf Stream itself is partly thermohaline-driven (Holland 1973); and finally, the thermohaline contribution to the western boundary current partly accounts for the faster throughflow rate.

Many geographical clues to the asymmetry of the thermohaline circulation in the two oceans do, of course, exist. The first and most obvious one is that the North Atlantic extends farther north than the North Pacific and has a deeper connection with the Arctic. Schmitt et al (1989) have further suggested that the narrower width of the Atlantic compared to the Pacific also causes a greater fraction of its area to be susceptible to the incursions of dry cold continental air that favor evaporation and heat loss at the sea surface. In situ measurements of heat transport across 24°N in the Atlantic (Hall & Bryden 1982) and Pacific (Bryden et al 1991) Oceans are 1.2 $\pm$ 0.3 PW (1 PW $=$ 10^{15} W) and 0.76 $\pm$ 0.3 PW, respectively. The larger zonal extent of the Pacific allows for larger east-west temperature contrasts so that gyre heat transport in the horizontal plane may be large enough to offset the need for a component in the meridional plane (Wang et al 1995). Finally, Reid (1961) has hypothesized that the poleward extension of South America compared with south Africa might impede the transport of fresh water out of the Pacific into the Atlantic by the Antarctic Circumpolar Current (ACC).

The sign of the atmospheric water vapor transport between the Atlantic and Pacific (e.g. Broecker et al 1990) is also important evidence, since this has been suggested to be the driving force for the global conveyor belt (Broecker et al 1985, Broecker 1991). The narrow Isthmus of Panama at low latitudes allows the export of water vapor from the Atlantic to the Pacific via the trade winds (Weyl 1968), whereas the Rocky Mountains along the west coast of North America block a similar export from the Pacific by the westerlies at higher latitudes. The result of these differences is that the net Atlantic-to-Pacific atmospheric freshwater flux in the present climate is thought to be about 0.45 Sv (Baumgartner & Reichel 1975). The circulation of fresh water within the boundaries of the Atlantic alone is equally important to the maintenance of the conveyor belt, since the formation of dense water at high latitudes depends on counteracting the atmospheric transport of water vapor from low to high latitudes (Broecker et al 1990) and the import of fresh water and sea ice from the Arctic (Aagaard & Carmack 1989).

The salty Mediterranean outflow at middepth contributes to the high salinity of the near-surface layer of the Norwegian Sea (Reid 1979), preconditioning it for deep convection. The only marginal sea that could provide a similar mechanism in the Pacific is the Sea of Okhotsk on the western side of the basin. Although the outflow of Mediterranean water does not contribute directly to NADW (since entrainment during descent along the continental slope reduces its density through mixing with thermocline water), it can influence the density within the Labrador Sea and Norwegian Sea overflow regions (Warren 1981). Vertical mixing of the saline waters may therefore contribute to the high-latitude Atlantic salinity (Reid 1979) and precondition waters for convection in the Labrador and GIN Seas. A recent series of hydrographic measurements within the Sea of Okhotsk (H Freeland, personal communication) found that water properties exhibit a maximum density (owing to cold winter outbreaks) at a salinity of 32.94 to 33.08 psu (densities of 26.51 to 26.62 σ_θ) which is not high enough to allow deep convection below the upper several hundred meters of the water column.

A number of climate modeling studies have been conducted to investigate the stability of the present North Atlantic thermohaline circulation with respect to freshwater export from the Atlantic. One of the first such studies (Manabe & Stouffer 1988) demonstrated multiple equilibria in a fully coupled ocean-atmosphere model. Their first state was similar to the present climate, with active NADW production, while their second state had upwelling in both the Atlantic and Pacific. The surface freshwater flux difference between the cases indicates a roughly 0.1 m/yr freshening of the entire North Atlantic. Stocker & Wright (1991) have also identified these states in a zonally averaged coupled ocean-atmosphere model. Through a sensitivity analysis of the Atlantic-to-Pacific atmospheric freshwater transport, they showed that a cessation of NADW production occurred in their coupled model when the net interbasin atmospheric freshwater transport dropped to about 0.03 Sv (from the Atlantic to the Pacific). Once stopped, it took an increase of interbasin freshwater transport to 0.36 Sv before NADW production once more resumed. A similar hysteresis behavior was found by Rahmstorf (1995) through his analysis of the effects of imposed surface freshening to the high-latitude North Atlantic Ocean. In particular, he found that the addition of a perpetual 0.06 Sv freshening (about a quarter of the discharge rate of the Amazon River) to the North Atlantic led to an irreversible shutdown of the conveyor. None of these studies, however, found the occurrence of NPDW, even upon the total shutdown of NADW.

Two notable modeling studies have been able to generate NPDW formation within their parameter sensitivity analyses. In the first of these, Stocker et al (1992), using a zonally averaged ocean model coupled to an energy balance model with a parameterization of the hydrological cycle, found the inverse conveyor belt (with NPDW formation and no NADW formation) as a possible equilibrium solution. Without the inclusion of parameterized wind forcing, this equilibrium tended to occur only under relatively large Atlantic-to-Pacific interbasin atmospheric freshwater transports and relatively large freshwater flux perturbations into the North Atlantic. When wind forcing was included, NPDW states were possible under a wider range of allowable interbasin freshwater transports, although similar large local freshening in the North Atlantic was also required. Hughes & Weaver (1994) also undertook a detailed sensitivity study using an ocean GCM to specifically address why the present-day Pacific Ocean did not form deep water. Under present-day geometry and forcing, there was a clear preference for their model to end up in a conveyor belt scenario (with NADW and AABW formation and no NPDW formation), starting from a variety of initial conditions. It was also possible to realize the second solution (with AABW formation and no NADW or NPDW formation) of Manabe &

Stouffer (1988). Studies of the sensitivity of model solutions to the width of the basin, the northern extent of the Atlantic, the length of southern tip of South America (relative to Africa), and a reduction in the Southern Hemisphere winds all allowed possible equilibria (under present-day boundary conditions) with NPDW formation. Although they were not able to point to a single factor in determining the cause of a lack of NPDW formation in the present climate, they were able to show quantitatively that in cases where NPDW did occur (under changed geometry or wind forcing), the meridional surface density and depth-integrated steric height gradients from the high-latitude North Pacific to the high-latitude South Pacific reversed.[2]

It is clear from this subsection that the question of why deep water forms in the Atlantic and not in the Pacific is far from being resolved. While it is apparent that in the present climate it is due to the fact that the North Pacific is much fresher than the North Atlantic, this in itself is not independent of the existing thermohaline circulation. Nevertheless, it is clear that NPDW can be initiated if processes occur that reverse the present-day basin-scale depth-integrated steric height gradient along the western boundary of the Pacific Ocean. That is, since the large-scale surface meridional flow in both the Pacific and Atlantic Oceans is down the basin-scale meridional depth-integrated steric height gradient (in a frictional western boundary layer) (Hughes & Weaver 1994), processes that can change this gradient control whether deep water formation occurs or does not occur in a particular basin. These processes could occur in either the Northern Hemisphere (e.g. changes in high northern latitude evaporation minus precipitation or Arctic freshwater discharge) or Southern Hemisphere (e.g. changes in the intensity of southern winds).

INTERACTION BETWEEN THE ARCTIC AND THE HIGH-LATITUDE OCEANS

In this section we review the interactions between the Arctic and the North Atlantic and Pacific. We begin by addressing oceanic interactions between the Arctic and the North Atlantic, through the exchange of waters through Fram Strait, Barents Sea, and the Canadian Archipelago and subsequently between the Arctic and the North Pacific via exchanges through the Bering Strait. We then discuss atmospheric interactions between the high-latitude oceans and the Arctic. Owing to the sensitivity of deep convection in the North Atlantic to changes in salinity, a whole subsection is devoted to the freshwater budget of the Arctic.

[2]Steric height gives a measure of the height difference between two isobaric surfaces in the ocean. Gradients in depth-integrated steric height (from an assumed level of no motion to the surface of the ocean) give a measure of upper-ocean mass transport.

Ocean Circulation

This section is partitioned into two parts specifically dealing with the interactions between the North Atlantic and the Arctic and the North Pacific and the Arctic. These interactions are summarized in Figure 2, which shows the oceanic exchanges between the Arctic and high-latitude oceans.

ARCTIC/NORTH ATLANTIC INTERACTIONS The deepwater formation regions of the northern North Atlantic are delicately poised in their ability to sustain convection (Aagaard & Carmack 1989). Because density is largely a function of salinity at low temperatures, a moderate freshening of the surface waters of the North Atlantic could impact deepwater formation and cause a freshwater capping of these regions, leading to a cessation or reduction of deep-water formation. This is one means by which the waters of the North Atlantic are modified through ocean interactions with the Arctic. Additionally, the North Atlantic has a large influence on the import of heat, mass, and salinity into the Arctic Ocean. This has consequences for the stability of the Arctic ice pack.

The Arctic and North Atlantic Oceans communicate directly through the relatively deep (approximately 2500 m deep) Fram Strait, which runs between northeast Greenland and Spitsbergen at approximately 80°N latitude. The West Spitsbergen Current (WSC) brings about 3 to 5 Sv of relatively warm, saline water, with a temperature of approximately 3°C and salinity of 35 psu, northward along the western side of Spitsbergen into the Arctic Ocean. Observations suggest that the maximum transport of the WSC occurs during winter (Morison 1991). As this warm water flows around the Greenland Sea, it is cooled dramatically. Approximately 350 W m^{-2} of heat is lost to the atmosphere, whereas another 200 W m^{-2} or so is lost in the process of melting sea ice which originated in the Barents Sea (Boyd & D'Asaro 1994). The relative strength of these processes influences the density of the water column, with atmospheric cooling resulting in cold and saline water that is capable of sinking to great depths and ice melt causing much fresher and hence stable surface waters.

The WSC appears to split into a number of different branches at the Yermak Plateau just northwest of Spitsbergen (e.g. Quadfasel et al 1987, Aagaard et al 1987). The western section of the current recirculates into the Fram Strait, whereas the remainder of the water moves northward along the shelf break into the Arctic Ocean. It is estimated that only 20 to 30 percent of the WSC continues to the Arctic Ocean, with the remaining 70 to 80 percent recirculating into the Greenland Sea (e.g. Buorke et al 1988, Manley 1995). This recirculated water is relatively dense and is likely an important contributor to the formation of deep water. The northward flowing branch sinks below the relatively fresh polar surface water and makes up a portion of the Atlantic layer, which is observed in the Arctic basin between 200 and 800 m. This layer represents a large reservoir of heat and is identified by a subsurface potential temperature maximum with relatively high salinity. The impact of this water on Arctic sea ice is mitigated by a layer of fresh surface water that stabilizes the water column and reduces the upward heat flux. As discussed below, this surface water is largely maintained by river runoff and the inflow of fresh water from the Pacific Ocean through the Bering Strait.

In addition to the WSC the Atlantic layer in the Arctic is fed by waters that flow into the Arctic through the Barents and Kara Seas (e.g. Rudels et al 1994). This water flows northward to the east of Spitsbergen over the relatively shallow (400 to 500 m) Bear Island Trough. It is modified on the shelves of the Barents Sea through the mixing of river runoff with shelf waters, heat loss to the atmosphere, and sea ice processes. As a result, the Atlantic layer is colder and fresher than the modified Atlantic water that flows through the Fram Strait. The two branches of Atlantic water meet at the shelf break of the Kara Sea. Recent modeling studies (Gerdes & Schauer 1997) suggest that the Barents Sea branch is the most important in terms of the inflow of heat, mass, and salinity into the Arctic Ocean. The strength and variability of this inflow have consequences for ice formation on the shelf regions and the water mass structure within the Arctic. Recent observations (e.g. Carmack et al 1995, McLaughlin et al 1996) indicate that the Atlantic layer within the Arctic Ocean has undergone large changes since 1990. These include a shift in the frontal structure, which separates different Atlantic layer water masses (from the Lomonosov ridge to the Mendeleyev ridge (Figure 2)), and a significant warming of the Atlantic layer. By 1994, this warming extended across the Nansen, Amundsen, and Makarov Basins. Swift et al (1998) show that these changes are likely caused by an increase in the temperature of the Atlantic waters that enter the Arctic Basin through Fram Strait. The anomalous warmth of these waters appears to be correlated with the North Atlantic Oscillation (NAO), which corresponds to relatively warm air temperatures in the Greenland Sea region and thus a reduction in oceanic heat loss. The temperature signal of these waters is transported into the Arctic Ocean by topographically steered boundary currents. It then enters the interior ocean through intrusive layers that extend laterally into the ocean basins (Carmack et al 1998). The question remains open as to where the Arctic waters displaced through the intrusion of the Atlantic layer went, although enhanced transport through the Canadian Archipelago is plausible. This would be consistent with recent observations of anomalous cold and fresh waters in the Labrador Sea since the late 1980s (Dickson et al 1996).

The East Greenland Current (EGC) represents the primary outflow of water from the Arctic Ocean. It transports sea ice as well as surface, intermediate, and deep water southward through Fram Strait along the eastern coast of Greenland. The surface waters of this current are made up of relatively cold ($<0^\circ$C) and fresh (salinity below 34.4 psu) polar water. The recirculated water from the WSC sinks below the surface water and moves southward with the EGC (e.g. Swift & Aagaard 1981). A wide range of transport estimates exists for the EGC (see Carmack 1990). At approximately 72°N a portion of the current branches eastward, resulting in a net cyclonic circulation within the Greenland Sea. The remainder of the EGC continues southward along the Greenland coast.

A complicated series of relatively shallow (less than 400 m) channels make up the Canadian archipelago. Fissel et al (1988) estimate that 1.7 Sv of Arctic surface water is transported through this region into Baffin Bay (see also Aagaard & Carmack 1989). This water is relatively fresh, and variability in its outflow may impact the deep-water formation that occurs in the Labrador Sea region, as discussed earlier.

ARCTIC/NORTH PACIFIC INTERACTIONS The high-latitude North Pacific exchanges surface waters with the Arctic Ocean through the shallow (50 m deep) Bering Strait. This strait (by way of the Arctic Ocean) is one of only two ways in which the Pacific and Atlantic Oceans are connected (the other being via the ACC), suggesting that changes in the relatively fresh Bering Strait inflow may influence NADW formation (e.g. Wijffels et al 1992). Box model studies (Shaffer & Bendtsen 1994) have suggested that the stability of the thermohaline circulation in past climates may have been influenced by the mean transport (or lack thereof) of Pacific waters through the Bering Strait.

In the present climate, approximately 0.8 Sv is transported northward through Bering Strait. This inflow is driven by an approximately 0.5 m decrease in sea level between the Pacific and North Atlantic (e.g. Stigebrandt 1984, Overland & Roach 1987). Its variability is largely tied to the meridional wind field in the region (Aagaard et al 1985a, Coachman & Aagaard 1988). For example, Roach et al (1995) show that under strong (10 to 12 m s^{-1}) sustained northerly winds the Bering Strait flow can temporarily reverse, bringing Chukchi Sea water southward. The Pacific water that is transported into the Arctic Ocean is relatively fresh, with a salinity of approximately 32.5 psu. Seasonal and interannual variability in the transport and salinity of this inflow is high (e.g. Roach et al 1995).

Atmospheric Circulation

The Arctic atmosphere acts as heat sink for the Northern Hemisphere through radiative cooling to space. According to Nakamura & Oort (1988), when averaged annually, 98 percent of the energy needed to balance radiative cooling at the top of the atmosphere from 70° to 90°N enters the Arctic through the poleward convergence of atmospheric heat transport. The remaining two percent is supplied by the underlying surface, suggesting that the poleward oceanic transport of heat and export of ice contribute little to the Arctic atmosphere's energy budget. Nakamura & Oort (1988) showed that the largest contribution to the atmospheric energy transport is from transient eddies, with stationary eddies also making a significant contribution during the winter. Their analysis was later extended by Overland & Turet (1994) and Overland et al (1996), who found the major pathways of heat transport are over the Greenland, Barents, and East Siberian Seas.

Although the annually averaged net flux of heat from the atmosphere to Earth's surface is small, the annual variations are at times comparable to the atmospheric energy transport into the Arctic region (Oort 1974, Nakamura & Oort 1988, Overland & Turet 1994). Based on calculations by Maykut (1982), the net flux from the atmosphere to the underlying surface (averaged over sea ice and open water) in the central Arctic ranges from 75 W m^{-2} in July to -25 W m^{-2} in November–December. Analysis of observations by Vowinckel & Orvig (1971) indicates that the surface energy budget is controlled by advection in the winter and by solar radiation in the spring and summer. Ultimately, the energy exchanged with the atmosphere at the surface influences the mass of sea ice, resulting in approximately a 1 m annual range of the area-averaged thickness in the Arctic Basin.

Recent interest has focused on how the high-latitude atmospheric circulation influences sea ice motion (e.g. Walsh & Johnson 1979, Overland & Pease 1982, Serreze et al 1989, Fang & Wallace 1994), including the export of ice from the central Arctic to the Greenland Sea (Mysak et al 1990, Walsh & Chapman 1990) and the distribution of heat and moisture at high latitudes (Serreze et al 1995, Overland et al 1996). The mean Arctic winter sea-level pressure (SLP) is characterized by high-pressure centers over the Beaufort Sea and Greenland and a ridge over Siberia (Figure 9*a*) associated with the extremely low temperatures in these regions (Figure 10). The mean summer SLP distribution is quite flat, with a weakened Beaufort high shifted toward Canada from its wintertime position. Ice motion, shown with SLP in Figure 9 follows the anticyclonic wind stress along the Beaufort gyre and transpolar drift from Siberia through Fram Strait and into the Greenland Sea. The interaction of the jet stream with the northern Rocky Mountains creates a stationary wave structure, resulting in the polar stratospheric vortex, which is seen extending into the troposphere at the 500-mbar level in Figures 9*b* and 9*d*.

Winter cyclonic activity is most common south of Iceland and often moves along the ice edge into the Norwegian, Barents, and Kara Seas (Whittaker & Horn 1984, Serreze et al 1993). Summer cyclones are generally weaker, less frequent, and more broadly distributed.

NORTH ATLANTIC Observations of strong decadal variability in the North Atlantic SST and atmospheric circulation have prompted many studies to seek

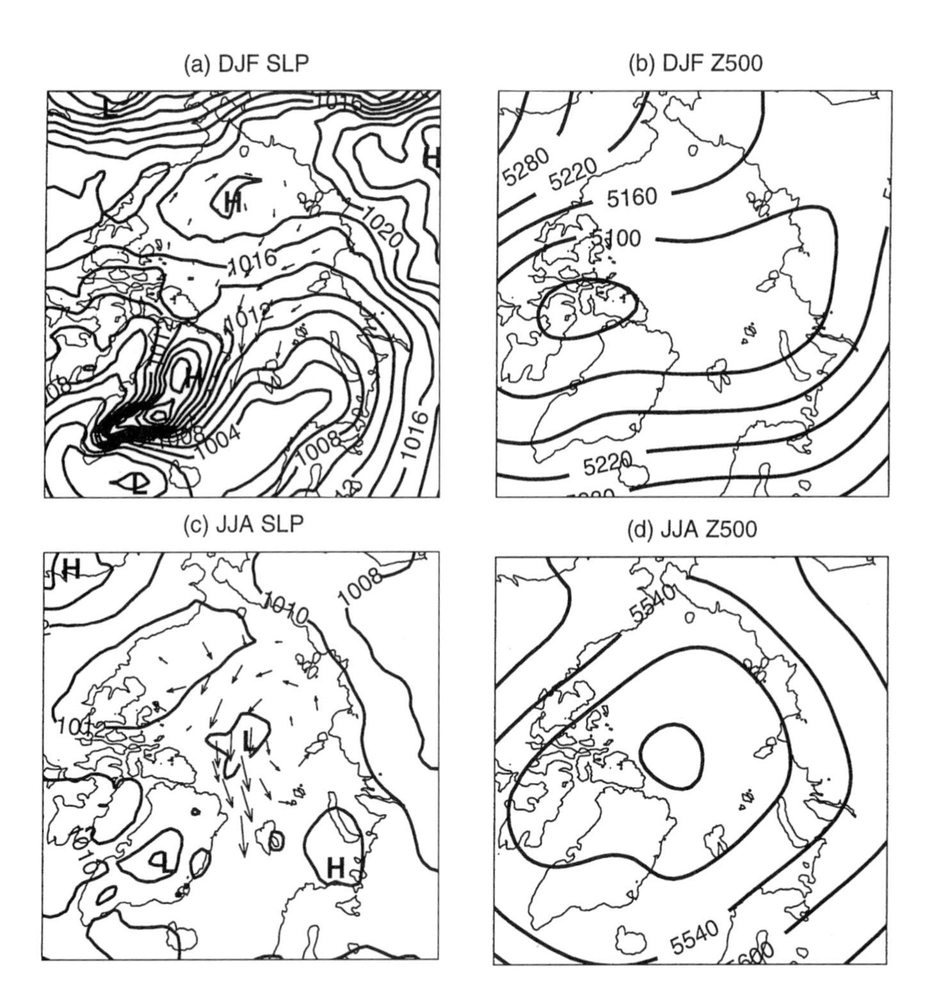

Figure 9 Mean sea level pressure (SLP) and 500 mbar height (Z500) for winter (December-January-February—DJF) and summer (June-July-August—JJA) from National Center for Atmospheric Research/National Center for Environmental Prediction (NCAR/NCEP) reanalysis data 1958–1997 provided through the NOAA Climate Diagnostics Center. Mean sea ice drift vectors for winter and summer 1979–1996 are shown with SLP. The drift speed in Fram Strait is 6.8 (3.0) cm/s in winter (summer). Ice drift data are provided by the International Arctic Buoy Programme.

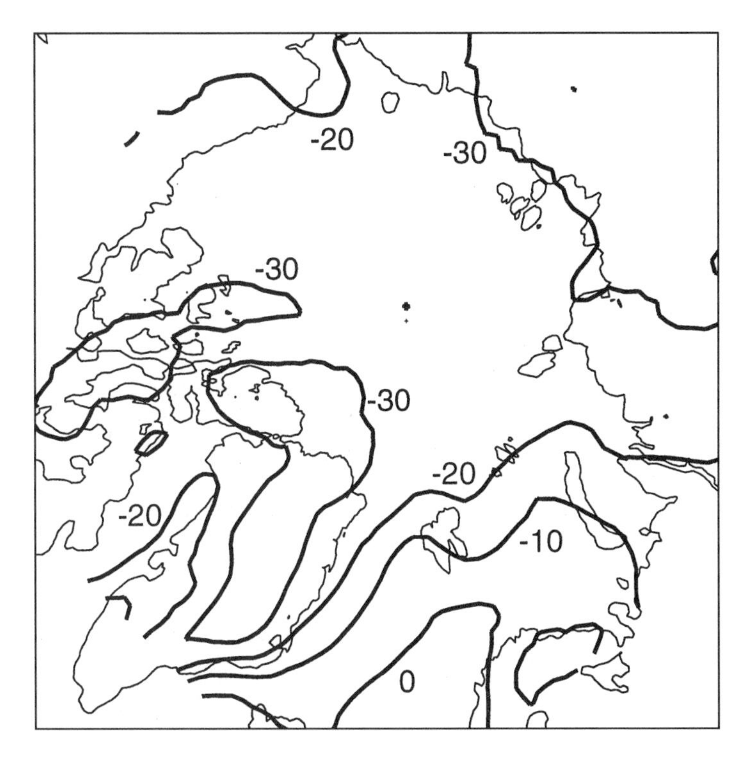

Figure 10 Winter (December-February) mean 2 m air temperature from the International Arctic Buoy Programme and the Polar Exchange at the Sea Surface Project (Martin & Munoz 1997).

linkages between the North Atlantic and Arctic climate systems. Here we focus on relationships established through the atmosphere. Investigations by Wallace & Gutzler (1981) and Barnston & Livezey (1987) showed that the dominant atmospheric circulation pattern in the North Atlantic is the North Atlantic Oscillation (NAO), a north-south oscillation in atmospheric mass with centers of action near Iceland and the Azores as defined by Walker & Bliss (1932). Fluctuations of the NAO have been linked to surface air temperature (SAT) anomalies in northern Europe and the northwestern Atlantic (van Loon & Rogers 1978), precipitation anomalies in Europe and over the Labrador Sea (Hurrell 1995), sea ice conditions in the Atlantic sector (Rogers & van Loon 1979, Walsh & Johnson 1979, Fang & Wallace 1994), sea ice export through Fram Strait (R. Dickson and D. Rothrock, personal communication), and the distribution of cyclones (Rogers 1990, Hurrell 1996, Serreze et al 1997). Figure 11 shows the NAO index and the signature of the NAO on SLP based on compositing years with strongly positive and negative NAO indices. The positive polarity of the NAO

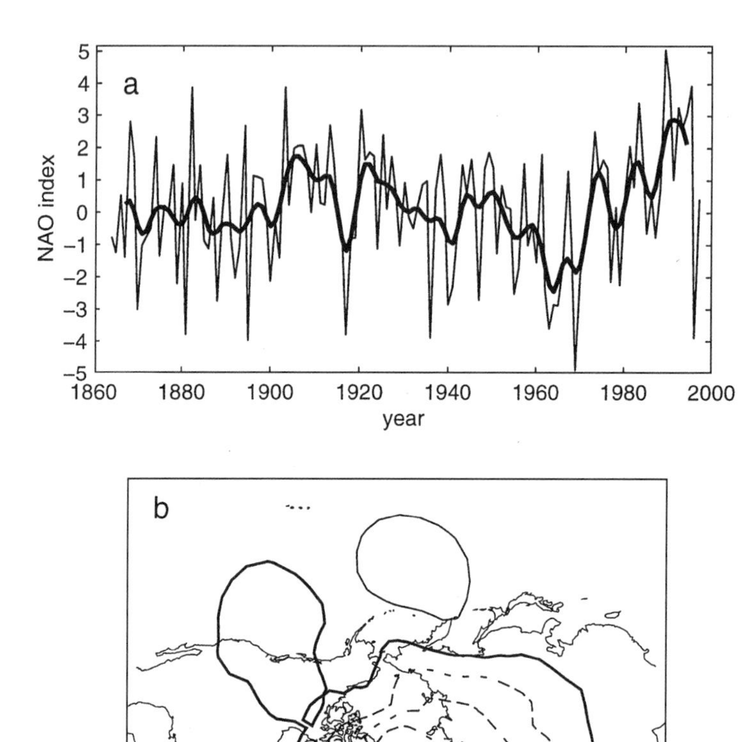

Figure 11 (*a*) Winter (December-March) index of the NAO based on the pressure difference between Lisbon, Portugal, and Stykkisholmur, Iceland, normalized by division of each seasonal pressure by the long-term (1864–1983) standard deviation. The heavy solid line shows the index filtered to remove variability with periods less than 4 years. (*b*) The SLP composite difference (high minus low) since 1899 for years with NAO index exceeding ± 1. Contours are drawn at 2 mbar intervals, negative contours are dashed, and the zero contour is a heavy solid line. (Figure 11*b* is modeled after Figure 1 in Hurrel 1995, with the addition of data through 1997.) The index is provided by the NCAR climate analysis section and differs from that of Hurrel (1995) only in the base period used for normalization (Hurrel used 1964–1994).

is characterized by a broad high-pressure center in the North Atlantic near 40°N and 30°W and a more spatially confined low centered over Greenland.

One of the primary ways in which the atmospheric circulation affects Arctic-North Atlantic interactions is via its direct and indirect forcing on sea ice in these regions. The links between atmospheric circulation and sea ice conditions in the Davis Straits and Labrador Sea regions are well known. Rogers & van Loon (1979) found that below normal temperatures in Greenland (or positive polarity of the NAO) are associated with heavy ice conditions in the Davis Strait and near Newfoundland, and light ice conditions in the Baltic Sea in winter and the following spring. Using an empirical orthogonal function (EOF) analysis of sea ice extent in the North Atlantic for all seasons from 1953 through 1977, Walsh & Johnson (1979) found the leading EOF corresponds to heavy ice in the Davis Straits/Labrador Sea and light ice in the Barents Sea. The pattern of sea ice extent is correlated with the leading EOFs of atmospheric variability from monthly anomalies of SLP, surface temperature, and 700 mbar heights and temperature. Lagged cross correlations suggest that during autumn the strength of forcing between sea ice and atmosphere is comparable. However, during summer, ice conditions are influenced by atmospheric forcing for a longer lead time.

Much has been written recently about a pattern of variability consisting of a dipole in the wintertime sea ice concentration between the Davis Straits/Labrador Sea region and the Greenland/Barents Sea (Mysak et al 1990, Fang & Wallace 1994, Slonosky et al 1997). For the winters of 1972 through1989, Fang & Wallace (1994) found that the sea ice dipole pattern is strongly coupled to the NAO such that the positive polarity of the NAO is observed with positive sea ice anomalies in the Davis Straits/Labrador Sea. They examined the data at seven-day intervals and determined that the relationship is strongest when the atmosphere leads the ice by two weeks. The negative polarity of the NAO is associated with blocking (diminished westerly flow) in the Greenland region, which Fang & Wallace (1994) link with a retreating (advancing) ice edge in the Davis Straits/Labrador (Greenland/Barents) Sea. Focusing on interannual variability, Slonosky et al (1997) show that a similar dipole pattern of wintertime sea ice anomalies for 1954–1990 is strongly correlated with atmospheric anomaly fields for simultaneous periods and for sea ice leading the atmosphere by one year. Positive sea ice anomalies are seen with high pressure over Greenland, Iceland, and the Canadian Archipelago.

Sea ice concentration anomalies in the Davis Straits/Labrador Sea region and Greenland/Barents Sea region are plotted in Figure 12*a*. Strong decadal features are present in the Labrador Sea as discussed by Mysak & Manak (1989) and Deser & Blackmon (1993). The tendency for area anomalies in the two regions to be out of phase is evident, particularly in the last half of the record. The cross correlation in Figure 12*b* reveals a significant negative correlation at zero

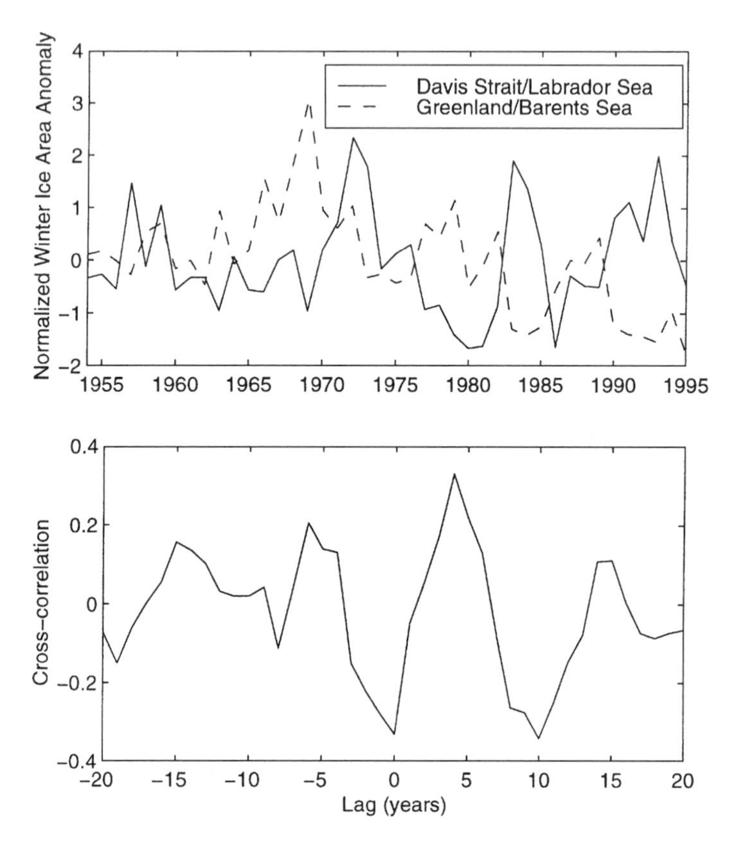

Figure 12 (*a*) Normalized wintertime (November to March) sea ice area anomalies in the Davis Straits/Labrador Sea region and Greenland/Barents Sea region and (*b*) cross correlation of the time series. The data are normalized by the standard deviation and are plotted in the year in which January occurs. Positive lag refers to anomalies in the Davis Straits/Labrador Sea lagging those in the Greenland/Barents Sea. The Davis Straits/Labrador Sea (Greenland/Barents Sea) region is defined as the area from 45°–65°N and 45°–70°W (65°–78°N and 25°W–60°E). Data were kindly provided by National Snow and Ice Data Center User Services and compiled by Walsh & Chapman.

lag. As noted by Mysak et al (1990), the strongest correlation occurs when area anomalies in the Davis Straits/Labrador Sea lag those in the Greenland/Barents Sea region by four years. The recurrence of a significant negative correlation at 10-year lag is evidence of the strong decadal variability relating ice anomalies in these regions.

Mysak & Manak (1989) discussed the propagation of sea ice anomalies from the Greenland Sea to the Labrador Sea over a period of approximately four years. In particular, they highlight the movement of the 1968 Greenland Sea anomaly associated with the Great Salinity Anomaly (GSA). Inspired by the strong decadal signal, Mysak et al (1990) gathered evidence in support of an interdecadal self-sustained climate cycle in the Arctic. The proposed negative feedback connects decreased cyclogenesis in the high-latitude atmosphere to positive sea ice anomalies in the Greenland Sea. Decreased storminess ultimately leads to higher salinity in the Arctic Ocean owing to a lack of precipitation and runoff. Finally, high-salinity water decreases the sea ice extent in the Arctic and Greenland Sea leading to the inverse process.

Wohlleben & Weaver (1995) proposed a modified version of the Mysak et al (1990) climate cycle that eliminates the connection between anomalous river runoff and sea ice extent in the central Arctic. Instead, they argued that deep convection in the Labrador Sea creates positive SST in the subpolar gyre, which enhances sea-level pressure over Greenland and weakens the Icelandic low (i.e. negative polarity of NAO) through the modification of storm tracks. Anomalous northerly winds enhance the transport of ice and fresh water as a result of the reduced east-west pressure gradient at high latitudes. Finally, these ice and freshwater anomalies are swept into the subpolar gyre and eventually reach the Labrador Sea, where they weaken convection and begin the inverse process.

It is possible that these feedback cycles may help explain why observations show that strong decadal variability in the sea ice extent in the Labrador Sea leads SST anomalies east of Newfoundland by two years (Deser & Blackmon 1993). Furthermore, these feedback cycles suggest that such diverse interactions as the GSA and the large-scale changes that have been seen in the northern North Atlantic are not simply explained. For example, there is observational (Walsh & Chapman 1990) and modeling (Häkkinen 1993) evidence to support the idea that ice and freshwater export were anomalously strong prior to the GSA of the late 1960s when the Icelandic Low was anomalously weak (negative polarity of the NAO, see Figure 11*a*). However, recent studies show that the relationship between NAO and ice export appears to have changed sign in the last two decades, such that export is positively correlated with the NAO (R. Dickson and D. Rothrock, personal communication). In the past two decades we have seen a strong positive NAO pattern leading to enhanced export and decreased ice concentration in the Greenland Sea. Hence the sea ice must have melted at an extraordinary rate in the Greenland Sea for the past 20 years. It is thought that this might contribute to the sustained weak convection seen in the Greenland Sea since the mid-1970s.

NORTH PACIFIC Interaction between the northern North Atlantic and the Arctic appears to be a two-way process, including the transport of heat and moisture to high latitudes by the atmosphere, the exchange of fresh water between the

Arctic and the North Atlantic, and the modulation of air-sea exchange by sea ice anomalies. However, studies of interactions in the northern North Pacific sector suggest primarily that large-scale atmospheric circulation forces a local response in sea ice anomalies.

Overland & Pease (1982) determined that the maximum wintertime sea ice extent in the Bering Sea is strongly correlated with the difference in the number of cyclone centers in the eastern and western parts of the northern Pacific Basin. Sea ice advance is inhibited by southerly winds associated with storms traveling along the western side of the Bering Sea. Rogers (1981) showed that the position of the ice edge in the Bering is linked to the strength of the north-south temperature gradient, which he associated with an atmospheric teleconnection pattern he called the North Pacific Oscillation after Walker & Bliss (1932). Analogous to their results in the North Atlantic, Fang & Wallace (1994) found a dipole pattern in the Pacific sector with opposing centers of action in the Bering Sea and the Sea of Okhotsk. They associated the advance (retreat) of ice in the Sea of Okhotsk (Bering Sea) with a blocking ridge in the Gulf of Alaska.

Freshwater Budget

Because water density is largely controlled by salinity at low temperatures, the freshwater exchange between the Arctic and high-latitude oceans is very important in determining the strength of the thermohaline circulation. Relatively small changes in the flux of freshwater to the convective regions are likely to significantly affect NADW formation (e.g. Aagaard & Carmack 1989). The freshwater flux between the Arctic and North Atlantic is made up of both ocean and sea ice transports. The Arctic freshwater budget is also important for the stability of the water column in the Arctic and the insulation of sea ice from the relatively warm subsurface Atlantic water.

The freshwater budget of the Arctic Ocean has been discussed by Aagaard & Carmack (1989). The primary freshwater input to the Arctic Ocean is river runoff. Approximately 10 percent of the world's river runoff, accounting for approximately 3300 $km^3\ yr^{-1}$, enters the Arctic Ocean, which occupies only 5 percent of the total ocean surface area and 1.5 percent of its volume.

Large seasonal and interannual variability exists in the river runoff (Cattle 1985). Bering Strait inflow represents the second largest freshwater source for the Arctic Ocean ($\sim$1670 $km^3\ yr^{-1}$), with precipitation minus evaporation ($\sim$900 $km^3\ yr^{-1}$) and the import of fresh water in the Norwegian coastal current ($\sim$330 $km^3\ yr^{-1}$) accounting for the remainder of the freshwater sources. These large inputs of fresh water allow the Arctic surface waters to remain relatively fresh ($<$34.4 psu), which results in a stable water column and the isolation of sea ice from the heat associated with the Atlantic layer of the Arctic.

The sources of fresh water for the Arctic are balanced primarily by the export of sea ice (which has a salinity of approximately 3 psu) in the East Greenland Current (EGC), which accounts for a freshwater loss to the Arctic and a gain by the Greenland Sea of approximately 2800 km^3 yr^{-1} (Aagaard & Carmack 1989). The exchange of water through the Canadian Archipelago and Fram Strait results in a loss of approximately 900 km^3 yr^{-1} and 820 km^3 yr^{-1} of fresh water, respectively. The variability in these outflows has important consequences for the deep-water formation that occurs in the GIN and Labrador Seas. For example, modeling results from Mauritzen & Häkkinen (1997) show that the thermohaline circulation increased by 10 to 20 percent in response to a decrease in sea ice export of 800 km^3. The relative strength of the freshwater sources to the GIN and Labrador Seas from the Arctic will also likely influence the preferred location and relative strengths of deep-water formation. From a relatively short time series (1979 to 1985), Steele et al (1996) show that simulated interannual variability in the outflow through the Canadian Archipelago is anticorrelated to the outflow through Fram Strait, with the Fram Strait anomalies leading the Canadian Archipelago anomalies by one year. This may explain why deep-water formation in the GIN and Labrador Seas has been observed to be out of phase in the past few decades.

The large changes that occur in Arctic/North Atlantic freshwater exchange are epitomized by the GSA of the late 1960s. This event freshened the upper 500 m of the northern North Atlantic with a freshwater excess of approximately 2000 km^3 (or 0.032 Sv over a two-year period). Dickson et al (1988) trace this freshwater anomaly as it was advected around the subpolar gyre for more than 14 years. It originated north of Iceland in the late 1960s, moving southwestward into the Labrador Sea (1971 to 1973) and then proceeding across the North Atlantic, returning to the Greenland Sea in 1981 to 1982.

Several studies have examined the cause of the GSA and have generally determined that it was a result of Arctic/North Atlantic interactions. Both modeling (Häkkinen 1993) and observational (Walsh & Chapman 1990, Wohlleben & Weaver 1995) studies concluded that strong northerly winds caused an increased sea ice export into the Greenland Sea. The large freshwater flux anomaly that was associated with this transport was likely enhanced by the relatively large advection of thick ice from north of Greenland. Additionally, as simulated by Häkkinen (1993), increased oceanic transport of fresh water from the Arctic occurred. This was caused by fresh anomalies within the Siberian Sea that were advected across the Arctic, entering the Greenland Sea approximately 4 years later. During the GSA the anomalous sea ice and oceanic freshwater transports were coincident, resulting in a significant and persistent freshening of the North Atlantic. This freshening appears to have resulted in a reduction of deepwater formation with winter convection in the Labrador sea limited to the upper 200 m (compared with 1000 to 1500 m for 1971 to 1973) (Lazier 1980).

DECADAL-INTERDECADAL VARIABILITY

In this section we discuss observational evidence concerning the existence of decadal-interdecadal variability in the North Atlantic and Pacific Oceans. An attempt is made to link analytical and numerical studies with observational data. More comprehensive reviews of this topic are given by Anderson & Willebrand (1996) and Latif (1998).

The Atlantic

North Atlantic SST records for the past century reveal slowly varying basin-scale changes including cold anomalies prior to 1920, warming from 1930 to 1940 and cooling again in the 1960s. Kushnir (1994) described the SST pattern associated with these long-term changes as unipolar (see Figure 13) with a strong maximum around Iceland and in the Labrador Sea and a weaker maximum in a band near 35°N across the central Atlantic. The atmospheric pattern associated with the cooling in the 1960s has a negative pressure anomaly to the east of positive SST anomalies (also see Deser & Blackmon 1993, who suggest that the pattern resembles the NAO). Because the SLP anomalies appear downstream of the SST anomalies, Deser & Blackmon suggest that the atmosphere is responding to the ocean on these timescales.

Evidence for changes in the subpolar North Atlantic Ocean over similar timescales, compiled by Dickson et al (1996), indicates that synchronous with the cooling in the late 1960s, convective activity reached a maximum in the Greenland Sea and a minimum in the Labrador Sea. These convective extremes occurred at the approximate time of the GSA. Since the early 1970s, the Greenland Sea has become progressively more saline and warmer through horizontal exchange with the deep waters of the Arctic Ocean. At the same time, the Labrador Sea has become colder and fresher as a result of local deep convection. Further signs of internal ocean dynamics can be seen in propagating signals of temperature and salinity. Hansen & Bezdek (1996) found temperature anomalies that remain coherent for 3 to 10 years circulating along the subpolar and subtropical gyres. Because the anomalies move slower than the near-surface ocean circulation, they speculated that the motion could be influenced by ocean wave activity. Reverdin et al (1997) explored patterns associated with salinity anomalies and found that a single pattern explains 70 percent of the variance of lagged salinity anomalies. The pattern represents a signal originating in the Labrador Sea that propagates from the west to the northeast in the subpolar gyre. The strong correlation between salinity and sea ice in the Labrador Sea

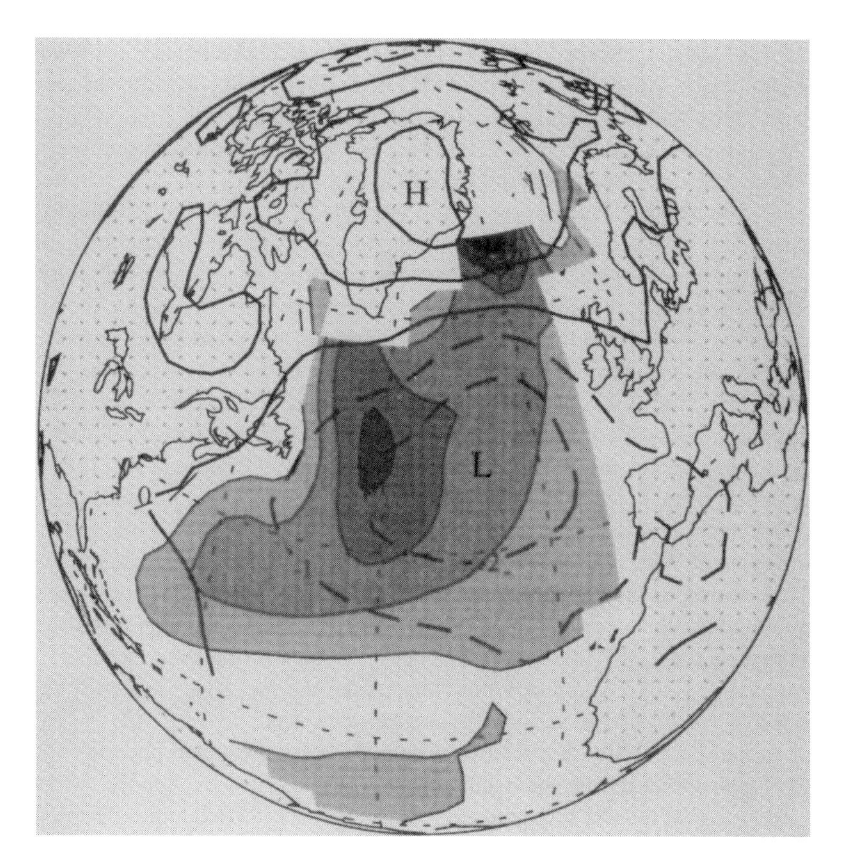

Figure 13 Composite differences of winters (December-April) 1950–1964 and 1970–1984. SLP is contoured with *heavy lines* at 1 mbar intervals, and SST is contoured with shaded regions separated by light lines at 0.2°C intervals. *Dashed lines* indicate negative values (from Kushnir & Held 1996).

led Reverdin et al (1997) to link the salinity anomalies to the export of Arctic fresh water.

The short and sparse record of oceanographic measurements, and the complicated nature of the interactions involved, motivate the use of models to clarify the processes associated with decadal variability. It is possible for ocean-only models (e.g. Weaver & Sarachik 1991) to exhibit internal variability on decadal timescales in their thermohaline structure when forced with mixed boundary conditions (i.e. SST is restored to a specified value and freshwater flux is prescribed). Low-frequency variability develops in these models when fluctuations in the thermohaline circulation are reinforced by salinity advection (a positive feedback) and defeated by heat transport (a negative feedback that is somewhat confounded by using a restoring condition on SST). Because salinity variations control the stability of the water column in the northern North Atlantic, instabilities arise when a sufficiently strong local minimum in the freshwater forcing is prescribed near the sinking region (Weaver et al 1991, 1993). In addition, models with weak SST restoring exhibit more stable thermohaline circulations (Zhang et al 1993). With a zonally averaged ice-ocean model, Yang & Neelin (1993) identified another feedback mechanism for long-term thermohaline variability. They showed that ice melt decreases the near-surface salinity, thereby weakening the thermohaline circulation. In turn, the poleward heat transport is reduced, which causes sea ice growth and so leads to the inverse process.

Delworth et al (1993) described the first coupled ocean-atmosphere GCM study of long-term thermohaline variability. They associate the variability primarily with oceanic processes. Later, Delworth et al (1997) found that salinity anomalies in the surface layer of the Arctic Ocean precede anomalies of the thermohaline intensity by 10 to 15 years. In agreement with the proposed climate cycle of Wohlleben & Weaver (1995), these Arctic freshwater anomalies are connected to the North Atlantic through sea-level pressure anomalies in the Greenland Sea resembling the pattern that Walsh & Chapman (1990) report preceded the GSA. Weaver & Valcke (1998) give further evidence that the Geophysical Fluid Dynamics Laboratory model thermohaline variability is a mode of the fully coupled atmosphere-ocean-ice system.

If decadal variability in the North Atlantic is truly a coupled phenomenon, then it is essential to our understanding to determine how the atmosphere responds to midlatitude SST anomalies. Palmer & Sun (1985) examined this problem by using an atmospheric GCM with anomalous SST specified in the region 40° to 50°N and 40° to 60°W. Although weaker than the response to SST anomalies of similar magnitude in the tropics, Palmer & Sun (1985) found a wave train emanating from this SST anomaly region such that positive height anomalies were found downstream of positive SST anomalies in the model. Palmer & Sun (1985) and Wallace & Jiang (1987) showed that observations of atmospheric circulations based on SST anomalies in the same region verify the structure simulated in the model, although the amplitude of the response is about half as great. Similar modeling studies that explore the atmospheric response to midlatitude SST anomalies by Pitcher et al (1988), Ting (1991), Kushnir & Lau (1992), Lau & Nath (1994), and Kushnir & Held (1996) yield results somewhat contradictory to those of Palmer & Sun (1985). Recent studies by Peng et al (1995, 1997) attribute some of the discrepancies between these studies to differences in the basic state of the atmosphere upon initialization. Model resolution is another factor that seems to influence the atmospheric response to SST anomalies. Latif & Barnett (1994) found that the atmospheric response to Pacific midlatitude

SST anomalies is much stronger than the response reported by those mentioned above. They attribute the response to an unstable air-sea interaction.

Bjerknes (1964) described how air-sea interactions differ on interannual and interdecadal timescales. Based on evidence that annual mean SST and wind speed anomalies are negatively correlated, Bjerknes argued that interannual SST variability is mainly a response to fluctuations of atmosphere-ocean latent and sensible heat transfer. He suggested that the longer timescale variability was ocean based by linking the SST anomalies to changes in the ocean currents.

Recent analysis of more than a century of data by Deser & Blackmon (1993) and Kushnir (1994) lends support to many of the ideas of Bjerknes (1964). Deser & Blackmon (1993) identified two prominent modes of SST variability in the North Atlantic (see Figure 14). The leading EOF represents the long-term changes that were described in the first paragraph of this section (Figure 14*a*) and resembles the pattern Kushnir (1994) obtained by compositing different time periods (Figure 13). Higher frequency variability is evident in the second EOF (Figure 14*b*) with strong roughly two- and 10-year timescales for SST anomalies with opposing centers of action east of Newfoundland and southeast of the United States. Kushnir (1994) obtained a similar SST pattern by compositing anomalies based on interannual fluctuations.

The atmospheric pattern associated with the higher frequency SST mode (see Figure 15) is very different from that of the lower-frequency mode shown in Figure 13. Like SST, the SLP has a dipole structure that is somewhat displaced to the northeast of the SST pattern. The atmospheric circulation has anomalous westerly winds over negative SST anomalies and anomalous southerly winds over positive SST anomalies. As noted by Deser & Blackmon (1993) and Hurrell (1995), the local nature of this relationship is governed by changes in the oceanic mixed layer that result from wind-induced heat flux anomalies, in agreement with Bjerknes (1964).

The Pacific

Interannual variability in the Pacific Ocean is well known to be dominated by El Niño/Southern Oscillation (ENSO) variability and its associated teleconnection through the atmosphere to the North Pacific (Bjerknes 1969, Weare et al 1976, Horel & Wallace 1981, Wallace & Gutzler 1981, Deser & Blackmon 1995, Zhang et al 1997). Recently, however, it has become apparent that the North Pacific possesses its own rich modes of decadal-interdecadal variability (see reviews of Trenberth 1990 and Nakamura et al 1997). One of the most notable events in the North Pacific was the apparent climate regime change that occurred from late 1976 through 1988. During this period the SSTs in the central and western North Pacific Ocean were cooler than normal, whereas SSTs along the western coast of North America were warmer than normal and the Aleutian low

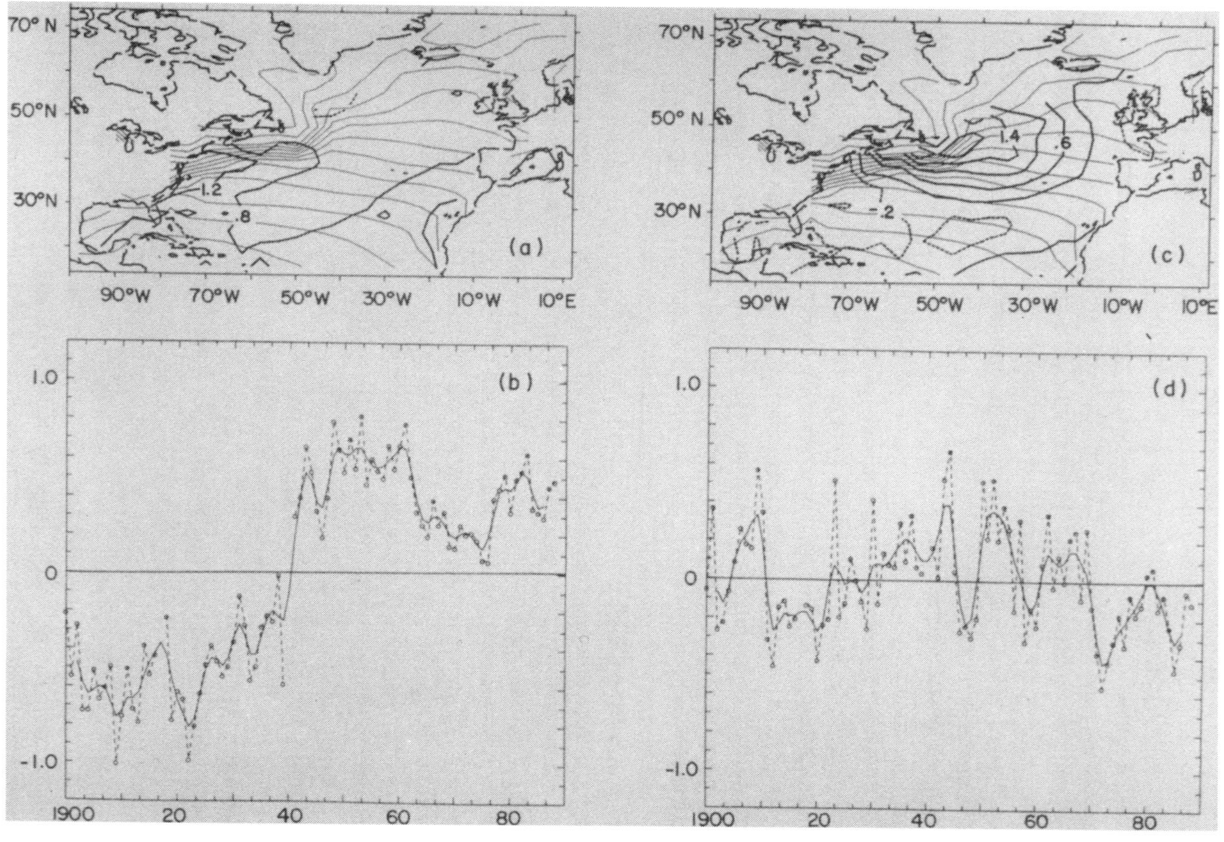

Figure 14 (*a*) Leading empirical orthogonal function (EOF) of winter mean (November–March) SST anomalies (*heavy lines*) and climatological winter mean (*light lines*) for the North Atlantic. (*b*) Expansion coefficient of EOF 1 (*dashed line*) and smoothed with a five-point binomial filter (*solid line*). (*c*) and (*d*) The same as Figures 14*a* and 14*b*, respectively, but for EOF 2. EOFs 1 and 2 account for 45 and 12 percent of the variance, respectively. (From Deser & Blackmon 1993.)

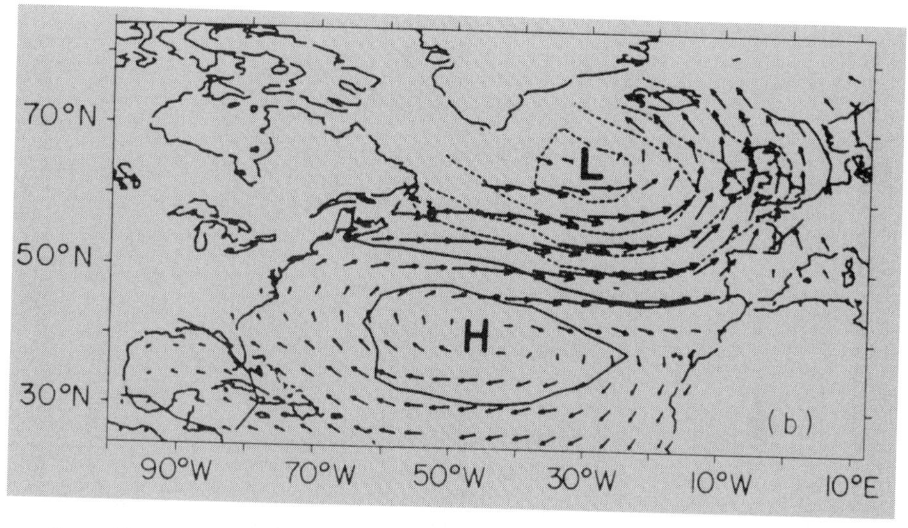

Figure 15 SLP and wind anomalies regressed on SST EOF 2, with opposite sign to the pattern shown in Figure 14*c* (from Deser & Blackmon 1993).

(see Figure 16) was deeper than normal (see Nitta & Yamada 1989, Trenberth 1990, Trenberth & Hurrell 1994). In attempting to explain this departure from normality, Trenberth (1990) pointed out that during this period there were three warm ENSO events with no intervening cold events (see Figure 17). Thus tropical temperatures were warmer than average, and so, through teleconnections to higher latitudes (Figure 16), the Aleutian low was deeper than normal on average. The end of this apparent regime change was marked by the strong cold event in 1989 (Figure 17). Subsequently, Trenberth & Hoar (1996) suggested that the 1990–1995 period was also anomalous in that tropical indices (see Figure 17) appeared to be locked into the warm phase. They described this "1990–1995 ENSO event" as the longest on record with a probability of occurrence about once every 2000 years. They further suggested the possibility that the ENSO changes may be partly caused by observed increases in greenhouse gases. Zhang et al (1997), on the other hand, provide compelling evidence to suggest that the 1976 "regime change" is by no means anomalous. They suggested that equally dramatic (but less studied) changes occurred around 1957 to 1958 (Figure 17). They further argued that the conditions during the 1977–1993 period were reminiscent of those that occurred from 1925 to 1942. Dettinger & Cayan (1995) and Zhang et al argued that regime changes are probably not the most informative way of characterizing decadal-interdecadal climate variability in the North Pacific.

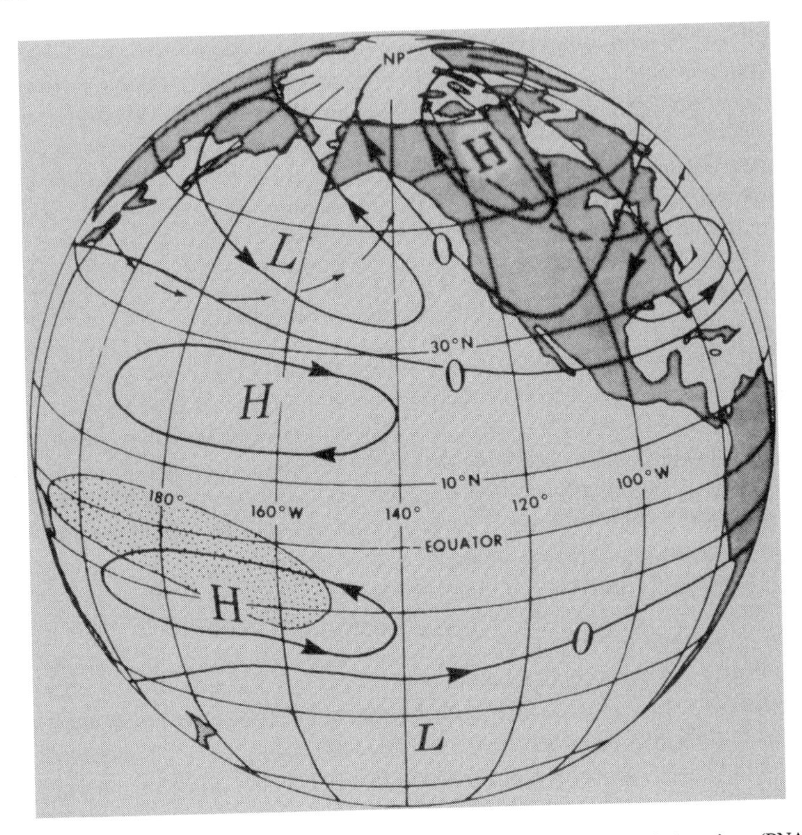

Figure 16 Diagram showing the Northern Hemisphere winter Pacific-North American (PNA) teleconnection pattern for warm SST anomalies in the equatorial Pacific Ocean (after Horel & Wallace 1981). The *wavy dashed line* with *arrows* lying between 30°N and 60°N shows a typical midtropospheric streamline that has been distorted by the anomaly pattern. The *shaded region* corresponds to where enhanced rainfall occurs in the equatorial band (redrawn from Mysak 1986).

Unlike the North Atlantic, the North Pacific has no region of deep-water formation so that internal fluctuations of the thermohaline circulation are not likely to be a source of low-frequency (decadal-interdecadal) climate variability there. Nevertheless, it is yet to be understood what effects (if any) low-frequency variability in convection and intermediate water production in the Sea of Okhotsk may have on the climate of the North Pacific. As pointed out by Hasselmann (1976), even a passive ocean, with its much higher heat capacity than the atmosphere, could integrate high-frequency atmospheric noise to produce a red response with variability at much lower frequencies. Such a response was indeed demonstrated in the model simulations of Hansen & Lebedeff (1987),

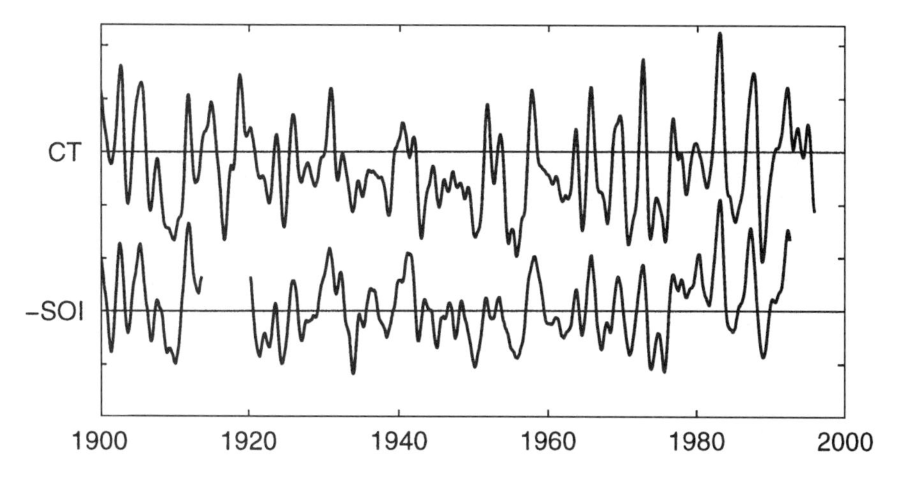

Figure 17 Cold tongue (CT) index based on SST in the equatorial Pacific (6°S–6°N, 180°W–90°W) and the Southern Oscillation Index (SOI) from pressure differences between stations at Darwin, Australia, and Tahiti with the sign reversed. Both time series have been smoothed with successive nine-month and five-month running mean filters. The interval between tick marks on the vertical axis is 1.0 standard deviation. These indices were kindly provided by Y Zhang & N Mantua.

where low-frequency climate variability was observed in a 100-year integration of the Goddard Institute for Space Studies atmospheric GCM coupled to a mixed layer ocean with specified ocean heat transport.

A number of theories exist to explain the observed decadal variability (Figures 18*b*, 19*b*) in the North Pacific. As seen in Figure 16, the Southern Oscillation Index (SLP anomalies at Tahiti minus Darwin, a statistical measure of ENSO) indicates that ENSO varies on decadal-interdecadal timescales (e.g. Trenberth 1990, Zhang et al 1997). Because ENSO is a nonlinear coupled tropical atmosphere-ocean phenomenon, it is possible that its decadal modulation and subsequent teleconnection to the North Pacific could explain the low-frequency variability observed there (Trenberth & Hurrel 1994). In this same area it may not be necessary to invoke a local nonlinearity of ENSO in the tropics. As pointed out by Gu & Philander (1997), a delayed negative feedback can be achieved through extratropical oceanic subduction of thermal anomalies (generated through the atmospheric teleconnection response to tropical SST anomalies) that slowly propagate along isopycnals toward the equator, where they reverse the sign of equatorial SSTs. Finally, Latif & Barnett (1994, 1996) have suggested a mode of decadal-interdecadal North Pacific variability solely involving midlatitude coupled atmosphere-ocean interactions and the strength of the subpolar gyre and its associated northward heat transport. As pointed

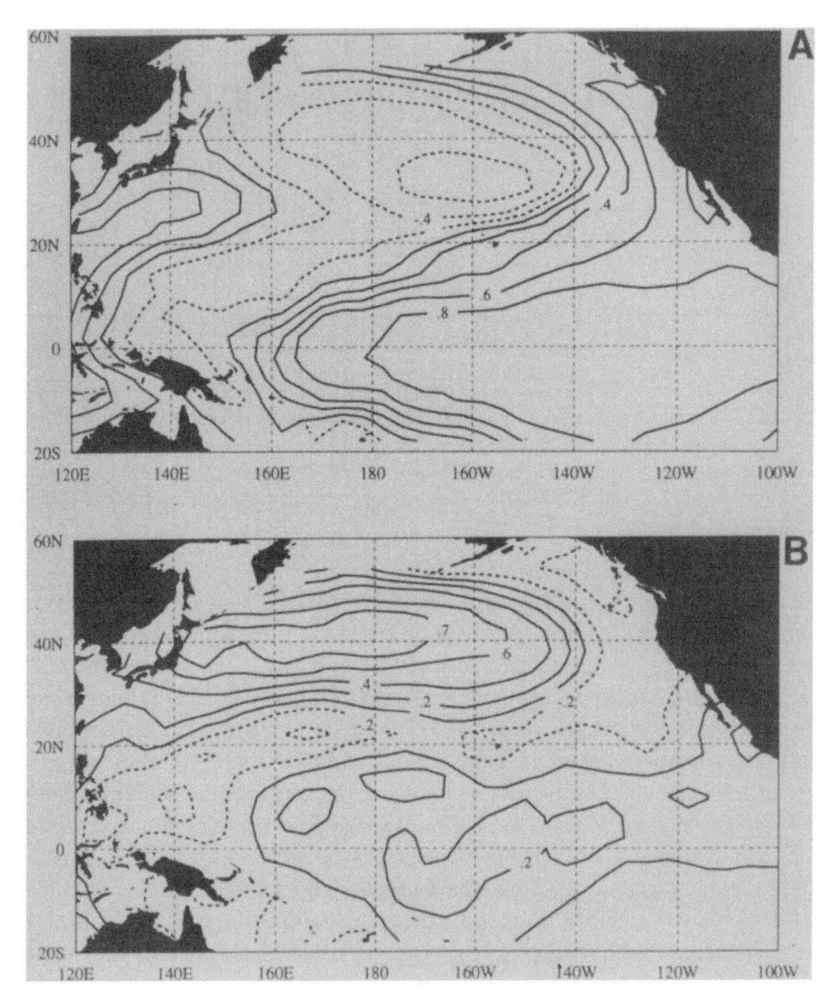

Figure 18 (*a*) Normalized EOF 1 (the interannual signal) of Pacific SSTs based on winter (November through March) anomalies during the period 1950–1951 to 1991–1992. The contour interval is 0.2, and negative contours are dashed. (*b*) As in Figure 18*a* but for EOF 2 (the decadal signal). The 0.7 contour is also shown. (From Deser & Blackmon 1995.)

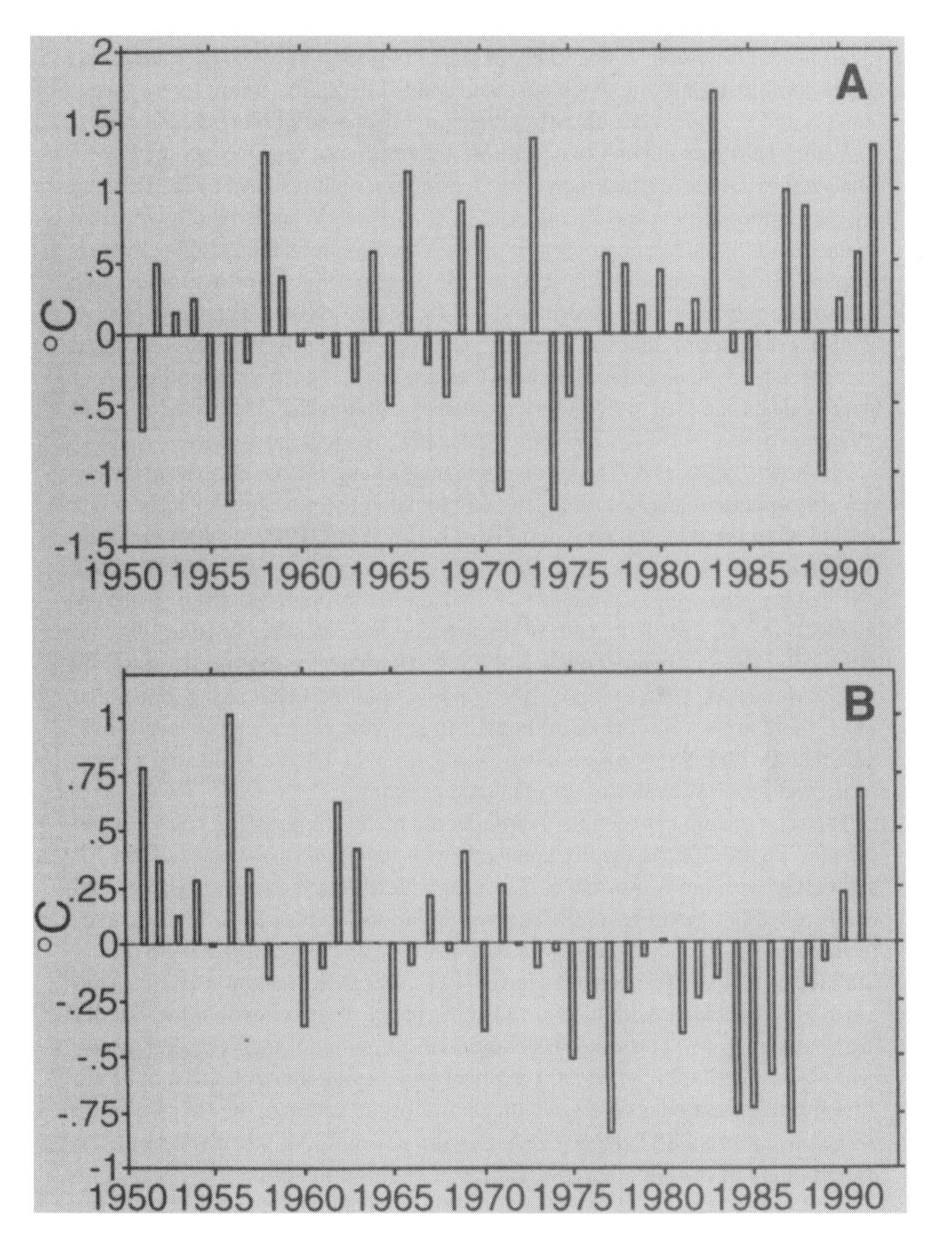

Figure 19 (*a*) Time series of SST anomalies averaged over the center of action of Figure 18*a*. (*b*) As in Figure 19*a* but for EOF 2 (Figure 18*b*) (From Deser & Blackmon 1995.)

out by Nakamura et al (1997), subtropical gyre SST variability on the decadal-interdecadal timescale is not solely explained through the tropical source, and some combination of the mechanisms may exist in reality. Below, these mechanisms and the evidence supporting them are discussed in more detail.

Figure 18 shows the first two EOFs of winter SST anomalies over the Pacific analyzed by Deser & Blackmon (1995). Similar structures have been identified by Kawamura (1994) and Zhang et al (1996, 1997). What is readily apparent is the distinctly different patterns expressed by the interannual (EOF 1) versus decadal (EOF 2) modes. The second EOF is primarily a North Pacific mode with maximum correlation centered on 40°N (the axis of the Kuroshio). It is also worth noting that the anomaly occurs as a linearly independent signal compared to tropical variability and is western intensified, perhaps indicative of gyre-scale as opposed to ENSO-modulated circulation changes (see Zhang et al 1996).

As shown by Deser & Blackmon (1995) and Zhang et al (1996), the preferential atmospheric teleconnection pattern associated with the North Pacific mode is similar to the well-known Pacific-North American (PNA) pattern (Wallace & Gutzler 1981). Trenberth & Hurrell (1994) devised a North Pacific index (NP) as the area-weighted mean SLP (November through March) over the region 30°–65°N, 160°E to 140°W. Figure 20*a* indicates the low-pass-filtered NP for the winter average, whereas Figure 20*b* indicates the spectrum for the time series 1924–1925 to 1990–1991. While not significant at the 95 percent level, there are several peaks in the two-to-six–year timescale (believed to be associated with ENSO), a deficit in power in the 7- to 12-year band, and a broad peak featuring timescales of 20 years and beyond.

The relationship between the North Pacific mode (Figures 18*b* and 19*b*) and the NP (Figure 20*a*) is readily apparent, with below normal values of the NP associated with below normal SST over the North Pacific (see also Trenberth & Hurrell 1994, their Figure 9). The main relationship is one in which changes in the atmospheric circulation are responsible for changes in the SST (e.g. Wallace et al 1990, Trenberth & Hurrell 1994, Deser & Blackmon 1995). Wind anomalies associated with the North Pacific index are maximum in the vicinity of the Aleutian low (Deser & Blackmon 1995), with stronger (weaker) westerlies associated with the cooler (warmer) SST. Along the west coast of North America the situation is reversed with below normal values of the NP associated with above normal SST due to the northward advection of warm waters on the eastern flank of the Aleutian low (Emery & Hamilton 1985).

While ENSO may modulate midlatitude thermal anomalies, a question arises as to how such anomalies persist through to the decadal timescale. Indeed, one would expect that thermal anomalies must ultimately be dissipated either

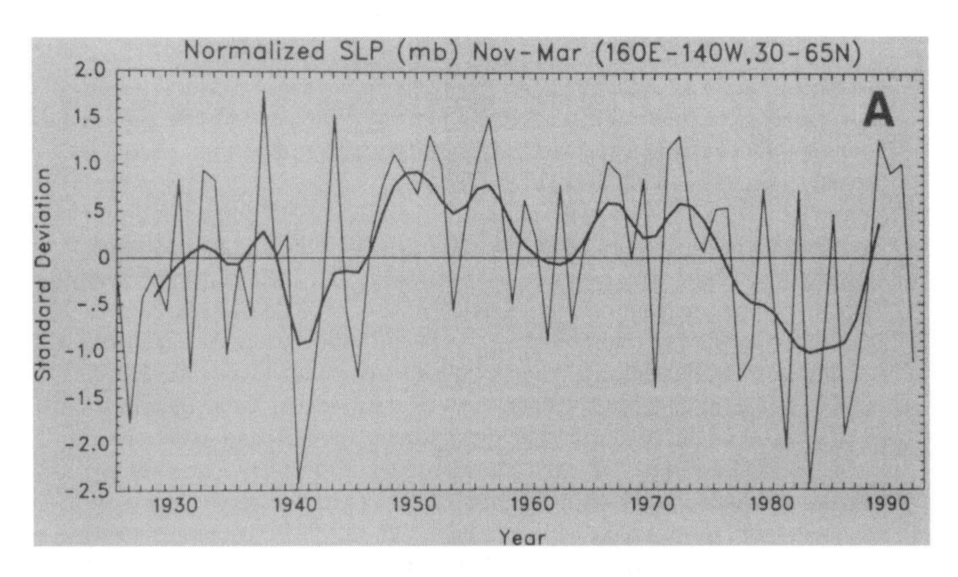

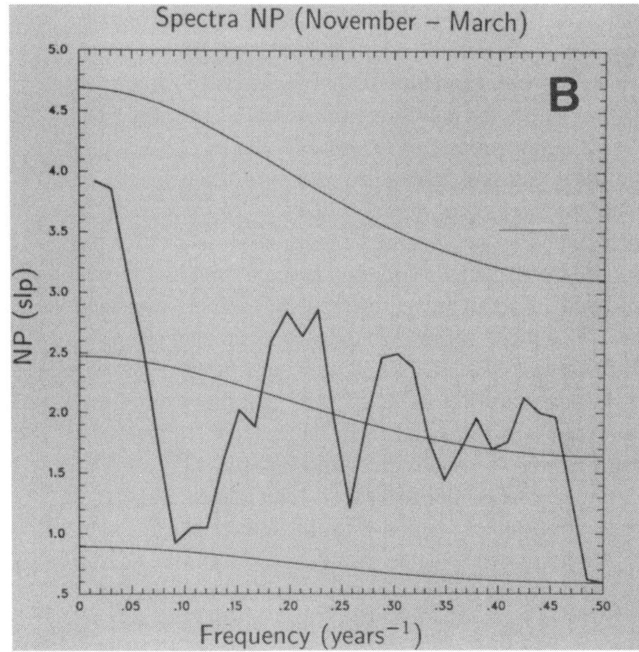

Figure 20 (*a*) Time series of the mean North Pacific SLPs for November–March. The total monthly mean value and the low-pass-filtered time series are shown. The filter employs seven weights (1, 4, 8, 10, 8, 4, 1)/36 across years to emphasize decadal time-scales. (*b*) Power spectrum of the NP index averages for the years 1924–1925 to 1990–1991. Also shown are the five and 95 percent confidence limits (*bottom and top smooth curves*), as well as the red noise spectrum (*middle smooth curve*) for the same lag one autocorrelation coefficient (0.91). (From Trenberth & Hurrell 1994.)

by ocean-atmosphere interaction (Alexander 1992*a*,1992*b*) (the most effective of which would be latent and sensible heat transfers) or by internal oceanic processes. Ocean surface anomalies created over the deep mixed layer in winter could be preserved below the summer thermocline and reappear at the surface during the following winter when the mixed layer deepens and intersects the anomaly (e.g. Namais & Born 1974, Alexander & Deser 1995). Persistent atmospheric features, such as the anomalous reoccurrence of intense Aleutian lows over decadal timescales, reinforce and further inject thermal anomalies into the deep mixed layer. These anomalies then subduct into the main pycnocline, where they flow southward along isopycnals (Deser et al 1996). Such thermal anomalies may then be transported (as a coherent structure) to later reemerge and alter the tropical Pacific SSTs (Gu & Philander 1997). In this manner, ENSO variability (e.g. more frequent occurrences of warm phases than cold phases or vice versa) could yield a decadal-interdecadal signal in the extratropical Pacific temperature, which in turn could give rise to a modulation of ENSO on a similar time scale and explain major phase shifts seen in the climate record (e.g. Trenberth 1990, Hastrom & Michaelson 1994, Minobe 1997, Mantua et al 1997, Nakamura et al 1997). Of course, such a phenomenon is inherently irregular owing to the ongoing effects of ENSO and midlatitude air-sea interactions, and additional irregularity could be imposed due to high-frequency weather.

Large-scale changes in heat content and transport of the Kuroshio and its extension can inject significant warm anomalies into the upper layer of the North Pacific, which extend with depth through to the core of the jet. In a similar fashion, reduction in Kuroshio transport favors expansion of the Oyashio and hence injection of cold thermal anomalies into the midlatitude gyre region (e.g. see Nakamura et al 1997, their Figure 1). Latif & Barnett (1994, 1996) (see also Latif 1998) describe the interdecadal vacillation as a gyre mode driven by midlatitude air-sea interactions without appealing to ENSO initiation or modification. Their coupled atmosphere-ocean modeling analysis suggested that this mode involved the weakening and strengthening of the subtropical gyre, which, through changes in heat transport in the western boundary through to the Kuroshio extension, creates SST anomalies in the central North Pacific (e.g. a strong gyre transports more heat into the central North Pacific via the Kuroshio extension, creating warm SST anomalies). A rather fascinating result was the subsequent strong atmospheric response to the central North Pacific SST anomalies in the PNA pattern, which acted locally to reinforce the anomalies. Changes in wind stress curl and the subsequent westward propagation of this signal in the ocean (via Rossby wave dynamics) then acted to spin down the subtropical gyre with the oscillation then entering the opposite phase. Support for the existence of such a mechanism in the observational record is given by Zhang & Levitus (1997), where they found subsurface thermal anomalies circulating clockwise around the subtropical gyre, as in the work by Latif & Barnett (1994, 1996). Jin (1997) also developed a simple conceptual model involving the central dynamics discussed above and arrived at a delayed-oscillator model which also captured the essence of this variability.

As an alternative hypothesis, Jacobs et al (1994) argue that the 1982–1983 El Niño spawned such intense poleward traveling Kelvin waves, and subsequently westward radiated Rossby waves, that it ultimately displaced the Kuroshio axis northward, giving rise to a warm anomaly extending across the Pacific basin (see also McPhaden 1994). The anomaly (which extended from Japan to the west coast of North America) then persisted over an 11-year period. Although they use satellite observations to provide support for their model results, a number of questions arise. Although Kelvin waves can communicate equatorial signals poleward, the Rossby radius of deformation diminishes rapidly along the coast and isopycnals begin outcropping. By the time these Kelvin waves reach midlatitudes they have little energy available. Furthermore, coastal waters are heavily stratified (due to freshwater input by rivers) so that anomalies are trapped at the surface and therefore subject to degradation by surface heat loss. Finally, the model of Jacobs et al (1994) used observed wind stress data over the period 1981 to 1993, which implicitly contained the PNA teleconnection signal in it. It is therefore more likely that the Rossby wave response seen in their model arose from the oceanic adjustment to atmospheric wind perturbations rather than through communication via the coastal wave guide.

SUMMARY

In this review we have attempted to describe the important differences between the thermohaline circulation in the North Atlantic and Pacific Oceans. Our emphasis throughout this review was the interaction between these regions and the Arctic climate system. We took this perspective because the role of the Arctic in climate change and variability is often overlooked. Rather than simply describing the annual mean climatology and fundamental physical processes of the thermohaline circulation, we also focused on decadal-interdecadal climate variability in the high-latitude regions and the role of the ocean in this variability.

The single most important problem in determining the underlying physical mechanisms to explain (and ultimately to predict) the observed variability in the high-latitude North Atlantic and Pacific is the lack of an adequate observational network. Many of the historical data records are too short in time and too sparse in space to provide conclusive validation of proposed mechanisms for variability on the decadal-interdecadal time scale. Consequently, further modeling studies in conjunction with and constrained by the observations must ultimately be carried out to unravel the underlying mechanisms.

One of the fundamental questions in the debate on climate change concerns the role of the oceans. By appealing to the results from both coupled atmosphere-ocean and atmosphere–mixed-layer ocean models we can begin to understand the differences between the North Atlantic and Pacific Ocean responses to increasing anthropogenic greenhouse gases. In the near term the ocean, through its large heat capacity, acts to reduce the effects of global warming relative to land. This is particularly noticeable in the high-latitude regions of the North Atlantic where a weakening of the thermohaline circulation and subsequent reduction in northward heat transport occurs. Both fully coupled atmosphere-ocean models (e.g. Gates et al 1992, Kattenberg et al 1996) and atmospheric models with only an ocean mixed layer (e.g. Mitchell et al 1990, Gates et al 1992) give similar responses in the Pacific Ocean, where there is a less active thermohaline circulation. Both classes of models further suggest that enhanced high-latitude warming is due to the reduction of sea ice cover and the accompanying decrease in surface albedo. Thus the Arctic is a prime location to look for a signal of anthropogenic climate change above the background natural climate variability.

Evidence for SAT changes over the Arctic Ocean is, however, inconsistent: Martin et al (1997) found significant warming in May and June from 1961 to 1990, whereas Kahl et al (1993) found cooling during autumn and winter over the western Arctic Ocean for 1950–1990. Trends in SAT over land are more apparent, as Chapman & Walsh (1993) report a significant increase over northwestern North America and central Siberia since 1960. A decrease in the Northern Hemisphere sea ice extent of about three percent per decade has also been observed (Chapman & Walsh 1993, Maslanik et al 1996, Cavalieri et al 1997).

Although it is apparent that we are only at the early stages of attempting to detect climate change at high latitudes and attributing this to anthropogenic effects, much progress will likely come in the near future. Of utmost importance is a deeper understanding of the mechanisms and structure of climate variability, as this is central to the detection and attribution issue. In addition, it is still unknown how the observed climate variability depends on the mean climatic state itself.

ACKNOWLEDGMENTS

This work was supported by NSERC strategic, operating, and Steacie Supplement grants, the Atmospheric Environments Service, the Canadian Institute for Climate Studies; and the NOAA Scripps Lamont Consortium on the Ocean's Role in Climate.

Literature Cited

Aagaard K, Carmack EC. 1989. The role of sea ice and other fresh water in the Arctic circulation. *J. Geophys. Res.* 94:14485–98

Aagaard K, Foldvik A, Hillman SR. 1987. The West Spitsbergen Current: disposition and water mass transformation. *J. Geophys. Res.* 92:3778–84

Aagaard K, Roach AT, Schumacher JD. 1985a. On the wind-driven variability of the flow through Bering Strait. *J. Geophys. Res.* 90: 7213–21

Aagaard K, Swift JH, Carmack EC. 1985b. Thermohaline circulation in the Arctic Mediterranean seas. *J. Geophys. Res.* 90:4833–46

Alexander MA. 1992a. Midlatitude atmosphere-ocean interaction during El Niño: Part I. the North Pacific Ocean. *J. Clim.* 5:944–58

Alexander MA. 1992b. Midlatitude atmosphere-ocean interaction during El Niño: Part II. the northern hemisphere atmosphere. *J. Clim.* 5:959–72

Alexander MA, Deser C. 1995. A mechanism for the recurrence of wintertime midlatitude SST anomalies. *J. Phys. Oceanogr.* 25:122–37

Anderson DLT, Willebrand J, eds. 1996. *Decadal Climate Variability: Dynamics and Predictability. NATO ASI Ser. I, Global Environ. Change*, Vol. 44. Berlin/New York: Springer-Verlag. 483 pp.

Armi L, Farmer DM. 1988. The flow of Mediterranean water through the Strait of Gibraltar. *Prog. Oceanogr.* 21:1–105

Barnston AG, Livezey RE. 1987. Classification, seasonality and persistence of low-frequency atmospheric circulation patterns. *Mon. Weather Rev.* 115:1083–126

Baumgartner A, Reichel E. 1975. *The World Water Balance*. Amsterdam/Lausanne/NY: Elsevier. 179 pp.

Berger WH. 1987. Ocean ventilation during the last 12,000 years: hypothesis of counterpoint deep water production. *Mar. Geol.* 78:1–10

Bjerknes J. 1964. Atlantic air-sea interaction. *Mon. Weather Rev.* 97:163–72

Bjerknes J. 1969. Atmospheric teleconnections from the equatorial Pacific. *Adv. Geophys.* 10:1–82

Borenäs KM, Lundberg PA. 1988. On the deep water flow through the Faroe Bank Channel. *J. Geophys. Res.* 93:1281–92

Boyd TJ, D'Asaro EA. 1994. Cooling of the West Spitsbergen Current: wintertime observations west of Svalbard. *J. Geophys. Res.* 99:22,597–618

Boyle EA. 1990. Quaternary deepwater paleoceanography. *Science* 249:863–70

Bretherton FP. 1982. Ocean climate modelling. *Prog. Oceanogr.* 11:93–129

Broecker WS. 1991. The great ocean conveyor. *Oceanography* 4:79–89

Broecker WS, Denton GH. 1989. The role of ocean-atmosphere reorganizations in glacial cycles. *Geochim. Cosmochim. Acta* 53:2465–501

Broecker WS, Peng TH, Jouzel J, Russel G. 1990. The magnitude of global fresh-water transports of importance to ocean circulation. *Clim. Dyn.* 4:73–79

Broecker WS, Peteet DM, Rind D. 1985. Does the ocean-atmosphere system have more than one stable mode of operation? *Nature* 315: 21–26

Bryden HL, Roemmich DH, Church JA. 1991. Ocean heat transport across 24 N in the Pacific. *Deep-Sea Res.* 38:297–324

Buorke RH, Weigel AM, Paquette RG. 1988. The westward turning branch of the West Spitsbergen Current. *J. Geophys. Res.* 93:14,065–77

Carmack EC. 1990. Large scale physical oceanography of Polar oceans. In *Polar Oceanography, Part A: Physical Science*, ed. WM Smith Jr., p. 171–222. New York: Academic. 406 pp.

Carmack EC, Aagaard K, Swift JH, MacDonald RW, McLaughlin FA, et al. 1998. Changes in temperature and tracer distributions within the Arctic Ocean: results from the 1994 Arctic Ocean section. *Deep-Sea Res.* 44:1487–502

Carmack EC, Macdonald RW, Perkin RG, McLaughlin FA, Pearson RJ. 1995. Evidence for warming of Atlantic water in the southern Canadian Basin of the Arctic Ocean: results from the Larsen-93 expedition. *Geophys. Res. Lett.* 22:1061–64

Cattle H. 1985. Diverting Soviet rivers: some possible repercussions for the Arctic Ocean. *Polar Rec.* 22:485–98

Cavalieri D, Gloersen JP, Parkinson CL, Comiso JC, Zwally HJ. 1997. Observed hemispheric asymmetry in global sea ice changes. *Science* 278:1104–06

Chapman WL, Walsh JE. 1993. Recent variations of sea ice and air temperature in high latitudes. *Bull. Am. Meteorol. Soc.* 74:33–47

Clarke RA. 1984. Transport through the Cape-Farewell-Flemish Cap section. *Rapp. Reun. Cons. Int. Explor. Mer.* 185:120–30

Coachman LK, Aagaard K. 1988. Transports through Bering Strait: annual and interannual variability. *J. Geophys. Res.* 93:15,535–39

Curry WB, Duplessy JC, Labeyrie LD, Shackleton NJ. 1988. Changes in the distribution of $\delta^{13}C$ of deep water ΣCO_2 between the last glaciation and the Holocene. *Paleoceanography* 3:317–41

Dean WE, Gardner JV, Hemphill-Halet E. 1989. Changes in redox conditions in deep-sea sediments of the sub-Arctic North Pacific Ocean: possible evidence for the presence of North Pacific Deep Water. *Paleoceanography* 4:639–53

Delworth T, Manabe S, Stouffer RJ. 1993. Interdecadal variations of the thermohaline circulation in a coupled ocean-atmosphere model. *J. Clim.* 6:1993–2011

Delworth T, Manabe S, Stouffer RJ. 1997. Multidecadal climate variability in the Greenland sea and surrounding regions: a coupled model simulation. *Geophys. Res. Lett.* 24:257–60

Deser C, Alexander MA, Timlin MS. 1996. Upper-ocean thermal variations in the North Pacific during 1970–1991. *J. Clim.* 9:1840–55

Deser C, Blackmon ML. 1993. Surface climate variations over the North Atlantic Ocean during winter: 1900–1989. *J. Clim.* 6:1743–53

Deser C, Blackmon ML. 1995. On the relationship between tropical and North Pacific sea surface temperature variations. *J. Clim.* 8:1677–80

Dettinger MD, Cayan DR. 1995. Large-scale atmospheric forcing of recent trends toward early snowmelt runoff in California. *J. Clim.* 8:606–23

Dickson RR, Brown J. 1994. The production of North Atlantic Deep Water: sources, sinks and pathways. *J. Geophys. Res.* 99:12,319–41

Dickson RR, Gmitrowicz EM, Watson AJ. 1990. Deep-water renewal in the northern North Atlantic. *Nature* 344:848–50

Dickson RR, Lazier J, Meincke J, Rhines P, Swift J. 1996. Long-term coordinated changes in the convective activity of the North Atlantic. *Prog. Oceanogr.* 38:205–39

Dickson RR, Meincke J, Malmberg SA, Lee AJ. 1988. The "great salinity anomaly" in the northern North Atlantic 1968–1982. *Prog. Oceanogr.* 20:103–51

Duplessy JC, Shackleton NJ, Fairbanks RG, Labeyrie L, Oppo D, Kallel N. 1988. Deepwater source variations during the last climate cycle and their impact on the global deepwater circulation. *Paleoceanography* 3:343–60

Ellis GE. 1868. *Memoir of Sir Benjamin Thompson, Count Rumford, with Notices from his Daughter*. Am. Acad. Sci. Philadelphia, PA: Claxton, Remsen Haffelfinger

Ellis H. 1751. A Letter to the Rev. Dr. Hales, F.R.S. from Captain Henry Ellis, F.R.S. dated Jan. 7, 1750–51, at Cape Monte Africa, Ship Earl of Halifax. *Philos. Trans. R. Soc. London* 47:211–14

Emery WJ, Hamilton K. 1985. Atmospheric forcing of interannual variability in the northeast Pacific Ocean: connections with El Niño. *J. Geophys. Res.* 90:857–68

Fang Z, Wallace JM. 1994. Arctic sea ice variability on a timescale of weeks and its relation to atmospheric forcing. J. Clim. 7:1897–913

Fissel DB, Loemon DD, Melling H, Lake RA. 1988. Non-tidal flows in the Northwest Passage. *Can. Tech. Rep. Hydrogr. Ocean Sci. 98*, Sidney, BC: Inst. Ocean Sci.

Foldvik A, Gammelsrøld T. 1988. Notes on Southern Ocean hydrography, sea-ice and bottom water formation. *Palaeogeogr. Palaeoclimatol. Palaeoecol.* 67:3–17

Ganopolski A, Rahmstorf S, Petoukhov V, Claussen M. 1998. Simulation of modern and glacial climates with a coupled global model of intermediate complexity. *Nature* 391:351–56

Gates WL, Mitchell JFB, Boer GJ, Cubasch U, Meleshko VP. 1992. Climate modelling, climate prediction and model validation. In *Climate Change 1992, the Supplementary Report to the IPCC Scientific Assessment*, ed. JT Houghton, BA Callander, SK Varney, p. 97–134. New York/London: Cambridge Univ. Press. 200 pp.

Gerdes R, Schauer U. 1997. Large-scale circulation and water mass distribution in the Arctic Ocean from model results and observations. *J. Geophys. Res.* 102:8467–83

Gordon AL. 1986. Interocean exchange of intermediate water. *J. Geophys. Res.* 91:5037-46

Grumbine RW. 1991. A model of the formation of high-salinity shelf water on polar continental shelves. *J. Geophys. Res.* 96:22,049–62

Gu D, Philander GH. 1997. Internal climate fluctuations that depend on exchanges between the tropics and extratropics. *Science* 275:805–7

Häkkinen S. 1993. An Arctic source for the Great Salinity Anomaly: a simulation of the Arctic ice-ocean system for 1955–1975. *J. Geophys. Res.* 98:16,397–410

Hall MM, Bryden HL. 1982. Direct estimates and mechanisms of ocean heat transport. *Deep-Sea Res.* 29:339–59

Hallock ZR, Teague WJ. 1996. Evidence for a North Pacific deep western boundary current. *J. Geophys. Res.* 101:6617–24

Hansen DV, Bezdek H. 1996. On the nature of

decadal anomalies in North Atlantic sea surface temperature. *J. Geophys. Res.* 101:8749–58
Hansen J, Lebedeff S. 1987. Global trends of measured surface air temperature. *J. Geophys. Res.* 92:13,345–72
Hasselmann K. 1976. Stochastic climate models: Part I. theory. *Tellus* 28:239–305
Hastrom L, Michaelson J. 1994. Long-term central coastal California precipitation variability and relationships to El Niño. *J. Clim.* 7:1373–87
Hasunuma K. 1978. Formation of the intermediate salinity minimum in the northwestern Pacific Ocean. *Bull. Ocean Res. Inst.* 9. 47 pp.
Holland WR. 1973. Baroclinic and topographic influences on the transport in western boundary currents. *Geophys. Fluid Dyn.* 4:187–210
Horel JD, Wallace JM. 1981. Planetary-scale atmospheric phenomena associated with the Southern Oscillation. *Mon. Weather Rev.* 109:813–29
Hughes TMC, Weaver AJ. 1994. Multiple equilibria in an asymmetric two-basin ocean model. *J. Phys. Oceanogr.* 24:619–37
Hurrell JW. 1995. Decadal trends in the North Atlantic Oscillation: regional temperatures and precipitation. *Science* 269:676–79
Hurrell JW. 1996. Influence of variations in extratropical wintertime teleconnections on Northern Hemisphere temperature. *Geophys. Res. Lett.* 23:665–68
Ivers WD. 1975. The deep circulation of the North Atlantic with especial reference to the Labrador Sea. PhD thesis. Univ. Calif., San Diego. 179 pp.
Jacobs GA, Hurlburt HE, Kindle JC, Metzger EJ, Mitchell JL, et al. 1994. Decade-scale trans-Pacific propagation and warming effects of an El Niño warming. *Nature* 370:360–63
Jacobs SS, Fairbanks RG, Horibe Y. 1985. Origin and evolution of water masses near the Antarctic continental margin: evidence from $H_2{}^{18}O/H_2{}^{16}O$ ratios in seawater. In *Oceanology of the Antarctic Continental Shelf, Antarct. Res. Ser.*, Vol. 43. ed. SS Jacobs, p. 58–86. Washington, DC: AGU
Jin F. 1997. A theory of interdecadal climate variability of the North Pacific ocean-atmosphere system. *J. Clim.* 10:1821–35
Johnson GC. 1990. *Near-equatorial deep circulation in the Indian and Pacific Oceans*. PhD thesis. Woods Hole, MA: Woods Hole Oceanogr. Inst.
Johnson GC, Toole JM. 1993. Flow of deep and bottom waters in the Pacific at 10°N. *Deep-Sea Res.* 40:371–94
Joyce TM, Warren BA, Talley LD. 1986. The geothermal heating of the abyssal subarctic Pacific Ocean. *Deep-Sea Res.* 33:1003–15
Kahl JK, Charlevoix DJ, Zaitseva NA, Schnell RC, Serreze MC. 1993. Absence of evidence for greenhouse warming over the Arctic Ocean in the past 40 years. *Nature* 361:335–37
Kattenberg A, Giorgi F, Grassl H, Meehl GA, Mitchell JFB, et al. 1996. Climate models—projections of future climate. In *Climate Change 1995, the Science of Climate Change*, ed. JT Houghton, LG Meira Filho, BA Callander, N Harris, A Kattenberg, K Maskell, p. 285–357. New York/London: Cambridge Univ. Press. 572 pp.
Kawamura R. 1994. A rotated EOF analysis of global sea surface temperature variability with interannual and interdecadal scales, *J. Phys. Oceanogr.* 24:707–15
Kawase M. 1987. Establishment of deep ocean circulation driven by deep-water production. *J. Phys. Oceanogr.* 17:2294–317
Keigwin LD. 1987. North Pacific deep water formation during the latest glaciation. *Nature* 330:362–64
Killworth PD. 1979. On "chimney" formations in the ocean. *J. Phys. Oceanogr.* 9:531–54
Killworth PD. 1983. Deep convection in the world ocean. *Rev. Geophys. Space Phys.* 21:1–26
Kirwan R. 1787. *An Estimate of Temperature at Different Latitudes*. London.
Krauss W. 1986. The North Atlantic Current. *J. Geophys. Res.* 91:5061–74
Kuo HH, Veronis G. 1973. The use of oxygen as a test for an abyssal circulation model. *Deep-Sea Res.* 20:871–88
Kushnir Y. 1994. Interdecadal variability in the Northern Hemisphere sea surface temperature and associated atmospheric conditions. *J. Clim.* 7:141–57
Kushnir Y, Held IM. 1997. Equilibrium atmospheric response to North Atlantic SST anomalies. *J. Clim.* 9:1208–20
Kushnir Y, Lau NC. 1992. The general circulation model response to a North Pacific SST anomaly: dependence on timescale and pattern polarity. *J. Clim.* 5:271–83
Latif M. 1998. Dynamics of interdecadal variability in coupled ocean-atmosphere models. *J. Clim.* 11:602–24
Latif M, Barnett TP. 1994. Causes of decadal climate variability over the North Pacific and North America. *Science* 266:634–37
Latif M, Barnett TP. 1996. Decadal climate variability over the North Pacific and North America: dynamics and predictability. *J. Clim.* 9:2407–23
Lau NC, Nath MJ. 1994. A modeling study of the relative roles of tropical and extratropical SST anomalies in the variability of the global atmosphere-ocean system. *J. Clim.* 7:1184–207

Lazier JRN. 1980. Oceanographic conditions at Ocean Weather Ship *Bravo*, 1964–1974. *Atmos-Ocean* 18:227–38

Leaman DK, Schott FA. 1991. Hydrographic structure of the convective regime of the Gulf of Lions: winter 1987. *J. Phys. Oceanogr.* 21: 575–98

Lee TN, Johns WE, Zantopp RJ, Fillenbaum ER. 1996. Moored observations of western boundary current variability and thermohaline circulation at 26.5°N in the subtropical North Atlantic. *J. Phys. Oceanogr.* 26:962–83

Manabe S, Stouffer RJ. 1998. Two stable equilibria of a coupled ocean-atmosphere model. *J. Clim.* 1:841–66

Manley TO. 1995. Branching of Atlantic water within the Greenland-Spitsbergen passage: an estimate of recirculation. *J. Geophys. Res.* 100:20,627–34

Mantua NJ, Hare SR, Zhang Y, Wallace JM, Francis RC. 1997. A Pacific interdecadal climate oscillation with impacts on salmon production. *Bull. Am. Meteorol. Soc.* 78:1069–79

Mantyla AW, Reid JL. 1983. Abyssal characteristics of the world ocean waters. *Deep-Sea Res.* 30:805–33

Martin S, Munoz E. 1997. Properties of the Arctic 2-meter air temperature for 1979–present derived from a new gridded data set. *J. Clim.* 10:1428–40

Martin S, Munoz E, Drucker R. 1997. Recent observations of spring-summer surface warming over the Arctic Ocean. *Geophys. Res. Lett.* 24:1259–62

Maslanik JA, Serreze MC, Barry RG. 1996. Recent decrease in Arctic summer ice cover and linkages to atmospheric circulation anomalies. *Geophys. Res. Lett.* 23:1677–80

Mauritzen C, Häkkinen S. 1997. Influence of sea ice on the thermohaline circulation in the Arctic-North Atlantic Ocean. *Geophys. Res. Lett.* 24:3257–60

Maykut GA. 1982. Large-scale heat exchange and ice production in the central Arctic. *J. Geophys. Res.* 87:7971–84

McCartney MS. 1992. Recirculating components to the deep boundary current of the northern North Atlantic. *Prog. Oceanogr.* 29: 283–383

McCartney MS, Bennett SL, Woodgate-Jones ME. 1991. Eastward flow through the Mid-Atlantic Ridge at 11°N and its influence on the abyss of the eastern basin. *J. Phys. Oceanogr.* 21:1089–121

McCartney MS, Talley LD. 1984. Warm-to-cold water conversion in the northern North Atlantic Ocean. *J. Phys. Oceanogr.* 14:922–35

McLaughlin FA, Carmack EC, Mcdonald RW, Bishop JKB. 1996. Physical and geochemical properties across the Atlantic/Pacific water mass front in the southern Canadian Basin. *J. Geophys. Res.* 101:1183–97

McPhaden M. 1994. The eleven year El Niño? *Nature* 370:326–27

Meincke J. 1983. The modern current regime across the Greenland-Scotland ridge. In *Structure and Development of the Greenland-Scotland Ridge*, ed. A Bott et al, p. 637–650, London/New York: Plenum

Minobe S. 1997. A 50–70 year climatic oscillation over the North Pacific and North America. *Geophys. Res. Lett.* 24:683–86

Mitchell JFB, Manabe S, Meleshko V, Tokioka T. 1990. Equilibrium climate change and its implications for the future. In *Climate Change The IPCC Scientific Assessment*, ed. JT Houghton, GJ Jenkins, JJ Ephraums, p. 131–172. New York/London: Cambridge Univ. Press. 365 pp.

Morison J. 1991. Seasonal variations in the West Spitsbergen Current estimated from bottom pressure measurements. *J. Geophys. Res.* 96:18381–95

Mysak LA. 1986. El Niño, interannual variability and the fisheries in the northeast Pacific Ocean. *Can. J. Fish. Aquat. Sci.* 43:464–97

Mysak LA, Manak DK. 1989. Arctic sea-ice extent and anomalies, 1953–1984. *Atmos. Ocean* 27:376–405

Mysak LA, Manak DK, Marsden RF. 1990. Sea-ice anomalies observed in the Greenland and Labrador Seas during 1901–1984 and their relation to an interdecadal Arctic climate cycle. *Clim. Dyn.* 5:111–33

Nakamura H, Lin G, Yamagata T. 1997. Decadal climate variability in the North Pacific during the recent decades. *Bull. Am. Meteorol. Soc.* 78:2215–25

Nakamura N, Oort AH. 1988. Atmospheric heat budgets of the polar regions. *J. Geophys. Res.* 93:9510–24

Namais J, Born RM. 1974. Further studies of temporal coherence in North Pacific sea surface temperatures. *J. Geophys. Res.* 79:797–98

Nitta T, Yamada S. 1989. Recent warming of tropical sea surface temperature and its relationship to the Northern Hemisphere circulation. *J. Meteorol. Soc. Jpn.* 67:187–93

Oort AH. 1974. Year-to-year variations in the energy balance of the Arctic atmosphere. *J. Geophys. Res.* 79:1253–60

Oppo DW, Fairbanks RG. 1990. Atlantic thermohaline circulation of the last 150,000 years: relationship to climate and atmospheric CO_2. *Paleoceanography* 5:277–88

Overland JE, Pease CH. 1982. Cyclone climatology of the Bering Sea and its relation to sea ice extent. *Mon. Weather Rev.* 110:5–13

Overland JE, Roach AT. 1987. Northward flow in the Bering and Chukchi Seas. *J. Geophys. Res.* 92:7097–105

Overland JE, Turet P. 1994. Variability of the atmospheric energy flux across 70°N computed from the GFDL data set. In *Polar Oceans and Their Role in Shaping the Global Environment, Geophys. Monogr. Ser.*, Vol. 85. ed. OM Johannessen, RD Muench JE Overland, p. 313–25. Washington, DC: AGU

Overland JE, Turet P, Oort AH. 1996. Regional variations of the moist static energy flux into the Arctic. *J. Clim.* 9:54–65

Palmer TN, Sun Z. 1985. A modeling and observational study of the relationship between sea surface temperatures in the northwest Atlantic and the atmospheric general circulation. *Q. J. R. Meteorol. Soc.* 111:947–75

Peng S, Mysak LA, Ritchie H, Derome J, Dugas B. 1995. The differences between early and midwinter atmospheric responses to sea surface temperature anomalies in the northwest Atlantic. *J. Clim.* 8:137–57

Peng S, Robinson WA, Hoerling MP. 1997. The modeled atmospheric response to midlatitude SST anomalies and its dependence on background circulation states. *J. Clim.* 10:971–87

Pitcher EJ, Blackmon ML, Bates GT, Munoz S. 1988. The effect of North Pacific sea surface temperature anomalies on the January climate of a general circulation model. *J. Atmos. Sci.* 45:226–46

Price JF, Baringer MO. 1994. Outflows and deep water production by marginal seas. *Prog. Oceanogr.* 33:161–200

Quadfasel D, Gascard JC, Koltermann KP. 1987. Large-scale oceanography in Fram Strait during the 1984 Marginal Ice Zone Experiment. *J. Geophys. Res.* 92:6719–28

Rahmstorf S. 1995. Bifurcations of the Atlantic thermohaline circulation in response to changes in the hydrological cycle. *Nature* 378:145–49

Reid JL. 1961. On the temperature, salinity and density differences between the Atlantic and Pacific Oceans in the upper kilometre. *Deep-Sea Res.* 7:265–75

Reid JL. 1979. On the contribution of Mediterranean Sea outflow to the Norwegian-Greenland Sea. *Deep-Sea Res.* 26:1199–223

Reid JL. 1997. On the total geostrophic circulation of the Pacific Ocean: flow patterns, tracers and transports. *Prog. Oceanogr.* 39:263–352

Reverdin G, Cayan D, Kushnir Y. 1997. Decadal variability of hydrography in the upper northern North Atlantic in 1948–1990. *J. Geophys. Res.* 102:8505–31

Rhein M. 1991. Ventilation rates of the Greenland and Norwegian Seas derived from the distributions of chlorofluoromethanes F11 and F12, *Deep-Sea Res.* 38:485–503

Rintoul SR. 1991. South Atlantic interbasin exchange. *J. Geophys. Res.* 96:2675–92

Roach AT, Aagaard K, Pease CH, Salo SA, Weingartner T, Pavlov V, Kulakov M. 1995. Direct measurements of transport and water properties through the Bering Strait. *J. Geophys. Res.* 100:18,443–57

Roemmich DH, Wunsch C. 1985. Two transatlantic sections: meridional circulation and heat flux in the subtropical North Atlantic Ocean. *Deep-Sea Res.* 32:619–64

Rogers JC. 1981. North Pacific Oscillation. *J. Climatol.* 1:39–57

Rogers JC. 1990. Patterns of low-frequency monthly sea level pressure variability (1988–1986) and associate wave cyclone frequency. *J. Clim.* 3:1364–79

Rogers JC, van Loon H. 1979. The seesaw in winter temperatures between Greenland and northern Europe: Part II. some oceanic and atmospheric effects in middle and high latitudes. *Mon. Weather Rev.* 107:509–19

Ross CK. 1984. Temperature-salinity characteristic of the 'overflow' water in Denmark Strait during "OVERFLOW 73," *Rapp. Reun. Cons. Int. Explor. Mer.* 185:111–19

Rudels B, Jones EP, Anderson LG, Kattner G. 1994. On the intermediate depth waters of the Arctic Ocean. In *The Polar Oceans and Their Role in Shaping the Global Environment, Geophys. Monogr. Ser.*, Vol. 85, ed. OM Johannessen, RD Muench JE Overland, p. 33–46. Washington, DC: AGU

Rudels B, Quadfasel D, Friedrich H. 1989. Greenland Sea convection in the winter of 1987–1988. *J. Geophys. Res.* 94:3223–27

Saunders PM. 1990. Cold outflow from the Faroe Bank Channel. *J. Phys. Oceanogr.* 20: 29–43

Saunders PM. 1994. The flux of overflow water through the Charlie-Gibbs Fracture Zone. *J. Geophys. Res.* 99:12,343–55

Saunders PM. 1996. The flux of dense cold overflow water southeast of Iceland. *J. Phys. Oceanogr.* 26:85–95

Schmitt RW, Bogden PS, Dorman CE. 1989. Evaporation minus precipitation and density fluxes for the North Atlantic. *J. Phys. Oceanogr.* 19:1208–21

Schmitz WJ Jr, 1995. On the interbasin-scale thermohaline circulation. *Rev. Geophys.* 33:151–73

Schmitz WJ Jr, McCartney MS. 1993. On the North Atlantic circulation. *Rev. Geophys.* 31:29–49

Schmitz WJ Jr, Richardson PL. 1991. On the sources of the Florida Current. *Deep-Sea Res.* 38, 1:S389–S409 (Suppl.)

Schmitz WJ Jr, Thompson JD, Luyten JR. 1992.

The Sverdrup circulation for the Atlantic along 24°N. *J. Geophys. Res.* 97:7251–56
Schott F, Visbeck M, Fischer J. 1993. Observations of vertical currents and convection in the central Greenland Sea during the winter of 1988–1989. *J. Geophys. Res.* 98:14,401–21
Serreze M, Barry R, McLaren AS. 1989. Seasonal variations in sea ice motion and effects on sea ice concentration in the Canadian Basin. *J. Geophys. Res.* 94:10,955–70
Serreze MC, Barry RG, Rehder MC, Walsh JE. 1995. Variability of atmospheric circulation and moisture flux over the Arctic. *Philos. Trans. R. Soc. London* 352:215–25
Serreze MC, Box JE, Barry RG, Walsh JE. 1993. Characteristics of Arctic synoptic activity, 1952–1989. *Meteorol. Atmos. Phys.* 51:147–64
Serreze M, Carse F, Barry R, Rogers JC. 1997. Icelandic low cyclone activity: climatological features, linkages with the NAO, and relationships with recent changes in the Northern Hemisphere circulation. *J. Clim.* 10:453–63
Shackleton NJ, Duplessy JC, Arnold M, Maurice P, Hall A, Cartlidge J. 1988. Radiocarbon age of last glacial Pacific deepwater. *Nature* 335:708–11
Shaffer G, Bendtsen J. 1994. Role of the Bering Strait in controlling North Atlantic Ocean circulation and climate. *Nature* 367:354–57
Siedler G. 1969. General circulation of water masses in the Red Sea. In *Hot Brines and Recent Heavy Metal Deposits in the Red Sea*, ed. ET Degen, DA Ross, p. 131–37. Berlin/New York: Springer-Verlag
Slonosky VC, Mysak LA, Derome J. 1997. Linking Arctic sea-ice and atmospheric circulation anomalies on interannual and decadal timescales. *Atmos. Ocean* 35:333–66
Steele M, Thomas D, Rothrock D, Martin S. 1996. A simple model study of the Arctic Ocean freshwater balance, 1979–1985. *J. Geophys. Res.* 101:20833–48
Stigebrandt A. 1984. The North Pacific: a global-scale estuary. *J. Phys. Oceanogr.* 14: 464–70
Stocker TF, Wright DG. 1991. Rapid transitions of the ocean's deep circulation induced by changes in surface water fluxes. *Nature* 351:729–32
Stocker TF, Wright DG, Mysak LA. 1992. A zonally averaged, coupled ocean-atmosphere model for paleoclimate studies. *J. Clim.* 5: 773–97
Stommel H. 1957. A survey of ocean current theory. *Deep-Sea Res.* 4:149–84
Stommel H, Arons AB. 1960. On the abyssal circulation of the world ocean: II. an idealized model of the circulation pattern and amplitude in ocean basins. *Deep-Sea Res.* 6:217–33
Swift JH, Aagaard K. 1981. Seasonal transitions and water mass formation in the Iceland and Greenland Seas. *Deep-Sea Res.* 28:1107–29
Swift JH, Jones EP, Aagaard K, Carmack EC, Hingston M, et al. 1998. Waters of the Makarov and Canada basins. *Deep-Sea Res.* 44:1503–29
Taft BA, Hayes SP, Frederich GE, Codispoti LA. 1991. Flow of abyssal water into Samoa Passage. *Deep-Sea Res.* 38, Suppl. 1: S103–28
Talley LD. 1991. An Okhotsk Sea water anomaly: implications for ventilation in the North Pacific. *Deep-Sea Res.* 38, Suppl. 1: S171–90
Talley LD. 1993. Distribution and formation of North Pacific Intermediate Water. *J. Phys. Oceanogr.* 23:517–37
Talley LD, McCartney MS. 1982. Distribution and circulation of Labrador Sea Water. *J. Phys. Oceanogr.* 12:1189–205
Talley LD, Nagata Y, Fujimura M, Iwao T, Kono T, et al. 1995. North Pacific Intermediate Water in the Kuroshio/Oyashio mixed water region. *J. Phys. Oceanogr.* 25:475–501
Thompson N, Count Rumford. 1798. The propagation of heat in fluids. Reproduced, 1870, In *The Complete Works of Count Rumford.* Vol. 1, p. 237–400. Boston, MA: Am. Acad. Sci.
Ting M. 1991. The stationary wave response to a midlatitude SST anomaly in an idealized GCM. *J. Atmos. Sci.* 48:1249–75
Trenberth KE. 1990. Recent observed interdecadal climate changes in the Northern Hemisphere. *Bull. Am. Meteorol. Soc.* 71:988–93
Trenberth KE, Hoar TJ. 1994. The 1990–1995 El Niño-Southern Oscillation event: longest on record. *Geophys. Res. Lett.* 23:57–60
Trenberth KE, Hurrell JW. 1994. Decadal atmosphere-ocean variations in the Pacific. *Clim. Dyn.* 9:303–19
van Loon H, Rogers JC. 1978. The seesaw in winter temperatures between Greenland and northern Europe: Part I. general description. *Mon. Weather Rev.* 106:296–310
Vowinckel E, Orvig S. 1971. Synoptic heat budgets at three polar stations. *J. Appl. Meteorol.* 10:387–96
Walker GT, Bliss EW. 1932. World weather V. *Mem. R. Meteorol. Soc.* 4:53–84
Wallace JM, Gutzler DS. 1981. Teleconnections in the geopotential height field during the Northern Hemisphere winter. *Mon. Weather Rev.* 109:784–812
Wallace JM, Jiang Q. 1987. On the observed structure of the interannual variability of the

atmosphere-ocean climate system. In *Atmospheric and Oceanic Variability*, ed. H Cattle, p. 17–43. London: R. Meteorol. Soc.
Wallace JM, Smith C, Jiang Q. 1990. Spatial patterns of atmosphere-ocean interaction in the northern winter. *J. Clim.* 3:990–98
Walsh JE, Chapman WL. 1990. Arctic contribution to upper-ocean variability in the North Atlantic. *J. Clim.* 3:1462–73
Walsh JE, Johnson CM. 1979. Interannual atmospheric variability and associated fluctuations in Arctic sea ice extent. *J. Geophys. Res.* 84:6915–28
Wang X, Stone P, Marotzke J. 1995. Poleward heat transport in a barotropic ocean model. *J. Phys. Oceanogr.* 25:256–65
Warren BA. 1973. Transpacific hydrographic sections at lats. 43°S and 28°S: the SCORPIO Expedition: II. deep water. *Deep-Sea Res.* 20:9–38
Warren BA. 1981. Deep circulation of the World Ocean. In *Evolution of Physical Oceanography*, ed. BA Warren, C Wunsch, p. 6–41. Cambridge, MA: MIT Press. 620 pp.
Warren BA. 1983. Why is no deep water formed in the North Pacific? *J. Mar. Res.* 41:327–47
Warren BA, Owens WB. 1988. Deep currents in the central subarctic Pacific Ocean. *J. Phys. Oceanogr.* 18:529–51
Watanabe T, Wakatsuchi M. 1998. Formation of 26.8–26.9 σ_θ water in the Kuril Basin of the Sea of Okhotsk as a possible origin of North Pacific Intermediate Water. *J. Geophys. Res.* 103:2849–65
Weare B, Navato A, Newell RE. 1976. Empirical orthogonal analysis of Pacific Ocean sea surface temperature. *J. Phys. Oceanogr.* 6:671–78
Weaver AJ, Eby M, Fanning AF, Wiebe EC. 1998. The climate of the Last Glacial Maximum in a coupled ocean GCM/energy-moisture balance atmosphere model. *Nature* 394:847–53
Weaver AJ, Marotzke J, Cummins PF, Sarachik ES. 1993. Stability and variability of the thermohaline circulation. *J. Phys. Oceanogr.* 23:39–60
Weaver AJ, Sarachik ES. 1991. Evidence for decadal variability in an ocean general circulation model: an advective mechanism. *Atmos.-Ocean* 29:197–231
Weaver AJ, Sarachik ES, Marotzke J. 1991. Freshwater flux forcing of decadal and interdecadal oceanic variability. *Nature* 353:836–38
Weaver AJ, Valcke S. 1998. On the variability of the thermohaline circulation in the GFDL coupled model. *J. Clim.* 11:759–67
Weyl PK. 1968. The role of the oceans in climatic change: a theory of the Ice Ages. *Meteorol. Monogr.* 8:37–62
Whittaker LM, Horn LH. 1984. Northern Hemisphere extratropical cyclone activity for four mid-season months. *J. Climatol.* 9:297–319
Wijffels SE, Schmitt RW, Bryden HL, Stigebrandt A. 1992. Transport of freshwater by the oceans. *J. Phys. Oceanogr.* 22:155–62
Wohlleben TMH, Weaver AJ. 1995. Interdecadal climate variability in the subpolar North Atlantic. *Clim. Dyn.* 11:459–67
Worthington LV. 1976. *On the North Atlantic Circulation. Oceanogr. Stud. 6, Baltimore, MD:* The Johns Hopkins Univ. 110 pp.
Wyrtki K. 1961. The thermohaline circulation in relation to the general circulation in the oceans. *Deep-Sea Res.* 8:39–64
Yang J, Neelin JD. 1993. Sea ice interactions with the thermohaline circulation. *Geophys. Res. Lett.* 20:217–20
Yasuda I. 1997. The origin of North Pacific Intermediate Water. *J. Geophys. Res.* 102:893–909
Yasuda I, Okuda K, Shimizu Y. 1996. Distribution and modification of North Pacific Intermediate Water in the Kuroshio-Oyashio interfrontal zone. *J. Phys. Oceanogr.* 26:448–65
Yin FL, Fung IY, Chu CK. 1992. Equilibrium response of ocean deep-water circulation to variations in Ekman pumping and deep-water sources. *J. Phys. Oceanogr.* 22:1129–57
Zhang RH, Levitus S. 1997. Structure and cycle of decadal variability of upper-ocean temperature in the North Pacific. *J. Clim.* 10:710–27
Zhang S, Greatbatch RJ, Lin CA. 1993. A reexamination of the polar halocline catastrophe and implications for coupled ocean-atmosphere modeling. *J. Phys. Oceanogr.* 23:287–99
Zhang Y, Wallace JM, Battisti DS. 1997. ENSO-like interdecadal variability: 1900-93. *J. Clim.* 10:1004–20
Zhang Y, Wallace JM, Iwasaka N. 1996. Is climate variability over the North Pacific a linear response to ENSO? *J. Clim.* 9:1468–78

Annu. Rev. Earth. Planet. Sci. 1999. 27:287–312

KUIPER BELT OBJECTS

David Jewitt
Institute for Astronomy, 2680 Woodlawn Drive, Honolulu, HI 96822;
e-mail: jewitt@ifa.hawaii.edu

KEY WORDS: Neptune, Pluto, planet formation, outer solar system, short-period comets

ABSTRACT

The region of the solar system immediately beyond Neptune's orbit is densely populated with small bodies. This region, known as the Kuiper Belt, consists of objects that may predate Neptune, the orbits of which provide a fossil record of processes operative in the young solar system. The Kuiper Belt contains some of the Solar System's most primitive, least thermally processed matter. It is probably the source of the short-period comets and Centaurs, and may also supply collisionally generated interplanetary dust. I discuss the properties of the Kuiper Belt and provide an overview of the outstanding scientific issues.

HISTORY OF THE KUIPER BELT

Edgeworth (1943) was the first to speculate on the existence of planetary material beyond the orbit of Pluto. Referring to the solar nebula, he wrote, "It is not to be supposed that the cloud of scattered material which ultimately condensed to form the solar system was bounded by the present orbit of the planet Pluto; it is evident that it must have extended to much greater distances." He also suggested that the trans-Plutonian region was the repository of the comets: "From time to time a member of this swarm of potential comets wanders from its own sphere and appears as an occasional visitor to the inner regions of the solar system." These qualitative ideas were later repeated by Edgeworth (1949) and echoed by Kuiper (1951), who gave no indication that he was aware of Edgeworth's papers. Rather, his motivation was in part to correct Oort's (1950) assertion that the comets formed near, and were ejected by, Jupiter and had a composition like that of the main-belt asteroids. Edgeworth's and Kuiper's remarks were purely speculative rather than predictions in the

0084-6597/99/0515-0287$08.00

accepted, scientific sense. Perhaps for this reason, the notion of a trans-Neptunian belt had a less immediate impact on the subsequent development of cometary science than did the contemporaneous but more quantitative work by Oort (1950) and Whipple (1951). Nevertheless, the possibility of a trans-Plutonian ring was clearly recognized before the middle of the 20th century. This concept was sustained by Whipple (1964) and others, while much later Fernandez (1980) explicitly re-proposed a trans-Neptunian ring as a source of the short-period comets.

Baum & Martin (1985) suggested (but apparently did not attempt) an early observational test of the comet belt hypothesis. Luu & Jewitt (1988) made a concerted but ultimately unsuccessful observational effort in which both photographic plates and an early charge-coupled device (CCD) were used to examine the ecliptic. In the same year, Duncan, Quinn, & Tremaine (1988) provided additional motivation for searches by showing that a flattened, disk-like source was required to fit the restricted range of inclinations of the short-period comets. Further ecliptic surveys by Kowal (1989), Levison & Duncan (1990), Cochran et al (1991) and Tyson et al (1992) proved negative. Observational success was achieved first with the discovery of 1992 QB1 (Jewitt & Luu 1993) and followed up with the rapid discovery of a growing number of trans-Neptunian bodies (Jewitt & Luu 1995, Irwin et al 1995, Williams et al 1995, Jewitt et al 1996, Luu et al 1997, Jewitt et al 1998, Gladman et al 1998). These discoveries have powered a veritable explosion of research on the Kuiper Belt in the past half decade. Indeed, Pluto itself is now considered to be the largest known Kuiper Belt Object.

The purpose of this review is to summarize the current observational constraints on the Kuiper Belt, and to discuss extant models and theories of its formation and evolution, all in a style suited to the diverse readership of the *Annual Review of Earth and Planetary Sciences*. The known population of the Kuiper Belt, as well as the important literature on this subject, changes on short timescales compared to the interval between reviews such as this. For the most recent information, the reader is referred to a list of orbital parameters maintained by Brian Marsden and Gareth Williams (http://cfa-www.harvard.edu/cfa/ps/lists/TNOs.html) and to a general site covering the Kuiper Belt maintained by the author (http://www.ifa.hawaii.edu/faculty/jewitt/kb.html).

OBSERVING THE KUIPER BELT

Kuiper Belt objects (KBOs) are best identified by their slow, retrograde (westward) motions when observed in the anti-solar direction. Then, the angular velocity, $d\theta/dt$ [arcsec hr^{-1}] is determined largely by parallax. Slow speeds indicate large distances, according to

$$d\theta/dt \approx \frac{148}{R + R^{1/2}} \tag{1}$$

where R[AU] is the heliocentric distance, and the geocentric distance is taken as $\Delta = R - 1$. The apparent red magnitude m_R, of an object observed in reflected light is given by

$$p_R r^2 \phi(\alpha) = 2.25 \times 10^{16} R^2 \Delta^2 10^{0.4(m_\odot - m_R)} \tag{2}$$

in which p_R is the geometric albedo, r[km] is the object radius, $\phi(\alpha)$ is the phase function and $m_\odot$ is the apparent magnitude of the sun (Russell 1916). At opposition we may take $\phi(0) = 1$. Equations 1 and 2 are plotted in Figure 1, assuming $p_R = 0.04$. The figure shows that 100 km sized objects beyond Neptune should exhibit apparent red magnitudes greater than 22 and characteristic slow retrograde motions (a few arcsec per hour). Successful Kuiper Belt

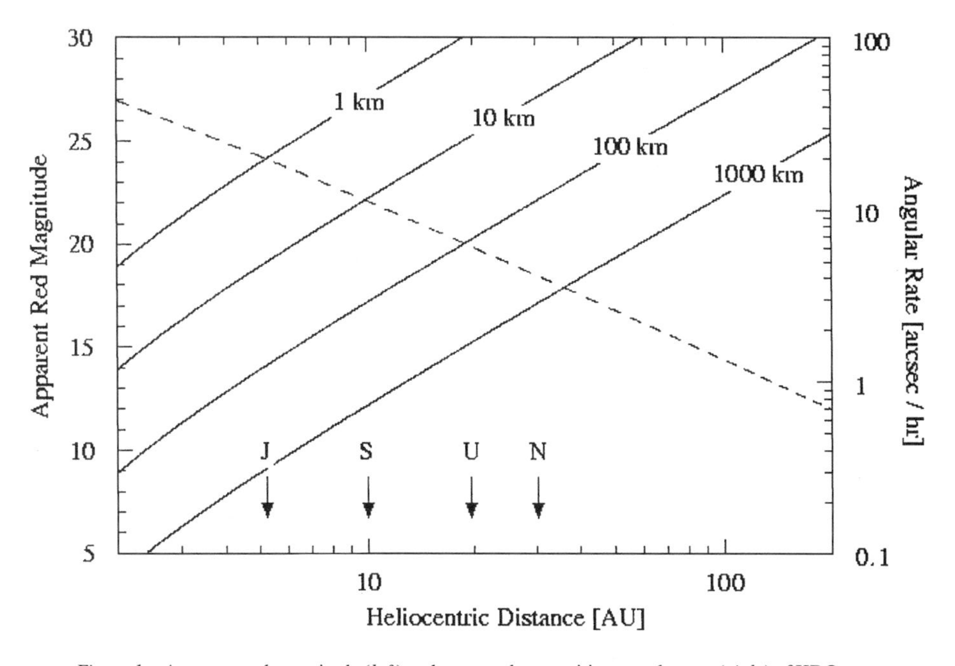

Figure 1 Apparent red magnitude (*left*) and retrograde opposition angular rate (*right*) of KBOs as a function of distance from the sun. Magnitudes were computed from Equation 2 assuming geometric albedo 0.04, radii as marked on the figure and observation at opposition (phase function = 1 and $\Delta = R - 1$). The angular rate is computed from Equation 1 (*dashed curve*). The locations of the major planets are indicated.

surveys have therefore been designed to reveal faint objects moving slowly with respect to the background stars. Increasingly, these surveys are automated (Trujillo & Jewitt 1998) so as to accommodate large data rates.

CONTENTS OF THE KUIPER BELT

At the time of writing (August 1998), ground-based surveys have revealed 69 KBOs, of which 43 possess relatively reliable multi-opposition orbits. The orbits of known KBOs naturally divide into three distinct categories.

Classical Objects

About two thirds of the observed KBOs have semi-major axes $42 \leq a \leq 47$ AU, and seem unassociated with resonances (Figures 2 and 3). These objects define

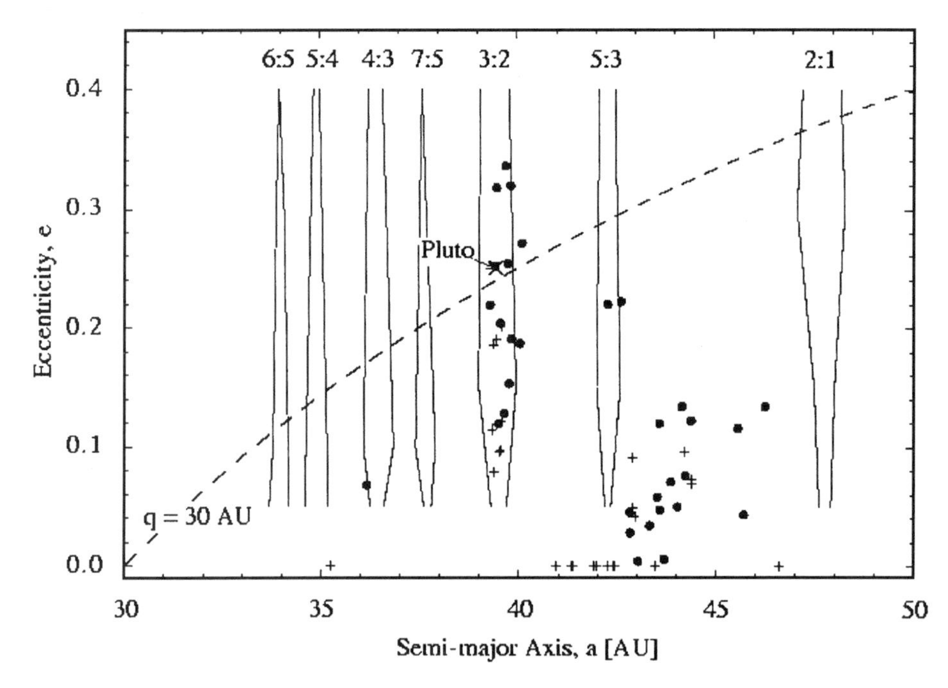

Figure 2 Semi-major axis versus orbital eccentricity for the known KBOs. Multi-opposition objects (*solid circles*) are distinguished from single-opposition objects (*pluses*) and Pluto (*cross*). The distribution is highly nonrandom and shows a clear correlation with mean motion resonances (approximate boundaries of these resonances are shown as *vertical bands*, taken from Malhotra 1995). The *diagonal dashed line* separates objects having perihelion inside Neptune's orbit (above the line) from those wholly exterior to that planet (below the line). Scattered KBO 1996 TL66 falls outside and is omitted from the plot.

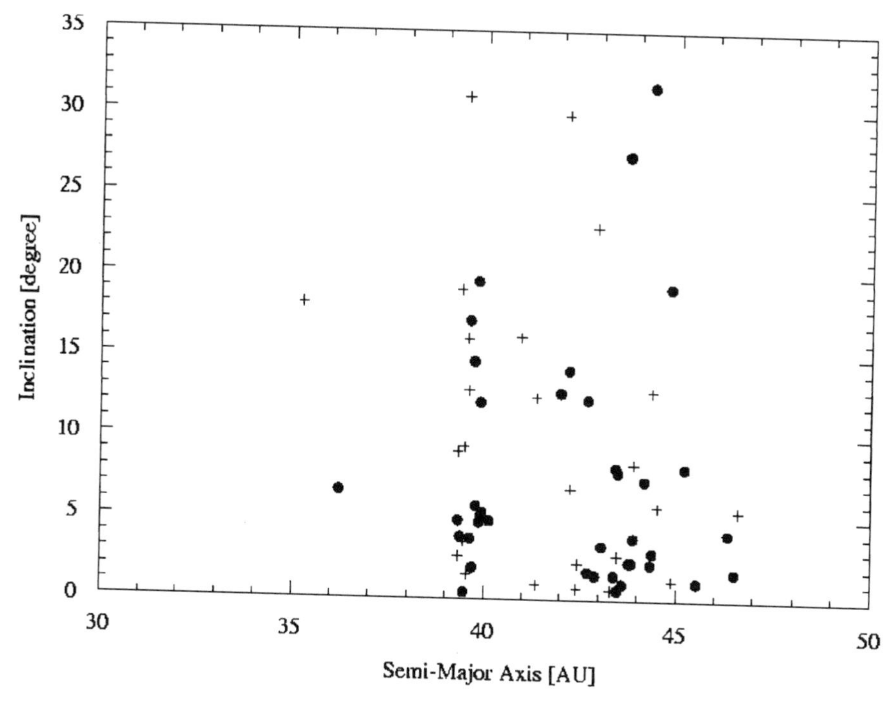

Figure 3 Distribution of semi-major axis versus orbital inclination for the known KBOs. Multi-opposition objects (*solid circles*) are distinguished from single opposition objects (*pluses*). Scattered KBO 1996 TL66 is omitted from the plot.

the "classical" Kuiper Belt. Classical KBOs have small eccentricities (median $e = 0.07$) that maintain a large separation (mostly >10 AU) from Neptune even when at perihelion. The inclinations occupy a wide range ($0 \leq i \leq 32°$).

Resonant Objects (Plutinos)

About one third of the known objects reside in the 3:2 mean motion resonance with Neptune at $a = 39.4$ AU (Figures 2 and 3). These bodies are collectively known as Plutinos ("little Plutos") to highlight the dynamical similarity with Pluto, which also resides in this resonance (Malhotra & Williams 1998). The apparent abundance of the Plutinos is affected by observational bias. Corrected for their smaller mean heliocentric distance (relative to classical KBOs), they probably constitute 10 to 15 percent of the population inside 50 AU (Jewitt et al 1998). The eccentricities ($0.1 \leq e \leq 0.34$) and inclinations ($0 \leq i \leq 20°$) of the Plutinos bracket the values of Pluto ($e = 0.25, i = 17°$). Presumably, the resonance provides immunity to destabilizing perturbations from Neptune.

Indeed, some of the Plutinos have perihelia inside Neptune's orbit (as does Pluto) and would be immediately ejected if not for the protection offered by the resonance. Other resonances (notably the 4:3 and the 5:3) may also be populated, although at a lower level than the 3:2. No objects have yet been found in the 2:1 resonance.

Scattered Objects

The observational sample includes one clearly deviant object, 1996 TL66 (a,e,i = 85 AU, 0.59, 24°; Luu et al 1997). The proximity of the perihelion (q = 35 AU) to Neptune (at 30 AU) suggests a weak dynamical involvement with that planet. It is likely that 1996 TL66 is the first detected member of a population of bodies scattered by Neptune (Fernandez & Ip 1983) and having dynamical lifetimes measured in billions of years (Duncan & Levison 1997). Such scattered KBOs may occasionally enter the planetary system, constituting a source of short-period comets separate from that of chaotic zones.

MODELS

Models of the Kuiper Belt have evolved in a somewhat piece-meal fashion. Most published dynamical models neglect mutual gravitational interactions and collisions among KBOs. This is reasonable in the present-day, low-density Kuiper Belt (but see Ip & Fernandez 1997). However, several lines of evidence now point to the existence of a once more massive Kuiper Belt, in which mutual interactions may have excited a collective (wave) response to perturbations (Ward & Hahn 1998) and in which collisions played an important role. Accordingly, models describing collisions have recently appeared (Farinella & Davis 1996, Stern & Colwell 1997), but have not yet been integrated with the dynamics. It seems likely that key features of the Kuiper Belt were imprinted at a time when collective effects dominated the transport of energy and angular momentum. Therefore, it is important to heed Ward & Hahn's (1998) warning that published simulations of Kuiper Belt dynamics, by omitting important physics, may be in serious error.

The division of the orbital parameters of the KBOs into three main groups was unexpected and has become the focus of recent theoretical attention. Observed properties worthy of explanation include:

(*a*) The general absence of KBOs with $a \leq 42$ AU, other than those trapped in mean motion resonances. Long-term numerical integrations show that clearing of the inner belt is a result of strong perturbations by Neptune (Holman & Wisdom 1993, Duncan et al 1995). This is seen in Figure 4, where most nonresonant orbits inside 42 AU have lifetimes $\ll$4 Gyr. Objects originally in this region were quickly scattered away or absorbed by

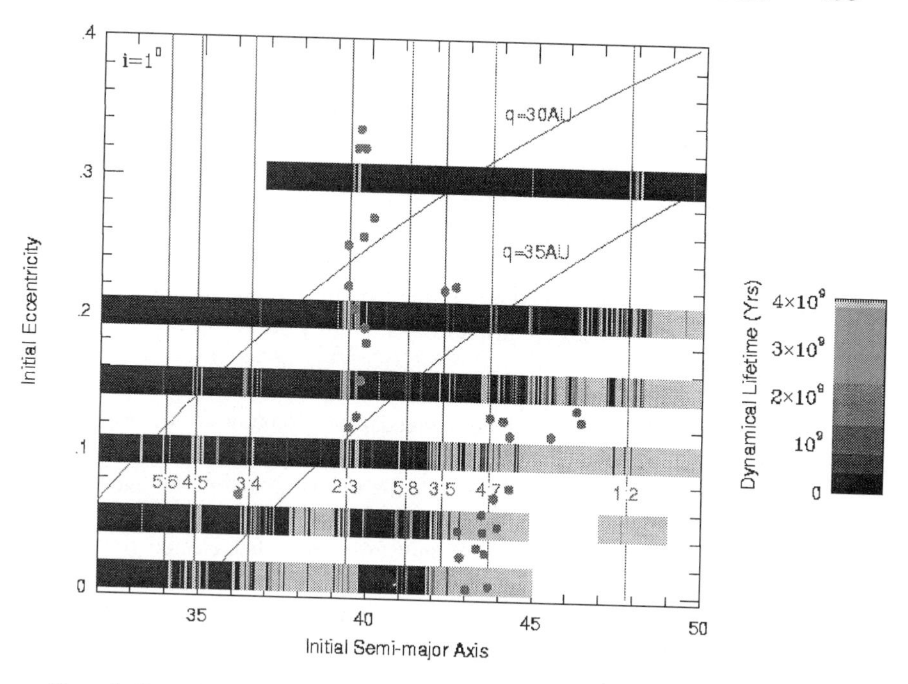

Figure 4 Dynamical simulations of test particle longevity compared with locations of real KBOs (*solid circles: see Figure 2*) in the semi-major axis versus eccentricity plane. Note that the simulations are for assumed inclinations of 1° while the real KBOs exhibit inclinations up to 32° (From Duncan et al 1995).

Neptune. A clear discrepancy exists between numerical simulations and the real Kuiper Belt. Low *e*, low *i* orbits in the 36 to 39 AU range are stable (Figure 4, Duncan et al 1995) and yet no KBOs have been found in this region (Figure 2). Objects in this region might have been removed by resonance sweeping (Malhotra 1995) due to planet migration, or by outward movement of the ν_8 secular resonance due to planet–disk interactions in an early, massive phase. A second empty zone, between 40 and 42 AU, is unstable due to overlapping secular resonances which induce chaotic motion (Knezevic et al 1991, Holman & Wisdom 1993). At greater distances, the influence of Neptune is diminished and nonresonant orbits are stable provided their perihelia remain ≥40 AU.

(*b*) The large number of objects trapped in the 3:2 resonance. The numerical integrations of Holman & Wisdom (1993) show that initially circular, coplanar orbits near the 3:2 resonance can develop moderate eccentricities and inclinations simply through the long-term action of planetary perturbations.

This was explored further by Levison & Stern (1995) as a way to explain the orbital properties of Pluto. They found that orbits excited in this way are unstable to unrealistically high libration amplitudes and they suggested that a dissipative process (possibly a collision) might have stabilized Pluto in the 3:2 resonance. Presumably, corresponding collisions would be needed to stabilize the many other Plutinos now known to be in the resonance. However, the distribution of orbital eccentricities predicted by this model is inconsistent with the orbits of Plutinos as determined from recent data (Figure 2). Whereas the majority of their simulated resonant objects have $e < 0.1$, precisely these eccentricities are missing from the observed Plutino population.

A second possible explanation is the resonance-sweeping hypothesis (Malhotra 1995). As originally envisaged, proto-Neptune scattered nearby planetesimals in the surrounding disk, exchanged angular momentum with them and, as a consequence, underwent a radial excursion (Fernandez & Ip 1984). Planetesimals that were scattered inward fell under the control of Uranus, while those scattered outward were either ejected from the planetary system, entered the Oort Cloud, or fell back to the region of the planets to be scattered again. This asymmetry in the fates of scattered objects caused the orbits of Saturn, Uranus and Neptune to expand (Fernandez & Ip 1984) while Jupiter, the innermost massive scatterer, migrated inward to provide the ultimate source of the angular momentum carried away by scattered planetesimals.

As Neptune migrated, its mean motion resonances swept through the planetesimal disk, trapping objects (Figure 5). Malhotra (1995) found that the maximum eccentricity reached by resonantly trapped particles is related to the distance of Neptune's migration. For a first order ($j+1$:j) resonance (where j is a positive integer),

$$e_{final}^2 \approx e_{initial}^2 + \left(\frac{1}{j+1}\right) \ln\left(\frac{a_{N,final}}{a_{N,initial}}\right) \tag{3}$$

where $e_{initial}$ and e_{final} are the starting and ending eccentricities of the planetesimal and $a_{N,initial}$, $a_{N,final}$ are the starting and ending semi-major axes of Neptune. For the 3:2 resonance ($j = 2$), with $e_{final} \gg e_{initial}$, the Plutinos (median eccentricity $e_{final} = 0.24$), give $a_{N,initial} \sim 25$ AU, or a total migration of about 5 AU. The 3:2 resonance would have swept from 33 AU to its present location at 39 AU and the more distant 2:1 resonance from 40 AU to its present location at 47.6 AU (Figure 5).

Numerical simulations of the sweeping process give partial agreement with the data (Figure 6). The ranges of eccentricity (and perhaps, inclination) of KBOs trapped in the 3:2 resonance are well matched by the model

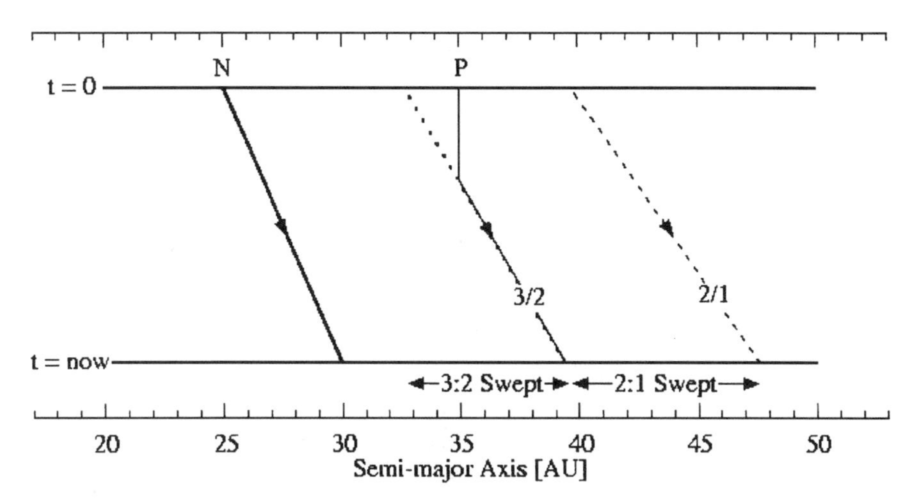

Figure 5 Radial migration of Neptune (*N*) and its 3:2 and 2:1 mean motion resonances. Trapping of Pluto (*P*) into the 3:2 resonance is symbolically illustrated. The zones swept by the 3:2 and 2:1 resonances are marked (Figure slightly modified from Malhotra 1995).

without the need for stabilizing collisions. The high population of 3:2 objects is also faithfully reproduced, because resonance trapping is efficient. Resonance sweeping also explains why apparently stable objects in the 36 to 39 AU region are not found: these bodies would have been swept by the outwardly moving 3:2 resonance. On the other hand, the simulations presented by Malhotra (1995) predict a substantial population of KBOs in the 2:1 resonance, whereas none are seen (Jewitt et al 1998). However, the trapping efficiency is a function of the speed and character of Neptune's migration, of the initial eccentricity and inclination of the planetesimals, and of proto-Neptune's mass (the trapping occurs during the late phases of the growth of the planet). Recently, Ida et al (1998) identified simulations of resonance sweeping in which the 2:1 resonance is left empty. Their simulations differ from Malhotra's in using extremely rapid Neptune migration timescales ($\leq 10^6$ yr) and sub-Neptune masses. This is too fast to be consistent with torques due to planetesimal scattering but might be appropriate for torques exerted by a massive gas disk. The physical significance of this result is not clear. Hydrogen and helium in Neptune (and Uranus) are depleted relative to their solar proportions. This is widely taken as evidence that Neptune grew slowly, reaching its final mass after the escape of the bulk of the gas disk (Lissauer et al 1995, Pollack et al 1996). In this case insufficient mass would be present in the gas disk to accelerate Neptune on 10^6 yr timescales.

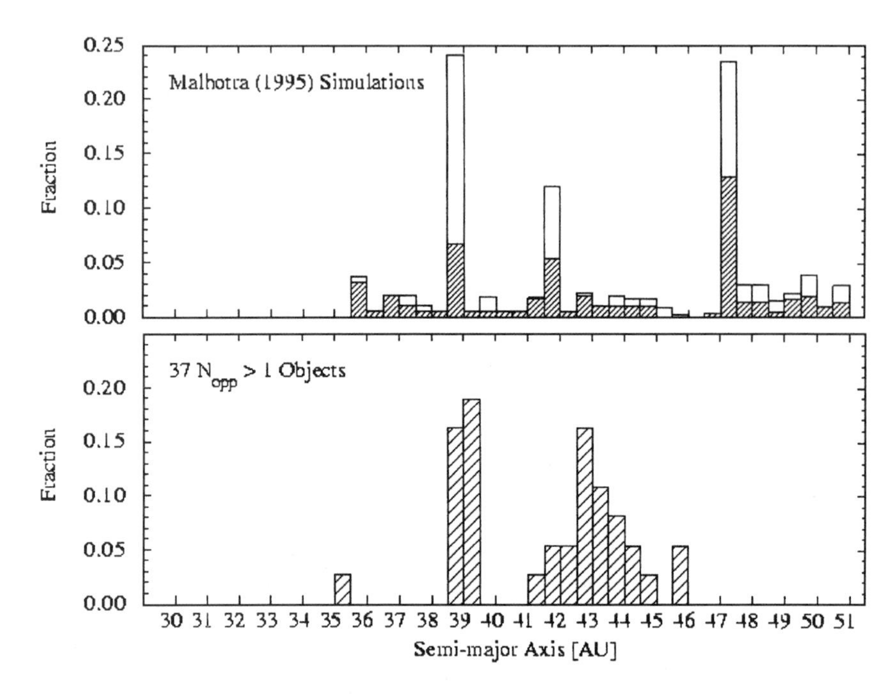

Figure 6 Distribution of semi-major axes of KBOs according to simulations of resonance sweeping due to planetary migration (*top panel*; adapted from Malhotra 1995) compared to the apparent distribution (*bottom panel*). The *shaded histograms* in the top panel indicate results from different sets of initial conditions. A Neptune migration timescale of 2×10^6 yr was adopted.

Whether resonance sweeping is responsible for the Plutinos is unclear. There is little doubt that planetary migration occurred. The gas giant planets must have exchanged angular momentum with the surrounding protoplanetary disk; only the amount and the timescale of the exchange are uncertain.

(*c*) The very wide range of inclinations of both classical and resonant KBOs. As noted, excited inclinations (and eccentricities) are natural consequences of resonance trapping (Holman & Wisdom 1993, Duncan et al 1995, Malhotra 1996). It is more surprising to observe very high inclinations in the classical Kuiper Belt (Figure 3). These inclinations and eccentricities correspond to relative collision velocities near $\Delta V = 1$ km s^{-1}. With a critical specific fragmentation energy near 10^3 J kg^{-1} (Fujiwara et al 1989), the velocity dispersion is large enough to ensure that collisions with objects having only 10^{-3} of the target mass (e.g. a 10 km projectile striking a 100 km target) cause fragmentation. Therefore, impacts in the present day Kuiper Belt are primarily erosive rather than agglomerative, and the KBOs cannot have grown in the present high-velocity regime (Stern 1996, Farinella & Davis 1996). The inescapable conclusion is that the velocity distribution among the KBOs has been amplified since the epoch of formation. The source of this amplification is a leading puzzle.

The disturbing influence of Neptune is largely restricted to the inner part of the Belt ($a \leq 42$ AU), and cannot explain the high mean e and i of the more distant classical KBOs (e.g. Figures 3 and 4 of Holman & Wisdom 1993 give median $e \sim 0.02$ and $i \sim 1^\circ$ for objects in initially circular, coplanar orbits between $42 \leq R \leq 50$ AU). Mutual gravitational interactions among KBOs may have pumped the velocity dispersion, especially if the early Kuiper Belt were very massive and contained large objects. There is good evidence for the temporary existence of short-lived, Earth-mass objects in the early outer solar system: large-body impacts are thought to account for the basic characteristics of planetary spin (Dones & Tremaine 1993) including the large obliquity of Uranus (Slattery et al 1992). Morbidelli & Valsecchi (1997) found that a handful of Earth-sized projectiles could excite the velocity dispersion of the KBOs and lead to extensive depopulation. In their model, the degree of damage inflicted on the Kuiper Belt is a sensitive function of the assumed masses and trajectories of the scattered objects. While the choice of these parameters is necessarily arbitrary, it appears that a few well-aimed Earth-mass bodies can disrupt the Belt and leave a dynamically excited remnant with traits similar to those observed (Petit et al 1998). It is not clear how to reconcile the massive-planetesimals hypothesis with the resonance-sweeping hypothesis for the population of the mean motion resonances. Massive scatterers capable of stirring up the velocity dispersion to 1 km s^{-1} would also dislodge most objects from resonance with Neptune. To argue that resonance sweeping occurred after the epoch of the massive scatterers is also problematic because the efficiency of trapping into resonance decreases as the initial eccentricity and inclination increase (Malhotra 1995): capture from nearly circular ecliptic orbits is much preferred. This dilemma currently evades resolution.

(*d*) The source of the short-period comets. The short-period comets have dynamical lifetimes of order 10^5 to 10^6 yr (Levison & Duncan 1994) and must be continually replenished if they are to maintain a steady-state population over the age of the solar system. There is a consensus that the Kuiper Belt is the source of the short-period comets, but the precise location of the source remains unidentified. Chaotic zones at the edges of resonances represent one plausible source (Morbidelli 1997). The chaotic zones may be populated and depopulated by mutual scattering as well as by dynamical chaos (Ip & Fernandez 1997). Randomly deflected scattered KBOs may also contribute to the short-period comet supply (Duncan & Levison 1997), as might the semi-stable 1:1 Lagrangian resonances with the major planets and the narrow, stable ring at 26 AU identified by Holman (1997). Potentially, all of these sources contribute to the flux of short-period comets.

CUMULATIVE LUMINOSITY FUNCTION, SIZE DISTRIBUTION

The cumulative luminosity function (CLF) is the number of KBOs per unit area of sky brighter than a given limiting magnitude, measured as a function of the magnitude. The CLF provides a measure of the size distribution and total number of KBOs and is therefore a quantity of particular observational interest. The CLF is well defined in the magnitude 20 to 26 range (Figure 7). At brighter magnitudes, the available constraints are based mainly on photographic data, and are difficult to calibrate and interpret. At fainter magnitudes, a measurement using the Hubble Space Telescope (Cochran et al 1995) has proved controversial. Brown, Kulkarni and Liggett (1997) found that the reported number of KBOs is, implausibly, two orders of magnitude beneath the sensitivity limits of the HST, a result that Cochran et al (1998) dispute. Independent confirmation of the measurement is desirable.

The CLF is described by

$$\log[\Sigma(m_R)] = \alpha(m_R - m_0) \tag{4}$$

where $\Sigma(m_R)$ is the number of objects per square degree brighter than red magnitude m_R, and α and m_0 are constants. Least squares fits to the CLF give $\alpha = 0.58 \pm 0.05, m_0 = 23.27 \pm 0.11$ ($20 \leq m_R \leq 25$; Jewitt et al 1998) and $\alpha = 0.54 \pm 0.04, m_0 = 23.20 \pm 0.10$ ($20 \leq m_R \leq 26.6$; Luu and Jewitt 1998b). Gladman et al (1998) used a different subset of the data (including some of the photographic constraints and the uncertain HST measurement) and a different (maximum likelihood) fitting method to find $\alpha = 0.76^{+0.10}/_{-0.11}, m_0 = 23.40^{+0.20}/_{-0.18}$ ($20 \leq m_R \leq 26$). Within the uncertainties, the various fits are consistent. In fact, the true uncertainties are likely to be larger than indicated because of hitherto ignored systematic errors. For example, it is probable that the surface density is a function of ecliptic longitude, owing to the nature of the Neptune-avoiding resonant objects, and yet no allowance for longitudinal effects has been attempted by the observers.

The gradient of the CLF is determined by the size and spatial distributions of the KBOs, and by the distribution of albedo among these objects. Simple models in which the size distribution is taken to be a power law with index $-\beta$, the spatial distribution is another power law with index -2, and the albedo is constant give $\beta \approx 4.0 \pm 0.5$ (Jewitt et al 1998, Luu & Jewitt 1998b) to $\beta \approx 4.8$

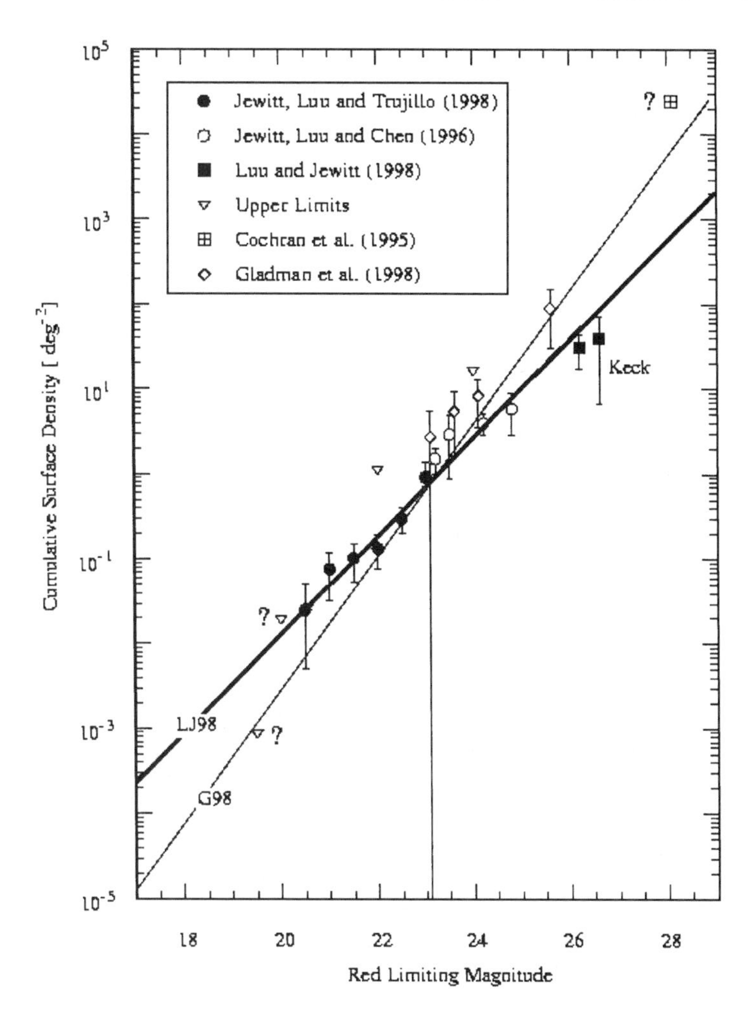

Figure 7 Cumulative luminosity function (CLF) of the Kuiper Belt. Lines through the data mark fits by Luu and Jewitt 1998b (*thick line*) and Gladman et al 1998 (*thin line*).

(Gladman et al 1998). Also, the albedo might be a systematic function of object size (gravity), adding another unmodeled effect to the CLF. In any event, the size distribution of KBOs (diameter greater than or approximately equal to 100 km) is probably steeper than the canonical $\beta = 3.5$ as produced by shattering collisions (Dohnanyi 1969).

It is not yet known whether the size distributions of the different dynamical classes in the Kuiper Belt are the same, but it is easy to think of reasons why they should not be. For example, impulses larger than a few tens of m s^{-1} (from collisions and mutual gravitational scattering) are capable of dislodging objects from resonance. Because small KBOs suffer larger average collisional impulse velocities than their massive counterparts, resonant objects may have a steeper size distribution than the classical Belt.

The total numbers of KBOs must be estimated by extrapolation from surveys of limited areas of the ecliptic. The extrapolations hinge on knowledge of the spatial extent of the Kuiper Belt, both radially and perpendicular to the ecliptic. Regions beyond $R \sim 50$ AU are poorly sampled by the available data, so that the number of objects is known only in the range $30 \leq R \leq 50$ AU. Furthermore, highly inclined KBOs spend a smaller fraction of each orbit near the ecliptic than KBOs of small inclination, and are thus subject to an observational bias. For this reason, the data provide only a lower limit to the intrinsic thickness of the Belt. The apparent thickness is $10^\circ \pm 1$ FWHM (Jewitt et al 1996) but, corrected for bias, the intrinsic thickness may be 30° or more, with no statistically significant difference between the classical and resonant populations.

The number of KBOs larger than 100 km in diameter (assuming 0.04 geometric albedo) in the 30 to 50 AU distance range is $\sim 10^5$ (Jewitt et al 1998). By extrapolation, assuming size index $\beta = 4$, the number of KBOs larger than 5 km diameter in the 30 to 50 AU region is 8×10^8, but with an uncertainty approaching an order of magnitude. Meaningful comparison with the number of comets required to replenish the short-period population for the age of the solar system is difficult because the sizes of the comets are not well known. The numbers seem of the right order, however (Duncan et al 1988, Duncan et al 1995). The total mass in observable (diameter $\geq$100 km) objects is of order $0.1 M_{Earth} (1 M_{Earth} = 6 \times 10^{24}$ kg; Jewitt et al 1996, Jewitt et al 1998, Luu & Jewitt 1998), based on an assumed bulk density 1000 kg m^{-3} and a power law differential size distribution with index $q = 4$. The mass is uncertain primarily because the KBO diameters are computed on the untested assumption of dark surfaces (albedo 0.04). The derived mass is consistent with the dynamical limit to the mass, $\sim 1 M_{Earth}$ (Hamid et al 1969).

PHYSICAL OBSERVATIONS

Optical (Luu & Jewitt 1996, Green et al 1997) and near-infrared colors (Jewitt & Luu 1998) of KBOs exhibit a diversity that suggests a range of surface types among KBOs. The best spectral discriminant is the V-J color index, which measures the ratio of the surface reflectances at V (approximately 0.55 μm) and J (approximately 1.2 μm) wavelengths. Figure 8 shows that V-J varies from 0.7 (1996 TO66) to 2.2 (1996 TP66) in the Kuiper Belt (and up to 2.6 in the Centaur 5145 Pholus). V-J can be measured to a 1σ accuracy of about

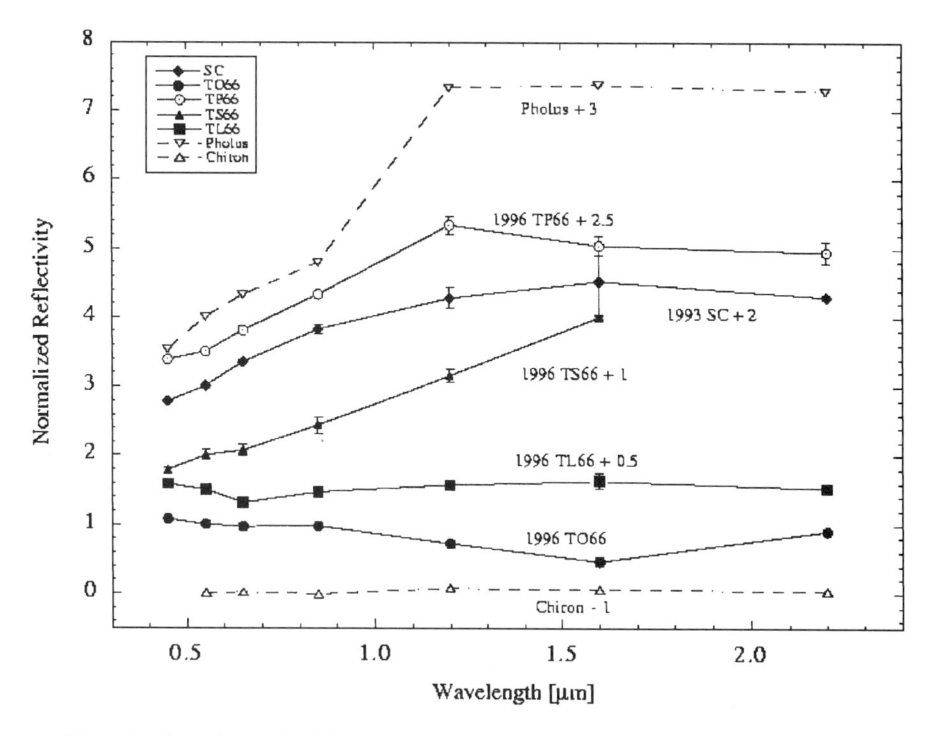

Figure 8 Normalized reflectivity versus wavelength for five KBOs. The KBOs are vertically offset for clarity, but preserve a fixed scale. Reflectivities of Centaur objects 2060 Chiron and 5145 Pholus are shown for comparison. (From Jewitt & Luu 1998).

0.1 mag., so that the color differences are highly significant ($\sim 20\sigma$). Tegler & Romanishin (1998) further claim that the KBOs are divided by optical color into two distinct classes with no intermediate examples, but the color separation of the groups exceeds the formal uncertainty of measurement by only a few times the 1σ accuracy of the colors. This intriguing result awaits independent confirmation.

Bombardment of simple ice mixtures (H_2O, CO_2, CO, NH3) by energetic particles (photons or atomic nuclei) is known to cause surface darkening and modification of the chemical structure of the surface to a column density of $\sim$100 g cm^{-2} (Johnson et al 1987, Moroz et al 1998). Incident particle energy is dissipated by breaking chemical bonds which then recombine to make new, structurally complex compounds. Hydrogen atoms liberated in this process are sufficiently small that they can escape from the irradiated material even at the low temperatures prevailing in the Kuiper Belt ($\sim$50 K). The resulting "radiation mantle" is a hydrogen-poor, carbon-rich (and therefore dark),

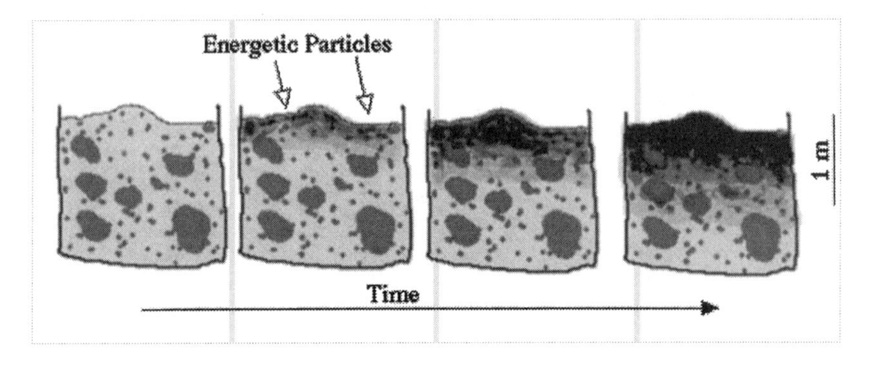

Figure 9 Development of an irradiation mantle. A KBO consisting of a mixture of refractory particles and common ices is bombarded by high-energy cosmic rays. Over time, a hydrogen-depleted crust of high molecular weight and of column density $\sim$100 g cm^{-2} develops. Carbon compounds in the mantle are responsible for the low albedo.

high-molecular-weight compound having low volatility (Figure 9). The timescale for saturation damage of the surface layer by cosmic rays is of order $10^{8\pm1}$ yrs (Shul'man 1972), but measurable surface darkening may occur much more rapidly. The surfaces of cometary nuclei are believed to be coated by such material, and it is natural to expect the same of KBOs. Within this context, it is difficult to see why the KBOs would vary in optical color from neutral to very red, or why the V-J index should be so variable from object to object.

A consensus regarding the origin of spectral diversity has yet to emerge. One possibility is that KBOs are occasionally resurfaced by collisionally generated debris (Luu & Jewitt 1996). The surface color would change because material excavated from depths greater than a few meters is unaffected by cosmic-ray irradiation. The instantaneous surface color would be determined primarily by the time since resurfacing. This remains a plausible hypothesis for color diversity only so long as the timescales for irradiation and resurfacing are of the same order. Unfortunately, neither timescale is well known. Should the color distributions really be bimodal, the resurfacing hypothesis would be immediately ruled out because partial resurfacing always produces intermediate colors.

A second possibility is that different KBOs have intrinsically different surface compositions, either because they formed differently, or because of long-term evolutionary effects. For example, large KBOs might be more thoroughly outgassed than small KBOs as a result of greater radioactive heating, and would be better at retaining surface frost deposits. However, at present, there is no compelling evidence for a color-size relation among the KBOs.

Energetic irradiation should produce a spectrally bland material corresponding to the destruction of all hydrogen bonds and the loss of hydrogen to space.

Spectral evidence is limited to near infrared spectra of KBOs 1993 SC and 1996 TL66. The heavily smoothed spectrum of 1993 SC is reported to show absorption bands reminiscent of (but not specifically identified with) complex hydrocarbon compounds (Brown et al 1997). The spectrum of 1996 TL66 is featureless and flat (Luu & Jewitt 1998). More spectra are urgently needed.

The albedos of KBOs have yet to be measured. The standard technique applied to main-belt asteroids is the simultaneous measurement of optical (reflected sunlight) and thermal (absorbed and re-radiated sunlight) radiation, from which both the cross section and the albedo can be measured. The Planck maximum for objects at 40 AU (temperature $\sim$50 K) falls near 60 μm wavelength, which is inaccessible from the ground. Measurements using the ISO orbiting telescope yielded marginal detections of 1993SC and 1996TL66, from which low (few percent) albedos may be inferred (Nick Thomas, private communication). Stellar occultations may soon be available from the TAOS project and others (Brown & Webster 1997), from which diameters and albedos might be directly inferred. At the present, however, all KBO diameters are computed on the assumption of a uniform red albedo $p_R = 0.04$. This value is adopted from measurements of the albedos of cometary nuclei and Centaurs (bodies in transition from the Kuiper Belt to the short-period comets), most of which tend to be very dark. For example, if the albedo should be higher by a factor of 10, the derived KBO diameters would decrease by a factor $10^{1/2}$, and the masses by a factor $10^{3/2}$. The diversity of surface types indicated by Figure 8, for example, certainly suggests that the albedos of KBOs may not all be the same. There is some evidence that the albedo is a function of object diameter (Figure 10). The largest objects have sufficient gravitational attraction to retain weak atmospheres from which albedo-enhancing surface frosts may be deposited. Conclusions about sizes and masses of KBOs are thus rendered uncertain.

DUST

Collisions between KBOs and the impacts of interstellar grains should provide a continuous source of dust. Liou et al (1996) presented numerical integrations of the equation of motion for dust particles released in the Kuiper Belt. They included radiation pressure, plasma drag, and Poynting-Robertson drag forces, as well as gravitational forces due to the planets (except Mercury and Pluto). Kuiper Belt dust particles that survive to cross the orbit of Earth do so with small eccentricities and inclinations, and would be difficult to distinguish from asteroidal dust grains on the basis of their orbital parameters alone. Liou, Zook, and Dermott found that $\sim$20 percent of the modeled 1-to-9-μm–diameter particles survive to reach the sun (the majority are ejected from the solar system

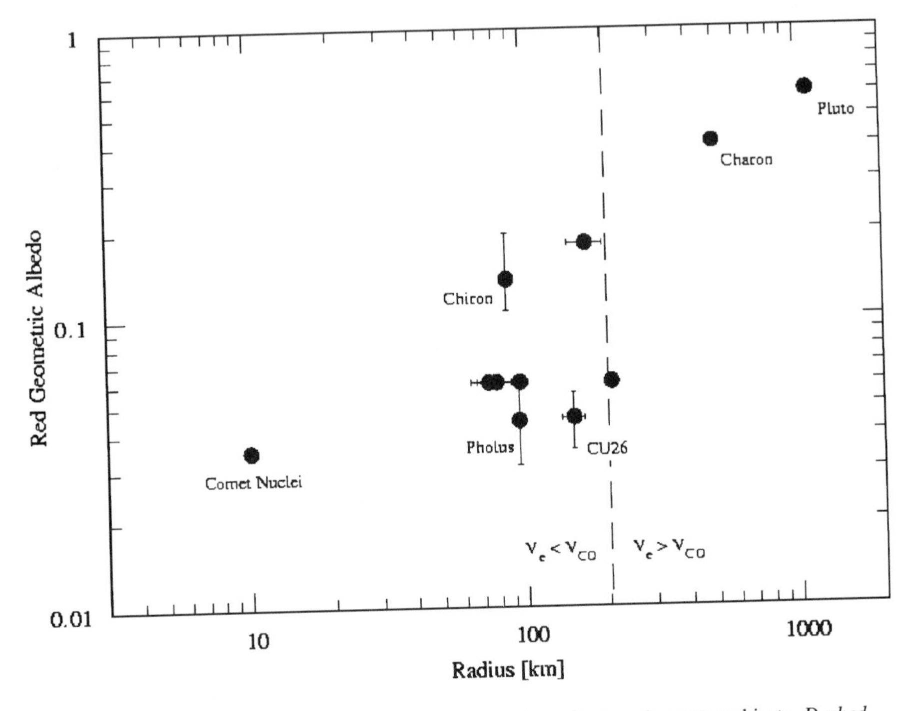

Figure 10 Red geometric albedo versus radius for a variety of outer solar system objects. *Dashed vertical line* marks $V_e = V_{CO}$, where V_e is the escape velocity (at assumed density $\rho = 1000$ kg m^{-3}) and V_{CO} is the thermal velocity of a CO molecule at $T = 40$ K. There is a trend for albedo to increase with object size. However, this trend is dominated by Pluto and Charon, objects both large enough to retain weak atmospheres.

by the gas giant planets). However, timescales for the collisional destruction of Kuiper Belt grains by interstellar dust particles (Figure 11) are less than the dynamical transport times (which cluster near 10^7 yr) for many of the small particles modeled by Liou et al (1996). Therefore, the survival probability of Kuiper Belt dust entering the inner solar system must be considerably smaller than 20 percent, and the size distribution of the surviving grains will be modified by shattering collisions with interstellar grains. Such particles might one day be identified in terrestrial stratospheric collections (Brownlee 1985) by their high solar wind track densities (due to long transport times from the Kuiper Belt).

I bracket the current dust production rate in the belt as follows. A lower limit is given by the rate of erosion due to interstellar dust impacts, $dM_d/dt \geq 4 \times 10^2$ kg s^{-1} (Yamamoto & Mukai 1998). An upper limit may be estimated by dividing the Kuiper Belt mass (approximately $0.1 M_{Earth}$; Jewitt et al 1998) by

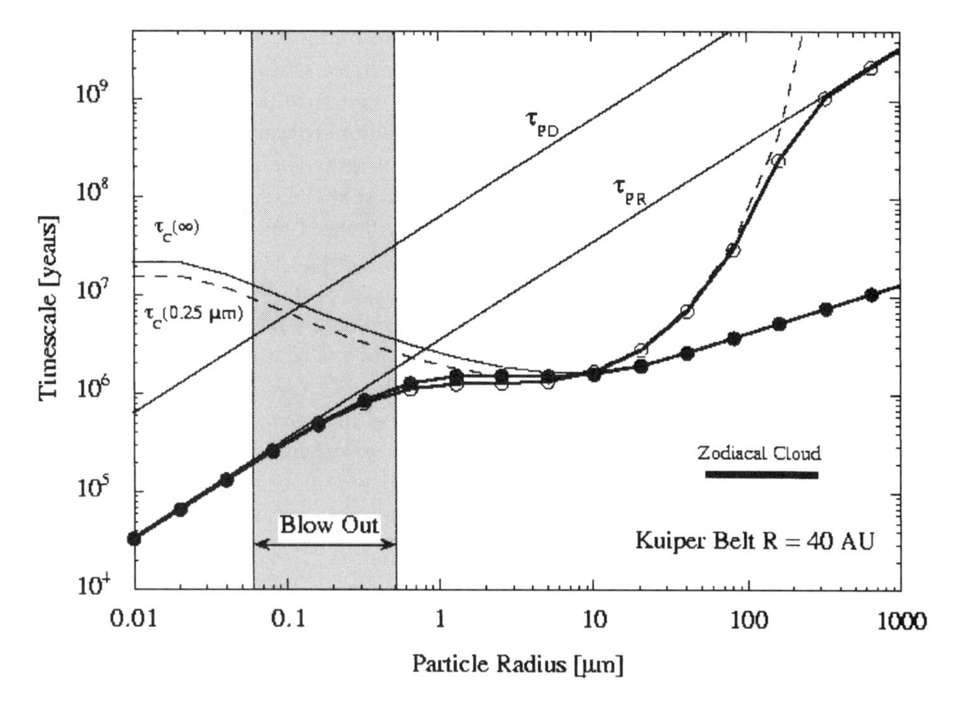

Figure 11 Timescales for Poynting-Robertson decay (τ_{pr}), plasma drag (τ_{pd}) and collision with interstellar dust (τ_c) all plotted as a function of radius for particles at 40 AU in the Kuiper Belt. *Shaded region* marks particles likely to be ejected from the solar system by radiation pressure. *Lines with circles* show the lifetimes to all three processes combined: *line with filled circles* corresponds to $a_0 = \infty$, and that with empty circles $a_0 = 0.25\ \mu$m. The interstellar dust particles were assumed to follow a power law size distribution with differential index -3.5, and with maximum particle radius a_0 (From Jewitt & Luu 1997).

the age of the solar system (4.6 Gyr), giving $dM_d/dt \leq 10^7$ kg s^{-1}. Thus:

$$4 \times 10^2 \text{ kg s}^{-1} \leq dM_d/dt \leq 10^7 \text{ kg s}^{-1} \tag{5}$$

is given for the current rate of dust production. For comparison, the production rate needed to sustain the Zodiacal Cloud is $\sim 10^4$ kg s^{-1} (Leinert et al 1983, Grün et al 1994).

With both foreground (Zodiacal Cloud) and background (galactic) confusion, it is perhaps not surprising that attempts to measure thermal emission from Kuiper Belt dust have failed (Backman et al 1995). However, there is one reported in situ detection of dust in the Kuiper Belt: Gurnett et al (1997) counted dust impacts using plasma wave analyzers on the Voyager 1 and 2 spacecraft. They determined an average number density of micron-sized grains

$N_1 = 2 \times 10^{-8}$ m^{-3} along the two Voyager trajectories, beyond 30 AU. The plasma wave analyzers are sensitive to only a narrow range of particle sizes: impacts below the threshold mass 1.2×10^{-14} kg (corresponding to particle radius 1.4 μm at unit density) do not excite measurable plasma waves (Gurnett et al 1997) while larger impacts are, presumably, rare. The average mass of a grain is taken as $m_d \sim 2 \times 10^{-14}$ kg (c.f. Gurnett et al 1997) and the volume of the Kuiper Belt (represented as an annular slab with inner and outer radii 30 AU and 50 AU, respectively, and a thickness of 10 AU) as $V \sim 2 \times 10^{38}$ m^3. The total mass of dust in micron-sized particles is then $M \sim m_d N_1 V \sim 8 \times 10^{16}$ kg. With a $\tau_c \sim 10^6$ yr dust lifetime, the implied production rate in micron-sized particles is of order $M/\tau_c \sim 3 \times 10^3$ kg s^{-1}, which is near the lower limit in Equation 5.

If the dust size distribution extends far above the micron size range, the Voyager impact measurements provide only a lower limit to the total rate of production of debris. In a Dohnanyi (1969) type power law distribution with index -3.5, micron-sized particles contain only ~1/16th of the mass, when measured to maximum radius 1 mm (~1/500th when measured to 1 m). The nondetection of centimeter-sized particles from Pioneer 10 (Anderson et al 1998) unfortunately does not provide a strong constraint on the dust size distribution.

CONSTRAINTS ON THE ORIGIN OF THE KUIPER BELT

Factors relating to the origin and evolution of the Kuiper Belt may be summarized in this manner:

(*a*) The surface density of the condensible part of the planetary mass obtained by smearing the masses of the giant planets is given roughly by $\sigma(R) \approx 10(10/R)^{3/2}$ [kg m^{-2}], where R is heliocentric distance measured in AU (Weidenschilling 1977, c.f. Pollack et al 1996). The mass obtained by integrating this surface density over the $30 \leq R \leq 50$ AU annulus is $\sim 25 M_{Earth}$, or about 10^2 times the estimated mass of the present Kuiper Belt. Either the extrapolation of $\sigma(R)$ beyond Neptune is invalid (i.e., the protoplanetary disk had an edge) or a large part of the mass initially present in the $30 \leq R \leq 50$ AU zone has been removed.

(*b*) Given only the present mass in the 30-to-50 AU region, the timescales for growth of the KBOs by binary accretion to 100 km and 1000 km scales are much longer than the age of the solar system (Stern 1995, Kenyon & Luu 1998). Such growth is therefore impossible, with the implication that either the KBOs formed elsewhere and were transported to their present locations by processes unknown or that the Kuiper Belt surface density was originally larger than it is at present.

(*c*) Perturbations from Neptune exert a major disturbing influence on the adjacent Kuiper Belt (Holman & Wisdom 1993, Levison & Duncan 1993, Duncan et al 1995), raising the velocity dispersion and decreasing the accretion rate. Therefore it is probable that the KBOs (in the inner part of the belt) were formed before Neptune (Stern 1996). However, assuming $35 M_{Earth}$ between 35 and 50 AU, Stern (1996) obtained Pluto growth times near 10^9 yr, which is longer than the 10^8 yr timescale for formation of Neptune (Lissauer et al 1995, Pollack et al 1996). Only recently have more compatible Pluto growth times have been obtained. Kenyon & Luu (1998) assert that the 10^9 yr timescales obtained by Stern were due to a binning error in the solution of the coagulation equation. Their numerical experiments show that 1000 km sized objects could form in $<10^8$ yr provided the initial Kuiper Belt mass is $10–30 M_{Earth}$ (Kenyon & Luu 1998). Again, the implication is that a substantial mass of material has been lost from the Kuiper Belt.

Several possibilities have been suggested to explain the depletion of mass from the Kuiper Belt. Long-term numerical integrations (e.g. Holman & Wisdom 1993; Duncan et al 1995) show that the decline in the number of KBOs is approximately logarithmic with time and therefore too slow to be the main cause of the depletion. The resonance-sweeping mechanism also appears incapable of losing 99 percent of the mass. Numerical experiments show that more than 50 percent of initially nonresonant objects are captured into mean motion resonances (Malhotra 1995). Admittedly, these experiments are simplistic, and the capture probability has not been assessed over a full range of initial orbital eccentricities. As noted by Fernandez & Ip (1984) and Malhotra (1995), the motion of Neptune is taken to be smooth and continuous. In reality, scattering of massive planetesimals would lend a stochastic character to Neptune's radial excursion, resulting in a decreased trapping efficiency. Stern & Colwell (1997) suggest that collisional grinding could have eroded the 30 AU–50 AU region from 10 to $35 M_{Earth}$ down to the observed $\sim 0.1 M_{Earth}$ in 10^9 yr. However, their model neglects velocity evolution and dynamics and is therefore of uncertain significance. Crucially, they note that sufficient erosion is obtained only if the mean eccentricity is somehow sustained above $e \geq 0.1$. In any case, the larger KBOs (namely, those detected in current ground-based surveys) cannot be destroyed by collisional grinding (Farinella & Davis 1996). For collisions to have removed 99 percent of the initial mass, one would have to postulate a steep initial size distribution ($q \sim 5$) in which objects large enough to escape comminution carry only one percent of the mass.

(*d*) The existence of the Pluto-Charon binary sets additional constraints on the Kuiper Belt. First, it is widely supposed (based on the large specific angular momentum of the planet-satellite pair) that Charon was formed by a giant impact into Pluto (McKinnon 1989, Dobrovolskis et al 1997). If so, this impact must have occurred prior to the capture of Pluto into the 3:2 resonance, for otherwise the recoil would have knocked the planet out of resonance (Hahn & Ward 1995). Second, the rate of impacts into Pluto large enough to eject Charon is presently vanishingly small. Accordingly, the existence of an impact-formed Charon suggests a much–higher-density Kuiper Belt, perhaps containing many Pluto-sized bodies (Stern 1991). Third, the eccentricity of the orbit of Charon should be tidally damped to a very low level ($e \ll 10^{-3}$) on timescales that are short compared to the age of the solar system. Surprisingly, Tholen & Buie (1997) have measured $e = (7.6 \pm 0.5) \times 10^{-3}$. If confirmed, this eccentricity will set important constraints on the impact rate in the modern-day Kuiper Belt.

SUMMARY

About 10^5 objects with diameters in excess of 100 km orbit the sun between 30 AU (the orbit of Neptune) and 50 AU (the practical limit of existing surveys). These objects obey a differential power law size distribution with index $q \sim -4$. The distribution seems to extend smoothly to Pluto (diameter ~2300 km) and allows the possibility that other Pluto-sized objects await discovery. The combined mass of these objects is about $0.1 M_{Earth}$. When extrapolated to small sizes, the $q = -4$ distribution predicts that the total number of trans-Neptunian objects larger than 1 km diameter is 10^{11}. However, the size distribution of bodies smaller than 100 km in diameter is obtained by extrapolation and has not been confirmed by direct observation.

The orbits of the KBOs are divided into three main groups. So-called classical KBOs orbit the sun with semi-major axes $42 \leq a \leq 50$ AU, possess small eccentricities and maintain a large separation from Neptune even at perihelion. Their orbits appear stable on timescales longer than the age of the solar system. About two thirds of the KBOs belong to the classical group. Most of the remaining objects reside in mean motion resonances with Neptune. The 3:2 resonance at $a = 39.4$ AU is particularly densely populated. Pluto is the largest of the tens of thousands of objects trapped in this resonance. The scattered KBOs follow large, eccentric orbits with perihelia near 35 AU, close enough to Neptune to permit weak dynamical control by that planet on billion year timescales. One example, 1996 TL66, is presently known. The inferred population is large, and may even dominate the total mass of trans-Neptunian objects.

The abundance of resonant objects (about 35 percent in the raw data and about 10 to 15 percent when corrected for the effects of observational bias) may provide important clues about the formation and early evolution of the Kuiper Belt. The resonance-sweeping hypothesis (in which Neptune's orbit expanded during planet growth) makes verifiable predictions of the inclination and eccentricity distributions and of the resonance population ratios. In its present guise, this hypothesis suggests a total Neptune excursion from 25 AU to the present 30 AU, on a timescale that was evidently short.

The velocity dispersion among KBOs is near 1 km s^{-1}, suggesting that collisions are largely erosive and that the velocity dispersion has been amplified since the formation epoch. Excitation of the orbits of the resonant KBOs is an expected consequence of trapping into the resonance. However, the non-resonant orbits of classical KBOs are equally excited, suggesting the action of a more general disturbing agent. One suggested possibility is that massive scattered objects were projected through the Kuiper Belt during the late stages of Neptune accretion, stirring up the velocity distribution of the KBOs as they repeatedly passed through.

The combined mass of KBOs in the 30 to 50 AU region is too small for these objects to have grown by accretion in the 10^8 yrs prior to the emergence of Neptune. Roughly $10M_{Earth}$ of material are needed to ensure growth of Pluto-sized objects on this timescale. By implication, the present Kuiper Belt is a mere shadow of its former self, with about 99 percent of the initial mass now lost. The loss mechanism remains unidentified.

In short, our understanding of the formation and evolution of the Kuiper Belt is currently incomplete, even confused. This is to be expected in a field that is still very young. As more and better observations of KBOs are obtained, existing models will become more tightly constrained and new and unexpected ones will suggest themselves.

ACKNOWLEDGMENTS

I thank the National Aeronautics and Space Administration for support.

NOTE ADDED IN PROOF

Subsequent to the completion of this review two KBOs were identified as possible residents of the 2:1 mean motion resonance. 1996 TR66 has semimajor axis $a \sim 48.2$ AU, eccentricity $e \sim 0.40$ and inclination $i \sim 12$ degree, while the corresponding quantities for 1997 SZ10 are $a,e,i = 48.3$ AU, 0.37, 12 degree (Brian Marsden, IAU Circular 7073, December 26, 1998). When combined with the factor $\sim$3 observational bias against detecting the more distant 2:1

objects (Jewitt, et al 1998), the data are statistically consistent with the hypothesis that the populations in the 3:2 and 2:1 resonances are of the same order. This, in turn, is compatible with a basic prediction of the resonance-sweeping hypothesis as presented by Renu Malhotra (1995).

Literature Cited

Anderson J, Lau E, Scherer K, Rosenbaum D, Teplitz V. 1998. Kuiper belt constraint from Pioneer 10. *Icarus* 131:167–70

Backman DE, Dasgupta A, Stencel RE. 1995. Model of a Kuiper belt small grain population and resulting far-IR emission. *Ap. J. Lett.* 450:35–38

Baum WA, Martin LJ. 1985. Testing the comet belt hypothesis. *Publ. Astron. Soc. Pac.* 97:892

Brown ME, Kulkarni SR, Liggett TJ. 1997. An analysis of the statistics of the Hubble Space Telescope Kuiper belt object search. *Ap. J. Lett.* 490:119–22

Brown MJ, Webster RL. 1997. Occultations by Kuiper belt objects. *MNRAS* 289:783–86

Brown R, Cruikshank D, Pendleton Y, Veeder G. 1997. Surface composition of Kuiper belt object 1993 SC. *Science* 276:937–39

Brownlee D. 1995. Cosmic dust: collection and research. *Annu. Rev. Earth Planet Sci.* 13:147–73

Cochran AL, Cochran WD, Torbett MV. 1991. A deep imaging search for the Kuiper disk of comets. *Bull. Am. Astron. Soc.* 23:131

Cochran AL, Levison HF, Stern SA, Duncan MJ. 1995. The discovery of Halley-sized Kuiper belt objects. *Ap. J.* 455:342–46

Cochran AL, Levison HF, Tamblyn P, Stern SA, Duncan MJ. 1998. Calibration of the HST Kuiper belt object search; setting the record straight. *Ap. J.* 503:L89–93

Dobrovolskis A, Peale S, Harris A. 1997. Dynamics of the Pluto-Charon binary. In *Pluto and Charon*, ed. A Stern, D Tholen, pp. 159–90. Tucson: Univ. Ariz. Press

Dohnanyi J. 1969. Collisional models of asteroids and their debris. *J. Geophys. Res.* 74:2531–54

Dones L, Tremaine S. 1993. On the origin of planetary spins. *Icarus* 103:67–92

Duncan M, Levison H. 1998. A disk of scattered icy objects and the origin of Jupiter-family comets. *Science* 276:1670–72

Duncan M, Levison H, Budd S. 1995. The dynamical structure of the Kuiper belt. *Astron. J.* 110:3073–81

Duncan M, Quinn T, Tremaine S. 1988. The origin of short-period comets. *Ap. J.* 328:69–73

Edgeworth KE. 1943. The evolution of our planetary system. *J. Br. Astron. Soc.* 53:181–88

Edgeworth KE. 1949. The origin and evolution of the solar system. *MNRAS* 109:600–9

Farinella P, Davis D. 1996. Short-period comets: primordial bodies or collisional fragments? *Science* 273:938–41

Fernandez J. 1980. On the existence of a comet belt beyond Neptune. *MNRAS* 192:481–91

Fernandez J, Ip W-H. 1983. On the time evolution of the cometary influx. *Icarus* 54:377–87

Fernandez J, Ip W-H. 1984. Some dynamical aspects of the accretion of Uranus and Neptune. *Icarus* 58:109–20

Fujiwara A, Cerroni P, Davis D, Ryan E, di Martino M, et al. 1989. Experiments and scaling laws for catastrophic collisions. In *Asteroids II*, ed. RP Binzel, T Gehrels, MS Matthews, pp. 240–65. Tucson: Univ. Ariz. Press

Gladman B, Kavelaars J, Nicholson P, Loredo J, Burns J. 1989. Pencil-beam surveys for faint trans-Neptunian objects. *Astron. J.* 116:2042–54

Green SF, McBride N, O'Ceallaigh DP, Fitzsimmons A, Williams IP, Irwin MJ. 1997. Surface reflectance properties of distant solar system bodies. *MNRAS* 290:86–92

Grun E, Gustafson E, Mann I, Baguhl M, Morfill GE, et al. 1994. Interstellar dust in the heliosphere. *Astron. Ap.* 286:915–24

Gurnett DA, Ansher JA, Kurth WS, Granroth LJ. 1997. Micron-sized dust particles detected in the outer solar system by the Voyager 1 and 2 plasma wave instruments. *Geophys. Res. Lett.* 24:3125–28

Hahn J, Ward W. 1995. Resonance passage via collisions. *Lunar Planet. Sci.* 26:541–42

Holman M. 1997. A possible long-lived belt of objects between Uranus and Neptune. *Nature* 387:785–88

Holman M, Wisdom J. 1993. Dynamical stability in the outer solar system and the delivery of short period comets. *Astron. J.* 105:1987–99

Ida S, Tanaka H, Bryden G, Lin D. 1998. Migration of proto-Neptune and the orbital distribution of trans-Neptunian objects. Preprint: U.C. Santa Barbara Press

Ip W-H, Fernandez J. 1997. On dynamical scattering of Kuiper belt objects in 2:3 resonance with Neptune into short-period comets. *Astron. Ap.* 324:778–84

Jewitt D, Luu J. 1993. Discovery of the candidate Kuiper belt object 1992 QB1. *Nature* 362:730–32

Jewitt D, Luu J, Chen J. 1996. The Mauna Kea-Cerro Tololo (MKCT) Kuiper belt and Centaur survey. *Astron. J.* 112:1225–38

Jewitt D, Luu J, Trujillo C. 1998. Large Kuiper belt objects: the Mauna Kea 8k CCD survey. *Astron. J.* 115:2125–35

Jewitt DC, Luu JX. 1995. The solar system beyond Neptune. *Astron. J.* 109:1867–76

Jewitt DC, Luu JX. 1997. The Kuiper belt. In *From Stardust to Planetesimals. ASP. Conf. Ser., 122*, ed. YJ Pendleton, A Tielens, p. 335–45

Jewitt DC, Luu JX. 1998. Optical-infrared spectral diversity in the Kuiper belt. *Astron. J.* 115:1667–70

Johnson RE, Cooper JF, Lanzerotti LJ, Strazzulla G. 1987. Radiation formation of a nonvolatile comet crust. *Astron. Ap.* 187:889–92

Kenyon SJ, Luu JX. 1998. Accretion in the early Kuiper belt. I. coagulation and velocity evolution. *Astron. J.* 115:2136–60

Knezevic Z, Milani A, Farinella P, Froeschle Ch, Froeschle Cl. 1991. Secular resonances from 2 to 50 AU. *Icarus* 93:316–30

Kowal C. 1989. A solar system survey. *Icarus* 77:118–23

Kuiper GP. 1951. On the origin of the solar system. In *Astrophysics*, ed. JA Hynek, pp. 357–424. New York: McGraw-Hill

Leinert C, Roser S, Buitrago J. 1983. How to maintain the spatial distribution of interplanetary dust. *Astron. Ap.* 118:345–57

Levison HF, Duncan MJ. 1990. A search for proto-comets in the outer regions of the solar system. *Astron. J.* 100:1669–75

Levison HF, Duncan MJ. 1993. The gravitational sculpting of the Kuiper belt. *Ap. J. Lett.* 406:35–38

Levison HF, Duncan MJ. 1994. The long-term dynamical behavior of short-period comets. *Icarus* 108:18–36

Levison HF, Stern SA. 1995. Possible origin and early dynamical evolution of the Pluto-Charon binary. *Icarus* 116:315–39

Liou J-C, Zook HA, Dermott SF. 1996. Kuiper belt dust grains as a source of interplanetary dust particles. *Icarus* 124:429–40

Lissauer J, Pollack J, Wetherill G, Stevenson D. 1995. In *Neptune and Triton*, ed. D Cruikshank, M Matthews, A Schumann, pp. 37–108. Tucson: Univ. Ariz. Press

Luu J, Jewitt D. 1996. Color diversity among the Centaurs and Kuiper belt objects. *Astron. J.* 112:2310–18

Luu J, Marsden BG, Jewitt D, Trujillo CA, Hergenrother CW, et al. 1997. A new dynamical class of object in the outer solar system. *Nature* 387:573–75

Luu JX, Jewitt DC. 1998a. Optical and infrared reflection spectrum of Kuiper belt object 1996-TL66. *Ap. J.* 494:L117–20

Luu JX, Jewitt DC. 1998b. Deep imaging of the Kuiper belt with the Keck 10-m telescope. *Ap. J. Lett.* 502:91–94

Malhotra R. 1995. The origin of Pluto's orbit: implications for the solar system beyond Neptune. *Astron. J.* 110:420–29

Malhotra R. 1996. The phase space structure near Neptune resonances in the Kuiper belt. *Astron. J.* 111:504–16

Malhotra R, Williams JG. 1998. Pluto's heliocentric orbit. In *Pluto and Charon*, ed. SA Stern, D Tholen, pp. 127–57. Tucson: Univ. Ariz. Press

McKinnon WB. 1989. On the origin of the Pluto-Charon binary. *Ap. J. Lett.* 344:41–44

Morbidelli A. 1997. Chaotic diffusion, and the origin of comets from the 3/2 resonance in the Kuiper belt. *Icarus* 127:1–12

Morbidelli A, Valsecchi GB. 1997. Neptune scattered planetesimals could have sculpted the primordial Edgeworth-Kuiper belt. *Icarus* 128:464–68

Moroz LV, Arnold G, Korochantsev AV, Wusch R. 1998. Natural solid bitumens as possible analogs for cometary and asteroid organics: 1. Reflectance spectroscopy of pure Bitumens. *Icarus* 134:253–68

Oort JH. 1950. The structure of the cloud of comets surrounding the solar system and a hypothesis concerning its origin. *Bull. Astron. Inst. Netherlands* 11:91–110

Petit J-M, Morbidelli A, Valsecchi G. 1998. Large scattered planetesimals and the excitation of the small body belts. *Icarus.* In press

Pollack J, Hubickyj O, Bodenheimer P, Lissauer J, Podolak M, Greenzweig Y. 1996. Formation of the giant planets by concurrent accretion of solids and gas. *Icarus* 124:62–85

Russell HN. 1916. On the albedo of the planets and their satellites. *Ap. J.* 43:173–95

Shul'man LM. 1972. The chemical composition of cometary nuclei. In *The Motion, Evolution of Orbits, and Origin of Comets, IAU Symp. 45*, ed. G Chebotarev, E Kazimirchak-Polonskaia, B Marsden. p. 265–70. Dordrecht, The Netherlands: Reidel

Slattery W, Benz W, Cameron A. 1992. Giant impacts on a primitive Uranus. *Icarus* 99:167–74

Stern SA. 1991. On the number of planets in

the solar system: the population of 1000 km bodies. *Icarus* 90:271–81

Stern SA. 1995. Collisional timescales in the kuiper disk and their implications. *Astron. J.* 110:856–68

Stern SA. 1996a. On the collisional environment, accretion time scales, and architecture of the massive, primordial Kuiper belt. *Astron. J.* 112:1203–11

Stern SA, Colwell JE. 1997. Collisional erosion in the primordial Edgeworth-Kuiper belt and the generation of the 30-50 AU Kuiper gap. *Ap. J.* 490:879–82

Tholen D, Buie M. 1997. The orbit of Charon. *Icarus* 125:245–60

Trujillo C, Jewitt DC. 1998. A semi-automated sky survey for slow-moving objects suitable for a Pluto express mission encounter. *Astron. J.* 115:1680–87

Tyson JA, Guhathakurta P, Bernstein G, Hut P. 1992. Limits on the surface density of faint Kuiper belt objects. *Bull. Am. Astron. Soc.* 24:1127

Ward WR, Hahn JH. 1998. Dynamics of the trans-Neptune region: apsidal waves in the Kuiper belt. *Astron. J.* 116:489–98

Weidenschilling S. 1977. The distribution of mass in the planetary system and solar nebula. *Astrophys. Space Sci.* 51:153–58

Whipple FL. 1951. A comet model. I. The acceleration of comet Encke. *Ap. J.* 111:375–94

Whipple FL. 1964. Evidence for a comet belt beyond Neptune. *Proc. Natl. Acad. Sci. USA* 51:711–18

Williams IP, O'Ceallaigh DP, Fitzimmons A, Marsden BG. 1995. The slow moving objects 1993 SB and 1993 SC. *Icarus* 116:180–85

Yamamoto S, Mukai T. 1998. Dust production by impacts of interstellar dust on Edgeworth-Kuiper objects. *Astron. Ap.* 329:785–91

Annu. Rev. Earth Planet. Sci. 1999. 27:313–58

STROMATOLITES IN PRECAMBRIAN CARBONATES: Evolutionary Mileposts or Environmental Dipsticks?

John P. Grotzinger
Department of Earth, Atmospheric and Planetary Sciences, Massachusetts Institute of Technology, Cambridge, MA 02139; e-mail: grotz@mit.edu

Andrew H. Knoll
Botanical Museum, Harvard University, 26 Oxford Street, Cambridge, MA 02138; e-mail: aknoll@oeb.harvard.edu

KEY WORDS: Archean, fossil, limestone, microbe, Proterozoic

ABSTRACT

Stromatolites are attached, lithified sedimentary growth structures, accretionary away from a point or limited surface of initiation. Though the accretion process is commonly regarded to result from the sediment trapping or precipitation-inducing activities of microbial mats, little evidence of this process is preserved in most Precambrian stromatolites. The successful study and interpretation of stromatolites requires a process-based approach, oriented toward deconvolving the replacement textures of ancient stromatolites. The effects of diagenetic recrystallization first must be accounted for, followed by analysis of lamination textures and deduction of possible accretion mechanisms. Accretion hypotheses can be tested using numerical simulations based on modern stromatolite growth processes. Application of this approach has shown that stromatolites were originally formed largely through in situ precipitation of laminae during Archean and older Proterozoic times, but that younger Proterozoic stromatolites grew largely through the accretion of carbonate sediments, most likely through the physical process of microbial trapping and binding. This trend most likely reflects long-term evolution of the earth's environment rather than microbial communities.

"If you can look into the seeds of time and say which grain will grow and which will not, speak then to me."

Macbeth, Act I, Scene III

0084-6597/99/0515-0313$08.00

INTRODUCTION

Stromatolites are among the most widespread and easily recognized components of Precambrian carbonate platforms. In terms of shape, stromatolites range from morphologically simple domes and cones to more complexly branched columnar structures (Figure 1). Stromatolitic facies characteristically range from thin sheets and lenses to major formation scale units that extend for hundreds of meters vertically and hundreds of kilometers laterally. Stromatolites commonly are interpreted to have formed in a spectrum of shallow-water environments, although available data occasionally suggest deeper-water settings. We know how stromatolites look and where they occur. However, despite almost 100 years of research, including the collection of taxonomic, micropaleontologic, and sedimentologic data, the origin and significance of these distinctive structures is still disputed. Several factors help to sustain this controversy: the routine use and acceptance of genetic rather than descriptive definitions; potential overemphasis on modern "analogs" that may have limited relevance for the interpretation of Precambrian structures—and which often are poorly understood themselves; too little emphasis on experimental and theoretical studies of stromatolite morphogenesis; and neglect of broader paleobiological and geological records that provide a necessary context for historical interpretation.

Our objective is to provide a framework for assessing the relative roles of different biologic and abiologic processes in stromatolite accretion. We adopt the nongenetic definition of stromatolites recommended by Semikhatov et al (1979), which states that a stromatolite is ". . . an attached, laminated, lithified sedimentary growth structure, accretionary away from a point or limited surface of initiation." This definition provides a concise statement of the basic geometric and textural properties of all stromatolites while also allowing for multiple or even indeterminate origins. Accepting this as a general definition makes it possible to evaluate objectively the various processes that may influence stromatolite development. A few specific but critical issues are chosen for discussion, with emphasis on processes that lead to the macroscopic forms recognizable in the field. We specifically address the roles of microbial, physical, and chemical processes in the development of stromatolite laminae and textures, and we consider how these processes collectively determine stromatolite morphology. At the heart of this discussion is the observation that stromatolite morphologies and textures change through time. Can we understand enough about the genesis of these structures to interpret secular variation in terms of evolutionary and/or environmental change?

Definitions

Probably the most potent source of controversy concerning the origin and broader significance of stromatolites lies in the implication of knowledge

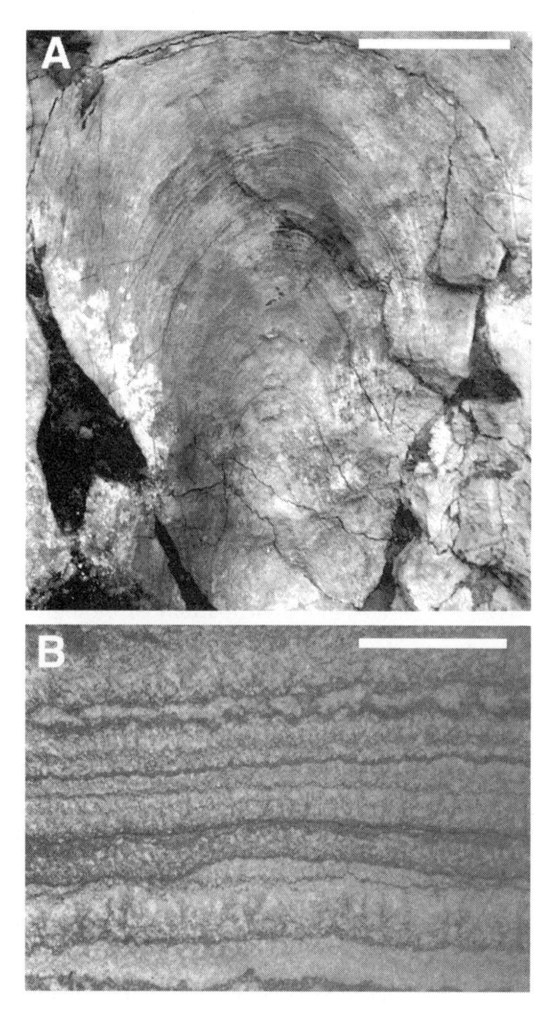

Figure 1 (*a*) Representative columnar stromatolite showing well-developed lamination that defines an upward-widening morphology during growth. Mesoproterozoic Debengda Formation, Olenek Uplift, Siberia. Scale bar is 5 cm. (*b*) Thin-section photomicrograph of representative stromatolite lamination, defined by alternations in finer, more clay-rich laminae (*dark layers*), and coarser laminae with internally mottled, possibly peloidal texture (*light layers*). Note the lack of preserved microfossils or filament molds. Paleoproterozoic Hearne Formation, Athapuscow Basin, northwest Canada. Scale bar is 2 mm.

inherent in genetic definitions of these features. This problem is immediately apparent in a consideration of Kalkowsky's original definition (Kalkowsky 1908): stromatolites are "organogenic, laminated calcareous rock structures, the origin of which is clearly related to microscopic life, which in itself must not be fossilized" (translated in Krumbein 1983). This seemingly paradoxical definition requires that microbial influence be interpretable from features of lamination and texture in the absence of direct paleontological evidence for mat organisms.

The controversy has changed little since Kalkowsky's day. A more recent and widely accepted definition indicates that "Stromatolites are organosedimentary structures produced by sediment trapping, binding and/or precipitation as a result of the growth and metabolic activity of micro-organisms, principally cyanophytes" (Walter 1976b). Injecting such a strong component of genesis into these definitions, without providing a clear basis for how biogenicity is to be demonstrated without recourse to direct observation of fossils, can retard useful description and attention to potentially important details. The point is not that this interpretation of ancient stromatolites is necessarily wrong—in many instances it is very likely correct—but, in assuming a set of accretionary processes, scientists have avoided developing specific process models that accurately describe stromatolite accretion dynamics. Such models might routinely predict the relative or even absolute contributions of biologic, physical, and chemical effects.

There are indeed some spectacular examples of stromatolites in which well-preserved microfossils document the specific roles of mat-building populations (Golubic & Hofmann 1976), but despite the attention that stromatolites have received since the discovery of microfossils in Precambrian sedimentary rocks more than 40 years ago, it is probably conservative to estimate that less than 1 percent of all stromatolites ever described have a fossilized microbiota associated with them. Note, as well, that the mere presence of fossils within ancient stromatolites does not demonstrate that these structures accreted as a direct result of microbial mat activities—just as the presence of skeletons in the La Brea tar pits does not obligate us to conclude that mammals secreted the tar. Preserved microfossils could simply have been trapped in accreting carbonates, either as plankton that settled out of the water column or as benthos that colonized surfaces between episodes of accretion. To sustain genetic definitions, one must be able to demonstrate via observations of the density and orientation of preserved populations that mat organisms trapped and bound or precipitated stromatolitic laminae. In the absence of such evidence, the role of biology in stromatolite accretion must be inferred by indirect methods. Insofar as laminated carbonates can accrete via processes other than those associated with mat biology, care must be taken in the interpretation of such morphological or textural observations. This suggests that Kalkowsky's insight of nearly a century ago should be regarded not as fact but as a working hypothesis. At a level of detail where stromatolitic morphologies and microstructures might be used to illuminate Earth history, we still have a great deal to learn.

There is much to be gained in stating our ignorance. Our biggest loss might be the proposed utility of stromatolites in providing a proxy record for the antiquity of life on earth. This loss would be small, however, because in recent years stromatolites have been eclipsed by the more compelling records of fossil microorganisms in (nonstromatolitic) early Archean rocks (Schopf 1992, Schopf & Packer 1987) and by the carbon-isotopic composition of Archean organic matter, which shows high fractionation consistent with photosynthesis (Hayes 1983, Strauss et al 1992). The future value of stromatolite research lies more in the potential of stromatolites to provide a basis for reconstructing ancient environments and to help us understand how benthic microbial communities interacted with these environments. This must necessarily involve a process-oriented approach to stromatolite morphogenesis in which the correct interpretation of carbonate recrystallization textures is as important as understanding microbial diversity in modern mats. The goal is to build an understanding of stromatolite development that stems not from definitional assertion but from rigorous, quantitative analyses of stromatolite form and lamina texture, including the deconvolution of diagenetic overprints to reveal primary textures diagnostic of specific microbial and sedimentary processes. Critical questions, which get at the core of texture and morphologic development, can then be addressed. For example, over what length and timescales do biological, physical, and chemical processes operate? Do any of these processes—which might be critical at microscopic scales—remain sensitive at larger scales? If not, at what scale does the transition in process response take place? Questions like these must be answered before we have a real understanding of what, for example, some fundamental property such as stromatolite shape signifies.

LAMINATION AND TEXTURE—STROMATOLITE BUILDING BLOCKS

Background

The most conspicuous feature of stromatolites is their lamination (Figure 1). Lamination at any scale in sedimentary rocks is a manifestation of the discontinuous nature of sedimentation itself (Sadler 1981), and stromatolites are no exception. Individual laminae are the building blocks of stromatolites and therefore comprise a time series of progressive, albeit incremental accretion. The morphology of any stromatolite is a function of how lamina shape, particularly its relief, evolves in time. Topographic anomalies that are reinforced in time give rise to greater relief for successive laminae, those that are stabilized give rise to greater inheritance of shape for successive laminae, and those that are damped result in diminished relief.

An interesting consequence of history, well discussed in Hoffman (1973), is that the first modern stromatolites to be studied carefully occurred entirely in freshwater environments and grew mostly through the precipitation of carbonate around algal or cyanobacterial filaments. On this basis, many geologists concluded that ancient stromatolites similarly grew through in situ precipitation and likely indicated freshwater to brackish environments (Bradley 1929, Eardley 1938, Walcott 1914). The subsequent discovery of microbial mats and columnar stromatolites growing in coastal marine environments primarily through microbial trapping and binding of carbonate mud and grains (Black 1933, Ginsburg & Lowenstam 1958, Logan 1961) caused a shift in interpretational emphasis: ancient stromatolites and stromatolitic facies came to be seen as products of the trapping and binding of loose sediment. Note that this interpretational shift did not stem from any new observations of the ancient stromatolites themselves. Rather, it grew from a new perception of what the most appropriate modern analog should be. Given that ancient stromatolites are generally associated with compelling evidence for growth in marine environments, it is understandable how nonmarine analogs could have been rejected in favor of marine analogs.

The logical prediction of this analogy is that ancient stromatolites should display lamination textures consistent with the trapping and binding of loose sediment. The lamination of ancient stromatolites is expressed as a physical attribute (Figure 1*b*), mostly related to variations in crystal size and orientation and to composition (Ca/Mg ratio, clay, silt, and organic content). Because the overwhelming majority of modern marine carbonates are precipitated as metastable

→

Figure 2 Examples of aggrading neomorphism in affecting stromatolite lamination textures in microdigitate-columnar and laminar-colliform stromatolites, Rocknest Formation, northwest Canada. (*a*) Polished slab showing microdigitate stromatolite columns and internal lamination. (*b*) Thin-section photomicrograph of three adjacent columns, showing crude lamination and palimpsest radial texture, interpreted to represent former fibers of aragonite now replaced by dolomite. Scale bar is 1 cm. (*c*) Same as in *b*, but texture now reflects aggrading neomorphism, which results in an increase in crystal size and loss of palimpsest fibrous texture except near center of photograph. Scale bar is 5 mm. (*d*) Thin-section photomicrograph of laminar-colliform stromatolites showing good preservation of lamination texture. Note that dark micritic layers infill rough microtopography on tops of lighter, more coarsely crystalline layers; light layers also have systematically smoother bases. Scale bar is 3 mm. (*e*) With aggrading neomorphism, the fabric in *d* is increasingly homogenized due to an increase in the crystal size of the darker, finer-grained layers, such that a distinction between the geometric attributes of the *dark* and *light layers* is no longer possible. Scale bar is 3 mm.

aragonite or high-Mg calcite (Bathurst 1975, Schroeder & Purser 1986), the occurrence of low-Mg calcite and dolomite in ancient stromatolites implies that at least some diagenesis of the laminae has occurred. Minimally, this would involve neomorphic inversion to more stable phases (Sandberg 1985), with associated grain size enlargement and, in former aragonite, loss of primary crystal orientation (Figure 2). More aggressive diagenetic regimes, especially dolomitization, can result in the loss of all primary variation in grain size and orientation (Sears & Lucia 1980, Zempolich et al 1988). In these instances, only a crude lamination is preserved, often expressed by traces of insoluble silt

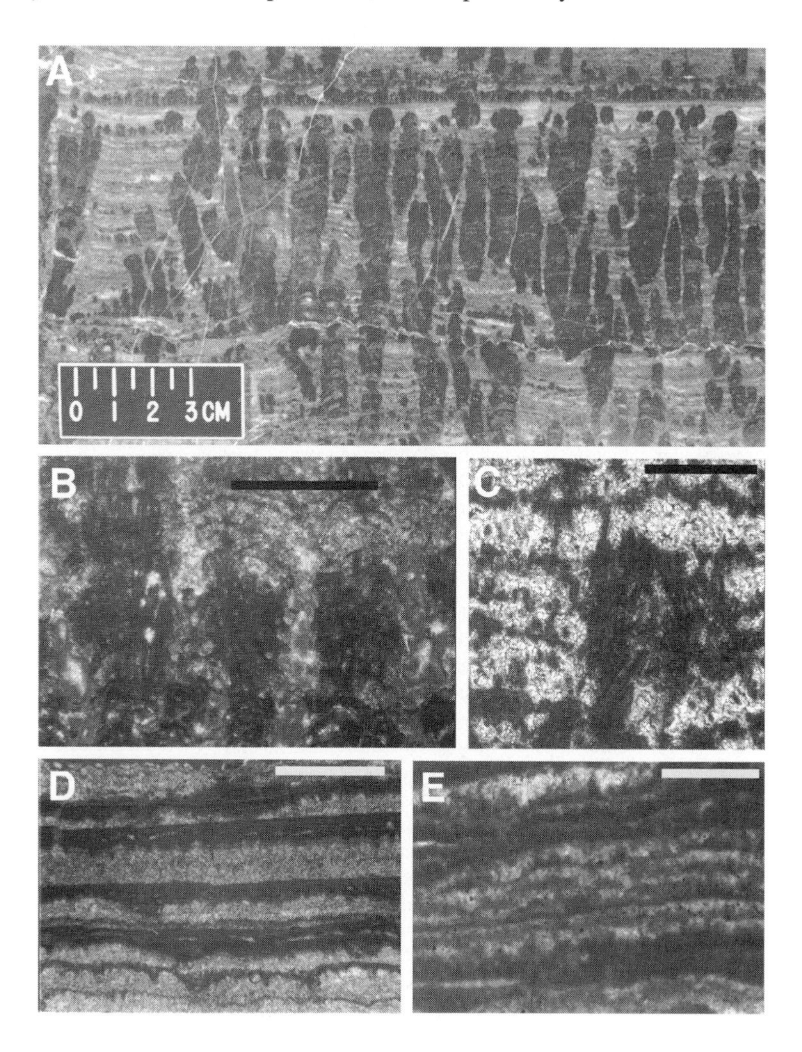

and clay preserved along badly altered laminar surfaces (see Figure 4*d* in Sears & Lucia 1980).

Many ancient stromatolites and stromatolitic facies, therefore, are sufficiently well recrystallized to preclude reconstruction of lamination processes. Fortunately, the record also includes numerous stromatolites that have suffered only minimal recrystallization so that primary textures are well preserved (Figure 3*a*, 3*b*, 3*c*), in some cases, with preservation of cyanobacterial filaments and molds (Figure 3*d*, 3*e*, 3*f*; see also Knoll & Semikhatov 1998, Semikhatov et al 1979, Walter et al 1988). In certain cases, fabrics are so well preserved that they imply neomorphic inversion of aragonite and high-Mg calcite directly to dolomite without a low-Mg calcite intermediary (Grotzinger & Read 1983). In rare instances, primary fabrics have been replaced by chert before neomorphic recrystallization, allowing clear distinctions to be made between laminae formed by sediment deposition, precipitation of sea-floor mineral crusts, and growth of microbial mats (Figure 4; see also Bartley et al 1999, Hofmann & Jackson 1987, Kah & Knoll 1996).

Much evidence has been supplied for the involvement of loose sediment in the formation of laminae (summarized in Semikhatov et al 1979); in contrast, the role of in situ precipitation has sometimes been viewed skeptically (e.g. Ginsburg 1991). Nonetheless, at least some ancient stromatolites were interpreted

→

Figure 3 Photomicrographs of stromatolite lamination textures, microbial filaments, and filament molds preserved in carbonate. (*a*) Small domal stromatolite with lamination defined by alternating layers of radiaxial dolomite, formed by in situ precipitation of calcite (replaced by dolomite). Note how some precipitated laminae pinch out in adjacent depression, filled by both precipitated laminae and peloidal grains. Paleoproterozoic Rocknest Formation, northwest Canada. Scale bar is 1.5 cm. (*b*) Stratiform stromatolite with lamination defined by alternating layers of radiaxial dolomite, formed by in situ precipitation of calcite (replaced by dolomite), and peloidal grainstone. Note how precipitation is localized on topographic highs formed by intraclasts and small peloid mounds. Mesoproterozoic Kyutingda Formation, Siberia. Scale bar is 450 μm. (*c*) Domal stromatolite with lamination defined by alternating micritic (*darker*) and microsparitic (*lighter*) layers. Lighter, microsparitic layers are highly discontinuous, consistent with an origin by sediment deposition. In contrast, darker layers are more continuous and organic rich and, through their draping, effectively bind sediment lenses in place. Mesoproterozoic Omachta Formation, Siberia. Scale bar is 20 mm. (*d*) Micrite-encrusted cyanobacterial filament sheaths. Note high abundance and recumbent position, both of which suggest preservation as a fossil mat. Neoproterozoic Ulukhan-Yuryakh Formation, Kolyma Massif, Russia. Scale bar is 500 μm. (*e*) Micrite-encrusted cyanobacterial filament sheaths. Note high abundance, and preservation of open framework created by intertwined filaments. Framework is lithified by drusy marine cement (*light gray, often lining filaments*), and residual pore space is filled by blocky spar (*bright white patches*). Texture is interpreted as a fossil mat, lithified before significant degradation of filaments. Neoproterozoic Chernya Rechka Formation, Igarka Uplift, Siberia. Scale bar is 500 μm. (*f*) Filament molds preserved in micritic stromatolite laminae. Stromatolites are completely enclosed in siliciclastic shales, thereby indicating in situ precipitation of micrite. Mesoproterozoic Svetli Formation, Uchuro-Maya Region, Siberia. Scale bar is 250 μm.

early on as the products of in situ precipitation, mostly on the grounds of indirect criteria that included fineness of lamination, oscillating Ca/Mg ratios between laminae, and bulk compositional differences between stromatolites and their enclosing sedimentary facies (Serebryakov & Semikhatov 1974). The last of these is perhaps most convincing, particularly where calcareous stromatolites are enclosed in entirely siliciclastic sediments (Figure 5; see also Hoffman 1976).

Petrographic studies of ancient stromatolites were slow to provide independent support for in situ precipitation models until criteria had been developed for the recognition of ancient cements based on distinctive crystallographic

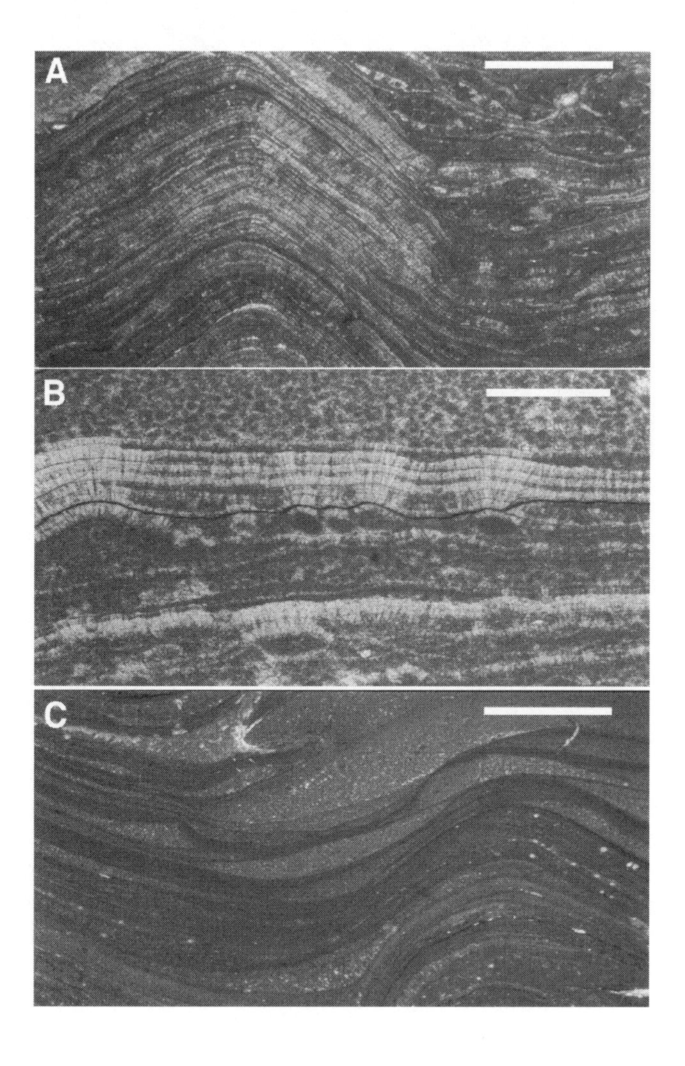

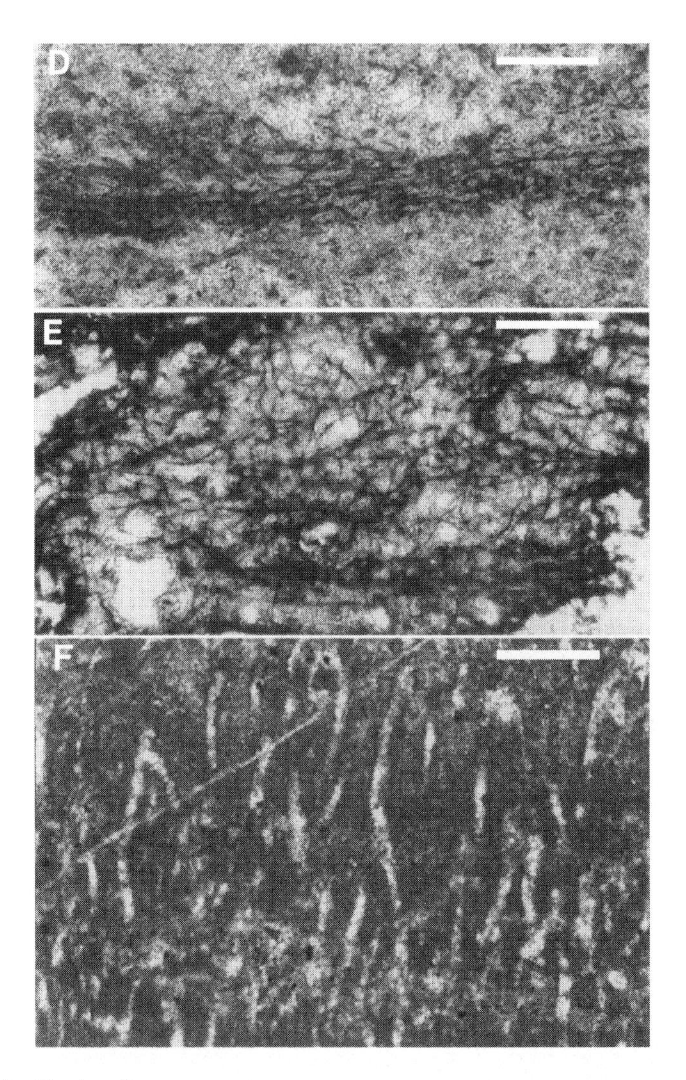

Figure 3 (*Continued*)

attributes.Based on recrystallization textures and analogy to Phanerozoic marine cements, it was argued that it was possible not only to recognize in situ precipitation in ancient stromatolites, but also to determine original mineralogy (Grotzinger 1986a, Grotzinger & Read 1983). Informative textures include radial fibrous (Figure 2*b*, 2*c*), radiaxial fibrous (Figure 3*a*, 3*b*), and fascicular optic calcite/dolomite (cf Kendall 1977, 1985; Kendall & Tucker 1973; Sandberg 1985). Strong support for this interpretation was provided by Hofmann and Jackson (1987), who discovered primary crystal textures that had been silicified before neomorphic inversion or recrystallization to more stable mineral phases. Comparable textures have subsequently been observed in a number of other Precambrian stromatolites (Bartley et al 1999, Kah & Knoll 1996, Knoll & Semikhatov 1998, Sami & James 1996), mostly of Paleo- and Mesoproterozoic age (Figure 4*c*, 4*d*). Another mineral texture consistent with in situ mineral precipitation, herringbone calcite, commonly encrusts thin films of organic matter interpreted as former microbial mats (Sumner 1997, Sumner & Grotzinger 1996a). Finally, distinctive textures formed exclusively of micrite create encrusting, isopachous sheets as well as branching, micritic tufas, analogous to those forming in modern thermal pools (Figure 6; Pope & Grotzinger 1999).

The recognition of these textures makes it possible to recognize in situ precipitation textures in many other previously described stromatolites (e.g. Horodyski 1975, Walter et al 1988). Over the past decade it has become increasingly clear that in situ mineral precipitation is an important accretion mechanism in ancient stromatolites (Figures 2–6; see also Bartley et al 1999, Grotzinger 1986a, Grotzinger & Read 1983, Hofmann & Jackson 1987, Kah & Knoll 1996, Knoll & Semikhatov 1998, Pope & Grotzinger 1999). In some remarkably well-preserved stromatolites of late Archean age, it can be observed that the only components in the stromatolite structures were microbial mats, early marine cement, and later porosity-occluding burial cement—sedimentary particles are completely absent (Sumner 1997). Furthermore, it appears that the distribution of stromatolites with precipitated laminae is time-dependent; stromatolites with precipitated textures are common in Archean and Paleoproterozoic carbonates, decline through the Mesoproterozoic era, and are represented principally by encrusted filament textures in late Neoproterozoic marine successions (Grotzinger 1989, 1990, 1994; Grotzinger & Knoll 1995; Knoll & Semikhatov 1998).

Processes

A proper understanding of stromatolite morphogenesis must begin with an analysis of the processes that create lamination. In general, it is thought that there are a small number of essential processes that control lamina accretion, including growth of microbial mats, sedimentation, and precipitation of minerals (Monty 1973, Semikhatov et al 1979). It is useful to consider in more detail how each of these processes contributes to accretion of laminae. Although a thorough review of mat biochemistry, crystal chemistry, and sediment transport dynamics is beyond the scope of this article, our goal here is to touch on the main points that relate specifically to the growth of the macroscopic structures so commonly observed in the field.

GROWTH AND DECAY OF MICROBIAL MATS The biological imprint on lamina texture is created through the orientation of filaments and unicells, the motility of major filamentous forms, and the adhesiveness and abundance of microbial sheath material (Semikhatov et al 1979). It is important to distinguish between lamination in mats, which results from microbial-community stratification, and lamination in stromatolites, which results from the interaction of mats with depositing sediment and/or precipitating minerals—only the latter process creates stromatolites. Lamination in mats is driven by steep vertical gradients in light intensity and redox conditions that cause segregation of different microorganisms into layers that are generally less than a few millimeters thick (D'Amelio et al 1989, Revsbech et al 1983). Primary producers, usually filamentous cyanobacteria, form the top layer, which is supersaturated with oxygen and which effectively acts as a light filter, causing abrupt attenuation of the higher frequency end of the visible-light spectrum (Jorgensen & Des Marais 1988). Both oxygen and light decrease along sharp vertical gradients within the mat. If anoxic microenvironments form within the photic zone, a distinct layer of photosynthetic bacteria may develop beneath the cyanobacterial surface community. In any event, ambient oxygen levels commonly fall to zero near the base of the photic zone, providing microenvironments for bacterial heterotrophs that participate in the decomposition of the overlying mat via sulfate reduction and other metabolic processes (Canfield & Des Marais 1993). Thus, over relatively short vertical distances (less than a few millimeters), the mat biota are stratified into discrete communities with distinctive metabolic attributes. The daily cycle of light intensity may result in an oscillation of pore fluid composition between oxygen supersaturation and millimolar concentrations of sulfide (Canfield & Des Marais 1993).

Although relatively common in flat microbial laminites, preservation of fossil mat populations is rare in domal, conoform, and columnar stromatolites, and

Figure 4 Photomicrographs of stromatolite lamination textures and microfossils preserved in chert, Mesoproterozoic Kotuikan Formation, Siberia. See Bartley et al (1999) for further details. (*a*) Submillimeter-scale lamination in precipitated stratiform stromatolites. Early preservation in chert reveals near absence of organic matter and complete lack of fossil mats. Scale bar is 2 mm. (*b*) Well-preserved, uncompacted, coccoid microfossil (*Myxococcoides*) preserved in stratiform stromatolites shown in *a*. Note that the scale of laminations is markedly smaller than most microfossils. Scale bar is 50 μm. (*c*) Microdigitate stromatolites formed of radiating acicular crystals (interpreted as former aragonite) draped with organic-rich laminae. Each radiating crystal fan corresponds to a single column observed in outcrop, similar to that seen in Figure 2. Scale bar is 5 mm. (*d*) Radial crystal fans interlaminated with organic-rich layers (*dark layers*), interpreted as sapropels produced from decaying mats. *Light layers* are silicified clastic carbonates. Nucleation is inferred to have taken place at or just below the organic-rich sediment-water interface, perhaps triggered by sulfate-reducing heterotrophic bacteria. However, once established, fans produced a slightly positive topographic anomaly and thus sustained themselves despite occasional draping by subsequent mats. Note that *light layers* lap out against the margins of fans but occasionally smother them, forcing renucleation. Scale bar is 5 mm.

in no case has a multilayered community structure been observed. Generally, only thin films or disseminated sheets of organic matter remain, giving rise to justifiable uncertainty as to the source of the organic materials. Former mats are but one of several possibilities—an obvious alternative is the accumulation of plankton from the overlying water column. Assuming that organic films do represent the vestiges of former mats, their poor state of preservation commonly is attributed to homogenizing effects associated with diagenetic recrystallization (e.g. Ginsburg 1991, Semikhatov et al 1979). However, it is probably more closely related to the high rates at which cyanobacteria are decomposed by

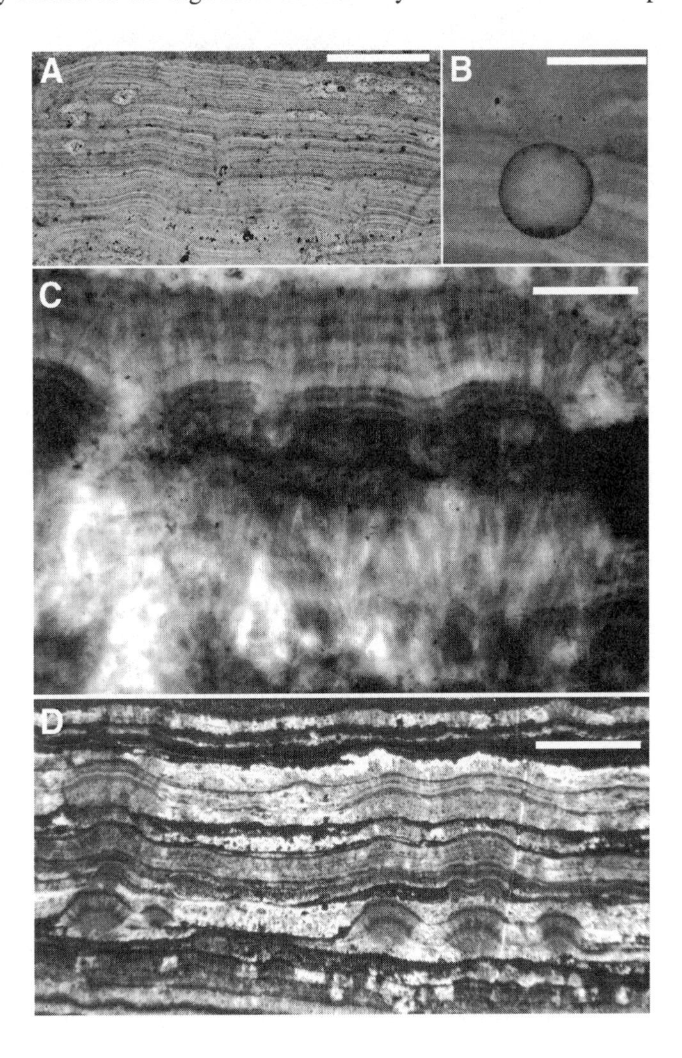

Figure 5 Stromatolites (*light*) within medium-grained to very-coarse–grained quartz sandstones (*dark*) of the Neoproterozoic Bildah Member where it onlaps the Witvlei Arch, central Namibia. Carbonates also form uncommon thin beds, but only in close proximity to stromatolites. These observations are taken to indicate in situ precipitation of fine-grained carbonate. Lens cap is 6 cm.

heterotrophic bacteria within the lower layers of mats (Doemel & Brock 1977, Golubic 1991). Bartley (1996) has estimated that significant degradation of both cyanobacterial trichomes and sheaths can occur in a matter of days to weeks, greatly exceeding the rates at which diagenetic recrystallization occurs (more than 10 years) (Constanz 1986). This helps explain the general rarity of preservation of recognizable microfossils and mats in ancient stromatolites, except under conditions of extremely early lithification (Bartley et al 1999). Interestingly, the early lithification process may in part be aided by heterotrophic bacteria, which raise alkalinity in the course of degrading cyanobacterial sheaths (Canfield & Raiswell 1991) and therefore induce carbonate precipitation on sheath surfaces (Chafetz & Buczynski 1992). Some recent studies of well-preserved ancient stromatolites support a bacterial nucleation process in early carbonate precipitation (Knoll & Semikhatov 1998), whereas in other cases it appears that early marine cements preferentially avoided nucleation on mats (Sumner 1997). The summary point is that the lower, heterotrophic component of the mat has several roles in the development of stromatolitic lamination. It drives mat decay, while at the same time facilitating cementation that may lead to preservation of the upper, cyanobacterial component of the mat. Last,

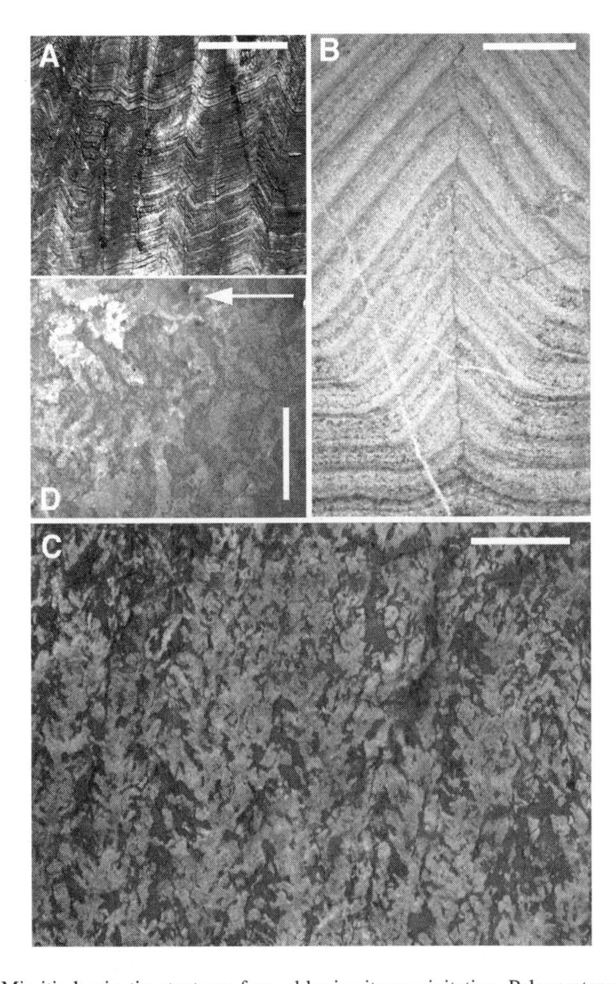

Figure 6 Micritic lamination textures formed by in situ precipitation, Paleoproterozoic Hearne Formation, northwest Canada. See Pope & Grotzinger (1999) for more details. (*a*) Outcrop photograph of very fine (submillimeter-scale), isopachous laminae defining laterally linked stromatolites. Scale bar is 10 cm. (*b*) Thin-section photomicrograph of lamination texture from stromatolites shown in *a*. Note extremely even, laterally continuous laminae with isopachous geometry. Isopachous geometry results from growth normal to the stromatolite surface, regardless of local curvature, typical of structures such as banded agates, which grow by in situ precipitation. Consequently, the stromatolite expands outwards (not just upwards) as it grows. Scale bar is 300 μm. (*c*) Outcrop photograph of dendritically branching tufa (*light*) which forms stratiform sheets on the scale of 1–2 cm. Branching structures are infilled with marine cement and micritic sediment. Scale bar is 5 mm. (*d*) Photomicrograph of a single dendrite structure with central thin stalk and numerous broad branches, both formed of micrite (*dark*). *Lighter material* is void-filling marine cement. Note arrow indicating growth orientation. Scale bar is 4 mm.

it generates high concentrations of dissolved sulfide that diffuse upward and stimulate mobility within the cyanobacterial layer.

MICROBIAL RESPONSE TO SEDIMENT FLUX The upper, cyanobacterial layer within a mat affects the development of stromatolitic layering and lamina growth in several important ways. As a prime example, this is where the physical activity is concentrated that results in the process known as "trapping and binding." Loose sediment deposited on the upper surface of the mat is tethered in place by the upward propagation of cyanobacterial sheaths through the sediment layer (Gebelein 1974). Surprisingly, since the rediscovery of the trapping-and-binding effect almost 40 years ago (Ginsburg & Lowenstam 1958, Logan 1961), there has been limited progress in elaborating the actual mechanics by which variable sediment flux and microbial response combine to produce stromatolite accretion. It is readily apparent that, physically, the microbiota must compete with the influx of sedimentary detritus to populate the depositional interface at densities sufficient to maintain a coherent mat. Under conditions of relatively small sediment influx, all constituents of the mat community are capable of rising through a given sediment layer (Des Marais 1995). Primary producers are displaced first, followed by an assemblage of anaerobic photobacteria and heterotrophs (Golubic 1991). If a higher sedimentation rate is sustained, the proportion of filamentous cyanobacteria in mats increases relative to coccoid forms, because the gliding motility of filamentous forms provides a selective advantage (Des Marais 1995). Logically, as the sedimentation rate increases past some (currently unknown) critical value, the sediment-stabilizing effect should drop off dramatically because sediment accumulation simply outpaces the maximum possible microbial response. The key point is that in natural systems there will be specific response times and scales for both microbial and sedimentation processes, and the growth of stromatolites will clearly be sensitive to how these processes balance. The end-member products of these interactions are clear (Monty 1976). In the absence of sedimentation, mats will decay and stromatolites will not be formed due to a lack of building material. On the other hand, stromatolites will not develop in the presence of critically high sediment fluxes because mat growth is not sustainable. This critical point will vary as a function of the biology of mat-building populations.

We consider a proper evaluation of the temporal and spatial scaling of these processes to be critical to the understanding of stromatolite morphogenesis. Additional studies of modern and experimental depositional systems are essential to address this problem because of the importance of quantifying sediment fluxes and mat growth rates. At this point, significant advances have been made in evaluating the temporal and spatial scales of biogeochemical cycling of elements in mats as it relates to mat layering, although these efforts have taken place at sites with little or no sedimentation (see summary in Des Marais 1995). On the other hand, recent work directed at understanding sediment accretion processes in regions where sediments significantly interact with mats has generated new information on the growth history of stromatolites, but not on the specific processes that lead to the accretion of sediment in the mats (Dill et al 1986, Dravis 1983, Macintyre et al 1996). Somewhat ironically, in the attempt to evaluate sediment accretion processes that take place on timescales of hours, days, and years, most studies of "modern" stromatolites have focused on the Holocene history and stratigraphy of stromatolitic facies. These studies seldom address sediment–mat interactions, in which microbial responses to sedimentation events might be monitored. For example, what is the minimum thickness of a layer of sediment that is required to smother and extinguish growth of an existing mat, so that it must recolonize? Alternatively, how frequently do tolerably thin layers have to be deposited before their integrated effect similarly results in termination of mat growth? Is either of these effects dependent on sediment grain size? How do these effects scale as a function of mat community structure? Only after these processes are fully characterized will it become possible to understand the significance of lamination in Holocene stromatolites formed by trapping and binding and, by extension, their equivalents in the more distant past.

EARLY LITHIFICATION A point on which all students of stromatolites seem to agree is that microbial mats and their associated sediments must be lithified early to be preserved in the record as stromatolites. Precisely when this happens and by exactly what mechanism are vigorously debated for both modern and ancient stromatolites. In modern (Holocene) stromatolites, it is clear that lithification occurs within a few centimeters of the depositional interface by the precipitation of aragonitic and high-Mg calcitic fibrous and micritic cements to form hard, current-resistant structures (Dill et al 1986, Dravis 1983, Logan 1961, Logan et al 1974, Macintyre et al 1996, Monty 1976). The case is clear cut because the stromatolites are obviously hard and primary marine cements can be observed petrographically to fill voids and cement particles. In ancient stromatolites, however, diagenesis characteristically has erased most evidence for early, pore-filling cements, and arguments for early lithification were, thus, traditionally based on indirect criteria such as the growth of steep or even overhanging margins (Donaldson 1976), the ability to withstand strong currents (Hoffman 1974), the chemical purity of lamination (Serebryakov & Semikhatov 1974), and the ability to form reefal escarpments with up to hundreds of meters of relief (Grotzinger 1986b). As mentioned previously, more recent studies supply direct petrographic evidence not only for early lithification, but also for growth of encrusting marine cement directly on the growing stromatolite, particularly for stromatolites of Mesoproterozoic and older ages (Bartley et al 1999, Grotzinger & Read 1983, Hofmann & Jackson 1987, Kah & Knoll 1996, Knoll & Semikhatov 1998, Pope & Grotzinger 1999, Sami & James 1996, Sumner 1997, Sumner & Grotzinger 1996a).

The processes of early lithification and growth of sea-floor crusts are poorly understood. Early lithified mats occur in modern marine and nonmarine environments and relate to calcification of the sheaths of primary, filamentous cyanobacteria which often form upward-divergent radially arranged bundles (Monty 1967, 1976). Consequently, when certain Precambrian stromatolites were discovered to contain laminae with palimpsest radial fabrics, these were attributed to the former growth of calcifying cyanobacteria (e.g. Bertrand-Sarfati 1976, Grey 1984, Walter et al 1988). However, subsequent detailed examination of petrographic textures revealed that these textures were more consistent with recrystallization of former radially arranged crystals than the micritic sheath coatings of cyanobacteria (Fairchild et al 1990, Grotzinger 1986a, Grotzinger & Read 1983, Hofmann & Jackson 1987). These observations favored a non-cyanobacterial interpretation, at least in accounting for both the domal morphology of related microdigitate or "asperiform" stromatolites and the development of an initially macrocrystalline (as opposed to micritic) texture. Of course, this interpretation does not deny a role for biologic processes in crystal nucleation. For example, it recently has been observed in some stromatolites that crystals were nucleated on mats only after the mats had decayed to thin organic laminae. This supports the hypothesis that heterotrophic bacteria played a role in nucleation of the crystal bundles (Bartley et al 1999, Knoll & Semikhatov 1998). Although they allow for the potential role of bacteria in catalyzing precipitation, these studies provide no evidence for cyanobacterial calcification of sheaths (cf Riding 1982). A number of studies of ancient stromatolites now indicate that early lithification of mats was simply a process of entombing microbes and mats within sheets and botryoids of marine cement. Cyanobacterial processes exerted little or no control over the morphology of crystal bundles; at best, heterotrophic bacteria may have helped trigger crystal nucleation. Even when Proterozoic stromatolites contain cyanobacterial sheaths preserved by carbonate encrustation, there is no direct evidence that precipitation was driven by cyanobacterial rather than heterotrophic bacterial metabolism (Knoll et al 1993, Knoll & Semikhatov 1998). Consistent with this, Chafetz & Bucyzinski (1992) succeeded in inducing the calcification of cyanobacterial sheaths in seawater only when experimental conditions included dead cyanobacteria and living heterotrophic bacteria.

Whatever the role of bacteria in aiding or impeding (Golubic 1991) carbonate precipitation and the genesis of crystalline crusts, it is clear that these crusts decline in importance through Proterozoic time (Grotzinger 1989, 1990, 1994;

Grotzinger & Kasting 1993; Grotzinger & Knoll 1995). This decline likely relates to a corresponding decrease in the calcium carbonate saturation of surface seawater, discussed in more detail below.

ARE STROMATOLITES JUST HARD GROUNDS? We close this section with one final point that addresses the importance of early lithification. An obvious but little discussed fact concerning the distribution of stromatolites is how profoundly biased they are toward formation in carbonate and, to a much lesser extent, other chemical sediments precipitated from seawater (iron, manganese, and phosphate, for example). Despite being the most common rock types in the sedimentary geologic record (Ronov 1968), siliciclastic sandstones and shales do not contain stromatolites made of sandstone or shale. If stromatolites are present their laminae are composed mainly of marly carbonate (Hoffman 1976, Serebryakov & Semikhatov 1974). Curiously, siliciclastic sediments are not devoid of features that suggest the former presence of mats (Hagadorn & Bottjer 1997, Pflueger 1997, Schieber 1986), the mats just simply didn't form stromatolites. Put another way, biology is ubiquitous and appears to have been so for more than three billion years; stromatolites are not. Therefore, stromatolite accretion requires conditions beyond those necessary for the establishment of microbial mats. Siliciclastic settings share many features in common with carbonate environments: they support well developed, ecologically diverse mats; they contain abundant grains of suitable size for trapping and binding; and they exhibit features such as shallow water depth, a spectrum of wave energies and salinities, and adequate nutrient supply known to be correlated with stromatolite development on carbonate platforms. One potential difference may be the degree of light penetration in the more turbid water column of siliciclastic settings, but this should place shallower limits on the depths at which photosynthetic mats could exist, rather than act as a fundamental barrier to mats being established within shallow-water settings. Another factor might be higher sediment mobility in siliciclastic settings, but stromatolites grow well in many modern carbonate environments with moderate to high sediment transport rates (Dill et al 1986, Macintyre et al 1996).

In our view, the fundamental difference is in the relative timing of early lithification. Trapping and binding by mats is insufficient to permanently stabilize sedimentary particles against currents and other destructive environmental processes (papers in Krumbein et al 1994) unless the particles are lithified very early with marine cement. The key is to recognize that this process is kinetically regulated and is therefore dependent on several factors besides the saturation of calcium carbonate in pore fluids. For the most part, the upper layer of the modern oceans is oversaturated with calcite and aragonite, regardless of latitude and whether siliciclastic or carbonate sediments are being deposited (Broecker

& Peng 1982, Broecker & Takahashi 1978, Li et al 1969). Similarly, microbial communities that might help catalyze carbonate precipitation (photoautotrophs and sulfate-reducing bacteria) are also ubiquitous in mats of all settings (Des Marais 1995, Golubic 1991). These important factors appear to be constant for both carbonate and siliciclastic settings. However, the suitability of substrate for catalyzing calcium carbonate precipitation is far greater for calcium carbonate versus quartz or clay. Therefore, the preference of mineral overgrowths for preexisting crystals will systematically select for early lithification of stromatolites made of carbonate grains rather than other sedimentary minerals. Viewed in this manner, the weight of the evidence argues strongly that crystal chemistry may have as much to do with the preservation of stromatolites as microbial processes have to do with forming them. Stromatolites are just another form of hard ground. For stromatolites formed by trapping and binding loose sediment, where biology begins and ends is comparatively easy to interpret. Without an active mat, sediments would not have accreted to form a topographic anomaly on the sea floor, whereas without grains of the correct composition, the structure would not have been lithified early and therefore preserved in the record. However, for stromatolites formed by mineral precipitation the problem of distinguishing uniquely biogenic and abiogenic processes becomes more difficult simply because the process of mineral precipitation can be dependent on only chemical processes. In the next section of this paper, we explore how the three lamina-forming processes (mat growth, sediment deposition, and mineral precipitation) interact to form stromatolites and their diverse morphologies.

STROMATOLITE MORPHOGENESIS

Background

In addition to lamination, the other distinguishing feature of stromatolites is their shape. A typical stromatolite is made up of numerous successive laminae that stack on top of each other to form domal, conoform, columnar, or branching columnar structures. Although laminae generally describe upwardly convex structures, they can also form upwardly concave or discrete conical structures. In general, it is thought that there is a broad but gradational variation in the forms of stromatolites encompassing several major morphological motifs (Semikhatov et al 1979). It has long been observed that stromatolite morphology varies as a function of facies. Thus there is broad agreement that physical environment plays a role in the generation of shape (papers in Walter 1976b).

At the level of process, however, there is no such guiding consensus, severely limiting our ability to understand either paleoenvironmental or stratigraphic

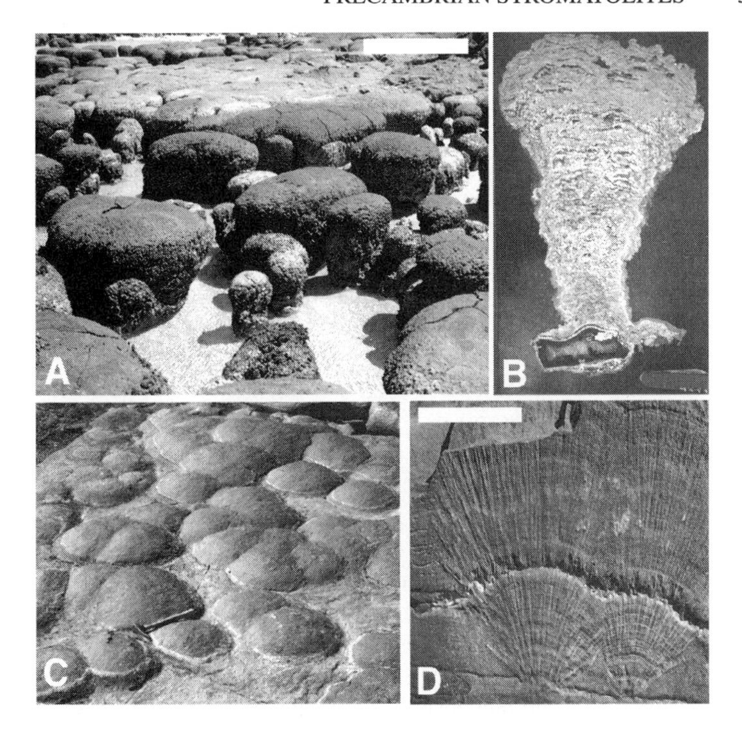

Figure 7 Stromatolites with similar shape but different origins. (*a*) Modern domal to columnar stromatolites from Shark Bay, Western Australia. Scale bar is 40 cm. (*b*) Stromatolites are formed by trapping, binding, and early lithification of loose carbonate sediment to form crude lamination. Knife is 7.5 cm long. (*c*) Domal stromatolites preserved in Neoarchean Campbellrand Subgroup, South Africa. Hammer is 35 cm long. (*d*) Stromatolites are formed by growth of crystal fans that trap sediment as it settles into the interstices of growing crystals. This produces a faint but relatively fine lamination along which preferential weathering has exposed domal shapes whose radii of curvature correspond to that of each radiating crystal fan. Scale bar is 20 cm.

variations in stromatolite form. As Hofmann (1987) stated, "... we still have no stromatolite theory, no model that shows which attributes changed in what way through time." Without a viable theory we are always at risk of misinterpreting the genetic significance of growth form. This is well illustrated in Figure 7, where it is obvious that the present is not the key to the past—modern domal and columnar structures formed by microbial mats (Figure 7*a*, 7*b*) are mimicked by Neoarchean domal structures formed by radiating crystal fans of calcitized aragonite (Figure 7*c*, 7*d*). Clearly, two very different sets of processes have acted over time to produce nearly identical structures. What is missing is a model in which lamina-scale morphology can be related to macromorphology through iteration of specific biological, physical, and chemical processes.

Accretion Mechanisms

One can easily list numerous factors that might influence stromatolite development, including light intensity, salinity, nutrient supply, current velocity, sediment grain size distribution, mat community diversity, and carbonate saturation, to name a few. In detail, stromatolite growth is dependent on many processes that are complexly interrelated. Not only are the processes mechanistically complex, but they evolve over long timescales that are difficult to reproduce experimentally or monitor in the field. Given these difficulties, our initial goal should be to construct a simple model of this complex system. Even this will be a difficult task, in need of studies that can serve to calibrate important model parameters (e.g. Jorgensen & Des Marais 1988, 1990). In principle, growth of stromatolites can be described as a simple system, depending only on three fundamental processes: growth and degradation of a microbial mat or biofilm, deposition of sediment, and precipitation of minerals. Interactions among these end-member processes should account for the bulk of stromatolites in the record.

As mentioned earlier, stromatolite growth depends on the iterative process of upward growth by mats or sea-floor crusts alternating with periods of sediment deposition. In addition, these processes must be balanced such that sediment does not overwhelm mats or crusts. On further inspection, an additional but critically important attribute of the iterative process is revealed. The growth of mats tends to produce an irregular, relatively rough surface, whereas the settling of sediment tends to create a smoother surface by filling in the microtopography of the underlying mat (Figure 8). The surface roughness of mats will vary depending on community. Gebelein (1974) notes that "Mats that have very smooth surfaces and little or no micro-relief are usually created by

→

Figure 8 Development of surface roughness in microbial mats and the consequences of sedimentation. (*a*) Irregular, mamillate growth surface produced by chroococcalean mat, Shark Bay, Western Australia. Knife is 7.5 cm long. (*b*) Tufted mat produced by *Phormidium*, Lake Hoare, Antarctica. Although these mats are often suggested as morphologic analogs of conoform stromatolites (e.g. Parker 1981), we note that the tufted morphology in this case is not preserved at depth in the mat due to rapid decay in the absence of early mineralization. (*c*) Cartoon illustrating how a building mat may interact with depositing sediment. If the sediment is thin and capable of spreading laterally, it will fill depressions, such as for the basal darker layer. Smothering of the mat in depressions may lead to fragmentation of the growing mat, resulting in growth of columns. However, if the sediment layer thickness is too great, as shown here, then the entire mat may be covered leading to damping of topography and forcing mat growth to start again. (*d*) Possible fossil mat (*light layer*), showing flat base and mamillate top, smothered by several successive layers of clastic carbonate. Note that the first set of laminae drape the growth topography, but the third layer infills depressions and thereby "resets" the growing interface back to a flat condition. Mesoproterozoic Omachta Formation, Siberia. Scale bar is 300 μm. Photograph courtesy of M Semikhatov.

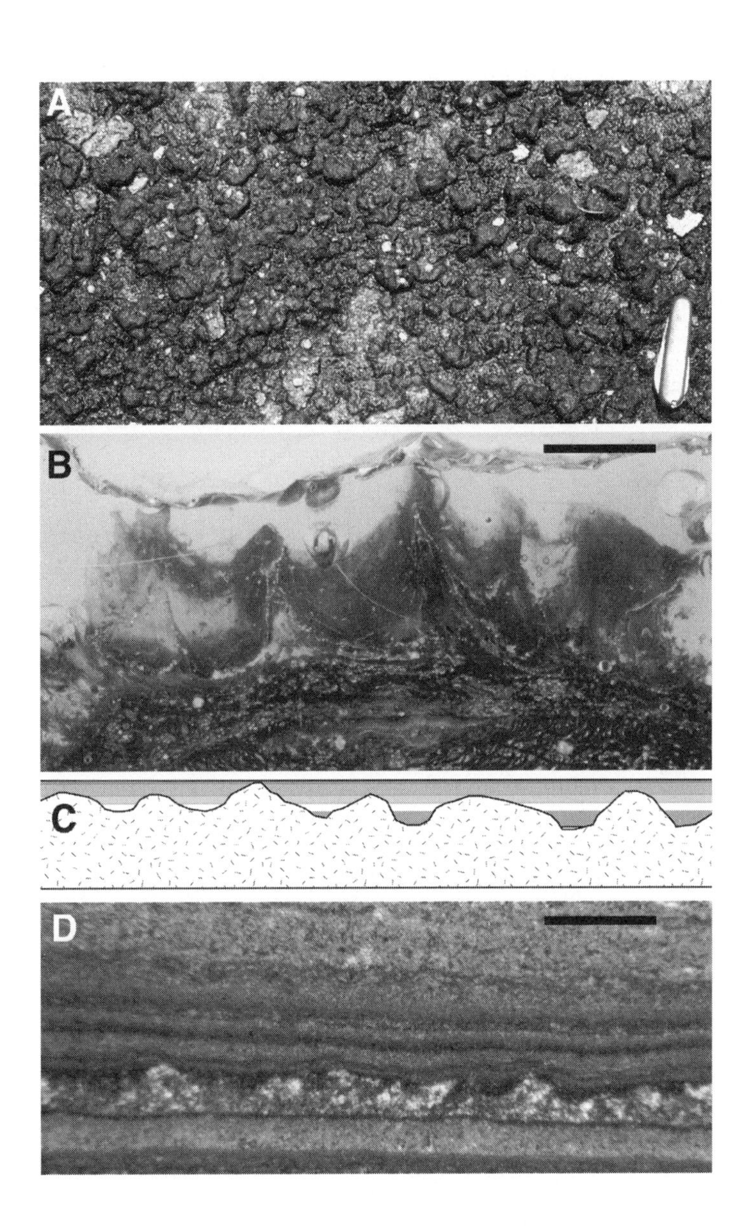
A
B
C
D

oscillatoriacean algae, which have thin sheaths. Mats that have a tufted or crenulated surface are created by oscillatoriaceans, which have thick mucilaginous sheaths [Figure 8*b*]. Mats that have an irregular, mamillate surface texture are created by chroococcalean algae, mainly of the family Entophysalidaceae" (Figure 8*a*). Therefore, over many iterations the surface roughness of a growing stromatolite may be enhanced or suppressed depending on the competitive processes of surface roughening by mats and surface smoothing through sedimentation (Figure 8*c*, 8*d*). Precipitation may preserve or dampen surface roughness depending on whether mineral emplacement occurs before or after significant mat decay. As used here, the "surface roughness" (or width, w) of any depositional (or erosional) interface is defined (Barabasi & Stanley 1995) as the root-mean-square fluctuation in the height of the interface

$$w(L,t) \equiv \sqrt{\frac{1}{L}\sum_{i=1}^{L}[h(i,t) - \bar{h}(t)]^2} \tag{1}$$

where L is the length of the interface, h is the interface height for any position along L, t is time, and i is an index. By definition, growth starts from a horizontal line; the interface is initially a simple straight line. As deposition occurs, peaks and valleys form and the interface gradually roughens. This quantity is easily measured and could potentially reveal a great deal about both mat biology and community structure, as well as the relative proportions of mat growth and sedimentation.

Unfortunately, even though it has been recognized for decades that mats have variable surface roughness in a qualitative sense, this attribute has never been quantified. In future studies of modern mats it would be sensible to measure, particularly where sedimentation also occurs. Here's why: theory, experiments, and observations in nature of dynamically evolving interfaces over the last 15 years have placed important constraints on the possible range of growth mechanisms for both microbial and abiotic systems. In almost all of these studies, the most commonly measured feature is the roughness of the interface, which is used as a basis for deduction of dynamical process and for taxonomic assignment to model class. It is worthwhile to review some of these advances and how they may relate to growth and interpretation of stromatolites. Studies of interface dynamics may possibly lead to a "stromatolite theory" of the sort envisioned by Hofmann (1987).

INTERFACE DYNAMICS AND STROMATOLITES The texture, particularly the roughness, of any depositional surface (interface) is subject to certain force balances and the presence of noise or randomness. In general, and regardless of whether growth is abiologic or solely microbial, there are two end-member models that control the morphology of an accretionary interface. These different models feature local versus nonlocal growth processes. When interface growth is controlled by local growth processes, the rate of growth is dependent only on the local properties of the interface, such as height of the interface and its nearest neighbors (e.g. Kardar et al 1986). Local growth models are consistent with accretion under equilibrium conditions, with "particles" (ions, sediment, or cells) rejecting attachment sites of the growing object until the most stable configuration is established. For growth by local processes, interface morphologies may be smooth or irregular, but always relatively compact (Figure 9*a*), whereas nonlocal processes lead to steady-state morphologies that are usually highly ramified, featuring columnar to dendritic branching patterns (Figure 9*b*).

A widely applied local growth model is represented by the KPZ equation (Kardar et al 1986), whose relevance to understanding stromatolite growth has been recently evaluated (Grotzinger & Rothman 1996). In that study, the following four mechanisms were considered important in controlling stromatolite growth: (*a*) fallout of suspended sediment; (*b*) diffusive smoothing of the settled sediment (i.e., sediment fills in depressions in underlying microtopography and moves downhill at a rate proportional to slope); (*c*) surface-normal mineral precipitation or mat growth, and (*d*) uncorrelated random noise representative of surface heterogeneity and environmental fluctuations. Under these circumstances, the KPZ equation can be slightly modified (Grotzinger & Rothman 1996) to predict a growth rate ($\partial h/\partial t$) for the stromatolite by

$$\frac{\partial h}{\partial t} = \nu_s + \kappa \nabla^2 h + \nu_p \sqrt{1 + (\nabla h)^2} + \eta(x, t) \tag{2}$$

where ν_s is the time-averaged rate of sedimentation, κ is an effective diffusion coefficient, ν_p is the time-averaged rate of surface-normal precipitation, and η is uncorrelated random noise with zero mean and variance η_0^2. The square-root factor is a geometric correction that acts to project the surface-normal growth along the h-axis. The smoothing represented by the second term of Equation 2 is equally representative of surface tension and diffusion. The net effect of these two processes is represented in the single coefficient κ. The validity of the KPZ equation and therefore the process of local growth in accounting for growth of these stromatolites were tested and tentatively confirmed by calculating the surface roughness of several stromatolitic laminae and comparing the obtained scaling exponent (and fractal dimension) to that predicted by the KPZ theory (Grotzinger & Rothman 1996). This growth model predicts smoothing and broadening of domes with time because particles (ions, nutrients, and sediment) that arrive at the surface have an equal probability of attaching to all sites, including those on the sides of bump in addition to the tops of bumps or the

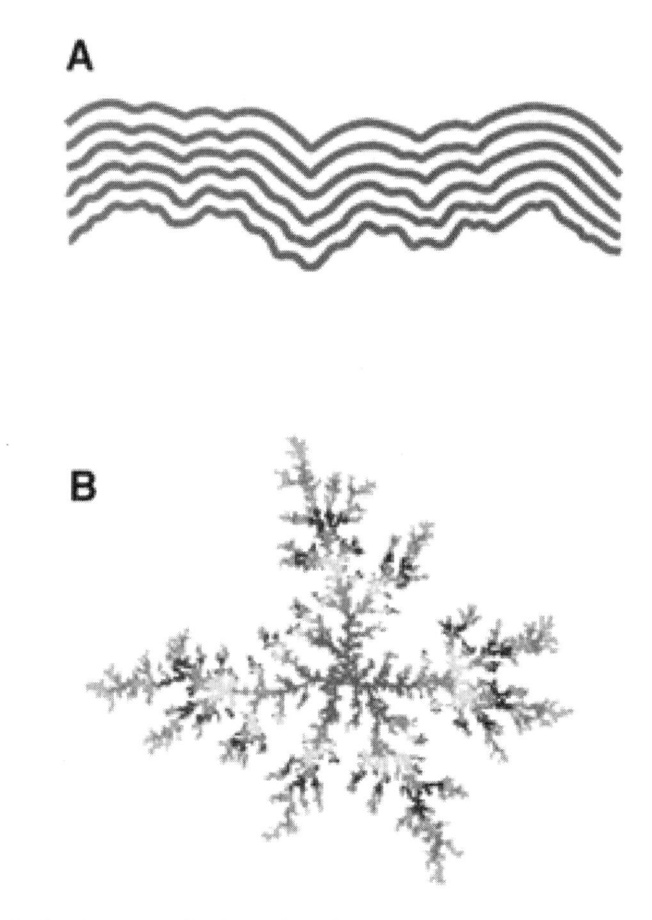

Figure 9 Local versus nonlocal growth models. (*a*) Local growth predicted by KPZ model, in which an initial surface with arbitrary roughness grows through time. Because of the strong influence of surface-normal growth, irregularities in the topography merge, resulting in selection of progressively fewer, progressively wider domes through time. Growing structures are relatively compact. Modified after Kardar et al (1986). (*b*) Nonlocal growth created by diffusion-limited aggregation. Growth begins at center as a result of accretion of particles undergoing random walks, which simulates the diffusive process. All particles stick on reaching the growth surface. Once a protruding branch develops, the probability of a random walker making it to an attachment site located near the branch point is dramatically reduced. This condition amplifies in time, resulting in preferential growth of tips over depressions and generation of a high degree of branching. Common shades of gray record equivalent times of growth. Model algorithm based on Witten & Sander (1983) and executed courtesy of K Chan.

depression between bumps. Thus, lateral growth may be as important as upward growth (Figure 9*a*). An initially rough surface grows with constant velocity normal to all local surfaces, and with time larger domes overtake smaller domes leading to a smoother interface with fewer broader domes, when compared to the initial condition (Figure 9*a*). Growth of this type may characterize many Precambrian stromatolites and also describes the geometry of layering in the walls of agates, botryoidal mineral clusters, travertines, and at least certain types of stromatolites, in a quantitative manner where seafloor precipitation is thought to have been important. For example, stromatolites formed immediately before precipitation of some of the world's largest evaporite deposits are characterized by fine, isopachous lamination, and internal textures consistent with in situ precipitation (Pope et al 1999).

In contrast, for other systems a number of nonlocal effects contribute to interface morphology and growth velocity, the most important of which is the presence of a diffusing field, which may reflect pressure, electric potential, temperature, and chemical or nutrient concentration (e.g. Witten & Sander 1983). For these models, growth rate ($\partial h/\partial t$) is simply related to a diffusion process and is given by

$$\frac{\partial h}{\partial t} \propto \frac{\partial c}{\partial t} = D\frac{\partial^2 c}{\partial z^2} \tag{3}$$

where D is the diffusion constant, h is the height of the interface, z is the vertical position in the overlying water column, c is the concentration of diffusing ions or nutrients, and t is time. This equation states that the increase (or decrease) in the growth rate of the interface in time is proportional to the increase (or decrease) of the concentration of diffusing ions or nutrients in time. In turn, the time rate of change in concentration is directly related to the gradients in concentration scaled by a diffusion constant. Basically, growth is fastest where concentration gradients are steepest. Nonlocal growth models are consistent with accretion under nonequilibrium conditions such that particles are accreted to the first site of attachment they encounter, and attachment sites that project farther into the diffusing field have a higher probability of being selected. The most common model used to account for dendritic growth, biologic or abiotic, is known as diffusion limited aggregation, or DLA (Witten & Sander 1983). As the name of the model implies, aggregation in DLA is controlled by the dynamics of diffusion away from the interface, rather than the kinetics of reaction at the interface—the essential difference between nonlocal and local growth dynamics. The key premise of the DLA model is that particles (ions, nutrients, and sediment) arrive at the site of deposition through a random walk that simulates Brownian motion (diffusion). Particles are successively released, each undergoing a random path, until a macroscopic cluster is formed

(Figure 9*b*). Numerical models have shown that the dendritic structures that are formed by DLA have fractal geometry, and experiments in both biological and abiotic systems have confirmed the essential role of diffusion in controlling the geometry. The complex branching of the dendritic structures best illustrates the nonlocality of the DLA model. As branches begin to develop, they create a screening effect, which makes it increasingly improbable (exponentially so) that new particles will ever find their way into the depressions between branches (Figure 9*b*). Thus, no matter what may occur at the interface itself, the growth process is fundamentally controlled through effects away from the interface (i.e. nonlocal effects), in this case diffusion of particles in a potential field. Furthermore, the growth rate at any given point depends on the entire geometry of the growing structure, not only on the local morphology.

Initial studies of the DLA model strongly suggest that it is also applicable to understanding the growth of certain stromatolites (Chan & Grotzinger, unpublished observations). Whereas DLA by itself predicts highly branched, dendritic structures, in the presence of incremental sedimentation events a simple model predicts many of the domal-, columnar-, and branching-columnar-stromatolite morphologies that are observed in the record (Figure 10*a*). In this model, an episode of upward growth by randomly attaching particles is taken to simulate growth of either mats or crystals and is followed by an episode of sediment settling in which the sediment is allowed to settle preferentially well into the microdepressions formed in the underlying mat or crystal layer. If thick enough and/or diffusive enough, the sediment may damp all of the initial topography created by the underlying layer. However, if antecedent topography remains, the next iteration of mat/crystal growth will result in preferential accretion on those topographic highs. The next layer of sediment now has a tougher

→

Figure 10 Stromatolite growth model of Chan & Grotzinger (unpublished observations) and comparison to ancient stromatolites. (*a*) Growth model based on diffusion-limited aggregation and episodic sedimentation. Initially the interface is allowed to grow through diffusion-limited aggregation, which simulates the growth of either microbial mats or precipitating minerals (*dark layers*). After some time, the interface has become rough, and sediment is allowed to settle down onto the rough surface (*light layers*). It is assigned a lateral mobility and therefore can migrate into depressions; in doing so it partially damps the preexisting topography. However, this process is incomplete, so the next interval of upward growth builds selectively on the remnant highs, reinforcing their topography. As long as the thickness of sedimentation events does not exceed some critical value, the growing interface eventually will produce branched columnar structures, similar to certain ancient stromatolites. Note that, in the late stages of growth, depressions are filled only by sediment. (*b*) Branching columnar stromatolites of the Paleoproterozoic Talthelei Formation, northwest Canada, showing strong similarity to model results. (*c*) Columnar stromatolites from Mesoproterozoic Debengda Formation, Siberia, also showing strong similarity to model results. See text for further discussion.

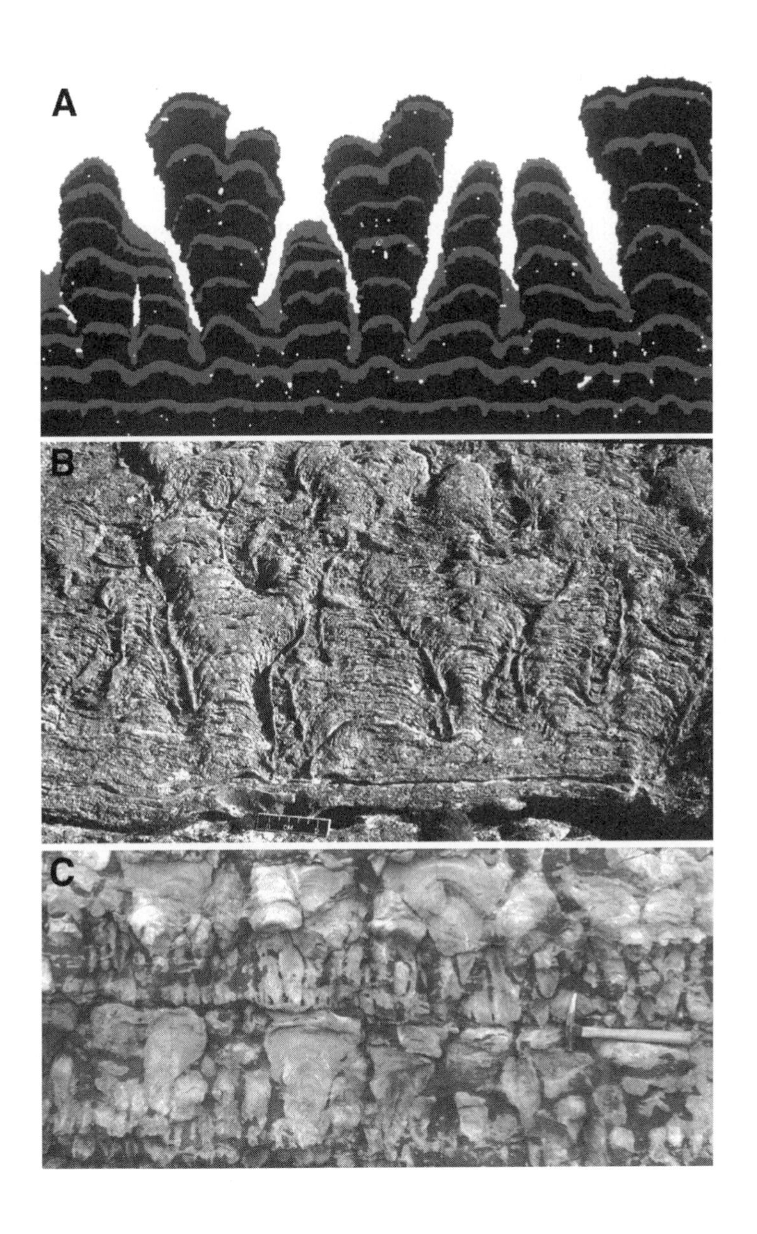
A
B
C

task to fill depressions, giving rise in the next iteration of mat/crystal growth to an even higher preference for growth on highs. This is the particular feature of DLA models, that small perturbations can be amplified in time to become dominant features of the structure itself. In this manner, no special conditions may be required to generate columns and branching columns in stromatolites—only time and the positive reinforcement of randomly produced protuberances. This type of growth may help account for the diversity of branched columnar stromatolites, which are common in the geologic record (compare Figure 10*a* with Figure 10*b* and 10*c*).

The fundamental point to be made is that growth of morphologically complex stromatolites is possible only within a fairly narrow range of environmental conditions, where mat growth, sediment flux, and mineral precipitation rates all optimally coincide (Chan and Grotzinger, unpublished observations). Sediment accumulation rates that are either too high or too low tend to force growth of stratiform stromatolites. Where too high, mats and/or precipitating crystals are blanketed in sediment, terminating stromatolite growth. At the other extreme, where sedimentation rates are too low, mats will decay before sediments can be bound into place to build larger structures.

Stromatolites formed by laminae precipitated in situ reflect a considerably different range of environmental conditions. Again, if the stromatolite is exposed to critically high sedimentation rates, growth will be terminated. However, the effect of negligible sedimentation rates is less severe because upward growth can be sustained by either early lithification of mats (preventing their decay and collapse of relief) or simply through crystal growth in the absence of mats. Consequently, development of precipitated stromatolites is predicted at sites where sedimentation rates are low, consistent with field data that indicate restricted peritidal settings for many such stromatolites (Bartley et al 1999, Grotzinger 1989, Kah & Knoll 1996).

Branching in stromatolites is apparently favored when topographic low points are preferentially filled, leaving high points as selected sites for continued growth (Figure 10). Although the temporal and spatial scales over which small topographic anomalies are amplified to generate branched, columnar structures may be dependent on the specific mat-building communities and/or precipitating minerals, from a dimensionless viewpoint the fundamental parameters are revealed as the mat/crystal growth rate and sedimentation rate, whose balance regulates both the onset and termination of branching and whose ultimate control is essentially environmental.

Conoform stromatolites appear to constitute a special case. Consistent with the empirical observations of Walter (1976a) of conoform stromatolites in silica-charged thermal pools at Yellowstone Park, K Chan & JP Grotzinger's unpublished observations suggest that both highly motile mat builders and penecontemporaneous mineral precipitation are required to generate and sustain these distinctive structures. Motile cyanobacterial filaments aggregate to form vertical tufts on mat surfaces, and these provide the template for conical accretion. However, unless the tufts are lithified early, their distinctive morphology can be lost quickly (see Figure 8*b*). Although filament tufts characterize mats found in a variety of environmental settings and presumably have done so for 2000 Ma or more, preserved vertical filament tufts are characteristic only of Mesoproterozoic and older successions (Kah 1997, Knoll & Sergeev 1995) more or less coincident with the stratigraphic distribution of conical stromatolites (Walter & Heys 1985). It is intriguing that the Miocene conoform stromatolites described recently by Feldmann and McKenzie (1997) are part of a transitional, evaporite-related facies assemblage formed when seawater may have strongly favored early lithification of mats.

BIOGENIC VERSUS ABIOGENIC GROWTH Although the biogenicity of stromatolites is seldom questioned outside of the hallways of academia, we tend to agree with Hoffman (1973) that "Something that haunts geologists working on ancient stromatolites is the thought that they might not be biogenic at all." Twenty-five years after its publication, we can profitably rephrase this statement as a question: how can we recognize biogenic stromatolites based on their morphology? The curious aspect of both local and nonlocal models is that examples of each can be found for microbial systems (compare Matsushita & Fujikawa 1990 with Ben-Jacob et al 1994), as well as purely abiotic systems (compare Kardar et al 1986 with Matsushita et al 1985) that have qualitatively identical appearances and often quantitatively similar scaling relationships (see summary in Barabasi & Stanley 1995). This similarity may be frustrating in the attempt to use the morphology of depositional surfaces as a parameter to ascertain the biogenicity of stromatolites (cf Grotzinger & Rothman 1996). On the other hand, it is fascinating in that it implies that microbial populations may behave almost atomistically in their self organization to form clusters, biofilms, and mats. The physical basis and rationale for this kind of microbial organization is well explained in Berg (1983).

Understanding, through the use of simple process models, that stromatolite growth may result from the competitive interaction of upward growth and surface roughness forced by mats and damping of surface relief by sediment settling, it becomes easy to see how the growth of abiotic marine crusts might substitute for mats and create the same end result. The good news is that we may now have a theory that can account for the growth of a remarkable range of stromatolites, but the bad news is that this theory predicts that we can no longer accept only morphological descriptions of stromatolites as evidence of their biogenicity. This does not mean that stromatolites may not have grown in the presence of biogenic influences; it means that morphology may well be a non-unique parameter—a point made clear by Grotzinger & Rothman (1996) but easily misunderstood. Biogenicity cannot be easily demonstrated by relationships observed at the outcrop scale; it is essential to examine lamination textures petrographically and demonstrate the presence of textures uniquely attributable to the presence of microbial mats or biofilms (Cady & Farmer 1996, Knoll & Semikhatov 1998). However, for many stromatolites this may not be possible due to an indecipherable level of diagenetic recrystallization.

After almost a hundred years of debate, it seems obvious that there is still a fundamental lack of agreement over where the roles of biology begin and end in many stromatolites, modern as well as ancient. Now, at least we may know why this controversy has been so long lived. As individual cells, microbes may react in similar fashion to many of the same influences that stimulate abiotic growth or vice versa. Consequently, we choose to reinforce the many valid points made by Semikhatov et al (1979) in formulating their definition. This definition specifies the principal textural and morphological attributes that make their identification in the field easy, but it does not make assumptions about their origin. Doing so allows for laboratory-based investigations that may reveal whether a given stromatolite is largely biogenic or abiogenic. If biogenic, it allows for the fact that it may be principally of bacterial or algal, as opposed to cyanobacterial, origin.

STROMATOLITE DIVERSITY—TRENDS AND SIGNIFICANCE

Background

In fairness to the many studies of Proterozoic stromatolites, the fundamental question of biogenicity is most critical in older Proterozoic and Archean (and, potentially, Martian) stromatolites. However, there is a second set of biological questions that is highly relevant to research on Proterozoic stromatolites. If we accept that most Proterozoic stromatolites accreted under the influence of microbial mats, can we assume further that the biological influence on accretion is so pervasive that specific stromatolites can be used as proxies for discrete microbial communities? Can we further view observed secular variations in stromatolite form and microstructure as the preserved record of microbial evolution?

In widespread, if long debated, practice, stromatolites have been classified by a quasi-Linnean system in which groups and forms—analogous to the genera and species of Linnean hierarchy—are recognized by the reconstructed shapes of columns, branching patterns, and lamination textures. The shorthand provided by this classification system has undeniably contributed to the important observation that at least some stromatolite morphologies and microstructures have distributions that are limited in time and space (summarized in Semikhatov

1991). It has also spawned a cottage industry in the analysis of stromatolitic diversity through time (Figure 11). Of course, the utility of such compilations depends greatly on the discreteness of the entities formally recognized as groups and forms.

Awramik (1971) was the first to compile Proterozoic stromatolite diversity data (Figure 11*a*). He recorded an increasing diversity of columnar forms through the Proterozoic Eon that culminated in a Neoproterozoic peak followed by a sharp decline in the latest Proterozoic and Early Cambrian. Awramik attributed this drop to the radiation of animals capable of disrupting mats by grazing and burrowing and outcompeting mat populations for space on the shallow seafloor; Garrett (1970) had earlier attributed the ecological restriction of modern mat populations to similar processes. Awramik (1971) did not take into consideration the unequal time intervals under consideration or the unequal representation of carbonate rocks in these time blocks, but his general conclusions have by and large been confirmed by subsequent analyses.

In 1985, Walter & Heys (1985) revisited the problem, using a much expanded database (Figure 11*b*). They explicitly addressed the issues of time and sampling, showing how various efforts to correct for these influenced diversity trends. Walter & Heys (1985) also added a parameter, "abundance," and defined it as the number of occurrences of individual forms in different basins, summed over the total number of forms documented for each time interval. Used this way, "abundance" does not have the meaning ascribed to it by ecologists. Rather, the term carries biogeographic (and both taxonomic and sampling) implications. Walter & Heys' (1985) summary diagram shows a broad diversity plateau that extends across the Meso- and Neoproterozoic Eras, followed by a decline comparable to that noted by Awramik (1971). Interestingly, Walter & Heys (1985) also plotted the diversity of conoform stromatolites separately and showed that such forms peaked in diversity during the Mesoproterozoic Era—with the unstated but necessary consequence that the diversity of columnar stromatolites was highest in the early Neoproterozoic, as claimed by Awramik (1971). Walter & Heys (1985) further suggested that overall stromatolitic diversity began to decline midway through the Neoproterozoic Era—perhaps 700 to 800 Ma—and hypothesized that this reflects the initial diversification of animals, as yet unrecorded by metazoan fossils or trace fossils. The Late Riphean interval (circa 1000 to 600 Ma) is not subdivided in their dataset, and so support for this conjecture cannot be drawn from their analysis. Further work by Awramik (1991, 1992) showed overall diversity trends (Figure 11*c*) comparable to those reported by Walter & Heys (1985).

The most comprehensive (and recent) attempt to interpret stromatolite diversity through time is that of Semikhatov & Raaben (Figure 11*d*; Semikhatov & Raaben 1993, 1994, 1996; Raaben & Semikhatov 1996). In this work, the

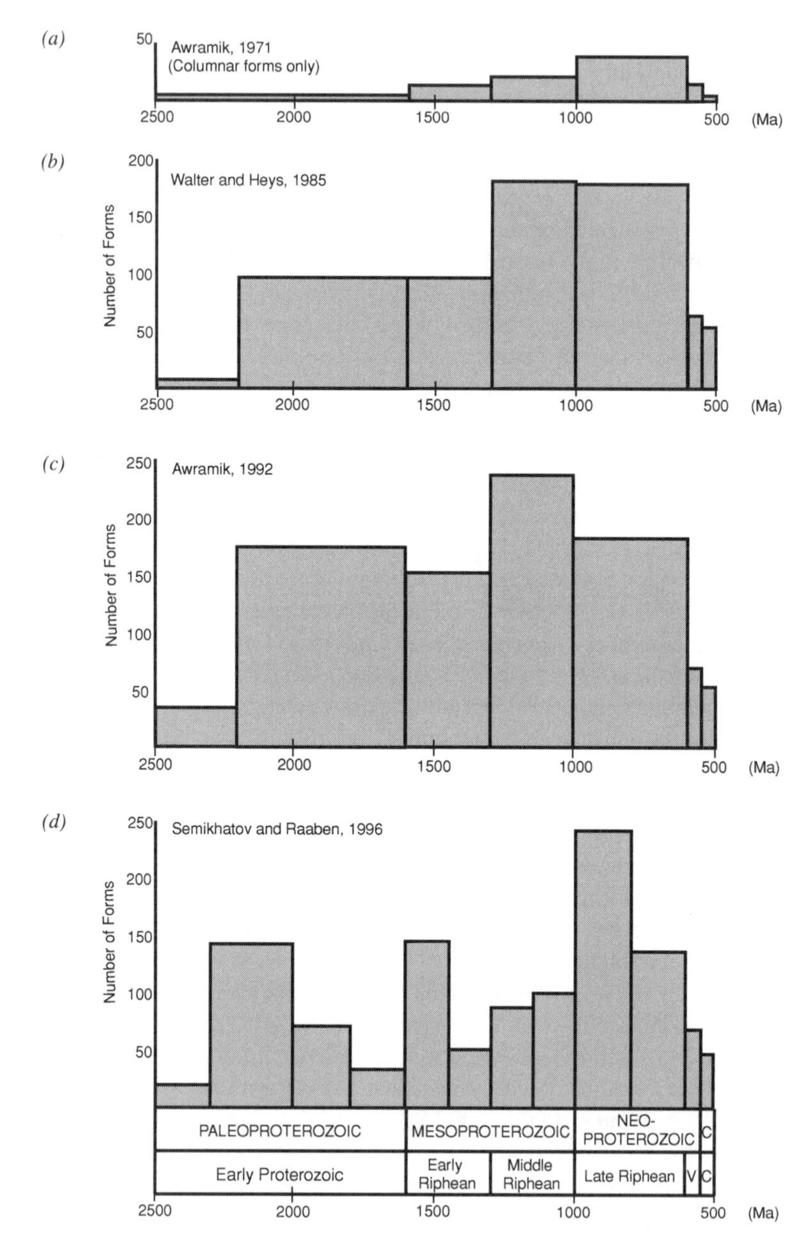

Figure 11 Four attempts to characterize the diversity history of stromatolites. Stratigraphic divisions reflect standard international usage (*upper bar*) and Russian stratigraphic scheme used in published compilations (*lower bar*). *V*, Vendian; *C*, Cambrian.

stratigraphic intervals used in previous studies were subdivided, and data were presented for both individual geographic regions and all major morphological classes of stromatolites. The effects of carbonate distribution are clearly seen in comparisons among continents, as are the influences of taxonomic practice (the three prominent peaks in their global diversity compilation correspond to peaks in the histograms presented for China). With this in mind, it is interesting to note that Raaben & Semikhatov (1996) support the hypothesis that stromatolite diversity began its decline within the early Neoproterozoic Era. Further, Semikhatov & Raaben (1996) confirmed earlier conclusions (e.g. Grey & Thorne 1985, Grotzinger 1989) that the microdigitate stromatolites considered by most workers to be pure precipitate structures are abundant in Paleoproterozoic carbonates, but rare thereafter. While acknowledging the potential role of animals in stromatolite decline, Semikhatov & Raaben stressed the roles of algal evolution and changing ocean chemistry.

Taken together these studies suggest that stromatolites are prominent features of carbonates throughout the Proterozoic, but are less common in later Neoproterozoic successions than earlier and much less common after the initial radiation of macroscopic animals. The initial decline in stromatolite abundance correlates in time with diversification of seaweeds capable of outcompeting microbial mats for space on the sea floor (Fischer 1965, Knoll & Swett 1990, Monty 1973). As well, it corresponds to a time of change in the chemistry of seawater and carbonate production (Grotzinger 1990, Semikhatov & Raaben 1996). The end-Proterozoic decline in stromatolites is accepted by most authors as a reflection of grazing and competition for space by metazoans (Awramik 1971, Garrett 1970, Walter & Heys 1985).

Further interpretation of these diagrams requires that we know what a stromatolite taxon is. Most biologists accept that species are discrete evolutionary units, with genera being more arbitrary groupings of closely related species. Species and genera are both components of a larger taxonomic hierarchy justified by evolutionary process. The information content of a stromatolitic form or group is less clear. It certainly denotes a shape or range in shape and may also be associated with a particular microstructure or limited range of petrographic textures; however, unlike the taxonomic hierarchy in biology, this classification scheme for stromatolites is not (or, at least, not yet) underpinned by process models that inform morphological interpretation.

There is broad agreement that stromatolite morphology reflects influences of environment as well as biology. Thus, unless we know what component of morphology is contributed by specific microbial associations, it is impossible to equate diversity of form with biological diversity of mat builders. Lamination texture is more commonly interpreted as a direct reflection of constituent mat builders, but as Knoll & Semikhatov (1998) have demonstrated, a single mat community can give rise to several distinct lamination textures depending on the relative timing of organic decay and carbonate precipitation. Textures in many stromatolites are principally the products of carbonate precipitation and diagenesis, obscuring mat community influence. Thus, whether defined by morphology or microstructure, stromatolite names do not provide a quantitatively reliable proxy for microbial diversity.

Analysis of diversity trends in stromatolites was inspired by studies of animal diversity through Phanerozoic time, and continuing research on secular variation in stromatolites might well borrow another approach from invertebrate paleontology. In efforts to move beyond taxon counting, paleontologists (e.g. Foote 1997) have turned to morphospace studies in which the shapes of fossils are quantified and displayed within a multidimensional space defined morphometrically. Such an approach has been slow to catch on in studies of stromatolites, with notable exceptions (Banerjee & Chopra 1986; Hofmann 1976, 1994; Zhang & Hofmann 1982). However, a weakness of some such studies is that the developmental (i.e., process) basis for interpreting position within the morphological space is unknown. Fortunately, morphospace studies of stromatolite form can potentially be integrated with process models (Grotzinger & Rothman 1996), enabling research to move beyond debate over taxa and their meaning. Given the possibility of defining and measuring dozens of attributes (Hofmann 1994), it is essential to identify which of these might reflect variability in the fundamental parameters controlling growth and, thus, form (Grotzinger & Rothman 1996; K Chan & JP Grotzinger, unpublished results).

Diversity of Microbes Versus Diversity of Stromatolites

Secular variation in stromatolites has commonly been interpreted as a reflection of evolution within constituent mat communities. In part this stems from analogies to the Phanerozoic record, where biological evolution is a principal vector in the changing composition of sedimentary rocks. In part, evolutionary inferences also reflect the observation that the mat communities observed along environmental gradients in modern environments, like Shark Bay, Australia (Golubic 1991), vary systematically in both taxonomic composition and texture. The inference is that since microstructure relates closely to taxonomic composition, stratigraphic changes in microstructure relate to evolutionary changes in mat communities. Logically, however, this inference is questionable. Observations of modern systems show that mat communities and microstructures vary as a function of environment, so stratigraphic changes could equally well reflect changes through time of some environmental variable, especially the chemistry of carbonate precipitation and diagenesis.

We can approach this question another way. Given that Proterozoic cherts and shales preserve a good record of ancient biology, do observed changes in stromatolite form or microstructure correlate with independently observed changes in biological diversity? This exercise is straightforward in principle—less so in practice. First of all, the most conspicuous changes in Proterozoic biological diversity are seen in nonstromatolitic fossils. Acritarchs (single-celled protists) and seaweeds document a major radiation of eukaryotic organisms beginning 1200–1000 Ma (Knoll 1992). It is conceivable that the influx of eukaryotes into mat communities might have affected stromatolite morphology, microstructure, or both. However, as better data become available, the stratigraphic correspondence between paleontologically inferred evolutionary radiations and changes in stromatolite morphology and microstructure is weakening (e.g. Xiao et al 1997).

Cyanobacteria have a rich fossil record that extends throughout the Proterozoic, and heroic attempts have been made to track their diversity through time (Schopf 1992). It is clear, however, that compiled diversity strongly reflects sampling and provides only limited insight into the details of cyanobacterial evolution. What we do know is that by the time that abundant and widespread cyanobacteria are first found in the fossil record (circa 2100 Ma), all major clades within this group had already diverged from one another (Golubic et al 1995). The second observation of relevance is that Proterozoic cyanobacteria that are morphologically and developmentally distinctive tend to have paleoenvironmental distributions closely approximating those of their modern equivalents (Knoll & Golubic 1992). This suggests that Proterozoic cyanobacteria were physiologically as well as morphologically comparable to their living descendants. In light of this evolutionary conservatism it is notable that silicified cyanobacterial assemblages actually change in composition near the Mesoproterozoic-Neoproterozic boundary (Kah & Knoll 1996, Knoll & Sergeev 1995). They do so, however, because of a change in the nature of carbonate substrates in coastal marine environments (Kah & Knoll 1996). Thus, even the observation that cyanobacterial representation in the fossil record changes through time can be traced to environmental rather than evolutionary causes. In any event, the cyanobacteria best represented in Proterozoic cherts come from peritidal laminites and not the domal, conoform, and branching columnar stromatolites that contribute to diversity curves.

In summary, any relationship between microbial evolution and stromatolite distributions in the Proterozoic record is indirect and difficult or impossible to substantiate based on known fossils. To the extent that evolution played an important role in influencing the Proterozoic stromatolite record, it may have done so through the effects of seaweeds and, later, animals in restricting the environmental distribution of stromatolite-building mat communities (Knoll & Semikhatov 1998). Certainly, the initial Neoproterozoic decline in stromatolites corresponds in time with the independently observed radiation of seaweeds

(Knoll 1994, Knoll & Golubic 1992). A role for micrometazoans in this initial decline is more speculative, but the Ediacaran and Cambrian radiations of large animals undoubtedly contributed to the further, terminal Proterozoic-Cambrian decline of stromatolites (Awramik 1971). Additional support for the role of algae and animals in limiting stromatolite-building mat communities to restricted coastal environments comes from carbonates deposited in the immediate aftermath of mass extinctions. When algal and animal biomass is low, stromatolites show a transient reprise in abundance and environmental distribution (Schubert & Bottjer 1992).

One potential example of microbialite response to evolutionary events within the microbenthos is provided by thrombolites—microbialites characterized by a clotted rather than laminated microstructure (Kennard & James 1986). Based on the observed link between clotted textures in modern Bahamian microbialites and green algal microbenthos, it can reasonably be hypothesized that the latest Proterozoic expansion of thrombolites reflects aspects of chrolophyte evolution (Feldmann & McKenzie 1998).

Stromatolite Diversity as a Record of Environmental Change

If the diverse patterns of form and lamination texture in Proterozoic laminates are not a record of microbial evolution, what do they signify? As discussed, accretion of sediments and/or cement crusts to form primary stromatolitic lamination results from biological, chemical, and physical processes at the sediment-water interface. However, the lamination expressed in ancient stromatolites also reflects the postdepositional processes of mat degradation and diagenetic recrystallization of metastable mineral phases. To a lesser or greater degree, primary lamination is variably modified to form a secondary lamination, and the range of textures observed in ancient stromatolites must surely reflect this spectrum. Recent research strongly suggests that stromatolite diversity, as recorded in the variability of these lamination textures, reflects closure of a primary facies and early diagenetic/taphonomic window at some point during Proterozoic time. For example, the radial fibrous textures of microdigitate stromatolites likely reflect the former presence of fibrous cement crusts (Grotzinger 1986a, Grotzinger & Read 1983, Hofmann & Jackson 1987); so-called "tussocky" texture is similarly reinterpreted as crusts of former acicular marine cements (Fairchild et al 1990; but see Bertrand-Sarfati & Pentecost 1992). It has been shown recently that several important stromatolite textures reflect variable degrees of mat degradation—the degree of preservation is closely correlated with the degree of early lithification—the sooner, the better (Bartley et al 1999, Kah & Knoll 1996, Knoll & Semikhatov 1998, Sumner 1997). If mats are not preserved, as is likely the case for most younger Proterozoic and Phanerozoic stromatolites, then lamination textures should be significantly more limited in their diversity.

It is likely that the literature of stromatolitic microfabrics seriously underestimates the importance of precipitation in the formation of stromatolites. For example, Semikhatov et al (1979) state, "In interpreting microstructures it is, of course, essential that only recognizably primary or penecontemporaneous ones be utilized. Superimposed microstructures of later diagenetic and metamorphic origin, commonly revealed as acicular or radiating microcrystallites or other doubtfully primary micromorphologies, must be excluded." However, as is argued here, textures that were once thought be associated with recrystallization are now regarded as palimpsest relicts of primary textures, particularly acicular or radiating microcrystallites. Studies of early silicified textures have been extremely important in demonstrating the precipitated origin of certain stromatolites (Bartley et al 1999, Hofmann & Jackson 1987, Kah & Knoll 1996), but, because these are few in number, interpreting the vast majority of stromatolite lamination textures will come from analysis of carbonate recrystallization fabrics (e.g. Fairchild et al 1990, Grotzinger & Knoll 1995a, Grotzinger & Read 1983, Knoll & Semikhatov 1998, Sumner 1997). Future investigations of Proterozoic stromatolites will have to consider the wide variety of possible fabrics, and their origins, that are associated with neomorphically recrystallized marine sediments and cements.

The role of in situ precipitation in the development of stromatolitic lamination is interpreted here to be a time-dependent process, correlated with changes in the carbonate chemistry of seawater. The transition through time is ultimately interpreted to be partially responsible for the decline of Proterozoic stromatolites as a result of their reduced capacity to accrete sediment. According to previous interpretations, saturation is proposed to have been highest in the early Proterozoic, decreasing through the middle and late Proterozoic (Grotzinger 1989, 1994; Grotzinger & Kasting 1993; Sumner & Grotzinger 1996b). Late Proterozoic levels are interpreted to be somewhat higher than Phanerozoic levels (Fairchild 1991, 1993; Knoll & Swett 1990). An obvious implication of this model is that the decline in diversity of Proterozoic stromatolite textures could, in part, be related to a global reduction in carbonate saturation through time (Grotzinger 1990). It is revealing that the microdigitate stromatolites (tufas) decline at the end of the early Proterozoic (Grey & Thorne 1985, Grotzinger 1989), followed by the conoform stromatolites and *Omachtenia*-like forms, which decline at the end of the middle Proterozoic, before the decline of most other stromatolite taxa (Walter & Heys 1985). Of all the stromatolite groups, the mechanisms for growth of these three can be most obviously related to direct precipitation (Grotzinger 1986a, Grotzinger 1990, Knoll & Semikhatov 1998). Whereas the microdigitate stromatolites formed on tidal flats, the conoform stromatolites are of subtidal origin. As carbonate saturation of seawater decreased through the Proterozoic, it can be expected that subtidal environments would have remained more productive, where a constant supply of calcium and bicarbonate was available.

Toward the end of Mesoproterozoic time, the abundance of textures associated with primary sea-floor encrustation and penecontemporaneous lithification of fossil mats declines sharply and so does the diversity of stromatolites. To a first-order approximation, this event signals the end of a period in Earth history when diverse stromatolite lamination textures could be developed owing to extremely early lithification, in many cases directly on the seafloor itself (Grotzinger 1990, Kah & Knoll 1996, Knoll & Semikhatov 1998). It is important that the decline of these textures also implies the loss of a powerful accretion mechanism for stromatolite growth, in that precipitation on the seafloor is extremely efficient (Grotzinger 1990). As discussed above, precipitation of carbonate on the sea-floor contributes directly to increasing local surface roughness and also accelerates the propagation of instabilities that may result in development of branching. In contrast, sedimentation either from suspension or by traction is a smoothing process that will contribute toward damping of small-scale surface irregularities. Consequently, branching stromatolites are predicted to occur more abundantly in regimes in which growth via sea-floor encrustation can occur. Thus, a decrease in stromatolite diversity, as measured either by stromatolite form or texture, is predicted if the time scale for early lithification is increased.

CONCLUSIONS

The Phanerozoic perspective on stromatolites is that they are unusual sedimentary features, commonly indicative of restricted environments or mass extinction. Rocks deposited during the first 85 percent of Earth history tell a different story, one in which stromatolites are the principal features of platform and shelf carbonates. We know that stromatolites generally reflect a spectrum of interactions among microbial-mat communities, sedimentation, and carbonate precipitation, but we remain in need of models and experiments that will enable us to deconvolve the sedimentary signals encrypted by each contributing process.

The common wisdom is that marine stromatolites result from the trapping and binding of micrite and calcisiltite by microbial-mat communities, and this does indeed provide the best explanation for a range of stromatolites seen in later Neoproterozoic successions. As we go back further in time, however, precipitated carbonates comprise a larger and larger proportion of the record, and the relationship between mat biology and lamination and microstructure becomes more difficult to interpret with confidence. At the extreme, in Archean rocks and (potentially) in sediments on Mars or other planetary bodies, the role, if any, played by biology can be difficult to ascertain. At no time in the last 3.5 Ga has the Earth's surface been sterile, so all stromatolitic structures surely accreted in the presence of biology. The question, then, is not whether organisms were in site as stromatolites accreted, but what roles they played in development and how those roles can be understood from preserved morphology and microstructure. On a sterile Earth, carbonates would still be removed from seawater, and they would likely form precipitated laminates in coastal environments.

The conclusion that biology played a role in the accretion of most stromatolites does not equate to a statement that secular changes in stromatolite form or microstructure reflect changes in the mat-building biota. Testable hypotheses about the role of evolution in driving stromatolite stratigrahy require that one must articulate the features of stromatolites that most directly reflect biology, and explain how evolution can account for observed changes through time. To date, this has not been accomplished. Whether longstanding interpretations of stromatolites as "evolutionary mileposts" can be sustained will require quantitative studies of stromatolite morphologies, to search for forms that can be shown to be uniquely biologic, coupled with detailed analyses of microfabric in which the influences of precipitation and diagenesis have been removed.

In contrast, we can be relatively confident in our assessment of how environmental change has contributed to the stratigraphic distribution of stromatolitic forms and textures. Thus, a promising avenue for continued research lies in the use of Proterozoic stromatolites as "environmental dipsticks"—as sensitive proxies for the evolution of seawater. Changing ocean chemistry undoubtedly contributed to the observed stromatolite record. The outstanding question is whether environmental change can account for all of the stratigraphic variation observed by geologists.

Whatever the outcome of current debates, it is clear that, insofar as stromatolites represent a conspicuous sedimentary manifestation of interactions among physical and biological processes, they will remain principal foci of research in the emerging discipline of geobiology.

ACKNOWLEDGMENTS

We are indebted to Kelvin Chan, Mike Pope, Dan Rothman, Misha Semikhatov, and Malcolm Walter for many helpful discussions and access to unpublished data. This research was supported by NASA Grant NAG5-6722 to JPG and by NASA Grant NAGW-783 to AHK. This paper is a contribution to the NASA Astrobiology program.

Literature Cited

Awramik SM. 1971. Precambrian columnar stromatolite diversity: reflection of metazoan appearance. *Science* 174:825–27

Awramik SM. 1991. Archean and Proterozoic stromatolites. In *Fossil Algae and Stromatolites*, ed. R Riding. New York: Springer-Verlag

Awramik SM. 1992. The history and significance of stromatolites. See Schidlowski 1992, pp. 435–49

Banerjee DM, Chopra J. 1986. Morphometric analysis of Proterozoic stromatolites from India—preliminary report on testing of a new technique. *Precambrian Res.* 33:265–82

Barabasi AL, Stanley HE. 1995. *Fractal Concepts in Surface Growth.* New York: Cambridge Univ. Press. 366 pp.

Bartley JK. 1996. Actualistic taphonomy of cyanobacteria: implications for the Precambrian fossil record. *Palaios* 11:571–86

Bartley JK, Knoll AH, Grotzinger JP, Sergeev VN. 1999. Lithification and fabric genesis in precipitated stromatolites and associated peritidal carbonates, Mesoproterozoic Billyakh Group, Siberia. See Grotzinger & James 1999.

Bathurst RGC. 1975. *Carbonate Sediments and Their Diagenesis.* New York: Elsevier Sci. 658 pp.

Ben-Jacob E, Shochet O, Tenenbaum A, Cohen I, Czirok A, Viczek T. 1994. Communicating Walker's model for cooperative patterning of bacterial colonies. *Nature* 368:46–49

Berg HC. 1983. *Random Walks in Biology.* Princeton, NJ: Princeton Univ. Press. 152 pp.

Bertrand-Sarfati J. 1976. An attempt to classify late Precambrian stromatolite microstructures. See Walter 1976b, pp. 251–38

Betrand-Sarfati J, Pentecost A. 1992. Tussocky microstructure, a biological event in upper Proterozoic stromatolites; comparisons with modern freshwater stromatolite builders. See Schidlowski 1992, pp. 468–77

Black M. 1933. The algal sediments of Andros Island, Bahamas. *Philos. Trans. R. Soc. London Ser. B* 222:165–92

Bradley WH. 1929. *Algae reefs and oolites of the Green River Formation. US Geol. Survey Prof. Pap.* 154:225–66.

Broecker WS, Peng TH. 1982. *Tracers in the Sea.* New York City: Lamont-Doherty Geophys. Observ. 690 pp.

Broecker WS, Takahashi T. 1978. The relationship between lysocline depth and in situ carbonate ion concentration. *Deep Sea Res.* 25:65–95

Cady SL, Farmer JD. 1996. Fossilization processes in siliceous thermal springs: trends in preservation along thermal gradients. In *Evolution of Hydrothermal Ecosystems on Earth (and Mars?)*, ed. MR Walter, pp. 150–73. New York: Wiley

Canfield DE, Des Marais DJ. 1993. Biogeochemical cycles of carbon, sulfur and free oxygen in a microbial mat. *Geochim. Cosmochim. Acta* 57:3971–84

Canfield DE, Raiswell R. 1991. Carbonate precipitation and dissolution: its relevance to fossil preservation. In *Taphonomy: Releasing the Data Locked in the Fossil Record*, ed. PA Allison, DEG Briggs, pp. 411–53. New York: Plenum

Chafetz HS, Buczynski C. 1992. Bacterially induced lithification of microbial mats. *Palaios* 7:277–93

Constanz BR. 1986. The primary surface area of corals and variations in their susceptibility to diagenesis. See Schroeder & Purser 1986, pp. 53–76

D'Amelio ED, Cohen Y, Des Marais DJ. 1989. Comparative functional ultrastructure of two hypersaline submerged cyanobacterial mats: Guerroro Negro, Baja California Sur, Mexico, and Solar Lake, Sinai, Egypt. In *Microbial Mats, Physiological Ecology of Benthic Microbial Communities*, ed. Y Cohen, E Rosenberg, pp. 97–113. Washington, DC: Am. Soc. Microbiol.

Des Marais DJ. 1995. The biogeochemistry of hypersaline microbial mats. *Adv. Microb. Ecol.* 14:251–74

Dill RF, Shinn EA, Jones AT, Kelly K, Steinen RP. 1986. Giant subtidal stromatolites forming in normal salinity waters. *Nature* 324:55–58

Doemel WN, Brock TD. 1977. Structure, growth and decomposition of laminated algal-bacterial mats in alkaline hot springs. *Appl. Environ. Microbiol.* 34:433–52

Donaldson JA. 1976. Paleoecology of Conophyton and associated stromatolites in the Precambrian Dismal Lakes and Rae Groups, Canada. See Walter 1976b, pp. 523–34

Dravis JJ. 1983. Hardened subtidal stromatolites, Bahamas. *Science* 219:385–86

Eardley AJ. 1938. Sediments of Great Salt Lake, Utah. *Am. Assoc. Pet. Geol. Bull.* 22:1305–1411

Fairchild IJ. 1991. Origins of carbonate in Neoproterozoic stromatolites and the identification of modern analogues. *Precambrian Res.* 53:281–99
Fairchild IJ. 1993. Balmy shores and icy wastes: the paradox of carbonates associated with glacial deposits in Neoproterozoic times. *Sedimentol. Rev.* 1:1–15
Fairchild IJ, Marshall JD, Bertrand-Sarfati J. 1990. Stratigraphic shifts in carbon isotopes from Proterozoic stromatolitic carbonates (Mauritania): influence of primary mineralogy and diagenesis. *Am. J. Sci.* 290A:46–79
Feldmann M, McKenzie J. 1998. Stromatolite-thrombolite associations in a modern environment, Lee Stocking Island, Bahamas. *Palaios* 13:201–12
Feldmann M, McKenzie JA. 1997. Messinian stromatolite-thrombolite associations, Santa Pola, Spain: an analogue for the Paleozoic? *Sedimentology* 44:893–914
Fischer AG. 1965. Fossils, early life, and atmospheric history. *Proc. Natl. Acad. Sci. USA* 53:1205–15
Foote M. 1997. The evolution of morphological diversity. *Annu. Rev. Ecol. Syst.* 28:129–52
Garrett P. 1970. Phanerozoic stromatolites; noncompetitive ecologic restriction by grazing and burrowing animals. *Science* 169:171–73
Gebelein CD. 1974. Biologic control of stromatolite microstructure: implications for Precambrian time stratigraphy. *Am. J. Sci.* 274:575–98
Ginsburg RN. 1991. Controversies about stromatolites: vices and virtues. In *Controversies in Modern Geology*, pp. 25–36. New York: Academic Press
Ginsburg RN, Lowenstam HA. 1958. The influence of marine bottom communities on the depositional environment of sediments. *J. Geol.* 66:310–18
Golubic S. 1991. Modern stromatolites—a review. In *Calcareous Algae and Stromatolites*, ed. R. Riding, pp. 541–61. New York: Springer-Verlag
Golubic S, Hofmann HJ. 1976. Comparison of modern and mid-Precambrian Entophysalidaceae (Cyanophyta) in stromatolitic algal mats: cell division and degradation. *J. Paleontol.* 50:1074–82
Golubic S, Sergeev VN, Knoll AH. 1995. Mesoproterozoic akinetes (Archaeoellipsoides): fossil evidence of heterocystous cyanobacteria. *Lethaia* 28:285–98
Grey K. 1984. Biostratigraphic studies of stromatolites from the Proterozoic Earaheedy Group, Nabberu Basin, Western Australia. *Geol. Surv. West. Aust. Bull.*
Grey K, Thorne AM. 1985. Biostratigraphic significance of stromatolites in upward shallowing sequences of the early Proterozoic Duck Creek Dolomite, Western Australia. *Precambrian Res.* 29:183–206
Grotzinger JP. 1986a. Cyclicity and paleoenvironmental dynamics, Rocknest platform, northwest Canada. *Geol. Soc. Am. Bull.* 97:1208–31
Grotzinger JP. 1986b. Evolution of early Proterozoic passive-margin carbonate platform: Rocknest Formation, Wopmay Orogen, N.W.T., Canada. *J. Sediment Petrol.* 56:831–47
Grotzinger JP. 1989. Facies and evolution of Precambrian carbonate depositional systems: emergence of the modern platform archetype. In *Controls on Carbonate Platform and Basin Development*, ed. PD Crevello, JL Wilson, JF Sarg, JF Read, Soc. Econ. Paleontol. Mineral. Spec. Publ. 44:79–106.
Grotzinger JP. 1990. Geochemical model for Proterozoic stromatolite decline. *Am. J. Sci.* 290A:80–103
Grotzinger JP. 1994. Trends in Precambrian carbonate sediments and their implication for understanding evolution. In *Early Life on Earth*, ed. S Bengtson, pp. 245–58. New York: Columbia Univ. Press
Grotzinger JP, James NP. 1999. *Precambrian Carbonates. Soc. Econ. Paleontol. Mineral. Spec. Publ.*
Grotzinger JP, Kasting JF. 1993. New constraints on Precambrian ocean composition. *J. Geol.* 101:235–43
Grotzinger JP, Knoll AH. 1995. Anomalous carbonate precipitates: Is the Precambrian the key to the Permian? *Palaios* 10:578–96
Grotzinger JP, Read JF. 1983. Evidence for primary aragonite precipitation, lower Proterozoic (1.9 Ga) dolomite, Wopmay orogen, northwest Canada. *Geology* 11:710–13
Grotzinger JP, Rothman DR. 1996. An abiotic model for stromatolite morphogenesis. *Nature* 383:423–25
Hagadorn JW, Bottjer DJ. 1997. Microbial carpeting of ancient siliciclastic seafloors: aid to preservation of early metazoan biosedimentary structures. *Geol. Soc. Am. Abstr. Prog.* 29:A193
Hayes JM. 1983. Geochemical evidence bearing on the origin of aerobiosis, a speculative hypothesis. In *Earth's Earliest Biosphere*, ed. JW Schopf, pp. 291–301. New York: Princeton Univ. Press
Hoffman PF. 1973. Recent and ancient algal stromatolites: seventy years of pedagogic cross-pollination. In *Evolving Concepts in Sedimentology*, ed. RN Ginsburg, pp. 178–91. Baltimore: Johns Hopkins Univ. Press
Hoffman PF. 1974. Shallow and deep-water stromatolites in lower Proterozoic platform-to-basin facies change, Great Slave Lake,

Canada. *Am. Assoc. Pet. Geol. Bull.* 58:856–67

Hoffman PF. 1976. Environmental diversity of Precambrian stromatolites. See Walter 1976b, pp. 599–612

Hofmann HJ. 1976. Stromatoid morphometrics. See Walter 1976b, pp. 45–54

Hofmann HJ. 1987. Precambrian biostratigraphy. *Geosci. Can.* 14:135–54

Hofmann HJ. 1994. Quantitative stromatolithology. *J. Paleontol.* 68:704–9

Hofmann HJ, Jackson GD. 1987. Proterozoic ministromatolites with radial-fibrous fabric. *Sedimentology* 34:963–71

Horodyski RJ. 1975. Stromatolites of the Lower Missoula Group (Middle Proterozoic), Belt Supergroup, Glacier National Park, Montana. *Precambrian Res.* 2:215–54

Jorgensen BB, Des Marais DJ. 1988. Optical properties of benthic photosynthetic communities: fiber optic studies of cyanobacterial mats. *Limnol. Oceanogr.* 33:99–113

Jorgensen BB, Des Marais DJ. 1990. The diffusive boundary layer of sediments: oxygen microgradients over a mat. *Limnol. Oceanogr.* 35:1343–55

Kah LC. 1997. *Sedimentological, geochemical, and paleobiological interactions on a Mesoproterozoic carbonate platform: Society Cliffs formation, Northern Baffin Island, Arctic Canda.* PhD thesis. Cambridge: Harvard Univ.

Kah LC, Knoll AH. 1996. Microbenthic distribution of Proterozoic tidal flats: environmental and taphonomic considerations. *Geology* 24:79–82

Kalkowsky VHE. 1908. Oolith and stromatolith im norddeutschen Buntsandstein. *Z. Dtsch. Geol. Ges.* 60:68–125

Kardar M, Parisi G, Zhang Y. 1986. Dynamic scaling of growing interfaces. *Phys. Rev. Lett.* 56:456–72

Kendall AC. 1977. Fascicular-optic calcite: a replacement of bundled acicular carbonate cements. *J. Sediment. Petrol.* 47:1056–62

Kendall AC. 1985. Radiaxial fibrous calcite: a reappraisal. In *Carbonate Cements.*, ed. N Schneidermann, PM Harris. Soc. Econ. Paleontol. Mineral. Spec. Publ. 36:59–78

Kendall AC, Tucker ME. 1973. Radiaxial fibrous calcite: a replacement after acicular carbonate. *Sedimentology* 20:365–89

Kennard JM, James NP. 1986. Thrombolites and stromatolites: two distinct types of microbial structures. *Palaios* 1:492–503

Knoll AH. 1992. The early evolution of eukaryotic organisms: a geological perspective. *Science* 256:622–27

Knoll AH. 1994. Proterozoic and Early Cambrian protists: evidence for accelerating evolutionary tempo. *Proc. Natl. Acad. Sci. USA* 91:6743–50

Knoll AH, Fairchild IJ, Swett K. 1993. Calcified microbes in Neoproterozoic carbonates: implications for our understanding of the Proterozoic/Cambrian transition. *Palaios* 8:512–25

Knoll AH, Golubic S. 1992. Proterozoic and living stromatolites. See Schidlowski 1992, pp. 450–62

Knoll AH, Semikhatov MA. 1998. The genesis and time distribution of two distinctive Proterozoic stromatolitic microstructures. *Palaios* 13:407–21

Knoll AH, Sergeev VN. 1995. Taphonomic and evolutionary changes across the Mesoproterozoic-Neoproterozoic boundary. Neues Jahrb. *Geol. Paleontol. Abh.* 195:289–302

Knoll AH, Swett K. 1990. Carbonate deposition during the later Proterozoic Era: an example form Spitsbergen. *Am. J. Sci.* 290A:104–32

Krumbein WE. 1983. Stromatolites: the challenge of a term in space and time. *Precambrian Res.* 20:493–531

Krumbein WE, Paterson DM, Stal LJ. 1994. Biostabilization of Sediments. Oldenburg, Ger: Bibl. Inf. Syst. Univerität Oldenburg. 526 pp.

Li TH, Takahashi T, Broecker WS. 1969. The degree of saturation of $CaCO_3$ in the oceans. *J. Geophys. Res.* 74:5507–25

Logan BW. 1961. Cryptozoan and associated stromatolites from the Recent, Shark Bay, Western Australia. *J. Geol.* 69:517–33

Logan BW, Hoffman PF, Gebelein CD. 1974. Algal mats, cryptalgal fabrics, and structures, Hamelin Pool, Western Australia. In *Evolution and Diagenesis of Quaternary Carbonate Sequences, Shark Bay, Western Australia*, ed. BW Logan, JF Read, GM Hagan. Am. Assoc. Petrol. Geol., Mem. 22:140–94

Macintyre IG, Reid RP, Steneck RS. 1996. Growth history of stromatolites in a Holocene fringing reef, Stocking Island, Bahamas. *J. Sediment. Res.* 66:231–42

Matsushita M, Fujikawa H. 1990. Diffusion-limited growth in bacterial colony formation. *Physica A* 168:498–506

Matsushita M, Hayakawa Y, Sawada Y. 1985. Fractal structure and cluster statistics of zinc-metal trees deposited on a line electrode. *Phys. Rev. A* 32:3814–16

Monty CLV. 1967. Distribution and structure of Recent stromatolitic algal mats. *Ann. Soc. Geol. Belg.* 90:55–100

Monty CLV. 1973. Precambrian background and Phanerozoic history of stromatolitic communities, an overview. *Ann. Soc. Geol. Belg.* 96:585–624

Monty CLV. 1976. The origin and development

of cryptalgal fabrics. See Walter 1976b, pp. 193–249
Parker BC, Simmons GM Jr, Love G, Wharton RA Jr, Seaburg KG. 1981. Modern stromatolites in Antarctic Dry Valley lakes. *BioScience* 31:656–61
Pflueger F. 1997. Matground structures: pinpoint facies indicators in inert sandstones. *Geol. Soc. Am. Abstr. Prog.* 29:A192
Pope M, Grotzinger J, Schreiber BC. 1999. Evaporitic subtidal stromatolites produced by in situ precipitation: textures, facies associations, and temporal significance. *Sedimentology* (In press)
Pope M, Grotzinger JP. 1999. Controls on fabric development and morphology of tufas and stromatolites, uppermost Pethei Group (1.8 Ga), Great Slave Lake, northwest Canada. See Grotzinger & James 1999.
Raaben ME, Semikhatov MA. 1996. Dinamika globalnogo raznoobraziya nadrodobich gryppirovok stromatolitov v proterozoye. *Dokl. Akad. Nauk.* 349:234–38 (in Russian)
Revsbech NP, Jorgensen BB, Blackburn TH. 1983. Microelectrode studies of the photosynthesis and O_2, H_2S and pH profiles of a microbial mat. *Limnol. Oceanogr.* 28:1062–74
Riding R. 1982. Cyanophyte calcification and changes in ocean chemistry. *Nature* 299:814–15
Ronov AB. 1968. Probable changes in the composition of sea water during the course of geologic time. *Sedimentology* 10:25–43
Sadler PM. 1981. Sediment accumulation rates and the completeness of stratigraphic sections. *J. Geol.* 89:569–84
Sami TT, James NP. 1996. Synsedimentary cements as platform building blocks, Paleoproterozoic Pethei Group, northwestern Canada. *J. Sediment. Res.* 66:209–22
Sandberg PA. 1985. Nonskeletal aragonite and pCO_2 in the Phanerozoic and Proterozoic. In *The Carbon Cycle and Atmospheric CO_2: Natural Variations Archean to Present*, ed. ET Sundquist, WS Broecker, p. 585–94. Washington, DC: Am. Geophys. Union
Schidlowski M, ed. 1992. *Early Organic Evolution: Implications for Mineral and Energy Resources.* New York: Springer-Verlag
Schieber J. 1986. The possible role of benthic microbial mats during the formation of carbonaceous shales in shallow Mid-Proterozoic shales. *Sedimentology* 33:521–36
Schopf JW. 1992. Paleobiology of the Archean. See Schopf & Klein 1992, p. 25–39
Schopf JW, Klein C, eds. 1992. *The Proterozoic Biosphere.* New York: Cambridge Univ. Press
Schopf JW, Packer BM. 1987. Early Archean (3.3-billion to 3.5-billion-year-old) microfossils from Warrawoona Group, Australia. *Science* 237:70–73
Schroeder JH, Purser BH. 1986. *Reef Diagenesis.* New York: Springer-Verlag. 455 pp.
Schubert JK, Bottjer DJ. 1992. Early Triassic stromatolites as post-extinction disaster forms. *Geology* 20:883–86
Sears SO, Lucia FJ. 1980. Dolomitization of northern Michigan Niagara reefs by brine refluxion and freshwater/seawater mixing. In *Concepts and Models of Dolomitization*, ed. DH Zenger, JB Dunham, RL Ethington. *Soc. Econ. Paleontol. Mineral. Spec. Publ.* 28:215–35
Semikhatov MA. 1991. General problems of the Proterozoic stratigraphy in the USSR. *Sov. Sci. Rev.* 1:1–192
Semikhatov MA, Gebelein CD, Cloud P, Awramik SM, Benmore WC. 1979. Stromatolite morphogenesis—progress and problems. *Can. J. Earth Sci.* 19:992–1015
Semikhatov MA, Raaben ME. 1993. Dynamics of the taxonomic diversity of Riphean and vendian stromatolites in northern Eurasia. *Stratigr. Geol. Correl.* 1:133–41
Semikhatov MA, Raaben ME. 1994. Dynamics of the global diversity of Proterozoic stromatolites. Article I: Northern Eurasia, China, and India. *Stratigr. Geol. Correl.* 2:492–513
Semikhatov MA, Raaben ME. 1996. Dynamics of the global diversity of Proterozoic stromatolites. Article II: Africa, Australia, North America, and general synthesis. *Stratigr. Geol. Correl.* 4:24–50
Serebryakov SN, Semikhatov MA. 1974. Riphean and recent stromatolites: a comparison. *Am. J. Sci.* 274:556–74
Strauss H, Des Marais DJ, Hayes JM, Summons RE. 1992. The carbon-isotopic record. See Schopf & Klein 1992, pp. 117–32
Sumner DY. 1997. Late Archean calcite-microbe interactions: two morphologically distinct microbial communities that affected calcite nucleation differently. *Palaios* 12:302–18
Sumner DY, Grotzinger JP. 1996a. Herringbone calcite: petrography and environmental significance. *J. Sediment. Res.* 66:419–29
Sumner DY, Grotzinger JP. 1996b. Were kinetics of Archean calcium carbonate precipitation related to oxygen concentration? *Geology* 24:119–22
Walcott CD. 1914. Pre-Cambrian Algonkian algal flora. *Smithson. Misc. Collect.* 64:74–156
Walter MR. 1976a. Hot-spring sediments in Yellowstone National Park. See Walter 1976b, pp. 489–98
Walter MR, ed. 1976b. *Stromatolites.* New York: Elsevier. 790 pp.
Walter MR, Heys GR. 1985. Links between the

rise of the metazoa and the decline of stromatolites. *Precambrian Res.* 29:149–74

Walter MR, Krylov IN, Muir MD. 1988. Stromatolites from Middle and Late Proterozoic sequences in the McArthur and Georgina Basins and the Mount Isa Province, Australia. *Alcheringa* 12:79–106

Witten TA, Sander LM. 1983. Diffusion-limited aggregation. *Phys. Rev. B* 27:5686–97

Xiao S, Knoll AH, Zhang Y, Yin L. 1997. Neoproterozoic fossils in Mesoproterozoic rocks? Chemostratigraphic resolution of a biostratigraphic conundrum from the North China Platform. *Precambrian Res.* 84:197–220

Zempolich WG, Wilkinson BH, Lohmann KC. 1988. Diagenesis of late Proterozoic carbonates: The Beck Spring Dolomite of eastern California. *J. Sediment. Petrol.* 58:656–72

Zhang Y, Hofmann HJ. 1982. Precambrian stromatolites: image analysis of lamina shape. *J. Geol.* 90:253–68

Annu. Rev. Earth Planet. Sci. 1999. 27:359–84

LINKING THERMAL, HYDROLOGICAL, AND MECHANICAL PROCESSES IN FRACTURED ROCKS[1]

Chin-Fu Tsang
Earth Sciences Division, Ernest Orlando Lawrence Berkeley National Laboratory, University of California, Berkeley, California 94720, CFTsang@lbl.gov

KEY WORDS: fractures, coupled processes, thermomechanical, hydromechanical

ABSTRACT

In this paper, an overview is presented of coupled processes linking thermo-hydro-mechanical (THM) effects in fractured rocks. A formulation is first presented to show the linkage mathematically, which can be used as a basis for numerical solutions and for further developments. Two simple examples of hydromechanical (HM) and thermo-hydro-mechanical (THM) coupled processes are discussed to convey physical insight into such couplings. Finally, three large-scale, long-term experiments currently under way are described. These are being conducted specifically to study coupled processes in situ.

BACKGROUND

During the last ten years or so, intense research has been performed on the processes that link thermal gradients, hydrologic flow, and mechanical deformation in fractured rock. Much of this work was summarized and discussed in a number of review papers and edited books (Tsang 1987, 1991; IJRMMS 1995; Stephansson et al 1996; Haijtink 1995). The main impetus for this intense activity has been the need to evaluate the impacts of these linked or coupled processes on the safety or isolation potential of a nuclear waste repository.

A nuclear waste repository is essentially an excavated underground cavity composed of tunnels and openings in which radioactive waste is emplaced. This waste releases heat, which causes expansion in the rock, possible closure of rock fractures, and disturbance of fluid flow in adjacent rock formations. Often, canisters containing the waste are surrounded by buffer or backfill materials after being emplaced in the repository. Coupling of thermal (T), hydro (H), and mechanical (M) effects also occurs in these materials. In order to give more definite examples of these coupled processes and point out relevant issues, I shall divide the repository problem into two stages: the construction stage, when heat-releasing waste has yet to be emplaced and only coupled HM processes in rock are expected; and the operation and containment stage, when waste, buffer, and backfill materials are present and coupled thermal-hydro-mechanical (THM) processes will occur.

During the first stage, the excavation of the repository causes a major perturbation of the rock mass. The impact depends on the initial stress field around the system, the nature of the excavation method, and the repository design. It is not easy to determine the in situ stress field of a region, especially in the presence of fracture sets forming a network, which could well be anisotropic and may not even follow a regular ellipsoidal angular distribution. The excavation will cause stress concentration around the opening, which in turn changes the local fracture apertures and permeability. Change in aperture and change in permeability of a fracture are functions of normal and shear stresses across the fracture. An increase in normal stress will cause the fracture aperture to close and the permeability will decrease as the square of the aperture. However, the aperture decrease will stop at an "irreducible" aperture after the normal stress reaches a certain high value. Changes in fracture permeability also occur with changes in shear stress; this process is more complex due to the influence of gouging, asperity deformation, and other possible effects. In these cases the permeability can easily change by one order of magnitude or more. In general, the question is how to determine the anisotropic change in hydraulic conductivity around the repository cavity. Excavation also represents a relatively sudden event, and as a result the normal and shear stress across nearby fractures may change in a short time, producing sudden aperture changes. This sudden change in stress may cause the pore pressure to rise quickly, before the water has had a chance to move and equilibrate. Such a transient coupled HM effect may cause local failures and create microfractures, as well as local hydraulic conductivity changes.

During the operation and isolation stage of a nuclear waste repository there is thermal input, giving rise to a series of THM processes. The heat from the nuclear waste will cause a rise in the temperature of the buffer/backfill and the rock, as the repository is being filled. The temperature will peak after a time period estimated to be a few hundred years. Thermally induced stresses are found around the repository, which may change the hydraulic conductivity, as discussed. The TM effects have been relatively well studied and much experience has been gained in its modeling and observation. However, what is interesting here is the progressive heating of the multiple media system, i.e. the waste canister, backfill materials (which could be bentonite), and then the surrounding rock, each of which has a different expansion coefficient. How they deform relative to each other and how the interfaces between them behave may make a difference in the hydraulic properties of these interfaces. In the worst-case scenario, these interfaces may form relatively higher hydraulic-conductivity paths for water flow. The heat will also induce convective flow in rock. Convective flow depends on thermal energy imparted to the fluid and would last much longer than the temperature peak time. In the very near field where the local temperature may be high, vaporization will occur. Vaporized water will move away from the repository and condense in cooler regions of the rock. The result is a complex and dynamic hydrologic system that requires a fully multiphase code for its analysis. How the thermohydraulic process affects stress, and thus the mechanical condition of the fracture-porous rock, is an open question.

MATHEMATICAL FORMULATION OF COUPLED PROCESSES LINKING THM EFFECTS

A number of authors have presented alternative formulations of coupled processes linking THM effects. Many of these approaches are summarized in Jing et al (1995). In this section I present, without development, the governing equations originating from the work of Noorishad and Tsang (1996), which can be consulted for details. The goal is to describe how the linkage of THM effects can be represented mathematically.

For saturated rock (i.e. rock saturated with water without the presence of air), the mathematical equations linking THM effects may be expressed in three governing and two constitutive equations. The first describes the combined mass and momentum balance law for moving fluid,

$$\frac{\rho_\lambda \partial e}{\rho_o \partial t} + \phi\beta_P \frac{\partial P}{\partial t} + \phi\beta_T \frac{\partial T}{\partial t} = \nabla \cdot \left[\frac{\rho_\lambda k_\lambda}{\rho_o \eta_\lambda}(\nabla P + \rho_\lambda g \nabla z)\right] \qquad (1)$$

where:

- ρ_λ is density of fluid in the rock.
- ρ_o is density of reference fluid.
- η_λ is liquid dynamic viscosity.
- k_λ is the local intrinsic permeability tensor.
- ϕ is the rock porosity.
- β_P is compressibility coefficient of the liquid.

β_T is the thermal volume expansion coefficient.
g is the gravity constant.
P is the liquid pressure.
T is the temperature.
e is the rock matrix volumetric strain.
z is the elevation.

The terms on the left side of Equation 1 describe the changes in rock volume due to strain, pressure, and temperature, respectively. The right side gives the liquid flow due to pressure gradient and gravity according to Darcy's Law. The second governing equation is the mechanical equation of motion,

$$\frac{\partial \tau_{ij}}{\partial x_j} + \bar{\rho}_s f_i = \rho \frac{\partial^2 U_i}{\partial t^2} \tag{2}$$

where:

τ_{ij} is the stress tensor.
f_i is the body force.
$\bar{\rho}_s$ is the average specific mass of the rock.
U_i is the displacement.
ρ is the density of the liquid-filled rock.

The third governing equation describes the conservation of energy,

$$(\rho C)_m \frac{\partial T}{\partial t} + T_o \gamma \frac{\partial}{\partial t}(\delta_{ij} e_{ij}) + \rho_\lambda C_{v\lambda} \frac{k_\lambda}{\eta_\lambda}(\nabla P - \rho_\lambda g \nabla z)\nabla T = \nabla \cdot K_m \cdot \nabla T \tag{3}$$

where:

$(\rho C)_m$ is the heat capacity of the liquid-filled medium.
K_m is the thermal conductivity of the liquid-filled medium.
T_o is the absolute temperature in the stress-free state.
δ_{ij} is 1 for $i = j$ and 0 for $i \neq j$.
e_{ij} is strain components of the rock.
$C_{v\lambda}$ is the liquid specific heat constant at constant volume.
γ $= (2\mu + 3\lambda)\beta$ with β being the isotropic linear solid thermal expansion coefficient and μ and λ being Lame's constants.

The first term on the left side of Equation 3 is the change in energy. The second term describes the deformation conversion energy, and the third term represents the convective energy due to heat carried by the moving liquid. The right side gives the temperature conduction in the liquid-filled rock medium.

The following two constitutive equations complete the mathematical description of the coupled processes linking T, H, and M,

$$\zeta = \frac{\rho_\lambda}{\rho_o} \times \alpha\delta_{ij}e_{ij} + \frac{1}{M}P + \frac{1}{M_T}T \tag{4}$$

$$\tau_{ij} = 2\mu e_{ij} + \lambda\delta_{ij}\delta_{kl}e_{kl} - \gamma\delta_{ij}T - \alpha\delta_{ij}P \tag{5}$$

where:

ζ is equivalent liquid production.
α is Biot's parameter.
M $= 1/\phi\beta_P$.
M_T $= 1/\phi\beta_T$.

Equation 4 describes the change in equivalent liquid strain due to rock strain, liquid pressure, and temperature represented by the three terms, respectively, on the right side. Equation 5 gives the stress as dependent on the strain tensor, represented by the first two terms on the right side, and the temperature and pressure as described by the third and fourth terms, respectively, on the right side.

Rock typically contains fractures that are cracks with two rough surfaces. These fractures may be filled partially or fully with clay or other materials. The constitutive models for rock fractures are summarized and reviewed by Ohnishi et al (1996). For the current purpose of representing a physical behavior, I describe the THM behavior of a fracture with its normal in the third direction by the following equations, in close analogy to equations for solid rock:

$$\tau_i = C_{ij}U_j + \alpha\delta_{i3}P \qquad i, j = 1, 3 \tag{6}$$

$$\zeta = \alpha\frac{U_3}{2b} + \frac{1}{M}P + \frac{1}{M_T}T \tag{7}$$

$$\frac{\partial\zeta}{\partial t} = \left[\frac{\rho_\lambda k^f_\lambda}{\rho_o\eta_\lambda}(\nabla P + \rho_\lambda g\nabla z)\right] \tag{8}$$

where:

C_{ij} is the deformation parameters for fractures (which could be stress or deformation-path dependent).
U_j is the relative displacement of the fracture surfaces in direction *j*.
$2b$ is the fracture aperture.
k^f is the fracture permeability for liquid flow.

Equations 6 and 7 are constitutive equations in terms of local values of normal and shear stresses and relative local deformation. The parameters α, M, and

M_T are defined similarly to those of the rock matrix and are dependent on fracture roughness and fracture infill materials. Because it is difficult to know the fracture roughness and the properties of infill materials, these quantities probably should be determined empirically.

Equation 8 describes fluid flow in the fracture plane. Equation 3 can be similarly applied for the flow of energy in the fracture. Because of the thinness of the fracture, no thermal gradient is assumed across the fracture aperture, and so heat is considered to flow only along the fracture plane.

Equations 1 through 8 can be extended to the case of unsaturated rocks. These equations were also developed by Noorishad and Tsang (1996). Other authors, instead of describing coupled processes in the rock matrix and fractures separately, have attempted to derive equivalent properties for a rock mass containing a large number of fractures. Thus the fracture rock is represented by an anisotropic elastic porous medium. Stietel et al (1996) present this approach and its validity in comparison with actual properties calculated using a discrete approach.

As shown by the equations provided, the linkage of T, H, and M occurs in two ways. First, most of these equations link the three effects through terms in the equations. For example, Equation 1 has the pressure compressibility effect (second term on the left side) and thermal expansion effect (third term on the left) interacting with the fluid flow represented by the right side. In Equation 3, thermal capacity, strain energy, and convective flow energy, represented by the three terms, respectively, on the left side, combine to equate the thermal energy diffusion term on the right. Equations 4 and 5 show the same types of linkage. The second way linkage occurs between T, H, and M is indirectly through the functional dependence of property coefficients. For example, the density of fluid (ρ_λ) is temperature and pressure dependent, and viscosity has a strong temperature dependence. Similarly the other coefficients may also have P, T, and τ dependence, though some to a lesser degree than others.

The solution to these equations with coefficients dependent on P, T, and τ is in general a major challenge. From the theoretical standpoint, the three processes have widely different characteristic time constants and spatial scales. The thermal gradient has a relatively large time constant and spatial scale because it is a function of the long heating cycle, and thermal dispersivity smoothes out the effects of local spatial property variations. Mechanical effects, on the other hand, propagate through the rock with the speed of sound waves, and the deformation is strongly affected by faults and fractures, though much less so by medium property variations. Finally, the hydrologic flow and transport are sensitive to smaller-scale medium heterogeneity, but with a timescale corresponding to the large solute transport times. Numerically these processes are handled by techniques such as finite difference methods, finite element methods, discrete element methods, and others; to combine these into an efficient model for simulating coupled THM processes in fractured rock is no easy task.

A number of numerical codes have been developed to solve the coupled problem linking thermo-hydro-mechanical effects. The main ones are well summarized by Jing et al (1995) and Stephansson et al (1996). One of the codes, ROCMAS, has been developed by Noorishad et al (1992) to solve the coupled Equations 1 through 8 based on a finite element model using mixed Newton-Raphson linearization within an incremental configuration. In the next section this code is used to study a coupled HM and a coupled THM problem, providing physical insight into the results of linking these different effects.

TWO EXAMPLES OF COUPLED PROCESSES

The first example is a coupled HM process in a constant injection test of a well that is transected by a fracture. Injection tests are often performed by bracketing the fracture between packers in the well to determine its permeability, which is usually quite low. In such cases, production tests would not be effective. If the injection pressure is high, e.g. above the lithostatic pressure, hydraulic fracturing will occur. However, it has been shown that at lower pressures, water pressure working against mechanical stress may enlarge the fracture aperture. This HM linkage was studied experimentally by Rutqvist (1995) and Rutqvist et al (1998). In their experiment, a section of a well containing a fracture is isolated by packers and pressurized in a step-wise manner (Figure 1*a*). At each step, as pressure attains a constant level, the flow into the well section also attains to a constant. The pressure is then increased to the next level. The results are plotted as pressure-versus-flow rates at the end of each pressure level (Figure 1*b*).

In this problem, a possible coupled HM effect occurs: as the well section is pressurized, the fracture opens in response. This opening will increase the permeability of the fracture and thus a lesser rise in pressure is required to induce an increase in flow rate (Figure 1*b*). Figure 1*b* shows two curves. The first is a straight line representing the linear relationship between pressure and flow rate to be expected for a non-deformable fracture, which is well known in conventional hydrology. The other is a curved line showing increasing steps of flow rate with increasing steps of pressure, which corresponds to the effect that the fracture opens mechanically (increased aperture) at higher fluid pressures.

To test this concept, an experiment was conducted (Rutqvist 1995) in a 500-m-deep vertical borehole 56 mm in diameter, located in a crystalline rock at Lulea, Sweden. A fracture at 417 m was isolated by packers in the well. In the experiment, pressure was increased in a step-wise manner. The increased fluid pressure reduced the effective stress across the fracture, allowing the fracture aperture to increase—this is called the unloading path of the fracture in

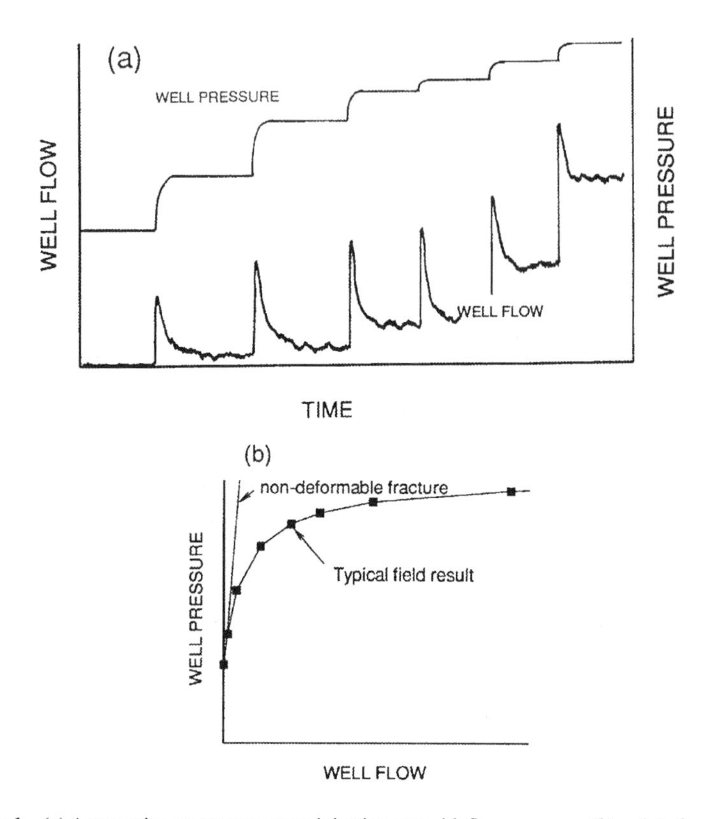

Figure 1 (*a*) A step-wise constant pressure injection test with flow response; (*b*) a plot of pressure against flow at the end of each successive step (from Rutqvist 1995).

a hydraulic fracturing process in rock mechanics. At the end of this step-wise pressure increase, the experiment is continued by reversing the process and decreasing pressure step by step—the loading part of the fracture. The results are shown in Figure 2. During the unloading period, the pressure-versus-flow curve follows the pattern of a deformed fracture (curved line in Figure 1*b*). It is interesting to note that during the subsequent loading period, in which fluid injection pressure is reduced, flow decreases quickly until there is a backflow from the fracture to the well. This is an indication that the fluid is being squeezed back into the well under mechanical stress, which reduces the fracture aperture.

The second example of a coupled THM process is designed to replicate the effect of heating of a waste canister in a rock cavity on flow in nearby fractures. The ROCMAS code is used to study the thermo-hydro-mechanical environment surrounding a 5-kW heater in granite at a depth of 350 m. A horizontal fracture is assumed to lie 3 m below the heater midplane and to

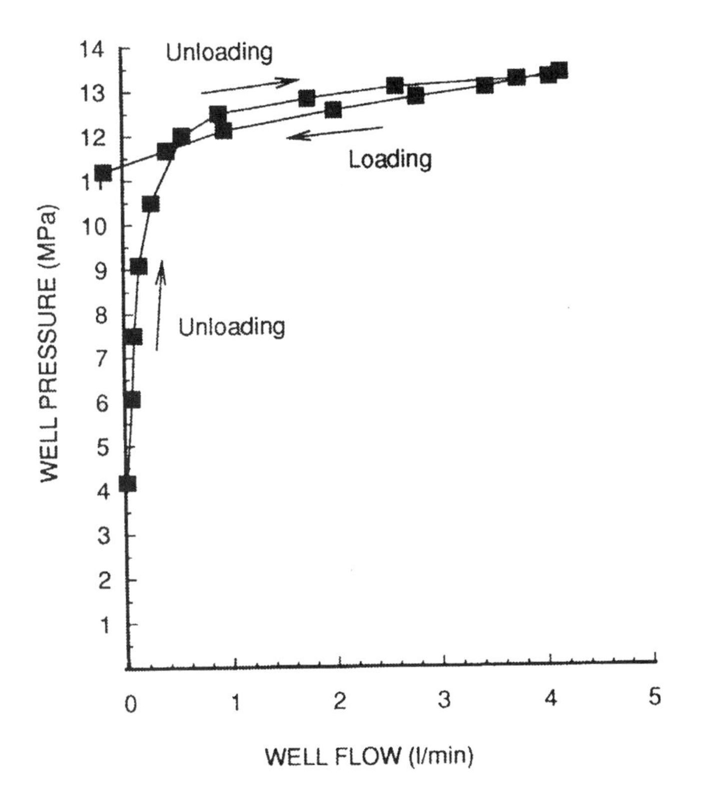

Figure 2 Field experimental results of pressure versus flow at the end of each step in a step-wise constant injection test. Unloading corresponds to increasing pressure when the effective stress is reduced. Loading corresponds to step-wise decrease of injection pressure (from Rutqvist 1995).

extend from the heater borehole to a hydrostatic boundary at a radial distance of 20 m from the borehole. The properties of rock and fracture are given in Table 1, and the two-dimensional axisymmetric (r, z) finite-element grid and some data are shown in Figure 3.

The heater drift, approximated by the cylindrical hatched area in Figure 3, is simulated by assigning a very low value of Young's modulus to the elements. Before the heater raises the temperature of a large volume of the rock, the flow of water from the hydrostatic outer boundary to the atmospheric (zero hydraulic pressure) borehole is high. Later, with the heated rock above the fracture expanding and the fracture aperture near the heater borehole closing, the flow decreases sharply (Figure 4).

The evolution of fracture aperture profile, together with the variations of the pressure and temperature distribution, is shown in Figure 5. The figure shows that at 0 day, before water in the fracture begins to flow, full hydrostatic pressure

Table 1 Properties of material used for the example application

Material	Parameter	Value
Fluid	Mass density, ρ_s	997 kg m^{-3}
	Compressibility, β_P	5.13×10^{-10} GPa^{-1}
	Dynamic viscosity at 20°C, η	10^{-3} N sm^{-2}
	Thermal expansion coefficient, β_T	3.17×10^{-4}°C^{-1}
	Specific heat, C_{Vw}	1.0 kcal kg^{-1}°C^{-1}
Rock	Mass density, ρ_s	2.6×10^3 kg m^{-3}
	Porosity, ϕ	5.0×10^{-2}
	Permeability, k	10^{-17} m^2
	Young's modulus, E	51.3 GPa
	Poisson's ratio, v	2.3×10^{-1}
	Biot's coefficient, α	1.0
	Biot's coefficient, M	5.0 GPa
	Thermal expansion coefficient (linear), β	8.8×10^{-5}°C^{-1}
	Specific heat, C_{Vs}	2.1×10^{-1} kcal kg^{-1}°C^{-1}
	Lumped thermal conductivity, K_M	3.18×10^{-3} KJ m^{-1} s^{-1}°C^{-1}
Fracture	Porosity, ϕ	1.0
	Initial aperture, 2b	10^{-4} m
	Initial normal stiffness, k_n	85 GPa/m
	Initial tangential stiffness, k_s	0.085 GPa/m
	Biot's coefficient, α	1.0
	Biot's coefficient, M	5.0 GPa
	Friction angle, ψ	30°
	Cohesion, C	0.0

prevails in the fracture. This pressure diminishes rapidly after the heater hole is opened to flow at atmospheric pressure and before major development of the thermal front, as shown by the curve labeled 0.25 days. However, as thermal stresses are established, the fracture starts closing. As a result, the pressure inside the fracture rises, leading to the establishment of full pressure in the fracture after 14 days, similar to the 0-day case. These results may provide a better understanding of some of the observations made in the in situ heater experiments in the Stripa granite (Witherspoon et al 1981). Similarly, aperture reduction leading to stoppage of water inflows into the heater boreholes (Nelson et al 1981) also can be explained by the same phenomenon.

THREE CURRENT THM FIELD EXPERIMENTS

Concern over the consequences of coupled THM processes as they relate to nuclear waste repository performance has stimulated much experimental study. A number of laboratory and small-scale field experiments have been carried

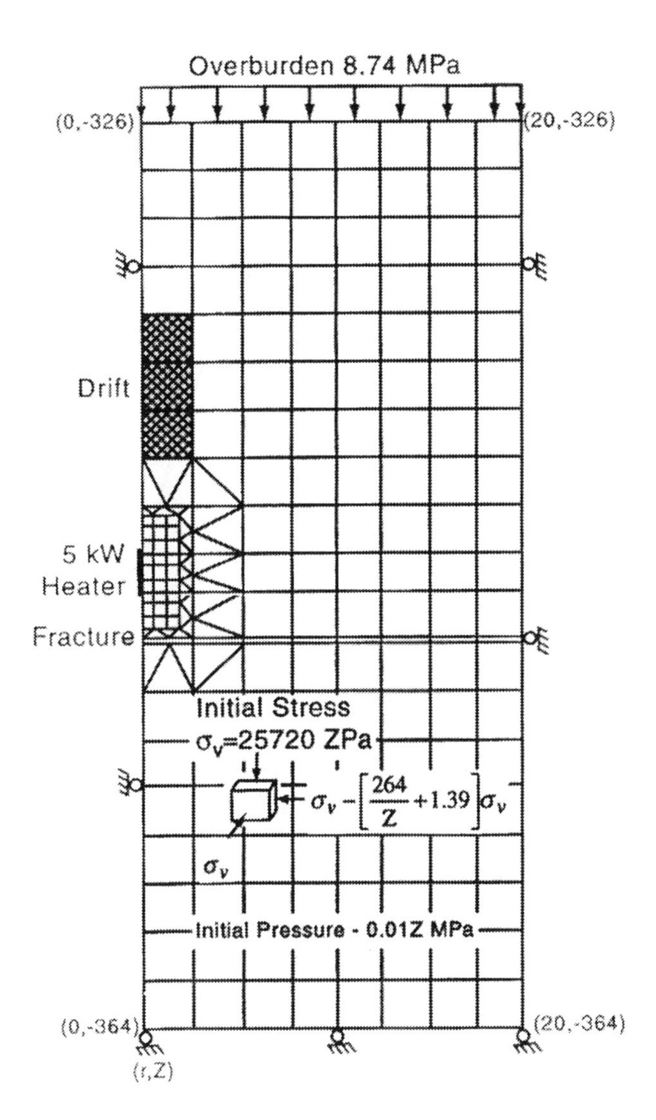

Figure 3 Vertical cross section of the finite element mesh around a heat source with a horizontal fracture just below the heater (from Noorishad and Tsang 1996).

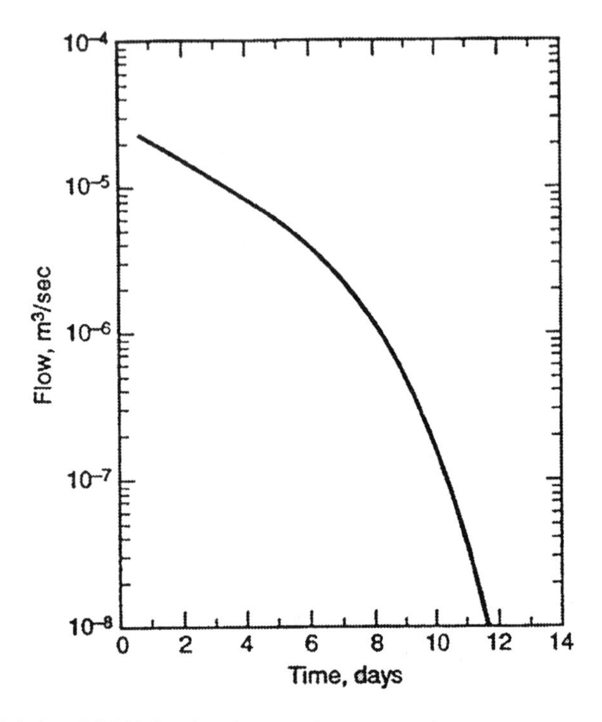

Figure 4 Variation of fluid inflow into the heater borehole as a function of time (from Noorishad and Tsang 1996).

out on some of these processes. Large-scale multiyear field experiments to specifically study coupled processes linking THM effects have been initiated only recently. Three major experiments that are ongoing or just completed are described in this section and summarized (Table 2).

Kamaishi Coupled THM Experiment

A THM experiment was conducted at the Kamaishi Mine in Japan from 1995 to 1998 by the Power Reactor and Nuclear Fuel Development Corporation (PNC). The purpose of the experiment was to study the mechanical effect of excavation on flow in fractured rock and the impact of a heater placed in an excavated pit with bentonite (clay) as a backfill material (PNC 1997). As Table 2 indicates, the Kamaishi experiment has concluded; however, not all results have yet been published.

Conducted at a depth of 250 m, the experiment simulated the condition of a waste canister storage pit in a potential nuclear waste repository in fractured hard rock (Figure 6). One objective of the test was to observe near-field coupled THM

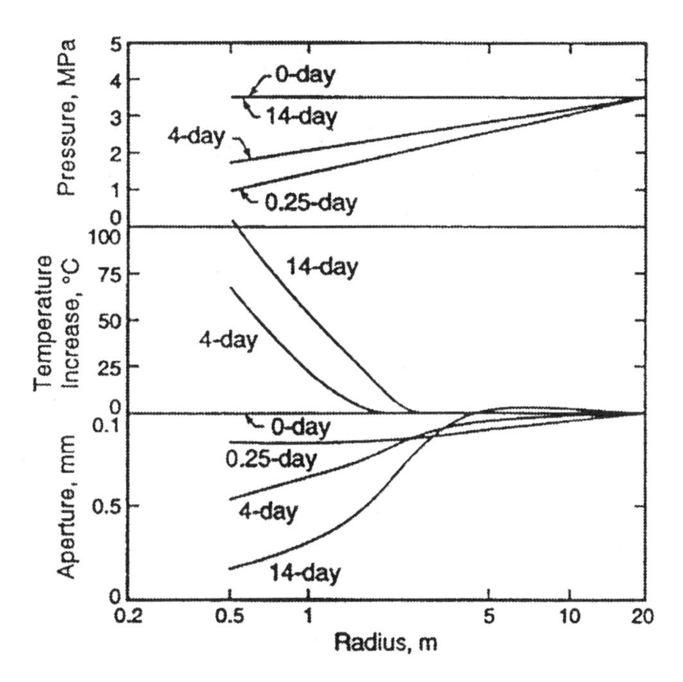

Figure 5 Pressure, aperture, and temperature profiles along the heater mid-plane for different times (from Noorishad and Tsang 1996).

Table 2 A summary of three major field tests to study THM processes

Experiment	Coupled THM experiment	FEBEX experiment	Drift scale heater test
Location	Kamaishi, Japan	Grimsel, Switzerland	Yucca Mountain, USA
Site	Kamaishi Mines	Grimsel Underground Research Laboratory	Exploratory Studies Facility (ESF)
Period (heating-cooling)	1.5 years	5.5 years	8 years
Heating start date	October 1996	February 1997	December 1997
Heater length	2 m	2 heaters: ~10 m	9 drift heaters and 25 wing heaters: 45 × 20 m
Maximum temp. expected	100°C	100°C	200°C
Remarks	Heater-bentonite system in vertical pit in fractured rock	Heater-bentonite blocks system horizontally in drift in fractured rock	Heater in drift with wing heaters in unsaturated volcanic tuff

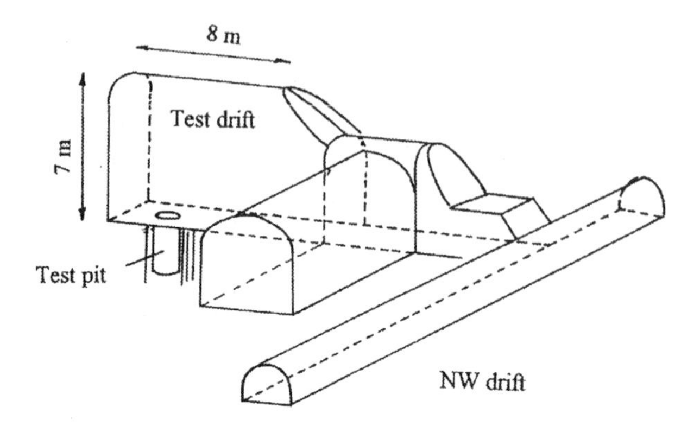

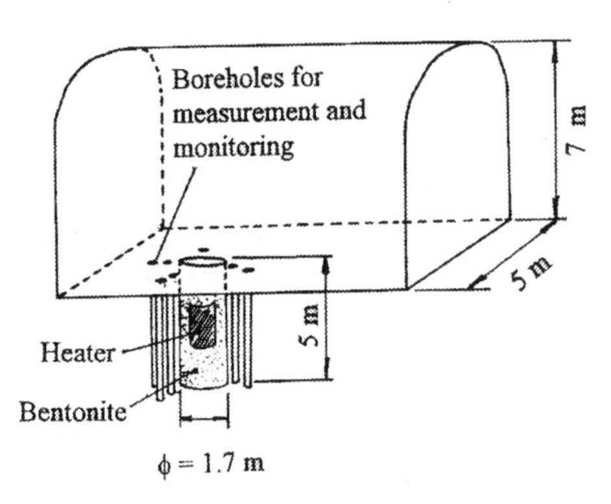

Figure 6 Arrangement of Kamaishi field experiment (from PNC 1997).

phenomena in situ and to build confidence in coupled mathematical models of these processes. The experiment was divided into several phases, starting with a hydromechanical evaluation of the effects of excavation and ending with a fully coupled THM analysis of the heater test.

To conduct the experiment (Fujita et al 1996), a test pit of 1.7 m in diameter and 5 m in depth was drilled in the floor of a 5-by-7 m alcove excavated off an existing drift at Kamaishi (Figure 6). An electric heater placed in the pit and surrounded by compacted bentonite powder simulated the heat output from a waste package in the test pit. Most of the measurements were taken from locations within an approximate 2 m radius from the center of the test pit, where most of the THM effects were expected to occur.

For the initial hydromechanical phase before emplacement of the heater and bentonite, measurements were focused on mechanical displacements of the test pit wall and along main fractures, and on the fluid pressure distribution and water inflow to the test pit. The fluid pressure distribution as a function of depth from the ground surface at Kamaishi does not correspond to the hydrostatic pressure gradient with depth due to continuous drainage in the existing drift system. At the horizon of the THM test the fluid pressure in the nearby 100-m area varies between 0.1 and 0.4 MPa, with an average pressure of about 0.3 MPa. Thus, the initial pore pressure at the site is much less than that expected at this depth.

The far-field properties and geometry affect the near-field behavior of fluid pressure and stress. In the rock mass at the level of the test drift, many fractures have northeast strikes and steep dip angles. The minimum principal stress is subvertical, and the maximum principal stress is oriented almost perpendicular to the axis of the test drift. The maximum hydraulic conductivity is oriented parallel to the dominating fracture orientation and perpendicular to the maximum principal stress orientation, indicating that the hydraulic conductivity is correlated to the fracture orientation rather than to the current in situ stress.

Fractures mapped on the floor of the drift have preferential northeast strikes with steep dip angles (Figure 7*a*). There are too many fractures to include discretely in a finite element model. On the other hand, there may not be enough fractures to treat them as an equivalent continuum near the test pit. However, three large shear fractures or faults adjacent to the test pit strike approximately

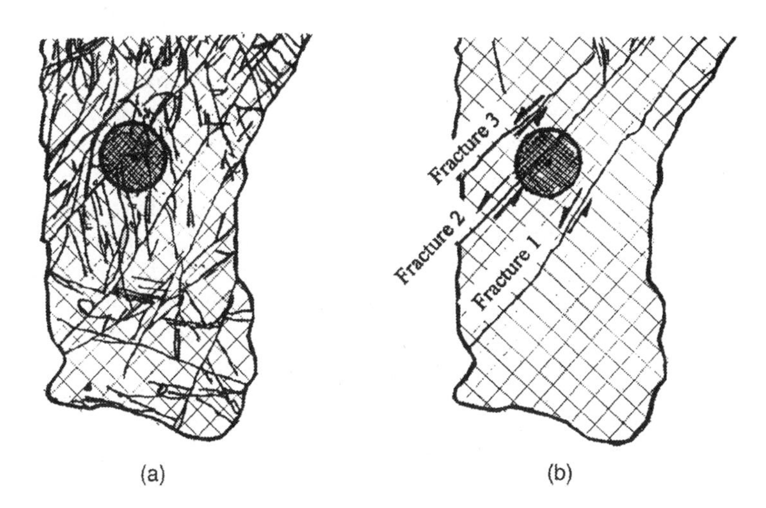

Figure 7 Fracture map (*left*) on the floor of the drift where the test pit is drilled, and a simplified map (*right*) with only the major fractures (from PNC 1997).

east-west (fractures 1, 2, and 3 in Figure 7*b*). These faults have been sheared and reported to be the most open fractures on the floor. They are up to 20 mm in thickness and contain some soft mineral filling. Similar fractures with the same strike and dip can be found on the roof, and are also found to dominate the inflow into the drift.

During excavation of the test pit, the rock was strained to allow the stresses to be redistributed. Most of the strain was concentrated within a distance of three diameters from the test pit. As mentioned, fluid pressure at the site was only 0.3 MPa and was not expected to cause any significant changes in fracture apertures. Thus the hydromechanical coupling can be considered a one-way coupling and the result of the disturbed stress field around the test pit. The disturbed stress field (disturbed due to excavation) deformed the shape of the test pit cross section and changed apertures—and hence permeabilities—of nearby fractures. The former observation was obtained by measuring distances between four points on the wall of the test pit near the floor of the alcove. The latter observations were estimated by borehole extensiometer measurements across fractures that were intersected by boreholes.

After the test pit was excavated, the rock near the pit surface was dehydrated because of ventilation: in the immediate neighborhood of the test pit, the rock is unsaturated and the local pore pressure can be negative. Nevertheless, water seeps into the test pit under hydraulic pressure. This seepage was measured by placing water-sorptive patches on the walls and measuring rates of water absorbed on these patches. The distribution of water rates around the pit wall is highly uneven, being largest along fracture traces on the wall. The total seepage rate into the test pit was two liters per day. To increase the seepage rate, a shallow water pool was imposed on the alcove floor around the test pit. The rate was then found to be 280 liters per day, and seepage distribution on the pit wall was again measured. These data, together with fracture distribution surveys on alcove floor and pit wall, were used to test computer models.

After the study of seepage into the test pit, a two-meter–long heater was emplaced and surrounded by bentonite. A one-year heating period was followed by a six-month cooling period, during which T, H, and M changes were carefully monitored all around the test pit. The unsaturated bentonite material underwent heating from the center of the test pit and wetting from the test pit wall. The heating caused water to vaporize and diffuse outwards, resulting in drying and possibly shrinkage near the center. The wetting from the rock saturated the bentonite and caused swelling and a corresponding increase in pressure near the pit wall. These processes depend on parameters that are both temperature and saturation dependent.

The coupled THM processes across the bentonite buffer are rather complex. In the rock, thermally induced coupled mechanical effects focused around

Table 3 Some of the coupled THM processes expected in the Kamaishi experiment

Linkage	Process description
HM-1	Interaction between deformability and permeability of rock matrix and rock fractures during test pit excavation
HM-2	Water flow under variable saturation conditions in bentonite–rock and bentonite–heater interfaces and also into bentonite, coupled with bentonite swelling
TH	Heat transfer through rock matrix, fractures, bentonite, and their interfaces coupled with flow under both unsaturated and saturated conditions
THM-1	Swelling pressure, multiphase flow with possible phase change (drying and wetting) in bentonite under a temperature gradient for variable-saturation conditions
THM-2	Interaction between deformability and permeability of rock matrix, rock fractures, bentonite, and their interfaces during heating and cooling periods, involving both saturated and unsaturated flows

fractures near the test pit on the alcove floor (Figure 7*b*) and at the bentonite-rock interface. The thermally induced coupled hydrologic effects included diffusive and convective flow in rock next to the test pit and in the bentonite with variable saturation. Although the experiment is concluded, the data have yet to be published. Some of the expected coupled processes in this experiment are summarized in Table 3.

The FEBEX Experiment

The Spanish nuclear waste management agency, Empresa Nacional de Residuos Radiactivos, S.A. (ENRESA), is conducting an in situ experiment as part of their FEBEX program (ENRESA 1996). The name FEBEX stands for Full-scale Engineering Barriers Experiment. The experiment (ENRESA, 1998) is located in a drift below the water table in the Grimsel Underground Laboratory in Switzerland operated by NAGRA, the Swiss agency for nuclear waste management. The drift is 70.4 m long and 2.28 m in diameter. A number of radial boreholes were drilled from the drift into the Grimsel fractured rock. Then two heaters, each 4.54 m long and 0.9 m in diameter, were placed in the drift along the drift's axis, separated from each other by 1 m and surrounded by packed bentonite blocks weighing 20–25 kg each. The radial boreholes are intensively instrumented to characterize the rock around the drift prior to heater and bentonite emplacement, as well as to monitor rock behavior during the heating and cooling phases. A total of 632 sensors of diverse types were installed in the bentonite blocks, the rock mass, the heaters, and the drift service zone to monitor temperature, humidity, total fluid pressure, mechanical displacement, water pressure, etc. Figures 8 and 9 show a sketch of the experiment and the dimension of the test design, respectively.

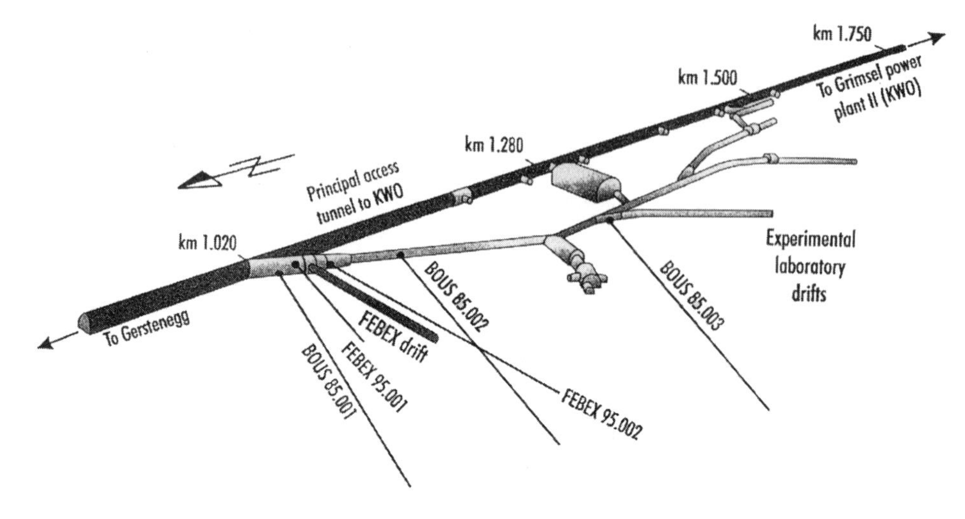

Figure 8 Location of the FEBEX experiment in the Underground Testing Facility in Grimsel, Switzerland. Scale is indicated in kilometers along the main tunnel (from ENRESA 1998).

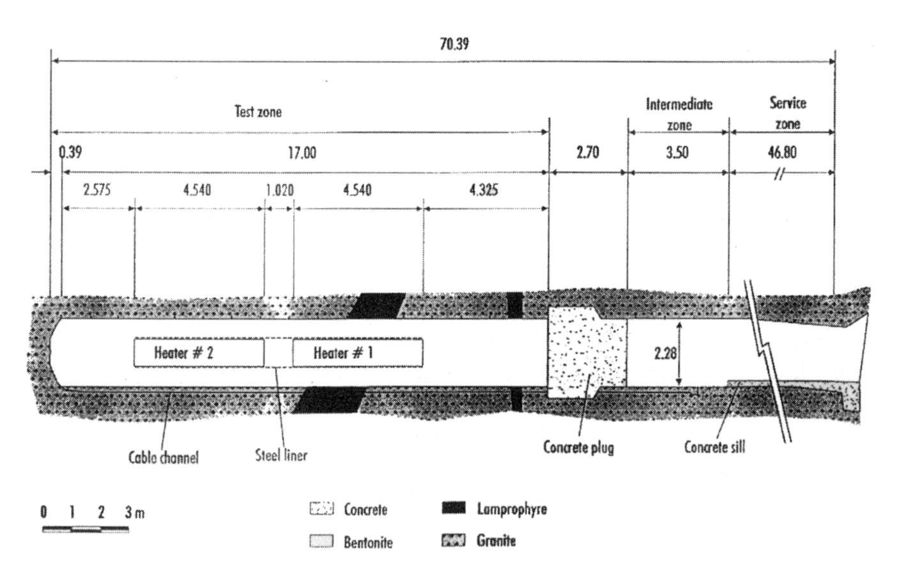

Figure 9 A closer view of the heater drift with two heaters; dimensions are in meters (from ENRESA 1998).

The heaters were activated in February 1997, and the current plans call for heating for five years, with an additional six-month cooling period. During the five-year period, the temperature will reach and be maintained at 100°C near the heaters. At the end of the five and a half years, the system will be dismantled to allow detailed sampling and observation for physical and chemical changes.

While the Kamaishi experiment focuses on a heater in a vertical deposition pit in an alcove, the FEBEX experiment involves a drift, along which two heaters are emplaced horizontally. In the Kamaishi case, the buffer material was bentonite powder that was compacted in the test pit. In the FEBEX case, the buffer is in the form of bentonite blocks arranged around the horizontal heaters. The scale of the FEBEX experiment is significantly larger, the length of the two heaters from end to end being 10 m, compared with the length of the single heater at Kamaishi, which was 2 m. Thus the two experiments complement each other, providing a range of data to understand coupled THM processes in heater-bentonite-fractured rock systems.

All the processes described in the previous section for the Kamaishi experiment are expected to occur in the FEBEX experiment. In addition, a number of other features are also important in the FEBEX case:

(*a*) Disturbed rock zone. During the excavation of the drift, the rock near the drift wall is disturbed by stress redistribution and by excavation methods, especially near fault zones. Small fractures may be created and existing fractures may be deformed. Once the drift is constructed, dehydration will occur in rock near the drift wall, creating an unsaturated flow condition. After the drift is closed, the rock will be resaturated and water will flow into the drift under regional hydraulic pressure. The role of the disturbed rock zone on coupled HM processes during the whole stress redistribution-dehydration-saturation process is very interesting, and the FEBEX experiment will be used to study it.

(*b*) Disturbed zone-bentonite interaction. The emplacement of the unsaturated bentonite introduces a material with high suction near the rock. If the suction rate is high relative to the inflow rate from the rock, the disturbed zone could undergo further dehydration, reducing its effective permeability. At the same time, as the bentonite is wetted, a swelling pressure will be imposed on the disturbed zone.

(*c*) Gaps between bentonite blocks and between bentonite blocks and the rock or heaters. Flow of water in the gaps under a temperature gradient imposed by the heaters will cause bentonite swelling and sealing of these gaps. If the sealing rate is slow, wetting and saturation of the blocks can occur from water in gaps between blocks. If the sealing rate is fast, saturation of the bentonite system will be uneven, with high saturation around block boundaries and low saturation in block centers. Because of this spatially variable saturation, the system will undergo a complex distribution of swelling pressure and deformation.

(*d*) Interference of the two heaters. The bentonite blocks between the two heaters will experience heating, vaporization, and drying from two sides as well as water wetting from the rock through interfaces.

(*e*) In the FEBEX drift, it was noticed that the part of the drift near one of the two heaters was much wetter than the other because of the presence of a fracture zone. This condition implies a possible uneven wetting of the bentonite blocks along the drift, unless the gap between the bentonite and the rock is so effective that water is quickly redistributed. The comparative THM behavior of the fractured rock in the two parts of the drift under the temperature gradient may be significant and could unveil the characteristics of coupled THM processes at the site.

Some of the coupled processes in the FEBEX experiments are summarized in Table 4. Actual experimental data will be published in the coming years, and

Table 4 Some of the coupled THM processes expected in the FEBEX experiment

Linkage	Process description
HM-1	Interaction between deformability and permeability of rock matrix and rock fractures during drift excavation
HM-2	Water flow under variable saturation conditions in bentonite-rock and bentonite-heater interface and also into bentonite, coupled with bentonite swelling
HM-3	Unsaturated water flow in gaps between bentonite blocks and bentonite swelling
HM-4	Role of disturbed zone in rock near the drift wall on mechanical and hydraulic properties during dehydration and resaturation
TH	Heat transfer through rock matrix, fractures, bentonite, and their interfaces coupled with flow under both unsaturated and saturated conditions
THM-1	Swelling pressure, multiphase flow with possible phase change (drying and wetting) in bentonite under a temperature gradient for variable-saturation conditions
THM-2	Interaction between deformability and permeability of rock matrix, rock fractures, bentonite, and their interfaces during heating and cooling periods, involving both saturated and unsaturated flows
THM-3	Interference between two heat sources on heating, water flow (wetting and drying), and swelling of bentonite blocks between the two heaters
THM-4	Comparison of THM processes in fractured rock in the wet and dry parts of the drift

Note: HM-1, HM-2, TH, THM-1, and THM-2 are identical to processes expected to occur in the Kamaishi experiment (Table 3), but the spatial and temporal scales are much larger in the FEBEX experiment.

a number of modeling efforts are already under way to make predictions and to perform evaluations of the experiment.

The Drift Scale Heater Test

To study technical issues related to the construction, operation, closure, and performance of a potential nuclear waste repository, the U.S. Department of Energy, Office of Civilian Radioactive Waste Management, is conducting a series of in situ experiments in the Exploratory Studies Facility (ESF) at Yucca Mountain, Nevada. One of these experiments, the Drift Scale Heater Test (DST), is intended to acquire an in-depth understanding of the coupled thermo-mechanical-hydrological-chemical processes anticipated in the rock mass around a repository.

The experimental arrangement (Peters et al 1997) is schematically shown in Figures 10 and 11. The drift to be heated is 47.5 m long and 5 m in diameter. Nine floor heaters are emplaced in the drift. In addition, 50 wing heaters have been placed in boreholes extending from the drift in a horizontal plane. Each of the floor heaters is 4.6 m long, and each of the wing heaters is 10 m long. The test is being conducted in unsaturated fractured tuff. The heaters were turned on in December 1997. Over a four-year heating period, the drift wall temperature will reach 200°C, and will be maintained at that temperature to ensure that coupled THM and chemical processes are properly observed. The

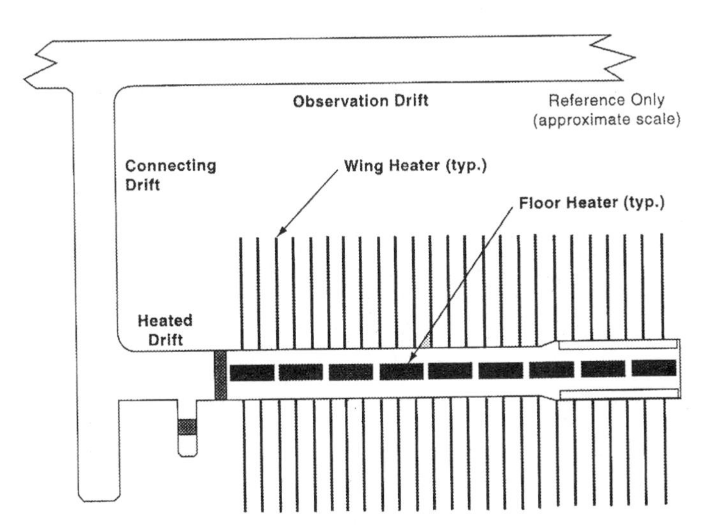

Figure 10 Arrangement of the Drift Scale Heater Test at the Exploratory Studies Facility at Yucca Mountain. The total length of the drift heater is about 45 m and each wing heater is 10 m (from Peters et al 1997).

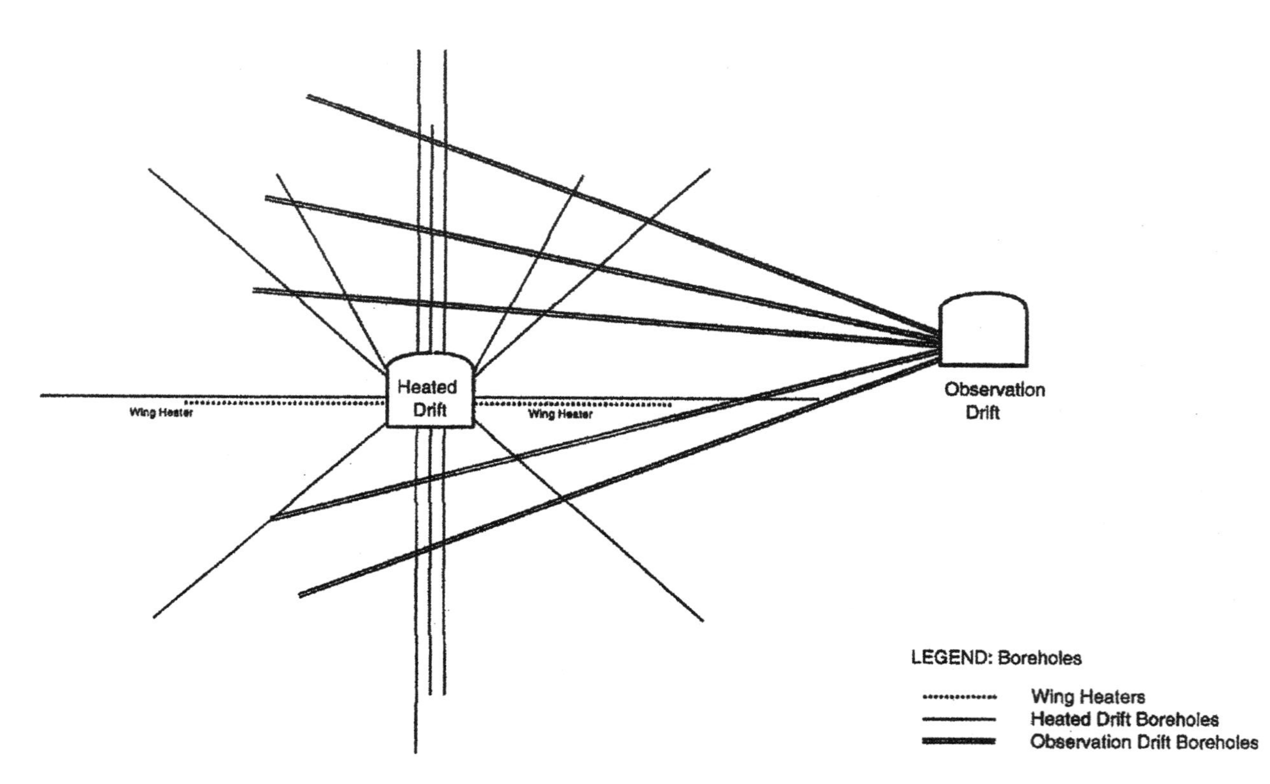

Figure 11 A cross section to illustrate the location of boreholes around the heater drift to monitor changes during the heater test. The drift is five meters in diameter (from Peters et al 1997).

heating period will be followed by a four-year cooling period. Approximately 3,500 measuring devices of various types have been installed in boreholes in the vicinity of the heated drift to monitor temperatures, rock displacements, rock moisture content, relative humidity, and pressures. Air permeabilities at various points will be measured at different stages of the experiment, and samples of water and gas will be collected periodically. Microseismic emissions, if any, will also be monitored.

Because the unsaturated volcanic tuff will be subject to temperatures as high as 200°C, significant evaporation and condensation is expected. In turn, temporal changes in saturation will strongly affect the effective permeability for fluid flow in the rock. Gravity also is an important factor in the downward flow of condensed water. Because the rock is highly fractured, the downward flow of condensed water will initially be in the fractures and will subsequently be imbibed into the rock matrix. Thus, the flow is highly complex and is dependent on the properties of the fracture–matrix interface, whose characteristics are still an open question. Mechanical strain in the neighborhood of the heater drift will be monitored, and water sampling will be carried out to obtain the chemical characteristics. The interlinking of strain, temperature, and flow in this test will be investigated. Water chemistry will also give indications of THM system behavior.

Table 5 summarizes some of the coupled processes present in the Yucca Mountain Drift Scale Heater Test. Because of unsaturation, the system has a very strong TH linkage. A temperature change will affect the water saturation level of the rock, which will change the capillary suction and effective

Table 5 Some of the coupled THM processes expected in the Yucca Mountain Drift Scale Heater Test

Linkage	Process description
TH-1	Effective permeability in an unsaturated fracture-porous medium as a function of temperature gradient, saturation, and fracture–matrix interaction
TH-2	Heat pipe effect in an unsaturated fracture-porous medium, with vapor produced near heat source, diffusing in fractures down thermal gradient, condensed and flowing back in matrix under capillary suction
HM	Opening of existing fractures during excavation and stress redistribution; flow in unsaturated fractures with modified apertures
THM-1	Change in fracture apertures under the high temperature gradient, unsaturated flow in fractures under gravity, and matrix imbibition, with the phenomenon of boiling and condensation
THM-2	Coupled THM processes in fractured porous medium under a high temperature gradient, with channelized flow because of fracture aperture variation and rock heterogeneity

permeability to water flow (TH-1). Further, a process called the heat pipe effect is expected to occur, which is a very effective heat transfer mechanism (TH-2). In this process, water near the heater is evaporated and the vapor moves along fractures down the thermal gradient and condenses in the cooler region. The condensed water then moves in the rock matrix under capillary suction toward the drier regions near the heat source. In general, vaporization (boiling) and condensation are expected to occur in and around the heater test region. Boreholes with packers and other methods have been put in place to detect these conditions and collect condensed water samples.

The Yucca Mountain ESF site is highly fractured, and fractures can open significantly due to stress redistribution during drift construction. Wang et al (1997) measured permeabilities of packed intervals in boreholes approximately 0.6 m above an alcove in the ESF some distance away from the Drift Scale Heater Test area. The measurements were made before and after the excavation, and an increase of one-and-one-half order of magnitude was found in the geometric mean of permeabilities. I expect this process to happen in the Drift Scale Heater Test area (HM).

Because of the high temperature (200°C) expected in the experiment, the THM linkage will include (*a*) fracture aperture changes, (*b*) convective and diffusive flow of gas and vapor as well as liquid under gravity, (*c*) boiling and condensation, and (*d*) matrix imbibition. All of these will interact and couple with each other in various ways (THM-1). Of particular interest in this THM linkage is the role of heterogeneity of the media—both heterogeneity in the fracture plane (i.e. variable apertures) and heterogeneity in the rock medium as a whole. It has been shown by Birkholzer and Tsang (1997) and Birkholzer et al (1998) that flow in unsaturated fractures is highly channelized, meaning that water flows along preferred fast paths. The effect is dependent on the degree of heterogeneity, the level of water saturation, and fracture–matrix interaction. The coupled THM linkage upon this kind of channelized flow will be very interesting and will display behavior not normally expected for homogeneous media or constant-aperture fractures (THM-2).

Initial data and information are beginning to emerge from the Yucca Mountain Drift Scale Heater Test. A parallel modeling program is also under way to predict and to understand the system behavior.

CONCLUSION

This paper presents an overview of coupled processes linking THM effects in fractured rocks. A formulation shows the linkage mathematically, which can be used as a basis for numerical solutions and for further developments. Two simple examples of HM and THM coupled processes convey physical insight into such couplings. Three current large-scale, long-term experiments are described. These tests are conducted specifically to explore and study coupled processes in situ.

After more than 10 years of research, the study of coupled THM processes has reached an important stage, cumulating in large-scale in situ experiments such as the three described in this paper. These experiments involve not only isolated coupled processes but a number of coupled processes occurring together. Because of the dimension of these experiments, the geologic structure of the experimental site may also be complex, presenting a particular challenge to data interpretation. Fortunately, the experimental design has allowed extensive monitoring with many sensors of diverse types. Hopefully, the data gathered in the monitoring program will help explain the behavior of the various coupled processes involved. I anticipate intense activity and significant advancement in the understanding of coupled processes in the next few years.

ACKNOWLEDGMENTS

Reviews and comments by Lanru Jing of the Royal Institute of Technology, Stockholm, Sweden, and by John Apps and Jonny Rutqvist of Ernest Orlando Lawrence Berkeley National Laboratory, Berkeley, California, are gratefully acknowledged. Collaboration and discussions with Jahan Noorishad, Jonny Rutqvist, Ove Stephansson, and participants of the International DECOVALEX project are very much appreciated. Thanks are also given to Julie McCullough for her careful technical editing. The paper is prepared with joint funding from the Swedish Nuclear Power Inspectorate (SKI), Stockholm, Sweden, and the Office of Energy Research, Office of Basic Energy Sciences, Engineering and Geosciences Division, of the U.S. Department of Energy, through Contract No. DE-AC03-76SF00098.

Literature Cited

Birkholzer J, Li G, Tsang CF, Tsang Y. 1998. Modeling studies and analysis of seepage into drifts at Yucca Mountain. *J. Contam. Hydrol.* (In press)

Birkholzer J, Tsang CF. 1997. Solute channeling in unsaturated heterogeneous porous media. *Water Resour. Res.* 33(10):2221–38

ENRESA. 1996. Test plan for FEBEX experiment, *Empresa Nac. Residuos Radiact.* S.A. Madrid, Spain

ENRESA. 1998. Febex-Full-scale engineered barriers experiment in crystalline host rock: Pre-operational stage, Summ. Rep. *Publ. Tec. Num 01/98. Empresa Nac. Residuos Radiact.* S.A. Madrid, Spain

Fujita T, Sugita Y, Sato T, Ishikawa H, Mano T. 1996. Coupled thermo-hydro-mechanical experiment at Kamaishi mine. *Plan. Power Reactor Nucl. Fuel Dev. Corp. Rep. TN8020*

Haijtink B, ed. 1995. Testing and modeling of thermal, mechanical and hydrogeological properties of host rocks for deep geological disposal of radioactive waste. *Proc. Brussels Workshop, Eur. Comm. Rep. EUR 16219,* 298 pp.

IJRMMS. 1995. Thermo-Hydro-Mechanical coupling in Rock Mechanics. *Int. J. Rock Mech. Min. Sci.* (Suppl.)

Jing L, Tsang CF, Stephansson O. 1995. DECOVALEX–An international cooperative research project on mathematical models of coupled THM processes for safety analysis of radioactive waste repositories. *Int. J. Rock Mech. Min. Sci.* 32(5):387–98

Nelson PH, Rachiele R, Remer JS. 1981. Water inflow into boreholes during the Stripa heater experiments. *Rep. LBL-12574.* Berkeley, CA: Lawrence Berkeley National Laboratories

Noorishad J, Tsang CF. 1996. Coupled thermo-hydroelasticity phenomena in variably saturated fractured porous rocks–formulation and numerical solution. In *Coupled Thermo-Hydro-Mechanical Processes of Fractured Media*, ed. O Stephansson, L Jing, CF Tsang, p. 93–134. Amsterdam: Elsevier Science Publishers. 575 pp.

Noorishad J, Tsang CF, Witherspoon PA. 1992. Theoretical and field studies of coupled behavior of fractured rocks–1. Development and verification of a numerical simulator. *Int. J. Rock. Mech. Min. Sci. Geomech. Abstr.* 29:401–9

Ohnishi Y, Chan T, Jing L. 1996. Constitutive models for rock joints. In *Coupled Thermo-Hydro-Mechanical Processes of Fractured Media*, ed. O Stephansson, L Jing, CF Tsang, p. 93–134. Amsterdam: Elsevier Science Publishers. 575 pp.

Peters M, Datta R, Wagner R, Boyle W, Yasck R. 1997. Progress in field testing and modeling activities at the Yucca Mountain site. *Proc. Field Test. Assoc. Modeling Workshop, Dec. 15–16, 1997,* Berkeley, CA: Lawrence Berkeley Nat Lab. p. 17–20

PNC. 1997. *DECOVALEX II Project, Definition for Test Case 2, Kamaishi,* DECOVALEX Secretariat, Stockholm: Roy. Inst. Technol.

Rutqvist J. 1995. Determination of hydraulic normal stiffness of fractures in hard rock from hydraulic well testing. *Int. J. Rock Mech. Min. Sci.* 32:513–23

Rutqvist J, Noorishad J, Tsang CF, Stephansson O. 1998. Determination of fracture storativity in hard rocks using high pressure injection testing. *Water Resour. Res.* 34(10):2551–60

Stephansson O, Jing L, Tsang CF. 1996. *Coupled Thermo-Hydro-Mechanical Processes of Fractured Media*, Elsevier Science B.V.: Amsterdam

Stietel A, Millard A, Treille E, Vuillod E, Thoravel A, Ababou R. 1996. Continuum representation of coupled hydromechanic processes of fractured media: Homogenization and parameter identification. In *Coupled Thermo-Hydro-Mechanical Processes of Fractured Media,* ed. O Stephansson, L Jing, CF Tsang. p. 135–64. Amsterdam: Elsevier Science Publishers. 575 pp.

Tsang CF. 1991. Coupled hydromechanical-thermochemical processes in rock fractures, *Rev. Geophys.* 29(4):537–55 (suppl.)

Tsang CF, ed. 1987. *Coupled Processes Associated with Nuclear Waste Repositories.* San Diego: Academic Press

Wang JSY, Trautz RC, Cook PC. 1997. Flow tests to quantify seepage into drifts. *Proc. Field Test. Assoc. Modeling Workshop, Dec. 15–16, 1997,* p. 21–22 Berkeley, CA: Lawrence Berkeley Nat. Lab.

Witherspoon PA, Cook NGW, Gale JE. 1981. Geologic storage of radioactive waste, field studies in Sweden. *Science* 211:894–90

Annu. Rev. Earth Planet. Sci. 1999. 27:385–415

IMPACT CRATER COLLAPSE

H. J. Melosh
Lunar and Planetary Laboratory, University of Arizona, Tucson, AZ 85721;
e-mail: jmelosh@lpl.arizona.edu

B. A. Ivanov
Institute for Dynamics of the Geospheres, Russian Academy of Sciences, Moscow,
Russia 117979

KEY WORDS: crater morphology, dynamical weakening, acoustic fluidization, transient crater, central peaks

ABSTRACT

The detailed morphology of impact craters is now believed to be mainly caused by the collapse of a geometrically simple, bowl-shaped "transient crater." The transient crater forms immediately after the impact. In small craters, those less than approximately 15 km diameter on the Moon, the steepest part of the rim collapses into the crater bowl to produce a lens of broken rock in an otherwise unmodified transient crater. Such craters are called "simple" and have a depth-to-diameter ratio near 1:5. Large craters collapse more spectacularly, giving rise to central peaks, wall terraces, and internal rings in still larger craters. These are called "complex" craters. The transition between simple and complex craters depends on $1/g$, suggesting that the collapse occurs when a strength threshold is exceeded. The apparent strength, however, is very low: only a few bars, and with little or no internal friction. This behavior requires a mechanism for temporary strength degradation in the rocks surrounding the impact site. Several models for this process, including acoustic fluidization and shock weakening, have been considered by recent investigations. Acoustic fluidization, in particular, appears to produce results in good agreement with observations, although better understanding is still needed.

INTRODUCTION

It is now generally accepted that impact cratering is an important geological process that has affected the surface of nearly every planet and satellite in the

0084-6597/99/0515-0385$08.00

solar system. Impact craters are nearly ubiquitous landforms on the ancient surfaces of, for example, the Moon, Mercury, and Mars. Even on Earth, with its highly active fluvial and tectonic recycling, some 150 impact craters are currently known (Grieve & Shoemaker 1994). Although nearly all fresh craters can be broadly described as "circular rimmed depressions," the detailed morphologies of impact craters show many variations. The size-morphology progression, first recognized for lunar craters by Gilbert (1893), ranges from small, simple, bowl-shaped craters through craters with central peaks and wreaths of terraces around the rims and up to basins with an internal mountainous ring and, on some bodies, exterior scarps that encircle the impact site.

Currently, it is believed that this morphologic diversity is not a direct result of the crater excavation process but develops only after most of the material has been expelled from the crater. The initial product of crater excavation is believed to be a circular, bowl-shaped cavity with a depth/diameter ratio between 1:4 and 1:3. The form of this initial crater is independent of its diameter, the impact velocity, impact angle (within limits), gravitational acceleration, and nearly every other property of the target or projectile. This "transient" crater then undergoes different degrees of modification as a result of gravitational instability and collapse. The final crater morphology is sensitive to conditions of the target planet, such as the acceleration of gravity, density, and disposition of the surface materials. This study affords both the opportunity of using impact crater morphology to learn about conditions in the target planet and the challenge of understanding how different conditions affect the course of crater collapse.

Transient crater excavation can be treated with relative accuracy using a variety of numerical methods based on a combination of simple Newtonian mechanics and a thermodynamic equation of state (Anderson 1987). Several two- and three-dimensional "hydrocodes" exist that, given simple characterizations of bulk material properties, accurately predict the formation and evolution of the crater formed by a high-speed impact or explosion when gravity is the main factor limiting crater growth. Further research is still needed regarding the role of material strength (or internal friction in granular materials) on limiting transient crater growth.

Crater collapse, however, depends on the essential details of the strength of material surrounding the crater. In particular, standard strength models used in conventional hydrocodes are not successful in describing crater collapse. The rock surrounding the site of an impact is broken, heated, and shaken by the forces that excavate the crater. Such material responds to the differential loads imposed by gravity in ways that are still not fully understood, although it is clear that some type of strength degradation mechanism must temporarily weaken the rock. Indeed, if the target rock retained its static strength properties, impact craters would not collapse at all, in flagrant contradiction to observations (Dent 1973).

The study of impact crater collapse thus requires a deeper understanding of the fundamentals of dynamic rock failure. A complete understanding of this collapse process does not currently exist. The present challenge in impact cratering studies is to use the observed morphology of extraterrestrial craters and the structure of terrestrial craters to infer the course of events during transient crater collapse. Coupling this phenomenology with a detailed mechanical analysis will, hopefully, bring us closer to a quantitative model for the morphologic variations observed in impact craters.

It may seem incredible that 50 years of study of the impact cratering process have not resulted in a predictive, quantitative model of crater formation. The fact that no such model yet exists, despite many attempts by many authors, indicates that we are still missing major pieces to the puzzle of how rocks respond to sudden shocks. Obviously, such a significant gap suggests that scientific treasures may be gained by further study. In this review we hope to point out the progress that has been made in the past and show some of the promising leads for future work.

IMPACT CRATER MORPHOLOGY

Dence (1965) first classified terrestrial impact craters as either simple or complex in structure. This structural classification, which was based on the study of craters exposed by erosion, also seems to apply to the morphology of fresh craters revealed by images of the surfaces of other planets and satellites. It is worth noting that data sets on terrestrial and planetary craters are highly complementary: because of erosion, geophysical investigations, and direct drilling, the subsurface structure of many terrestrial craters can be explored in detail. The surface morphology, which is usually missing, is often displayed in exquisite detail by images of fresh craters on the surfaces of other planets or satellites.

Simple Craters

Simple craters are circular, bowl-shaped depressions with raised rims and approximately parabolic interior profiles. They exhibit few other internal topographic features with the exception of occasional trails where boulders have rolled from their steep rims into their interiors. The rim-to-floor depths of a large sample of simple craters on the moon are roughly 1:5 of the rim-to-rim diameter (Pike 1977).

Simple craters are widely distributed in the solar system. Most of the craters on small solar system bodies such as asteroids are of this variety. The largest simple crater currently known is 90 km in diameter, on Jupiter's moon Almathea.

Most lunar craters smaller than about 15 km in diameter are simple, as are most terrestrial craters smaller than approximately 4 km in diameter.

Drilling at such simple craters as the 4-km-diameter Brent in Ontario, Canada, reveals a lens of broken rock debris and formerly shock-melted rock underlying the crater floor. This breccia lens in turn lies in a bowl of fractured country rock. The breccia lens thickness is about half of the crater rim-to-floor depth, so the depth from the rim to the bottom of the breccia lens is about one-third of the crater diameter. The volume of the breccia lens is roughly one-half the volume of the crater itself.

The distribution of melt in the breccia lens suggests that the lens is created by the collapse of the steep outer walls of the transient crater (Grieve et al 1977). The breccia lens at Meteor Crater, Arizona, is a mixture that contains clasts from all of the rock units intersected by the crater (Shoemaker 1963), suggesting that the debris first surged outward along the wall of the growing cavity and then drained back into the crater interior.

These facts indicate that a simple crater forms by the relatively straightforward collapse of the rim of the transient crater immediately after it forms. Because the rim is composed of broken rock debris and forms close to the angle of repose of loose rock debris, this process is mechanically plausible and does not present any special difficulties to understand or analyze.

Complex Craters

Complex craters, as their name implies, possess a much more complicated structure than simple craters. On Earth they exhibit central structural uplifts, rim synclines, and outer concentric zones of mainly normal faulting. Images from spacecraft show extraterrestrial craters with single or multiple central peaks, flat inner floors, and terraced rims. The depths of complex craters increase with increasing diameter, but they increase much more slowly than the depths of simple craters. Pike (1977) showed that the depths of complex lunar craters increase as approximately the 0.3 power of their diameter, a result consistent with the depths of complex craters on Mercury and Venus (McKinnon et al 1997).

Complex craters range in diameter from a few kilometers on Earth to a monster crater on asteroid 4 Vesta that is 460 km across, so large that its rim almost encircles the 530 km-diameter equator of Vesta, although its central peak juts out of Vesta's south pole (Thomas et al 1997). Even the icy satellites exhibit this type of crater. In addition to the many complex craters on Ganymede and Callisto, the 130 km-diameter "death star" crater Herschel on Saturn's moon Mimas is a classic complex crater, as is the isostatically relaxed 400 km-diameter crater Odysseus on Tethys. Recent Galileo images indicate that the 25 km-diameter crater Cilix on Europa is also a classic complex crater.

The transition between simple and complex craters occurs over a relatively narrow diameter range on any given solar system body and seems to scale as the inverse power of the surface gravity, g. Thus, the simple-complex transition is relatively well determined on the Moon at about 15 km diameter. On Mercury and Mars the transition occurs at about 7 km diameter, and on Earth it drops to 3 to 5 km diameter (the diameter range seems to depend on whether the crater forms in sedimentary or crystalline rocks). The large crater on Vesta corresponds in nearly every way to a gravity-scaled 62 km-diameter crater on the Moon (Thomas, et al 1997). With Vesta's low surface gravity of 0.22 m s^{-2}, the $1/g$ relation is extended to approximately more than two orders of magnitude in gravitational acceleration.

The simple-complex transition has not been determined accurately for the icy satellites around Jupiter and beyond because Voyager images lack sufficient resolution to accurately determine the depths of small craters. Early studies (Schenk 1991) suggest that the transition takes place at much smaller diameters on icy satellites than on silicate bodies. Thus ice appears to be weaker, in some sense, than silicates, but the exact quantification of this result must await analysis of the high-resolution Galileo data.

As crater size on any one body increases further, the central peak complex in a complex crater begins to break up and form an inner ring of mountains. In sufficiently large craters the ring appears at about one-half the rim diameter. Such craters are termed "peak ring" craters and have been observed on Earth, Venus, Mars, Mercury, and the Moon (Melosh 1989). Study of large Venusian craters shows that this transition is actually gradual (Alexopoulos & McKinnon 1994), with a small inner ring expanding to half of the rim radius as the crater diameter increases. The transition between central peak and peak ring craters also appears to scale as $1/g$, although the range of g for which we have data is more limited than for the simple-complex transition. Peak rings are uncommon on icy satellites: their place in the size-morphology sequence is taken by central pit craters, whose formation may be related to the unusual properties of water ice (Schenk 1993), but is otherwise not understood at the present time. Central pit craters also occur on Mars.

Multi-Ring Basins

The very largest impact craters on some bodies exhibit many more rings. The classic basin is Orientale, just over the Moon's western limb. First recognized on rectified lunar photographs in 1962 (Hartmann & Kuiper 1962), Orientale possesses at least five circular rings that form inward-facing scarps up to 6 km high. A second variety of multi-ring basin was discovered on the Jovian satellite Callisto during the Voyager encounters in 1979–1980 (Passey & Shoemaker 1982). Typified by the Valhalla structure, this type of multi-ring basin exhibits a central bright patch surrounded by a system of concentric ridges. These ridges are surrounded by dozens of grabens or outward-facing rings that may extend thousands of kilometers from the impact point. Multi-ring basins of both types appear to form as a tectonic response of the target's lithosphere to the cavity created by the impact (Melosh & McKinnon 1978, McKinnon & Melosh 1980). As such, the formation of multiple rings indicates the presence of a low-viscosity or low-strength layer below the surface. The extent of the ring system provides an indication of the strength and thickness of the lithosphere, with extensive rings forming in thin, weak lithosphere (Melosh 1982b).

Not all planets possess multi-ring basins. Despite some assertions to the contrary (Spudis 1993), it does not appear that any crater on Mercury possesses an external ring scarp (Wood & Head 1976), even including the 1300 km-diameter Caloris basin. The Moon possesses about nine multi-ring basins, and Venus supports four multi-ring basins (Alexopoulos & McKinnon 1994). The 180 km-diameter Chicxulub impact crater on Earth is now definitely identified as a multi-ring basin (Morgan et al 1997), and more circumstantial evidence suggests that the 200 km-diameter Sudbury structure might also be a multi-ring basin (Spray & Thompson 1995). The case for Mars is less clear because of erosion: neither the 2000 km-diameter Hellas basin nor the 1200 km-diameter Argyre basin shows multi-ring scarps. The highly degraded impact structure Isidis may have rings at 1100 and 1900 km diameter (Wood & Head 1976), but interpretation is hampered by extensive postimpact alteration. Ganymede possesses both an Orientale-type basin in the 550 km-diameter ring around the crater Gilgamesh and evidence for an enormous Valhalla-type structure in the furrows that cross Galileo Regio (Schenk & McKinnon 1987). Besides Valhalla itself, Callisto also possesses at least two other smaller structures of the same type. The recent Galileo images of Europa have revealed two classic Valhalla-type structures, Callanish and Tyre, neither of which is more than about 40 km in diameter.

Multi-ring basins are apparently formed by a type of collapse qualitatively different from the collapse that yields complex central peak or peak ring craters. The transition to multi-ring scarps does not scale as $1/g$ and seems to depend a great deal on the rheological conditions near the surface of the planet on which they form (specifically, a weak subsurface layer that can flow on the timescale of crater collapse). For these reasons, and in the interest of brevity, we exclude multi-ring basins from further consideration in this review, referring the reader to Melosh (1989, chapter 9). There is still a great deal of controversy about how these largest of impact structures form, and much work remains to be done before their formation is fully understood.

MECHANICS OF IMPACT CRATER FORMATION

Principal Stages of Impact Crater Formation

The impact process as a whole can be described as a kind of explosion, in which the initial kinetic energy of the projectile does work on the target to create a hole—the crater—as well as heating the material of both projectile and target. In most cases of planetary interest, the crater is much larger than the original projectile.

The course of events in a high-velocity impact may be separated into several sequential stages. Each of these stages is dominated by a specific set of major physical and mechanical processes. It is conventional to distinguish three main stages of an impact event (Melosh 1989). These stages are contact and shock compression, transient cavity growth by crater material ejection, and, finally, transient cavity modification (slumping or collapse). These stages do not have strict boundaries and are used for convenience in the analysis of impact processes because they highlight the dominant mechanisms acting at any given time.

Although the first stage, contact and compression, has little to do directly with crater collapse, it lays the foundation for the subsequent crater formation events, so we begin with a brief description of this stage. The subsequent stages of crater excavation and modification will be discussed in more detail.

Contact and Compression—Shock Wave Generation

In the first stage of an impact the energy released by deceleration of the projectile results in the formation and propagation of shock waves away from the point of impact. A shock wave propagates through the projectile and into the target. This results in a redistribution of the projectile's initial kinetic energy into kinetic and internal energy of all colliding material. The residual kinetic energy is spent ejecting material and opening the transient cavity. The internal energy heats both the projectile and target. For sufficiently strong shock waves this may result in melting or vaporization of material near the impact site.

Contact and compression compose the briefest of the three main stages, lasting only as long as it takes the projectile to enter the target and deposit its energy. If the projectile is approximated as a sphere of radius a that strikes at velocity v and at an angle to the horizontal θ, the duration of this stage is given by $t_{cc} = a/(v \sin\theta)$. For a typical 1 km-diameter impactor striking at 15 km s^{-1} and a 45° angle, this stage lasts literally for only the blink of an eye—about 0.1 seconds.

During this stage, shock pressures are of the order of the stagnation pressure ρv^2, where ρ is the smaller of either the projectile or target density. At speeds above a few kilometers per second these pressures greatly exceed the strengths of any known materials, and the process can be treated hydrodynamically.

Owing to the rapidly changing geometry of the projectile and target during this stage and the highly nonlinear equations of state that must be used to describe the thermodynamic properties of common materials at high pressure and temperature, contact and compression must be studied numerically. Fortunately, numerical hydrocodes are now highly developed for both two-dimensional (axisymmetric, corresponding to vertical impacts) and three-dimensional (required for oblique impacts, which may possess only bilateral symmetry) geometries (Johnson & Anderson 1987). Full three-dimensional computations at high resolution still tax modern computers but are now well within the range of solution on workstation-class machines.

Excavation and Transient Crater Growth

After the shock wave forms in the target it expands away from the impact site, compressing and accelerating the material it encounters. Although the pressure drops toward zero after the shock wave passes, the particle velocity drops to about 1/5 of the peak velocity in the shock wave. This residual velocity, the existence of which is attributed to the thermodynamic irreversibility of the shock wave, eventually acts to open the crater (Melosh 1985).

In the first moments after the shock passes over it, material in the target moves outward along directions that are approximately radial to the point of impact. Because of the presence of the target's free surface, however, pressure gradients behind the shock tend to deflect the particle trajectories toward the surface. The mutual action of these pressure gradients and the inertia of the initial pulse motion result in curved trajectories of target material. The complex material motion away from the point of impact opens the growing cavity, the "transient crater." Roughly equal volumes of material are either ejected ballistically from the crater or displaced by plastic flow downward into the target. In most experimentally and numerically modeled impacts the growing transient crater is initially shaped like a hemisphere. Owing to greater resistance with increasing depth in nonductile targets, the transient crater's depth stops growing at a time when the crater radius continues to increase. The transient cavity eventually reaches a maximum volume. This moment may be defined as the effective time of transient cavity formation (Melosh 1989).

The general timescale for the growth of a transient crater is defined by the basic principles of mechanics: a high-velocity impact transfers a specific amount of energy and momentum into the target material. In response, the material begins to move ("to flow") away from the point of impact. This motion is essentially an inertial one: the transfer of energy and momentum from the shock waves to the target occurs very quickly in comparison to the time of the cratering excavation flow. The initial kinetic energy of the cratering flow is spent working against strength or friction forces and gravity. The simple balance of these factors allows us to integrate the equation of motion and to estimate the moment when the excavation flow should stop.

Thus, if the transient crater growth is halted by gravity alone, as would be the case for impacts into liquid water, the timescale is of order $\sqrt{H_f/g}$, where H_f is the final crater depth and g is the acceleration of gravity. If crater growth is halted by elastic-plastic material strength Y, the timescale is given by $H_f\sqrt{Y/\rho}$. For more complicated failure laws or more accurate estimates of excavation time, detailed numerical studies with more elaborate material models are required.

The maximum depth and diameter of the transient crater are determined by the properties of the target, which may include density, strength, and acceleration of gravity. On the other hand, the ratio of depth to diameter seems to be nearly independent of size, strength, gravity, etc. Although some early centrifuge experiments and numerical computations suggested that large transient craters might be shallower than small ones, there is no longer support for this theory (O'Keefe & Ahrens 1993). Very recent studies of the crustal thickness beneath large lunar basins also support this hypothesis of "proportional scaling" for the dimensions of the transient crater up to craters at least 500 km in diameter (Wieczorek & Phillips 1998). For complex craters the validity of proportional scaling does not, however, imply that at some single point in time the crater had the form of a bowl-shaped depression with depth-to-diameter ratio of 1:3 to 1:4. In many simulations the floor of the transient crater achieves its maximum depth and begins to rise into a central peak while the diameter is still increasing (Melosh 1989).

The size of the transient crater can be estimated by a variety of methods, ranging from laboratory-scale experiments, numerical computations, and, most conveniently, scaling laws. The use of scaling laws for estimating the size of impact craters has reached a high level of sophistication (Holsapple & Schmidt 1982). Based on the idea that early-time phenomena described by projectile parameters are related to late-time cratering phenomena by a single, dimensional, "coupling constant," scaling laws link nondimensional crater descriptors by a variety of power laws. For well-studied cases such as impacts into water or into ductile metals whose strength is described by an elastic-plastic yield law, scaling relations give accurate descriptions of the overall dimensions of the transient crater (Holsapple & Schmidt 1982, O'Keefe & Ahrens 1993).

Although the strength of materials such as water (strengthless) or ductile metals seems to be reasonably well understood, the strength response of brittle materials such as rock or ice is much less clear. Ductile metals have nearly the same strength Y in tension as in compression, whereas rock is much weaker under tension than under compression. One consequence of this dependence of strength on pressure is that sufficiently large impacts may shatter the rock in tension long before the crater has time to grow, effectively destroying any strength the target rock may possess long before the crater opens. Asphaug & Melosh (1993) noted this effect in a study of the formation of the Stickney crater on Phobos. In this computation of an impact on an initially strong basalt (or ice) asteroid, the outward radial displacement of the target just behind the shock wave produced strong hoop tension that shattered the rock shortly after it was released from compression. This effect has also been observed experimentally in explosions on transparent media (Fourney et al 1984).

Target strength degradation by stress waves was studied in more detail by Nolan et al (1996), who concluded that, for craters on large bodies such as Earth, any intrinsic target strength was destroyed by tensile failure in craters produced by projectiles larger than a few meters in diameter. Transient crater growth in these cases proceeds in a mass of shattered rock, and, although rock friction may play a role in reducing crater size (Ivanov & Kostuchenko 1998), strength probably does not. Much more work needs to be done before the role of rock strength is fully understood, but it appears that the assumption long recommended by such crater-scaling experts as Schmidt & Housen (1987), of transient crater growth in a strengthless medium, may give the best approximation to the truth. Further work should concentrate on the mechanism of strength degradation and how to incorporate this mechanism in the growth of the transient crater.

Elastic Rebound

For many years, geologists used the terms "rebound" or "elastic rebound" to describe their observations in terrestrial impact structures. However, the exact mechanical nature of this phenomenon was not discussed in detail. From the detailed study of underground nuclear explosions and small-scale laboratory explosion experiments it is known that, far enough from the explosion point, rocks are not crushed but instead respond elastically to applied forces. This elasticity causes the distant rocks to experience only reversible deformation: after the explosion impulse, they return to their pre-event state. Such reverse motion of material caused by rebound in the elastic zone has been documented by experiments on contained explosions (Rodionov et al 1971).

In cratering events the presence of the free surface decreases the rebound amplitude in comparison to that observed in underground explosions because the deformational rebound of the elastic zone tends to be discharged in the direction of the free surface. Ivanov et al (1982) conducted experiments with an exotic target material consisting of small pieces (0.5 to 0.8 mm) of rubber. Mixtures of this "granulated" rubber with sand and chalk permitted control of the interparticle friction. Explosion cratering in this medium showed very strong true elastic rebound, producing a central peak in a crater only 20 cm in diameter. However, it is unclear how to scale these experiments to events several kilometers in size.

Numerical simulations also permit investigation of this elastic rebound. Ivanov and Kostuchenko (1998) presented results from a pure Lagrangian calculation of a 5 km s^{-1} impact into a rock target at a scale corresponding to a 40 km-diameter impact crater. The rebound of the elastic zone is obvious, but the maximum rebound velocity was only about 50 m s^{-1}. If the surface rocks were launched vertically at this velocity, they could rise only approximately 125 m. This is not sufficient to uplift the crater floor substantially in a crater of this size (Figure 1), for which the expected uplift is about 4 km (Equation 1b). One may suppose that a 50 m s^{-1} rebound could be enough to make a detectable uplift in craters of smaller scale. However, for smaller-scale events the role of lithostatic pressure, which keeps the rocks in an elastic state, is also decreased, and the damaged zone increases in size (Dabija & Ivanov 1978, Ivanov et al 1997). The larger relative size of the damaged zone decreases the role of the elastic rebound. Elastic rebound has been observed in experimental explosive cratering (Ullrich et al 1977) but only in specialized circumstances involving very strong rocks.

Although elastic rebound is not affected directly by gravitational acceleration, it is difficult to see how the various morphometric transitions in impact craters can be a simple inverse function of g unless gravity, not elastic forces, controls the process. The systematic change of crater morphology with increasing size (simple bowl-shaped to central peaked to double ringed) is found on all planetary bodies with a solid crust. For terrestrial-type planets (Mercury, Venus, Earth, the Moon, Mars and, now, the asteroid Vesta) the critical diameters of these morphological changes are inversely proportional to the surface gravity (Pike 1980). This strongly supports the idea that gravity is the main force driving transient crater collapse. We will henceforth neglect the role of elastic rebound in most large impact events, but it may play some role in special circumstances.

Modification of the Transient Crater

In most large craters, the modification or collapse of the transient cavity produces the final crater shape. The ultimate driving force is gravity: transient crater collapse results in a shallower crater geometry that is more stable in a gravity field (Quaide et al 1965). Depending on the size of the cratering event, modification may include the slumping of the crater walls in the form of landslides, stepped terrace formation, uplift of the crater floor, and central peak formation. The inward and upward material motion during the modification stage results in a complex intermixture of breccia and impact melt inside the crater

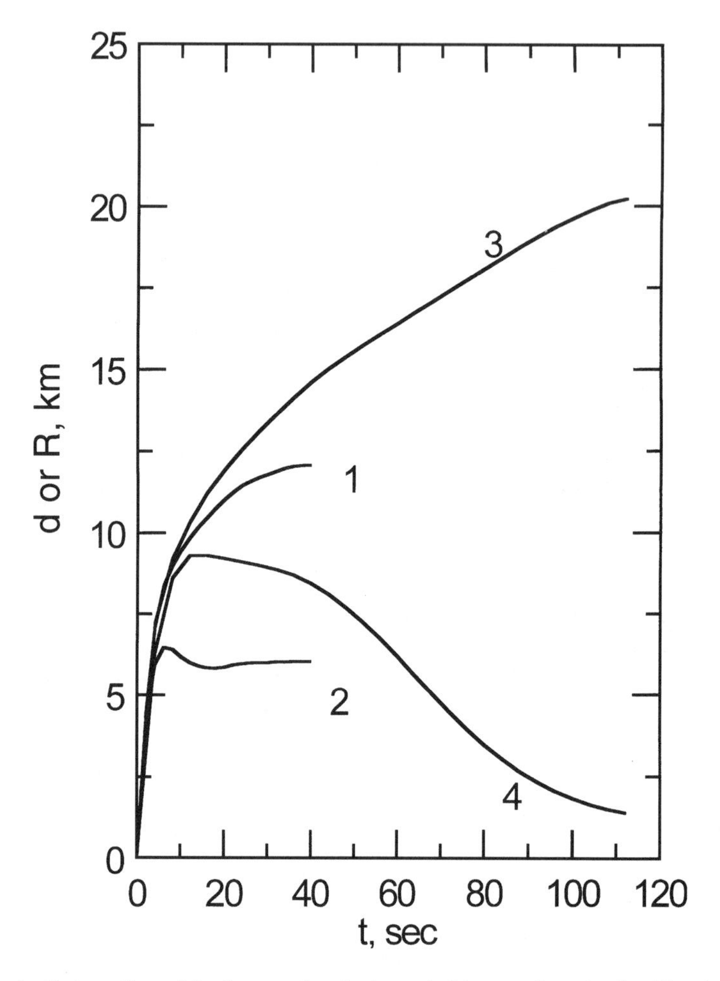

Figure 1 Crater radius and depth versus time for the vertical impact of a projectile with an impact velocity of 5 km s^{-1} and a diameter of 6 km in a terrestrial gravity field. Curves 1 and 2 show radius and depth, respectively, for dry friction. Curves 3 and 4 show radius and depth, respectively, for a model of acoustic fluidization. An elastic rebound for the case of dry friction is seen at $t = 8$ s (curve 2). Note the drastically increased transient depth owing to acoustic fluidization. After Ivanov & Kostuchenko (1997).

depression. The mechanical style of modification depends on the gravity field of the planet and the strength of near-surface rocks. A full mechanical model of crater collapse is still under construction. However, it is clear that to explain the observed dependence of final crater morphology on crater diameter (ranging from simple craters to craters with a central peak to double-ring craters), some type of extreme strength degradation must occur in the rocks surrounding the crater. One very specific (and perhaps somewhat exotic) mechanism that has been proposed is acoustic fluidization (Melosh 1979, 1989, Ivanov & Kostuchenko 1997). Another model that is patterned after thermal weakening of metals (see below) is by O'Keefe & Ahrens (1993, 1998).

FORMATION OF SIMPLE CRATERS

In media such as ductile metals, plasticene, or wet clay, plastic flow occurs after the deviatoric stresses exceed some threshold Y. This flow does not dramatically change the final shape of the transient cavity. So long as gravitationally induced stresses following excavation do not exceed the yield stress Y, the shape of craters in these materials remains close to a hemisphere. Thus the main morphometric parameter, the ratio of crater depth to crater diameter, is close to 1:2. Craters of this kind, commonly produced at laboratory scales, can be conveniently thought of as "frozen" transient craters.

In media such as sand or fragmented rocks, two mechanisms modify the transient cavity shape. The first mechanism is dry friction: the action of dry friction leads to the continued growth of the transient cavity diameter after the moment when the cavity depth reaches its final value. The second mechanism is the slumping of the steep transient cavity walls under the action of gravity. These two mechanisms result in similar crater shapes that are close to a paraboloid of revolution with a depth-to-diameter ratio in the range from 1:5 to 1:4. Both mechanisms generally act together, but typically dry friction is more important in laboratory-scale impact and 0.1- to 1-m-diameter explosion craters. Small lenses of mixed breccia produced by wall slumping appear in missile impact craters in the size range of 2 to 10 m on Earth (Moore 1976). Wall slumping dominates for natural craters with diameters above approximately 100 m.

Wall slumping is the main modification process for simple craters. The result of slumping is a breccia lens that overlies the true floor of the crater. Grieve & Garvin (1984) developed a qualitative model of the mass balance attributed to crater wall slumping. The true floor is the remnant of the transient cavity floor, and it may be recognized by the presence of melted rock that once lined the transient crater and by the fractured but unmixed rocks beneath. For example, Arizona's well-known Meteor Crater, with a diameter of 1.2 km, has a depth from the level of the preimpact surface to the true floor (called true depth) of approximately 300 m (true depth-to-diameter ratio of 1/3). The true floor is covered with a 150 m-thick breccia lens. The resulting apparent depth of the crater is (300 − 150) = 150 m below the original target surface. Finally, the ratio of the crater depth measured from the uplifted rim crest to the visible (apparent) crater floor of Meteor Crater is 0.19 (Roddy et al 1975).

FORMATION OF COMPLEX CRATERS

Complex craters have smaller depth-to-diameter ratios than simple craters. The geological study of terrestrial complex craters shows that rock strata beneath the center of the crater are uplifted above the pre-impact level (Dence et al 1977). At the center of such a crater this uplift creates a central mound or central peak. A ring depression (or circular trough or rim syncline) surrounds the central mound. This ring depression is filled with fragmented material (allogenic breccia) and impact melt. On other planets one can see only the summit of the central mound rising above a relatively level plain that is underlain by a mixture of breccia and melt. The central mound is a manifestation of the main modification mechanism for complex crater formation: the uplift of the transient cavity's floor. This uplift is accompanied by subsidence of the crater rim. Overall, the process is referred to as transient crater collapse.

MECHANICS OF TRANSIENT CRATER COLLAPSE

Uplift of Deep Strata

Numerous observations of the geology of complex terrestrial impact craters demonstrate that deep target strata are uplifted above the pre-impact level due to the collapse of the transient cavity. In some cases the boundary between sedimentary and basement rocks traces this uplift. Eroded impact structures clearly expose deep layers of basement rocks (for example, at Puchezh-Katunky and Kara in Russia or Vredefort in South Africa). For impact craters entirely formed in sedimentary targets (for example, Steinheim in Germany and Gosses-Bluff in Australia) one can trace the structural uplift using identifiable horizons of different kinds of sedimentary rocks.

Grieve et al (1981) summarized geological and geophysical stratigraphic data on the structural uplift with

$$SU = 0.06D^{1.1} \tag{1a}$$

where SU is the stratigraphic uplift (km) and D is the final crater diameter (km). The data set used here includes 14 structures from 3 to 30 km in diameter and the deeply eroded Vredefort structure. Without the point for Vredefort the data for 14 craters may be also presented as

$$SU = 0.1D \tag{1b}$$

to the same level of accuracy (Ivanov et al 1982, Basilevsky et al 1983).

An independent estimate of the stratigraphic uplift may be derived from the observations of unmelted shocked rocks in the central uplift. Rocks in the central uplifts of terrestrial craters exhibit a level of shock metamorphism characteristic of shock pressures from 30 to 50 GPa. This is well below the melting pressure for typical crustal materials. The physics of high-velocity impacts indicate that this range of pressure is much less than the maximum pressure during contact and compression. Consequently, the presence of these rocks close to the surface shows that they were shocked at greater depth and then uplifted to their present position during crater modification. The depth at which the shock pressure falls to 30 to 50 GPa gives an independent estimate of the stratigraphic uplift. Not surprisingly, this estimate is close to that derived from the geological data in Equations 1a and 1b (Ivanov et al 1982, Basilevsky, Ivanov et al 1983).

Timescale of Crater Collapse

Observational data on the geology of terrestrial craters give some limits on the timescale for complex crater formation. In most known terrestrial craters, overlying sediments younger than the crater itself are not deformed. This means that the crater modification time is short in comparison with the rate of accumulation of sediments. In some craters the presence of a massive impact melt sheet is well documented. The best examples are Boltysh (Ukraine, $D = 25$ km) and Manicouagan (Quebec, Canada, $D \sim 70$ km). In both cases the impact melt sheet surrounds the central uplift (Onorato et al 1978, Masaitis et al 1980). If the melt sheet had solidified before the central mound formed, the uplift should deform the impact melt sheet. In contrast, the observed geology corresponds far better to melt solidification after the central mound formed. Thus the impact melt solidification time gives an upper limit to the duration of central uplift formation.

Impact melt solidification is not, unfortunately, a simple process. Impact melts differ sharply from "normal" volcanic melts in two respects: (*a*) the wide range of initial temperatures ranging from the melting point to the boiling point (volcanic melts are mostly separated at the liquidus), and (*b*) the widely variable clast content, which is a result of mixing by the flow during crater excavation. Based on investigations of the Manicouagan melt sheet (Onorato et al 1978), the cooling history of this massive impact melt consists of two main phases. In the first phase, local heat exchange (millimeter to centimeter scale) results in the attainment of a local temperature equilibrium between melt and clasts. In most cases the temperature drops down to the liquidus in 100 seconds. After the first 100 seconds the impact melt viscosity becomes relatively large—some big boulders may be trapped before they reach the melt pool bottom. The local thermalization time depends on the clast temperature. On Venus the local thermalization time may be 10 times longer than on Earth due to Venus' inherently high crustal temperature (Ivanov et al 1992).

In the second phase of melt cooling (after local thermalization is complete), thermal conductivity controls the cooling rate of the entire mass. The thicknesses of hot melt/suevite bodies vary widely (suevite is the name given to a mixture of clasts and a quantity of impact melt). In the range of crater diameters from 10 to 100 km, this thickness h ranges from 200 m to 2 km. The cooling time of such a mass is given by the simple formula $t^* \sim h^2/\kappa$, where κ is the thermal diffusivity and t^* is the characteristic cooling time. For a typical rock, κ is of order 10^{-6} m^2 s^{-1}, and for h of 1 km, $t^* \sim 3 \times 10^4$ years. So 30,000 years is a first-order estimate for a 1 km layer, and it varies as h^2. This estimate may be dramatically changed by water circulation where conduction is outstripped by hydrothermal heat transfer.

The first phase of melt cooling establishes the best limit on the timescale for crater collapse. Since the viscosity of the melt mass increases greatly after local thermalization, it is clear that the central uplift in both Boltysh and Manicouagan craters formed within about 100 seconds of their excavation. Young lunar craters such as Copernicus and Tycho also exhibit flat, locally irregular floors that are believed to represent solidified impact melt units. Because these floors lap up against both central peaks and terraces in the outer walls, their solidification must postdate most of the crater collapse process.

A minimum timescale for transient crater collapse may be derived from the collapse of a crater in a strengthless, inviscid fluid. Numerical calculations as well as laboratory experiments give a timescale of the order of $\sqrt{H_f/g}$, where H_f is the maximum transient cavity depth. For transient depths in the range of 1 to 10 km (final crater diameter range 10 to 100 km) the gravity collapse time is 10 to 30 seconds. This is also approximately the same time required for the transient cavity to attain its maximum depth in a strengthless target.

By comparing all these estimates, we see that for strengthless targets the transient cavity has enough time to collapse and to form a prominent central uplift before the solidification of impact melt (timescale of several minutes). The subsequent cooling of melt/suevite bodies occurs over a timescale of thousands of years.

On a still longer timescale, slow viscous relaxation of a crater that is initially out of isostatic equilibrium will flatten it still further. The viscous (creep) relaxation time t_R is given roughly by

$$t_R = \frac{4\pi\eta}{\rho g D} \tag{2}$$

where η is an effective target viscosity, ρ is the target density, g is the acceleration of gravity, and D is crater diameter (Scott 1967). Upper crustal rocks on silicate planetary bodies have viscosities well above 10^{25} Pa s, so for craters with diameters in the range 10 to 100 km the viscous relaxation time is in the range of 10^{10} to 10^9 years. This time is comparable to or much larger than the age of the Solar System. Viscous relaxation may occur faster on bodies whose surface rocks are closer to their melting points or in craters that are large enough to respond to the flow of rocks deep in the interior of a planet. One of the best examples of a viscously relaxed crater is Odysseus on Tethys.

Gravity Collapse: Strength Properties, Friction, Thermal Softening, and Creep

As described, in large-scale events, gravity collapse results in uplift of the crater floor in place of (or in addition to) a simple landslide from the wall. This phenomenon is what one observes during cratering in water or milk (Worthington 1963) or any strengthless fluid. The analogy with a fluid was widely used to create a set of "splash" or "tsunami" models of complex crater formation (see e.g. Baldwin 1981). Quantitative analyses of transient cavity collapse in a gravity field show, however, that the process is better described by the collapse of a crater in a material that behaves as a Bingham fluid. This type of rheology has also recently been found to provide an excellent phenomenological description of the flow of large masses of rock debris in large rock avalanches (Dade & Huppert 1998).

A Bingham fluid responds elastically to an applied stress until some strength limit is reached, the "Bingham yield stress" Y_B, after which it flows as a viscous fluid (Bingham 1916). Although it was initially unclear just how this peculiar rheology is realized in the rock surrounding an impact event, Melosh (1977) showed that the observed morphology of collapsed craters is well described by a fluid with a Bingham yield stress. The required yield stress is in the vicinity of 30 bars and must be accompanied by an effective angle of internal friction below five degrees (McKinnon 1978). This model is also quantitatively consistent with the formation of slump terraces in lunar (Pearce & Melosh 1986) and Mercurian craters (Leith & McKinnon 1991). The assumption of an appropriate viscosity even gives a quantitative explanation of how peak rings form, by sloshing of the fluidized debris surrounding the transient crater (Melosh 1982a, Alexopoulos & McKinnon 1994).

In spite of the great difference between the behavior of a Bingham fluid and the familiar static rheology of rock material (Jaeger & Cook 1969), the Bingham rheology does seem to provide a good phenomenological description of the behavior of rock debris in the vicinity of an impact (Melosh 1989). Because the rock debris needs to exhibit this strange behavior only for as long as the crater takes to collapse, we seek a mechanism that causes only temporary fluidization. The nature of this fluidization is poorly understood at present (see the review of hypotheses, Melosh 1989). Melosh (1979) suggests that this fluidization is fundamentally caused by acoustic waves (strong vibrations) in the broken rock debris surrounding the freshly excavated crater. This approach is discussed further in the next section.

Although agreeing that strength degradation is necessary to correctly model crater collapse, O'Keefe & Ahrens (1993, 1998) propose a model apparently inspired by existing models for thermal weakening of metals. Numerical hydrocode modelers have long recognized that, as the melting point of a material is approached, its strength must decrease toward zero (Anderson 1987). A standard procedure is to multiply the elastic-plastic yield strength Y by a factor proportional to $(1 - E/E_m)^n$, where E_m is the internal energy at melting (hydrocode computations do not yield temperature directly; they yield only internal energy, which is the reason for this representation), and E is the specific internal energy of the material. The exponent n is usually taken as either 1 or 2. O'Keefe & Ahrens (1993), using $n = 1$, replace E_m with a new energy E_{sw} that is about two orders of magnitude smaller than E_m. E_{sw} is interpreted as the energy necessary for shock waves to shatter the rock. Thus, they are able to derive the same threshold for the simple to complex transition as derived by using the Bingham model discussed above, with the same $1/g$ dependence on surface gravity, although the spatial distribution of strength is different in the two models, as is the dependence on crater size.

The main mechanism investigated numerically by O'Keefe & Ahrens (1993, 1998) is the shock heating of the target material and the corresponding decrease of strength. Thermal softening is especially efficient for large-scale events: whereas gravity tends to decrease the transient cavity size in large events, the zone of intensive shock heating becomes larger in comparison with the zone of the excavation flow. Scaling laws show that the excavation volume for a terrestrial crater with a diameter of 300 km is approximately equal to the impact melt volume (p. 123 and Fig. 7.7, Melosh 1989). However, for smaller craters ($D < 200$ km), shock heating seems to influence too small a volume, and the flow of relatively cold rocks must control the excavation and modification stages of impact cratering.

Ivanov & Deutsch (1998) constructed a numerical model of the Sudbury impact crater's formation. They used a thermal-softening strength model for the rock target similar to the model used by O'Keefe & Ahrens (1993, 1998). In addition, Ivanov et al (1996) introduced a typical continental geotherm in the target. Nevertheless, the results for a Sudbury-scale event show that dry friction dominates the modification of the transient cavity. With thermal softening alone, the simulations show an avalanchelike collapse of the transient crater walls. Crater floor uplift was the main form of modification only when the friction coefficient was artificially decreased (Figure 2). The magnitude of this friction reduction is comparable with values previously estimated by McKinnon (1978). One possible mechanism for this friction reduction is discussed in the next section.

ACOUSTIC FLUIDIZATION

Original Model

Melosh (1979) originally proposed his model as a short-wavelength extension of existing models of earthquake-induced landsliding (Seed & Goodman 1964). It was well known that explosions (and, by extension, impacts as well) induce strong ground motions (Cooper & Sauer 1977). Analysis of a series of 10-ton TNT explosion tests revealed stress fluctuations exceeding the overburden stress near the crater (Gaffney & Melosh 1982). This suggested that the strong shaking produced during crater excavation might play a major role in affecting the rheology of the debris surrounding the crater (Melosh & Gaffney 1983).

The fundamental idea of acoustic fluidization relies on the fact that for a coulomb material the yield stress is, to first order, a linear function of the overburden pressure with a proportionality coefficient μ, the coefficient of internal friction. Portions of the material that are normally under too high an overburden pressure to fail may nonetheless flow plastically if the ambient vibrations temporarily reduce the overburden pressure below the coulomb threshold. Note that, in this context, "acoustic" refers to elastic (sound) waves in the rock debris, not in any adjacent atmosphere.

The strain rate $\dot{\varepsilon}$ in acoustically fluidized debris is given by a rather complicated function of applied shear stress τ, overburden pressure p, and the variance of the pressure fluctuations σ. By following common engineering practice (Crandall & Mark 1973), the amplitudes of the pressure fluctuations are assumed to be distributed according to a Gaussian law. The strain rate also depends on the density of the debris ρ, the S-wave velocity in the debris β, and the dominant wavelength of the acoustic field λ. In terms of these variables the strain rate of vibrated granular debris is

$$\dot{\varepsilon} = \frac{\tau}{\rho\lambda\beta}\left\{\frac{2}{\mathrm{erfc}[(1-\Omega)/\Sigma]} - 1\right\}^{-1} \tag{3a}$$

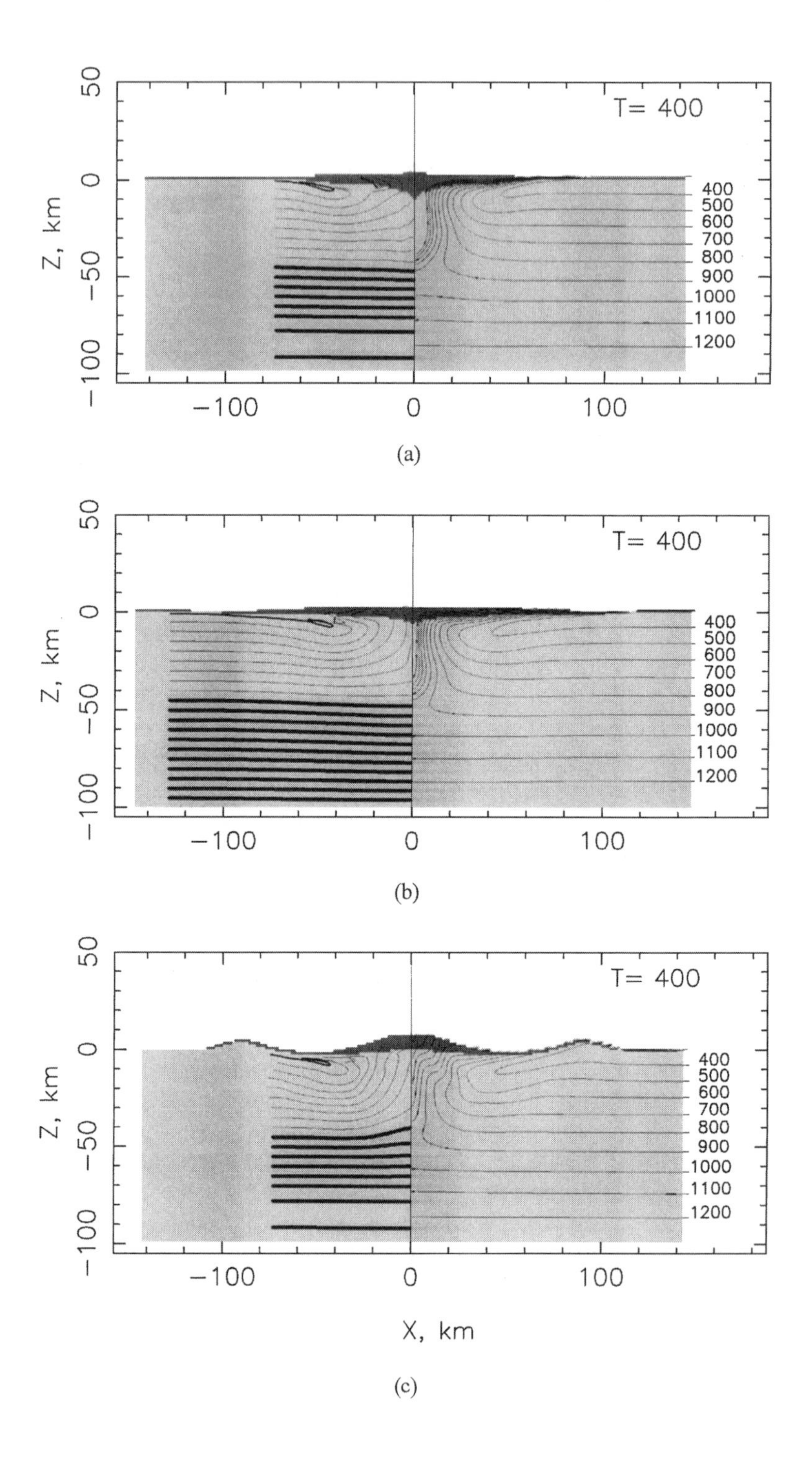
T= 400
Z, km
X, km
400
500
600
700
800
900
1000
1100
1200
(a)
(b)
(c)

where erfc is the complementary error function, Ω is a dimensionless measure of the driving stress, and $\Omega = \tau/\tau_{static}$, where $\tau_{static} = \mu p$ is the stress required to initiate failure when no vibrations are present. Ω ranges from 0 to 1. Σ is a dimensionless measure of the amplitude of the vibrations $\Sigma = \sigma/p$. A somewhat more compact way to represent this flow law is to combine the dimensionless expression into one term, $X = (1 - \Omega)/\Sigma$.

$$\dot{\varepsilon} = \frac{\tau}{\rho\lambda\beta}\left\{\frac{2}{\operatorname{erfc}(X)} - 1\right\}^{-1} \tag{3b}$$

where X ranges from 1 to 0. The utility of this expression will be seen when we compare it with the block model.

The behavior of this rheological law is shown in Figure 3, where Equation 3b is plotted versus the dimensionless driving stress Ω. Although the strain rate is finite for any $\Omega > 0$, the strain rate becomes large only for $\Omega > (1-\Sigma)$, thus exhibiting a kind of Bingham yield behavior in which the yield stress is a function of the amplitude of shaking Σ. As the driving stress approaches the static limit $\Omega = 1$, the strain rate approaches a constant value $\dot{\varepsilon} = \tau_{static}/\rho\lambda\beta$. The stress-strain rate curve is otherwise complex and exhibits substantial non-Newtonian curvature. Complex as this flow law may seem, it has recently received experimental verification in a laboratory study of vibrated sand (Melosh & Girdner 1995).

The original model of acoustic fluidization was entirely a continuum model: the lengths of the elastic waves were assumed to be larger than any intact rock fragments in the material. A fundamental limitation of this model is that it does not predict the wavelength of the vibrations dominating the flow and thus cannot be used to make quantitative predictions of the rheology of the material without further assumptions. The model also does not produce a Bingham rheology in a straightforward way, although it does predict a sharp increase in creep rate above a threshold that depends on the mean amplitude of the vibrations. Questions such as how strong the shaking is and how long it goes on after impact are

Figure 2 A hydrocode simulation of the Sudbury event using a modified version of the SALE code (Amsden et al 1980). A cylindrical projectile with height and diameter both equal to 12.5 km impacts the granite surface at a velocity of 20 km s^{-1}. Model friction coefficients, μ, are equal to (*a*) 0.25, (*b*) 0.125, and (*c*) 0.0625. Vertical (depth) and horizontal (radial distance) scales are in kilometers. The axis of symmetry (zero radial distance) divides each frame in two halves. (*left*) Distortion and displacement of originally horizontal marked layers are shown in the target at $T = 400$ s. Thick lines correspond to layers below 45 km—an estimated Moho depth. The dark gray tone marks cells with densities less than 2000 kg m^{-3} but above 200 kg m^{-3}. (*right*) Displacement and distortion of the isotherms are shown. Isotherms are labeled in kelvins. After Ivanov & Deutsch (1998).

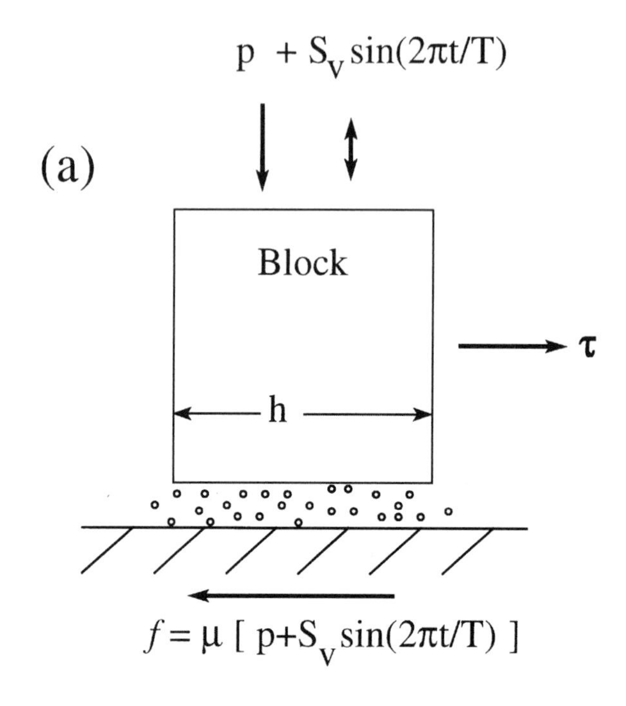
p + S_V sin(2πt/T)
(a)
Block
τ
h
f = μ [p+S_V sin(2πt/T)]

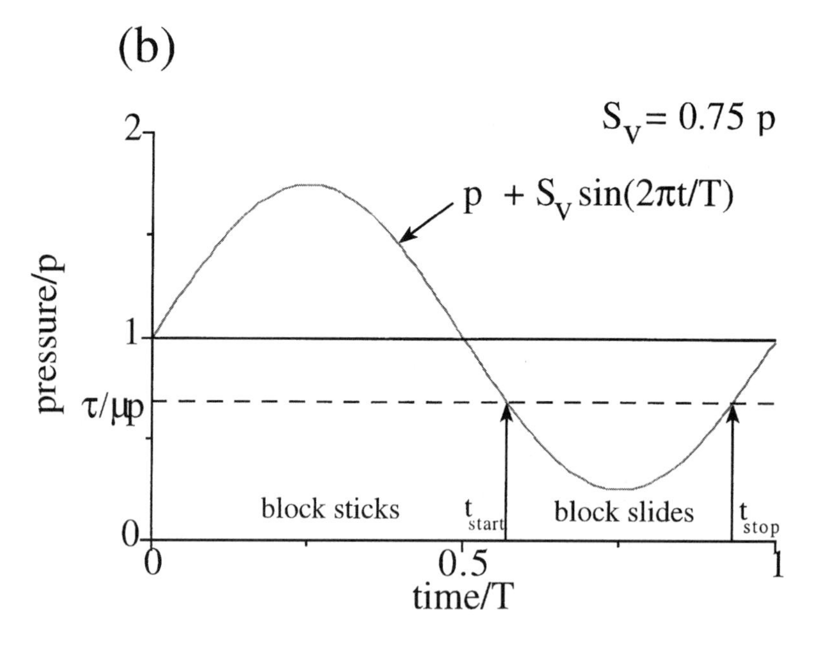
(b)
S_V = 0.75 p
p + S_V sin(2πt/T)
pressure/p
2
1
τ/μp
0
block sticks
t_start
block slides
t_stop
0
0.5
1
time/T

important for estimating the degree of collapse of the crater, but previous theories gave little information on these parameters. More recent studies (Melosh 1996) provide a rationale, still within the context of a continuum model, for estimating the rate at which acoustic (vibrational) energy in a mass of rock debris is generated, propagates elsewhere, and decays. In the future we hope that such modeling, carried out in conjunction with hydrocode computations, may yield deeper insight into how sub–mesh-scale vibrations might affect the gross strength properties of rock debris in the vicinity of an impact.

Block Model

One of the characteristic features of rock deformation is that the rock medium deforms not as a plastic metallike continuum but as a system of discrete rock blocks. Deep drilling of 40 km-scale impact craters reveals a system of rock blocks ranging in size from 50 to 200 m, with an average size of about 100 m (Ivanov et al 1996). A model of these block oscillations may be used to formulate an appropriate rheological law for the subcrater flow during the modification stage.

For this model to be valid, the time for sound to cross a block must be short compared with the period of vibration of one block against another. In other words, the sound speed of the matrix between the blocks must be much smaller than that of the intact rock. A soft interblock breccia layer of about 10 to 20 percent of the block's thickness (10–30 m) with a sound speed approximately 500 to 1000 m s^{-1} provides a plausible rationale for this model.

In a simple, one-dimensional approximation, the "acoustic fluidization" equations describe a single block sliding along the surface (Figure 3*a*). Imagine that the block is under a normal stress p that creates a dry friction force μp. This force prevents the block from moving under the traction stress τ. Let the block oscillate vertically with a period T. The oscillation creates a sinusoidal variation of the normal stress with amplitude S_v. Under this assumption the block is static (velocity $v = 0$) whenever the vertical stress $p + S_v \sin(2\pi t/T)$ creates a friction force larger than the traction. However, for the time period t_{free} (see Figure 4*b*), friction is less than the traction, and the block begins to move ($v > 0$). A minimal requirement for this motion is that $\tau > \mu(p - S_v)$, a condition which defines a kind of Bingham yield stress, $Y_B = \mu(p - S_v)$. If τ is

Figure 3 (*a*) Schematic drawing of the one-dimensional model for a block sliding along the underlying surface. The block, of size h, is under a static pressure p and a traction stress τ. The friction stress is proportional to the overburden pressure with a friction coefficient μ. The pressure oscillates around its static value p with amplitude S_v. (*b*) Time variation of the normal stress. Owing to oscillations, the friction force is below the strength limit during the period t_{free}. During this time the block accelerates under the applied traction. See text for a more detailed description.

less than Y_B, the block does not slide. If τ is greater than this threshold, there is a time during the cycle when the block is accelerated by the difference between the traction stress τ and the frictional resistance $\mu[p + S_v \sin(2\pi t/T)]$. When the normal stress increases back to the friction limit, the block stops. Figure 4*b* shows that this occurs between the third and fourth quadrant of the sine function at times given by

$$t_{start,stop} = \frac{T}{2\pi} \sin^{-1}\left(\frac{\tau - \mu p}{\mu S_v}\right) \tag{4}$$

During the next period of oscillation, the block moves again. This simple scheme allows us to construct a nonlinear rheological law similar to the acoustic fluidization equations proposed by Melosh (1979).

By integrating the acceleration and velocity of the block to obtain its displacement δ per cycle, it is easy to show that the strain rate of the block is given by $\dot{\varepsilon} = \delta/Th$ where h is a characteristic dimension of the block. The

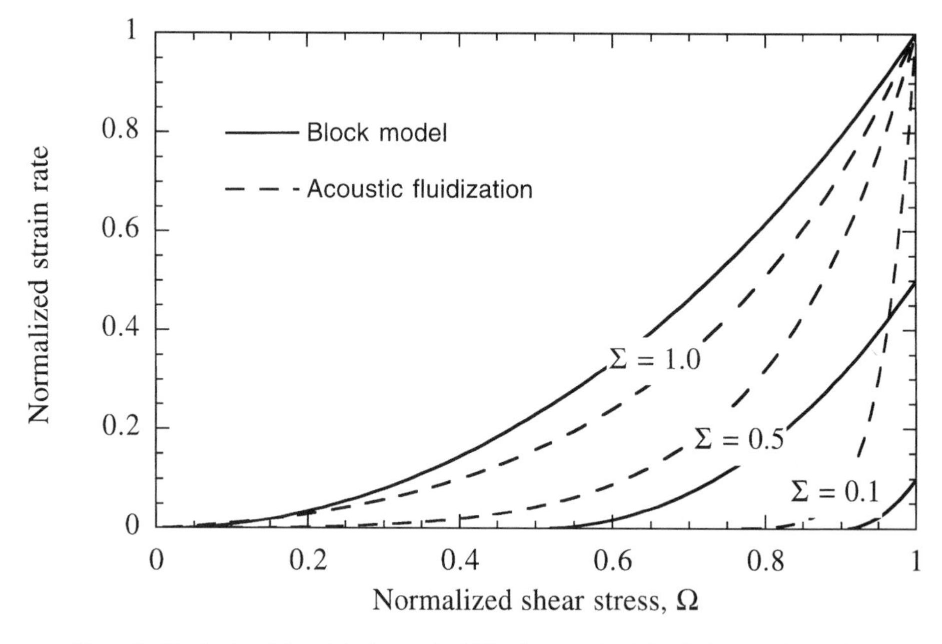

Figure 4 Rheologies of the original acoustic fluidization model and the block model for the flow of strongly vibrated rock debris. The plot illustrates the relation between strain rate, normalized by the maximum strain rate, and applied stress, normalized by the static sliding stress derived from Equations 3b and 5 in the text. The different curves are labeled by Σ, the dimensionless ratio between the amplitude of the vibrations and the overburden pressure. Although the two models differ in detail, their predictions for the rheology of strongly shaken rock debris are quite similar overall.

integration to derive δ yields a complex functional form, but the overall behavior is simply described. In analogy to Equation 3b the block model yields the rheological law

$$\dot{\varepsilon} = \frac{(\tau - Y_B)T}{2\pi^2 \rho h^2} \left\{ \sqrt{\frac{1+X}{1-X}} - \frac{X}{1-X} \cos^{-1} X \right\} \cos^{-1} X \tag{5}$$

where in this case $\Sigma = S_v/p$ and $X = (1 - \Omega)/\Sigma$, as before. Equation 5 applies to a model in which the block velocity drops to zero as soon as the frictional force exceeds the driving force. A more sophisticated numerical treatment accounts for inertial sliding for a short time after this, but the results differ little from Equation 5.

Figure 4 shows that this implies a rheological behavior very similar to that of acoustic fluidization. The maximum strain rate occurs when $\Omega = 1, \dot{\varepsilon} = (\tau_{static} - Y_B)T/(4\pi\rho h^2)$. Because Y_B is a function of the amplitude of vibration, the maximum strain rate depends on Σ, unlike the case for acoustic fluidization. There is a true Bingham threshold: flow does not begin until τ exceeds Y_B. After yielding, it flows with an effective viscosity given by $2\pi h^2/\rho T$. In spite of these differences the two flow laws are very similar because the ultimate fluidization mechanisms are quite similar. An advantage of the block model is that the length scale h is determined from an observable, whereas in the acoustic fluidization model the wavelength λ was undetermined. On the other hand, the block model itself does not define the period T of the dominant vibrations, which must be determined from other assumptions or observations.

Numerical Modeling: Central Peak, Double Ring

Ivanov and Kostuchenko (1997) published an example of numerical cratering simulations with a simplified acoustic fluidization model. The one-dimensional block oscillation model was implemented in a MAC-type free-surface Lagrangian numerical code (Welsh et al 1966). To simplify the model, they assumed initial conditions of a hemispherical transient crater cavity with initially flat layers deformed in accordance with the Z-model kinematic description of cratering (Maxwell 1977). The initial block oscillation intensity (measured as the amplitude of the velocity of a block oscillation) is assumed to be a constant fraction of the particle velocity behind the shock front (calculated with another type of a hydrocode). The oscillation intensity decays spatially as the inverse square of the distance from the impact point. The oscillation's decay in time follows an exponential law.

To define the scaling law, the block size was assumed to be proportional to the transient crater diameter. The quality factor Q for the oscillation's decay was assumed to be the same for all crater diameters. By definition, Q is the ratio between the energy stored per cycle and the energy lost over the same period. The frequency of the block oscillation needs to be in the range of several hertz,

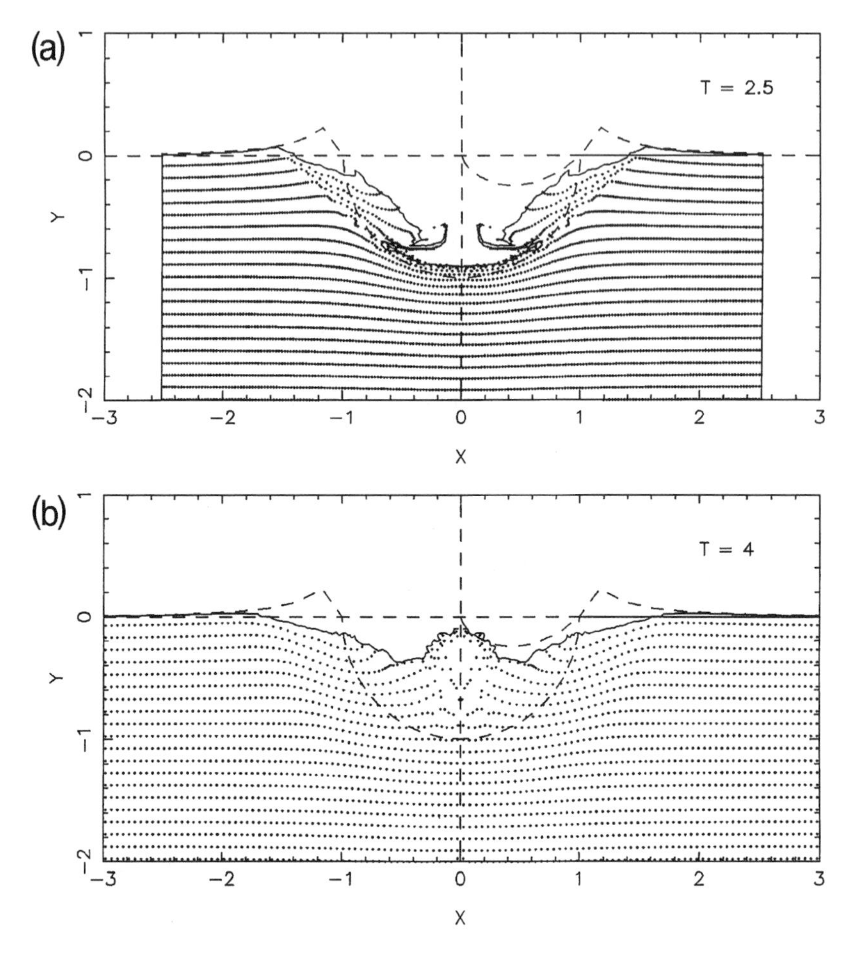

Figure 5 The final shape of collapsed craters with a diameter of ~2 km (*a*), ~20 km (*b*), ~40 km (*c*), and ~80 km (*d*) in the terrestrial gravity field. *Dashed lines* show the initial contour of the transient cavity and (*right*) the ejected volume. Initially flat layers (*dotted lines*) were distorted according to the Z model to form the transient cavity and then followed the crater collapse motion. The length unit is equal to the maximum transient cavity depth. The time unit is equal to the free-fall time from a height equal to the transient cavity radius, $(H_f/g)^{0.5}$.

and Q needs to be of the order of 10 to 100 to fit the observed crater profile. With these parameters we calculated the collapse of various size craters to study the morphology of the final crater. Selected results are shown in Figure 5, which illustrates the change of crater morphology with increasing crater diameter for complex craters.

The numerical computations are most efficiently done using nondimensional, or normalized, values for the viscosity and decay time parameters. Distances

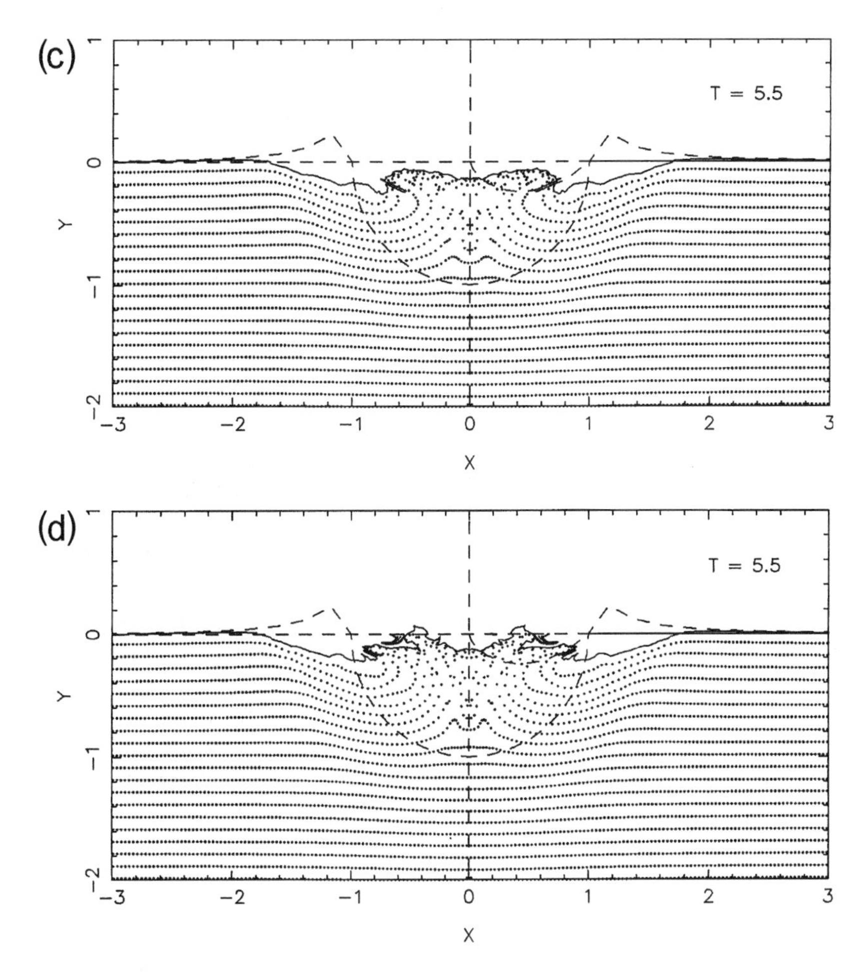

Figure 5 (*Continued*)

are normalized by the transient crater depth H_f. The normalized viscosity is $\eta^* = \eta\sqrt{H_f^3/g}$ and the normalized oscillation decay time is $t^* = t\sqrt{H_f/g}$, where η and t are the dimensional viscosity and decay time, respectively. The numerical computations indicate that a good fit with observations can be achieved with $\eta^* = 0.028$ and $t^* = 1.4$. For a 10 km-deep crater this corresponds to an actual viscosity of 2.4×10^8 Pa, an oscillation decay time of 45 seconds, and Q in the range of 10 to 200, depending somewhat on the assumed block size and the elastic modulus of the block system. For a Q of 10 the block size needs to be about 300 m, whereas a Q of 200 corresponds to 70 m blocks.

In small craters the rapid decay of the vibrations prevents floor uplift, and only a simple inward sliding of the crater walls modifies the transient cavity. Only for crater diameters above a critical threshold do block oscillations facilitate uplift of the crater floor before the prominent wall sliding. This uplift creates a central mound. The model approximately reproduces the well-known depth/diameter law for complex craters: the maximum depth of the final collapsed crater, d, grows with the final crater rim diameter, D, as $d \sim D^{1/3}$.

In this simulation the central uplift for very large craters experiences a substantial overshoot—the upper part of the growing central mound lifts well above the pre-impact ground level, then falls down before finally coming to rest. As the size of the modeled event increases, the further subsidence of the central mound following the overshoot begins to produce a pit at the top of the central uplift. The final topography strongly resembles a peak ring structure (Figure 5*d*). However, the resolution of the currently available numerical models is not yet high enough to permit quantitative analysis because the final model's topography is comparable with the size of computation cells.

The large block reformulation of the original Melosh acoustic fluidization model with proper parameterizations thus reproduces the main features of the impact crater collapse (*a*) the existence of a critical diameter below which no collapse occurs, and (*b*) a gradual change of crater morphology with increasing crater diameter.

The block size is an observable parameter, at least for terrestrial impact craters. Future work should incorporate some form of acoustic fluidization from the very beginning of transient crater growth. This may change the simple assumption of the hemispherical transient cavity used here.

CONCLUSIONS

Mechanical analysis of the collapse of impact craters has proved surprisingly difficult in comparison with the static analysis of the stability of big holes in the ground. Observations of impact crater collapse reveal that the dynamic strength properties of rock debris are strikingly different from their static properties. The very existence of a simple-complex crater transition requires the operation of some process that greatly degrades the strength of the material in the vicinity of an impact. This strength degradation is evidently transient, lasting little more than the few minutes required for craters in the 10 to 100 km range to collapse.

The study of large impact craters has thus revealed new and fundamental aspects of the behavior of Earth materials at scales much larger than we are accustomed to modeling in the laboratory. The information gleaned from studying impact crater collapse may have applications in other, apparently disparate, areas of Earth science. The physics of both earthquakes and long runout landslides have aspects in common with that of collapsing impact craters (Melosh 1983). A deeper understanding of the formation of impact craters may thus have implications far outside the domain of cratering studies.

Within this domain, future progress in understanding the motions that occur in the target subsequent to a large impact will require numerical studies using increasingly sophisticated and realistic strength models. It will also require field and remote studies of impact craters to define better what really happens during collisions of very large Solar System objects.

ACKNOWLEDGMENTS

Boris Ivanov thanks the Russian Foundation for Basic Science for support of his work on impact cratering (grant 96-05-64167). Jay Melosh has been supported by NASA grant NAGW-428 of the Division of Planetary Geology and Geophysics.

Literature Cited

Alexopoulos JS, McKinnon WB. 1994. Large impact craters and basins on Venus, with implications for ring mechanics on the terrestrial planets. In *Large Meteorite Impacts and Planetary Evolution*, ed. BO Dressler, RAF Grieve, VL Sharpton, Spec. Pap. 293:29–50. Boulder, CO: Geol. Soc. Am.

Amsden AA, Ruppel HM, Hirt CW. 1980. *SALE: A Simplified ALE Computer Program for Fluid Flow at All Speeds. LA-8095, Los Alamos Natl. Lab.*, Los Alamos, NM

Anderson CE. 1987. An overview of the theory of hydrocodes. *Int. J. Impact Eng.* 5:33–59

Asphaug E, Melosh HJ. 1993. The Stickney impact of Phobos: A dynamical model. *Icarus* 101:144–64

Baldwin RB. 1981. On the tsunami theory of the origin of multi-ring basins. See Schultz & Merrill 1981, pp. 275–88

Basilevsky AT, Ivanov BA, Florensky KP, Yakolev OI, Fel'dman VI, et al. 1983. *Impact Craters on the Moon and Other Planets.* Moscow: Nauka. 200 pp. (Engl. tech. transl. 1985. *NASA TM 77667*)

Bingham EC. 1916. An investigation of the laws of plastic flow. *Bull. Bur. Stand.* 13:309–53

Cooper HF, Sauer FM. 1977. Crater-related ground motions and implications for crater scaling. See Roddy et al 1977, p. 1245–60

Crandall SH, Mark WD. 1973. *Random Vibration in Mechanical Systems.* New York: Academic. 166 pp.

Dabija AI, Ivanov BA. 1978. Geophysical model of meteorite craters and some problems of cratering mechanics. *Meteoritika* 37:160–67. In Russian

Dade WB. 1998. Long runout rockfalls. *Geology* 26:803–6

Dence MR. 1965. The extraterrestrial origin of Canadian craters. *Ann. NY Acad. Sci.* 123:941–69

Dence MR, Grieve RAF, Robertson PB. 1977. Terrestrial impact structures: Principal characteristics and energy considerations. See Roddy et al 1997, p. 247–75

Dent B. 1973. Gravitationally induced stresses around a large impact crater. *EOS* 54:1207

Fourney WL, Holloway DC, Barker DB. 1984. Model studies of fragmentation. In *Mechanics of Oil Shale*, ed. KT Chang, JW Smith, p. 337–88. New York: Elsevier

Gaffney ES, Melosh HJ. 1982. Noise and target strength degradation accompanying shallow-buried explosions. *J. Geophys. Res.* 87:1871–79

Gilbert GK. 1893. The moon's face: A study of the origin of its features. *Bull. Philos. Soc. Wash.* 12:241–92

Grieve RAF, Dence MR, Robertson PB. 1977. Cratering processes: As interpreted from the occurrence of impact melts. See Roddy et al 1977, p. 791–814

Grieve RAF, Garvin JB. 1984. A geometric model for excavation and modification at terrestrial simple craters. *J. Geophys. Res.* 89:11561–72

Grieve RAF, Robertson PB, Dence MR. 1981. Constraints on the formation of ring impact structures. See Schultz & Merrill 1981, p. 37–57

Grieve RAF, Shoemaker EM. 1994. Record of past impacts on Earth. In *Hazards Due to Comets and Asteroids*, ed. T Gehrels, p. 417–62. Tucson: Univ. Ariz. Press

Hartmann WK, Kuiper GP. 1962. Concentric structures surrounding lunar basins. *Commun. Lunar Planet. Lab.* Tucson: Univ. Ariz. 1:51–66

Holsapple KA, Schmidt RM. 1982. On the scaling of crater dimensions–2. Impact processes. *J. Geophys. Res.* 87:1849–70

Ivanov BA, Basilevsky AT, Sazonoya LV. 1982. Formation of the central uplift in meteoritic craters. *Meteoritika* 40:60–81. In Russian. (Engl. tech. transl. 1986. *NASA TM-88427*)

Ivanov BA, Deniem D, Neukum G. 1997. Implementation of dynamic strength models into 2D hydrocodes: Applications for atmospheric breakup and impact cratering. *Int. J. Impact Eng.* 20:411–30

Ivanov BA, Deutsch A. 1998. Sudbury impact event: Cratering mechanics and thermal history. In *Proc. Sudbury Conf.*, ed. Dessler. Houston: Lunar Planet. Inst. In press

Ivanov BA, Kocharyan GG, Kostuchenko VN, Kirjakov AF, Pevzner LA. 1996. Puchezh-Katunki impact crater: Preliminary data on recovered core block structure. *Lunar Planet. Sci. Conf.* 27:589–90

Ivanov BA, Kostuchenko VN. 1997. Block oscillation model for impact crater collapse. In *Lunar Planet. Sci. Conf. 27th, Abstr. 1655*. Houston: Lunar Planet. Inst. (CD-ROM)

Ivanov BA, Kostuchenko VN. 1998. Impact crater formation: Dry friction and fluidization influence on the scaling and modification. In *Lunar Planet. Sci. Conf. 29th, Abstr. 1654*. Houston: Lunar Planet. Inst. (CD-ROM)

Ivanov BA, Nemchinov IV, Svetsov VA, Provalov AA, Khazins VM, Phillips RJ. 1992. Impact cratering on Venus: Physical and mechanical models. *J. Geophys. Res.* 97: 16167–81

Jaeger JC, Cook NGW. 1969. *Fundamentals of Rock Mechanics*. London: Chapman & Hall. 515 pp.

Johnson WE, Anderson CE. 1987. History and application of hydrocodes in hypervelocity impact. *Int. J. Impact Eng.* 5:423–39

Leith AC, McKinnon WB. 1991. Terrace width variations in complex Mercurian craters, and the transient strength of the cratered Mercurian and lunar crust. *J. Geophys. Res.* 96:20923–31

Masaitis VL, Danilin AN, Maschak MS, Raikhlin AI, Selivanovskaya TV, Schadenkov EM. 1980. *Geology of Astroblems*. Leningrad: Nedra. 231 pp. In Russian

Maxwell DE. 1977. A simple model of cratering, ejection, and the overturned flap. See Roddy et al 1977, pp. 1003–8

McKinnon WB. 1978. An investigation into the role of plastic failure in crater modification. *Proc. Lunar Planet. Sci. Conf.* 9:3965–73

McKinnon WB, Melosh HJ. 1980. Evolution of planetary lithospheres: Evidence from multiringed basins on Ganymede and Callisto. *Icarus* 44:454–71

McKinnon WB, Zahnle KJ, Ivanov BA, Melosh HJ. 1997. Cratering on Venus: Models and observations. In *Venus II*, ed. SW Bougher, DM Hunten, RJ Phillips, pp. 969–1014. Tucson: Univ. Ariz. Press

Melosh HJ. 1977. Crater modification by gravity: A mechanical analysis of slumping. See Roddy et al 1977, p. 1245–60

Melosh HJ. 1979. Acoustic fluidization: A new geologic process? *J.Geophys. Res.* 84:7513–20

Melosh HJ. 1982a. A schematic model of crater modification by gravity. *J. Geophys. Res.* 87:371–80

Melosh HJ. 1982b. A simple mechanical model of Valhalla Basin, Callisto. *J. Geophys. Res.* 87:1880–90

Melosh HJ. 1983. Acoustic fluidization. *Am. Sci.* 71:158–65

Melosh HJ. 1985. Impact cratering mechanics: Relationship between the shock wave and excavation flow. *Icarus* 62:339–43

Melosh HJ. 1989. *Impact Cratering: A Geologic Process*. New York: Oxford Univ. Press. 245 pp.

Melosh HJ. 1996. Dynamic weakening of faults by acoustic fluidization. *Nature* 379:601–6

Melosh HJ, Gaffney ES. 1983. Acoustic fluidization and the scale dependence of impact crater morphology. *J. Geophys. Res.* 88(Suppl. A):830–34

Melosh HJ, Girdner KK. 1995. Rheology of vibrated granular materials: Application to long runout landslides (Abs). *Eos* 76 (Suppl. F):270

Melosh HJ, McKinnon W. 1978. The mechanics of ringed basin formation. *Geophys. Res. Lett.* 5:985–88

Moore HJ. 1976. Missile impact craters (White Sands Missile Range, New Mexico) and applications to lunar research. *US Geol. Surv. Prof. Pap. 812-B*

Morgan J, Warner M, Group TCW. 1997. Size and morphology of the Chicxulub impact crater. *Nature* 390:472–76

Nolan MC, Asphaug E, Melosh HJ, Greenberg R. 1996. Impact craters on asteroids: Does strength or gravity control their size? *Icarus* 124:359–71

O'Keefe JD, Ahrens TJ. 1993. Planetary cratering mechanics. *J. Geophys. Res.* 98:17011–28

O'Keefe JD, Ahrens TJ. 1998. Complex craters: Relationship of stratigraphy and rings to the impact conditions. *J. Geophys. Res.* In press

Onorato PIK, Uhlmann DR, Simonds CH. 1978. The thermal history of the Manicouagan impact melt sheet, Quebec. *J. Geophys. Res.* 83:2789–98

Passey QR, Shoemaker EM. 1982. Craters and Basins on Ganymede and Callisto: Morphological indicators of crustal evolution. In *Satellites of Jupiter*, ed. D Morrison, MS Matthews, pp. 379–434. Tucson: Univ. Ariz. Press

Pearce SJ, Melosh HJ. 1986. Terrace width variations in complex lunar craters. *Geophys. Res. Lett.* 13:1419–22

Pike RJ. 1977. Size dependence in the shape of fresh impact craters on the Moon. See Roddy et al 1977, p. 489–509

Pike RJ. 1980. Control of crater morphology by gravity and target type: Mars, Earth, moon. *Proc. Lunar Planet. Sci. Conf.* 11:2159–89

Quaide WL, Gault DE, Schmidt RA. 1965. Gravitative effects on lunar impact structures. *Ann. NY Acad. Sci.* 123:563–72

Roddy DJ, Boyce JM, Colton GW, Dial AL. 1975. Meteor Crater, Arizona, rim drilling with thickness, structural uplift, diameter, depth, volume, and mass-balance calculations. *Proc. Lunar Sci. Conf.* 6:2621–44

Roddy DJ, Pepin RO, Merrill RB, eds. 1977. *Impact and Explosion Cratering*. New York: Pergamon

Rodionov VN, Adushkin VV, Kostuchenko VN, Nikolaevsky VN, Romashov AN, et al. 1971. *Mechanical Effects of An Underground Explosion*. Moscow: Nedra. 221 pp. In Russian. (Engl. transl. 1972. *USAEC UCRL-Trans-10676*. Los Alamos, NM)

Schenk PM. 1991. Ganymede and Callisto: Complex crater formation and planetary crusts. *J. Geophys. Res.* 96:15635–64

Schenk PM. 1993. Central pit and dome craters: Exposing the interiors of Ganymede and Callisto. *J. Geophys. Res.* 98:7475–98

Schenk PM, McKinnon WB. 1987. Ring geometry on Ganymede and Callisto. *Icarus* 72:209–34

Schmidt RM, Housen KR. 1987. Some recent advances in the scaling of impact and explosion cratering. *Int. J. Impact Eng.* 5:543–60

Schultz PH, Merrill RB, eds. 1981. *Multiring Basins*. New York: Pergamon

Scott RF. 1967. Viscous flow of craters. *Icarus* 7:139–48

Seed HB, Goodman RE. 1964. Earthquake stability of slopes of cohesionless soils. *Proc. Am. Soc. Civ. Eng.* 90 (SM-6):43–73

Shoemaker EM. 1963. Impact mechanics at Meteor Crater, Arizona. In *The Moon, Meteorites and Comets*, ed. BM Middlehurst, GP Kuiper, 4:301–36. Chicago, IL: Univ. Chicago Press

Spray JG, Thompson LM. 1995. Friction melt distribution in a multi-ring impact basin. *Nature* 373:130–32

Spudis PD. 1993. *The Geology of Multi-Ring Impact Basins*. Cambridge, UK: Cambridge Univ. Press 263 pp.

Thomas PC, Binzel RP, Gaffey MJ, Storrs AD, Wells EN, Zellner BH. 1997. Impact excavation on asteroid 4 Vesta: Hubble Space Telescope results. *Science* 277:1492–95

Ullrich GW, Roddy DJ, Simmons G. 1977. Numerical simulations of a 20-ton TNT detonation on the Earth's surface and implications concerning the mechanics of central uplift formation. See Roddy et al 1977, p. 959–82

Welsh JE, Harlow FH, Shannon JP, Daly BJ. 1966. The MAC method. A computing technique for solving viscous, incompressible, transient fluid-flow problems involving free surfaces. *Tech. Rep. TID-4500, Los Alamos Sci. Lab.*, Los Alamos, NM. 146 pp.

Wieczorek MA, Phillips RJ. 1999. Lunar multiring basins and the cratering process. *Icarus* (In press)

Wood CA, Head JW. 1976. Comparison of impact basins on Mercury, Mars and the Moon. *Proc. Lunar Planet. Sci. Conf.* 7:3629–51

Worthington AM. 1963. *A Study of Splashes*. New York: Macmillan. 169 pp.

Annu. Rev. Earth Planet. Sci. 1999. 27:417–62

WESTERN UNITED STATES EXTENSION: How the West was Widened

Leslie J. Sonder
Department of Earth Sciences, Dartmouth College, Hanover, New Hampshire 03755; e-mail: leslie.sonder@dartmouth.edu

Craig H. Jones
Department of Geological Sciences and CIRES, University of Colorado, Boulder, Colorado 80309; e-mail: cjones@mantle.colorado.edu

KEY WORDS: extensional tectonics, Basin and Range, geodynamics, driving forces

ABSTRACT

Cenozoic extension in the western United States presents a complex interrelation of extension, volcanism, and plate boundary tectonics that defeats simple notions of "active" or "passive" rifting. Forces driving extension can originate at plate boundaries, through basal traction, basal normal forces, or from buoyancy forces internal to the crust and lithospheric mantle. The latter two are most responsible for driving extension where it is observed in the Basin and Range. The complex evolution of the northern Basin and Range probably represents removal or alteration of mantle lithosphere interacting with buoyancy stored in the crust. In contrast, crustal buoyancy forces combined with a divergent plate boundary between about 28 and 16 Ma to drive extension in the southern Basin and Range. The central Basin and Range most likely extended as a result of boundary forces external to itself but arising from buoyancy forces elsewhere in the western United States.

1. INTRODUCTION

A major portion of western North America has experienced large-scale distributed extension during the Cenozoic. Compared with most other tectonically similar areas in the world (e.g. the Aegean Sea, the North Sea), which are near or below sea level, western North America is superb in the extent of

0084-6597/99/0515-0417$08.00

aridity-aided exposure it affords. Thus the extensional areas of western North America provide some of the best opportunities in the world for studying processes of continental extension.

There is an extensive body of recent literature about Cenozoic extension and magmatism in western North America. Reflecting this level of professional interest, at least seven reviews or syntheses of the geology or geophysics of these extensional provinces have been published in the 1990s alone (Armstrong & Ward 1991, Axen et al 1993, Best & Christiansen 1991, Jones et al 1992, Parsons 1995, Ward 1991, Wernicke 1992). Notwithstanding the complexity of the Cenozoic plate tectonic setting and the few constraints on mantle and lower crustal structure and thermal state, more is known about the timing, amount, and spatial variation of extension and related volcanism in western North America than is known in any other region. Thus western North America presents the best opportunity to resolve the driving forces for extension. This review focuses on critically evaluating such forces by identifying the geological and geophysical observations that place quantitative limits on these forces, as well as the observations that can test models postulating such forces.

We begin by providing a broad overview of Cenozoic extension and magmatism in the United States. We concentrate on extension that occurred in the Basin and Range province of the western United States (Color Plate 1). This approach sets the stage to discuss possible driving forces for extension, their geologic consequences, and the observations that provide tests of their significance. Finally we present our view of the most likely combination of forces driving extension. In doing so, we hope to motivate further studies that will provide critical data and geodynamic understanding needed to test our scenario and others.

2. SUMMARY OF EXTENSION AND MAGMATISM IN THE WESTERN UNITED STATES

2.1 *Plate Tectonic Framework*

Throughout the late Mesozoic and early Tertiary, the entire length of the western edge of North America was a convergent plate boundary, with North America overriding the subducting Farallon plate (Color Plate 2) (Atwater 1970, Engebretson et al 1984, Stock & Molnar 1988). About 52 Ma the Farallon plate broke into the Farallon and Vancouver plates, with the Vancouver plate west of the future Basin and Range (Rosa & Molnar 1988, Severinghaus & Atwater 1990, Stock & Molnar 1988).

A spreading ridge separated the western edges of the Farallon and Vancouver plates from the Pacific plate (Color Plate 2). Because the rate of subduction exceeded the half-spreading rate at the ridge, the ridge gradually neared the

North American continent. Around 28 Ma the Farallon plate began to break up into smaller fragments near its boundary with the Vancouver plate, between about 27° and 31°N (relative to modern North America) (Atwater & Stock 1998, Lonsdale 1991, Severinghaus & Atwater 1990). Between about 27 and 16 Ma the plate boundary to the west of the Southern Basin and Range experienced a mixture of slow subduction and strike-slip motion as the small plate fragments were either subducted under North America or captured by the Pacific plate (Lonsdale 1991). Because the motion of the Pacific plate was sufficiently parallel to the western edge of North America, subduction was replaced by right-lateral transform faulting as the Pacific and North American plates came into contact (Atwater 1970; Severinghaus & Atwater 1990). This boundary gradually lengthened as the Mendocino triple junction moved north and the Rivera triple junction moved south to their present positions. The present-day remnant of the Vancouver plate is the Juan de Fuca plate subducting under Washington and Oregon. Remainders of the Farallon plate are the Rivera and Cocos plates subducting under Central America.

2.2 *History*

The geologic history of Cenozoic extension in North America has been extensively reviewed (Armstrong & Ward 1991, Axen et al 1993, Burchfiel et al 1992, Eaton 1982, Hamilton & Myers 1966, Jones et al 1992, Parsons 1995, Stewart 1978, Thompson & Burke 1974, Ward 1991, Wernicke 1992, Wernicke et al 1988, Zoback et al 1981). From these and other more recent papers we summarize the background necessary to evaluate possible driving forces for extensional deformation in the western United States.

Cenozoic extensional deformation occurred over a wide area of western North America, ranging south from Canada into Mexico and east from California to Utah, Montana, and New Mexico. Extended areas can be divided from north to south (Wernicke 1992) into four distinct regions (Color Plate 1): the Omineca extensional belt in northern Washington and British Columbia; the Rocky Mountains Basin and Range in Montana, Idaho, and Wyoming; the Basin and Range in Nevada, Utah, Arizona, California, and northern Mexico; and the Rio Grande Rift, which although often considered a separate extensional province, merges at its southern end with the Basin and Range.

The Basin and Range can be subdivided into the Northern, Central, and Southern Basin and Range (Jones et al 1992, Wernicke 1992). The Northern Basin and Range (NBR) has the highest average elevation, highest heat flow, and thinnest crust (Color Plates 1 and 3). The Southern Basin and Range (SBR) has the lowest average elevation of the three subprovinces, lowest heat flow, and is tectonically least active. The Central Basin and Range (CBR) has the greatest local relief and represents a transition from the NBR to the SBR, with large north-south gradients in regional topography, heat flow, and gravity; it is tectonically active, particularly in its western half.

A critical piece of evidence in identifying the forces driving extension is to know the timing of extension, especially its initiation. But the process of identifying initiation of extension presents certain pitfalls that can be illustrated using the southern Sierra Nevada. Some minor normal faulting is observed at the surface (e.g. the Durrwood Fault, Bergquist & Nitkiewicz 1982) and extensional faulting has produced substantial swarms of earthquakes (Jones & Dollar 1986). Deeper levels of the crust and upper mantle have probably been profoundly tectonized and thinned over several million years (Jones 1987, Jones & Phinney 1998, Jones et al 1992, Wernicke 1992). However, the absence of preserved syntectonic sedimentation would preclude inferring from typical geological criteria that extension had initiated. In contrast, some of the minor volcanism and associated feeder dikes erupted or emplaced in the past 12 Myr in the southern Sierra (Bergquist & Nitkiewicz 1982, Moore & Dodge 1980) would probably be preserved if the range is tectonically dismembered, perhaps suggesting a lengthier pre-extension history of volcanism than in fact existed. Thus geologic estimates of the timing of initiation of extension, especially if based only on data from surficial extension, can be later than the actual initiation.

While the history of extension and volcanism across the Basin and Range is strongly variable in timing and magnitude, some generalizations apply. Nearly all of this area of Cenozoic extension was previously subjected to compression and crustal thickening during a procession of Mesozoic contractional orogenies (Allmendinger 1992). The oldest extension that is possibly associated with the formation of the Basin and Range dates to the Cretaceous. This event has been inferred principally from decompression of middle crustal rocks and some ductile extensional shear zones active at about the same depth (Applegate & Hodges 1995, Camilleri & Chamberlain 1997, Hodges & Walker 1992, Wells 1997). Coeval supracrustal extension is not generally observed. Rather, early Tertiary strata conformably overlie Mesozoic or Paleozoic strata in some large parts of the hinterland of the Sevier orogen, reinforcing the idea that surficial extension did not occur over much of the region (Armstrong 1972, Gans et al 1989). Contractional tectonism continued to the east and locally in the hinterland at this time (Camilleri & Chamberlain 1997, Wells 1997). Thus extension of this age probably reflects adjustments within the Sevier and Laramide orogens and does not contribute to later (Cenozoic) wholesale thinning of crust and mantle lithosphere. Wernicke and Getty (1997) present a provocative interpretation of this deformation as a result of "intracontinental subduction."

Extension not kinematically linked to compression appears to have initiated earliest in the north, in the Omineca extensional belt and the Rocky Mountain Basin and Range in Eocene time, about 55 to 45 Ma. Extension in this time frame occurs at least as far south as the Ruby Mountains of northeastern Nevada and might include some poorly dated faulting farther south in eastern Nevada (Axen et al 1993, Lee 1995, Mueller & Snoke 1993). The southern areas apparently lack any sedimentary record of surficial faulting, although broad lake basins existed in eastern Nevada through much of this time (Dubiel et al 1996), suggesting relatively mild extension. To the north, extension occurred in a volcanic setting, with extension directions initially nearly arc-parallel (Janecke 1992). This extensional event is also found near the eastern thrusts of the Sevier fold-and-thrust belt and follows the last contractional faulting by less than 10 Myr (Constenius 1996).

By late Paleogene time, large-scale extension and sedimentation occurred through most of the eastern two-thirds of the NBR, extending up into the Rocky Mountain Basin and Range (Axen et al 1993, Constenius 1996, Janecke 1994). Ages of extension may decrease from north to south in this interval, because extension in Idaho started no later than 45.5 Ma (Janecke 1992), is tightly bound at the surface at ca. 39 Ma in eastern Nevada (MacCready et al 1997, Mueller et al 1999, Mueller & Snoke 1993, Potter et al 1995), and seems no earlier than about 35 Ma in southern Nevada (Axen et al 1993). However, cooling ages from the Snake Range might indicate some extensional tectonism before 41 Ma (Lee 1995). A north-to-south or north-to-southwest migration of volcanism has been well documented across the NBR from about 40 to 20 Ma (Best & Christiansen 1991), terminating to the south at a Precambrian boundary in the mantle (Farmer et al 1989, Jones et al 1992).

At present there is no evidence that the western part of the NBR (the Walker Lane belt) was involved in extension until approximately 25 Ma as volcanism became active in the region (Dilles & Gans 1995, and references therein). However, we note that it is possible that the absence of Paleogene sedimentary or volcanic rocks, complexity of late Paleozoic and Mesozoic deformation, and relatively shallow levels of exposure might have conspired to prevent identification of earlier extension, as was noted above for the southern Sierra Nevada.

Unlike the NBR, where extension occurred within or outboard of the tectonically thickened Paleozoic continental margin, extension in the SBR was superimposed on thickened cratonic crust (Color Plate 2). Extension began at or before about 25 Ma and appears to have migrated northward with time over the whole region (Glazner & Bartley 1984), although not in a smooth sweep when examined at smaller scales (Spencer et al 1995). Volcanism is generally closely tied to extension in the SBR (Glazner & Bartley 1984) although volcanic centers rarely correlate with highly extended terrains (Spencer et al 1995), an observation discussed in more detail below.

The CBR exhibits the latest initiation of extension of all, beginning in middle Miocene time, about 16 to 14 Ma (Beard 1996, Gans & Bohrson 1998). Except in the Eldorado Mountains area at its eastern edge, volcanism was generally minor. A clear progression of surficial tectonism from east to west has been documented (Wernicke 1992, Wernicke et al 1988). In addition, this area provides the only displacement estimate across the whole Basin and Range based on a palinspastic reconstruction (Wernicke et al 1988, Wernicke & Snow 1998).

The sense of direction of early extension was generally aligned from west-northwest to east-southeast in the NBR and east-northeast to west-southwest in the SBR with temporally varying directions in the CBR. Stress directions inferred from dike orientations appear more variable through time (Best 1988), and might be related to some unusual crustal transport directions (e.g. north-south) present in some areas, such as the Ruby Mountains area (MacCready et al 1997, Mueller et al 1999, Mueller & Snoke 1993).

2.3 *Magnitude of Strains and Strain Rates*

Total extension across the CBR has been put at 250 ± 50 km, a 250 percent strain (Wernicke et al 1988, Wernicke & Snow 1998). No equivalent estimate spans the NBR, though extension across eastern Nevada and western Utah has been placed at about 120 to 150 km (Coogan & DeCelles 1996, Gans 1987, Wernicke 1992) and small vertical axis rotations inferred for the Sierra Nevada (Bogen & Schweickert 1985, Frei 1986) suggest that the NBR extended a distance comparable to the CBR. Total extension across the SBR is more difficult to estimate, in part because westernmost portions probably now lie west of the San Andreas Fault (Bohannon & Geist 1998, Hamilton 1987); one transect traversing most of the SBR east of the San Andreas Fault contains 80 to 90 km of extension (Spencer & Reynolds 1991).

Strain is strongly variable across the extensional provinces. Some uplifted ranges are relatively unextended, with strains less than approximately 10%, but other areas have extended by several hundred percent (Wernicke 1985, Gans 1987). Variation of strain with depth is also highly likely because the crust is the same thickness in extended and unextended terrains (Block & Royden 1990, Gans 1987). Although the mechanisms of such heterogeneity are of great significance, here we are most interested in strains averaged across broader regions.

Strain rates, like strains, are strongly variable both spatially and temporally. It is difficult to accurately reconstruct the strain rate history of a region because it requires knowing both extension rate and width of an extending region as functions of time. The CBR offers the best prospect at this point. Early extension rates appear to have been more than 20 mm yr^{-1} in the interval ~8 to 16 Ma when the region was between ~70 and ~230 km in width, giving a temporally and spatially averaged strain rate of $\sim 5 \times 10^{-15}$ s^{-1}. Rates as high as 2×10^{-14} s^{-1} have been suggested (Wernicke 1992). After ~8 Ma, extension rates slowed to ~15 mm yr^{-1} (Wernicke & Snow 1998), implying that the average strain rate was $\sim 2 \times 10^{-15}\ s^{-1}$ and suggesting a temporal decrease in strain rate by a factor of ≥ 2 to 3.

In the absence of constraints on displacements across the NBR with time (information that would be profoundly valuable in understanding the causes of extension), it is reasonable to assume that the extension rate was temporally constant when averaged across the entire region. If so, the 200-to-300 km of extension occurring since 35 to 45 Ma gives a rate of extension of 4 to 9 mm yr^{-1}. If the NBR was initially 600 to 700 km wide, time- and space-averaged strain rates would have been 2–4 $\times 10^{-16}\ s^{-1}$.

The 8-to-9 mm yr^{-1} average extension rate between $\sim$25 Ma and $\sim$15 Ma across a 120-to-140 km wide part of the SBR (Spencer & Reynolds 1991) gives an average strain rate of 2–5 $\times 10^{-15}\ s^{-1}$. A higher value is possible because the extension rate was derived from a section initially 10 to 20 km wide at the surface and involving crust approximately 50 km wide when extension initiated (Spencer & Reynolds 1991). Although strain rates were undoubtedly variable, in the absence of other estimates of strain rates, we take this as representative of the SBR as a whole.

2.4 *Volcanism and Extension*

Considerable controversy has surrounded the relationship, if one exists, of volcanism to tectonism. While a general correlation exists at the province level, several studies have pointed out the apparent absence or reduced presence of extensional tectonism in areas of peak volcanism (Anderson 1971, Best & Christiansen 1991, Burke & Axen 1997, Gans & Bohrson 1998, Sawyer et al 1994, Spencer et al 1995). This circumstance is frequently interpreted to mean that extension ceased across the Basin and Range during peak volcanism. Best and Christensen (1991) rejected extension through diking because such dikes were not at the surface; they considered magmatic additions to the crust to be in the form of sills. Gans & Bohrson (1998) interpreted the onset of extensional strain after volcanism to indicate that the onset of extension interfered with and terminated volcanism.

Given the long-term history of extension, we suggest that the reverse is equally plausible: as volcanism waned of its own accord, extension was accommodated by extensional faulting and stratal tilting rather than an earlier combination of diking, caldera emplacement and collapse, and minor faulting accompanying peak volcanism. In addition, some faulting or diking was probably covered by late volcanics or distributed on poorly dated faults within the volcanic regions. This suggestion arises in part from the observation of coeval extension in less volcanically active regions. It also stems from first-order mechanical considerations. Given that areas of volcanism are likely to be underlain by warm mantle and crust, they should be mechanically weak.

If extension in these regions did indeed slow or stop during volcanism, one is forced to postulate a significant reduction, by perhaps as much as an order of magnitude, in the forces driving extension in these regions. Given the possible driving forces (see next section), we find it difficult to envision how they could be sufficiently reduced, either spatially or temporally, to produce the apparent lack of correlation of volcanism and extension.

The most relevant modern analogue is the Long Valley caldera region of California. The 1980 Mammoth Lakes earthquakes produced well-defined displacement on the Hilton Creek Fault outside the caldera. Within the caldera, slip occurred over a wide region with smaller offsets (Clark et al 1982). Even more compelling, Bursik and Sieh (1989) documented a shift from surficial fault displacement to volcanic accommodation in the Mono Craters area just to the north of Long Valley. A similar mode of deformation has been proposed for the Snake River Plain (Parsons et al 1998). Thus we envision more a change in style than magnitude of extension during peak volcanic periods in the Basin and Range.

2.5 *Active Deformation*

2.5.1 SEISMICALLY DETERMINED STRAIN RATES Seismicity in the Basin and Range is usually described as being in two bands: the Central Nevada Seismic Belt/ Eastern California Shear Zone (Wallace 1984) on its west margin and the Intermountain Seismic Belt (ISB) on the east margin (Sbar et al 1972, Smith & Sbar 1974, Smith & Arabasz 1991). This view may be biased by the concentration of seismic stations and settlements at the margins of the NBR and the few large historic earthquakes in the region. Plotting events greater than magnitude 4 occurring since 1980 reveals that seismicity has occurred nearly uniformly across the bulk of the NBR, with a noticeable gap in western Utah and a greater concentration along the westernmost edge (Color Plate 3).

Eddington et al (1987) and Greensfelder et al (1970a, 1970b) estimated strain rates in the Basin and Range by summing seismic moments of earthquakes in subregions $\sim$100–250 km wide. In the ISB (excluding the Yellowstone area), strain rates are generally between 10^{-17} and 1–2×10^{-16} s^{-1}. The direction of maximum extension is more or less east-west. On the west side of the NBR, strain rate estimates are generally about an order of magnitude higher, from 10^{-16} to 3–4×10^{-15} s^{-1}, with maximum extensional strain rates oriented east-west to northwest-southeast. The greatest source of uncertainty is in the conversion of earthquake magnitudes to seismic moments, which makes strain rates uncertain to a factor of $\sim\pm 3$ (Eddington et al 1987). In addition, because the time interval of observation is short in relation to the recurrence intervals of large earthquakes on active faults, seismicity may not provide a reliable indicator of long-term geological rates of deformation, especially for smaller areas within the Basin and Range.

2.5.2 GEODETIC ESTIMATES Geodetic determinations of strain rates have been obtained from conventional surveys in small networks covering distances of tens of kilometers to a few hundred kilometers, as well as space-based networks spanning the entire Basin and Range. Coverage is best across the NBR, where the integrated rate of motion of the Sierra Nevada relative to North America is about 10 to 12 mm yr^{-1} in a 290-to-330° azimuth (Argus & Gordon 1991, Bennett et al 1998, Clark et al 1987, Dixon et al 1995).

In terms of strain rates, the NBR appears to be extending in an east-west direction at a fairly uniform rate of $3 \pm 0.3 \times 10^{-16}$ s^{-1} (Bennett et al 1998); a similar rate of 5×10^{-16} s^{-1} has been reported for the eastern third (about 160 km width) of the NBR (Dixon et al 1995). It is possible that strain rates might be locally higher near the eastern edge of the NBR in Utah. Martinez et al (1998) report strain rates of $\sim$1.6–1.7×10^{-15} s^{-1} across a 55 km wide region including the Wasatch fault. Whether this reflects a transient condition related to loading of the Wasatch fault is unclear.

In the western 200 to 300 km of the NBR, in addition to the abovementioned east-west extension, right-lateral shearing is occurring at a rate of about $8 \pm 2 \times 10^{-16}$ s^{-1} on northwest striking planes (Bennett et al 1998).

Rates of deformation are not as well constrained across the CBR and SBR. The western side of the CBR in the Death Valley and Owens Valley areas is active. GPS data suggest a northwesterly shear strain rate of at least $1\text{–}2 \times 10^{-15}$ s^{-1} across a 100 km wide zone including Death Valley (Bennett et al 1997). South of the Garlock fault there is very little evidence of ongoing extension (note absence of seismicity in Color Plate 3). Yuma (in the center of the SBR) moves with sites on the southern Colorado Plateau (Gordon et al 1993) and is frequently assigned to be part of stable North America.

Extension rates across the Rio Grande Rift are sufficiently slow that they cannot statistically be distinguished from zero. Savage et al (1980) measured a principal strain rate of $3 \pm 5 \times 10^{-16}$ s^{-1} in a N84E ± 15° direction (upper bound of about 1 mm yr^{-1} across a 70 km wide zone). Argus and Gordon (1996) report that the Colorado Plateau might be rotating clockwise such that displacements across the southern Rio Grande Rift might be a few mm per year, but distributed over a broader area than just the Rift itself.

3. FORCES ACTING ON THE LITHOSPHERE

Fundamental physical laws require a balance of forces. Because deformation in the earth is extremely slow, inertial forces (those due to accelerations) are negligible. Hence the force equilibrium in the lithosphere becomes a balance between driving forces, of which we identify three basic types, and resisting forces. The driving forces are boundary forces (those acting on the edges of a deforming region), basal forces (those applied at the base of the lithosphere), and buoyancy forces (those arising from horizontal gradients in density). Resisting forces result from the viscosity of ductile materials and the elasticity or yield stress of brittle materials. The resulting strain rate is that which makes the resisting forces balance the driving forces.

To resolve the resisting forces, we need to be able to estimate the rheology of the deforming material. With varying degrees of uncertainty, this can be accomplished by extrapolating laboratory observations to conditions and materials present in the crust and upper mantle inferred from geological observation, providing a constraint on the sum of the driving forces. However, to resolve the relative magnitudes of the three types of driving forces is more difficult because one needs to know two of them independently in order to constrain the third. As a step in that direction we attempt here to estimate the possible range in magnitude of each type of driving force, to see if each is independently capable of accounting for western United States extension.

Many models for western United States extension invoke one or several of these driving forces. In many instances, support for these models derives not from quantitative consideration of the physics, but from temporal and spatial correlations of implied consequences of the driving forces with the geological record. These models are summarized in Table 1 and are discussed below, along with the observations that provide support or provoke doubt. By separating forces from models in the discussion below, we hope to isolate observations that are useful in constraining the forces from those that allow only differentiation between models.

3.1 *Plate Boundary Forces*

3.1.1 MODELS BASED ON PLATE BOUNDARY KINEMATICS Plate boundary forces have been invoked to explain both the integrated strain in the western United States over the last few tens of millions of years as well as the deformation occurring today.

PAST EXTENSION Because of the approximate parallelism of the northwest direction of extension in western North America with the relative velocity of Pacific-North American plate motion, several researchers (Atwater 1970, Atwater & Molnar 1973, Livaccari 1979) have proposed that extension in western North America after 30 Ma resulted from very broadly distributed shear driven by right-lateral transform motion of the Pacific plate. Because Vancouver plate convergence with North America was slightly oblique (Stock & Molnar 1988), it is conceivable that large-scale shearing could also have driven extension before 30 Ma.

A second model derives from the observation that Pacific-North American plate motion has been slightly divergent during much of the time that the plates

Table 1 Models proposed for extension in western North America

Model	Observational tests and consequences	Success[a]			Selected references
		N	C	S	
	Boundary Forces				
Shear due to Pacific-North America transform plate motion or tangential component of Farallon-North America motion	• General westerly increase in shear strain rates	+	+	0	Atwater (1970), Atwater & Molnar (1973), Livaccari (1979)
	• Temporal changes in strain rates should correlate with changes in plate velocity	0	+	0	
	• No correlation of strain rates with surface uplift	–	–	–	
Stretching due to divergent component of Pacific-North American motion	• General westerly increase in extensional strain rates	0	~	–	Bohannon & Parsons (1995), Dokka & Ross (1995)
	• Temporal changes in strain rates should correlate with changes in plate velocity	0	+	–	
	• No correlation of strain rates with surface uplift	–	–	–	
	• Inception of extension should track northward with Mendocino triple junction	0	–	+	
Triple junction instability	• General westerly increase in strain rates	0	0	–	Ingersoll (1982), Dokka & Ross (1995), Glazner & Bartley (1984)
	• Temporal changes in strain rates should correlate with changes in plate velocity	0	+	0	
	• No correlation of strain rates with surface uplift	–	–	–	
	• Extension should track northward with Mendocino triple junction	0	0	+?	
	• Magnitude of deformation should decrease southward from triple junction	0	0	0	

(*Continued*)

Table 1 *(Continued)*

Model	Observational tests and consequences	Success[a] N	C	S	Selected references
	Basal Shear Forces				
Basal shear due to active East Pacific Rise under North America	• W to E progression of extension and volcanism as ridge subducts to east	0	0	0	Menard (1960), Dixon & Farrar (1980)
	• Spatial link to modern East Pacific Rise	0	0	0	
Basal shear due to mantle flow induced by subducting slab	• Strain rates should correlate with changes in subduction velocity or slab dip, but not with lithospheric geometry	0	0	0	McKenzie (1978) (for Aegean)
	Basal Normal Forces				
Slab gap/window permitting upwelling of hot asthenosphere	• Region of extension, volcanism, surface uplift should track northwards with Mendocino triple junction	0	0	+	Severinghaus & Atwater (1990), Dickinson & Snyder (1979a), Stewart (1978)
	• No extension before initiation of slab window	0	+	+?	
	• Asthenospheric source for magmas erupted through gap	~	~	~	
	Basal and Buoyancy Forces				
Detachment/rollback/delamination of flat Farallon slab	• Compression (or at least lack of extension) until slab detaches	–	–	–	Dickinson & Snyder (1978) Bird (1988), Humphreys (1995)
	• If slab detaches from E to W with no change in strike: E to W initiation of extension, volcanism, uplift as asthenosphere fills in gap left by slab	0	+?	0	
	• If slab detaches by buckling or peeling off along E-W axis; extension, volcanism, and uplift should migrate and converge in N-S direction to common axis or center	+?	~	+	

Mechanism	Predictions	N	C	S	References
	• Synextensional magmas should increase in asthenospheric component as slab detaches	+	+	+	
Mantle convective instability (detachment of blob)	• Gap in time between end of Sevier/Laramide thickening and beginning of extension should correlate with lithospheric strength and inversely with amplitude of convective instability	+	+	+	Houseman et al (1981), Sonder et al (1987), Conrad & Molnar (1996), Houseman and Molnar (1997)
	• Rapid surface uplift in region of previous compression, followed by extension and volcanism	+?	+?	+?	
Mantle plume	• Radial spreading of region of extension and volcanism as plume head contacts base of lithosphere	0	0	0	Parsons et al (1994), Saltus & Thompson (1997), Pierce & Morgan (1992), Fitton (1991)
	• NE progression of extension as continent moves over plume head	0?	0	0	
	• Increase in elevation and volcanism should precede or be synchronous with extension	~	+?	+?	
	• Asthenospheric source for magmas	~	~	–	
Mantle upwelling—various causes including upwellings mobilized above subducted slab	• Uplift and volcanism precede or are simultaneous with extension	~	+?	+?	Many authors, e.g. Gans et al (1989) Scholz et al (1971)
	Buoyancy Forces				
Release of potential energy stored in thickened crust or buoyant lithospheric mantle	• High surface elevations precede extension	+?	+?	+	Sonder et al (1987), Jones et al (1996), Jones et al (1998); Coney & Harms (1984); many others
	• Maximum strain rates should generally correlate with areas of greatest potential energy	+	+	+	
	• Volcanism should follow initiation of extension	~	+	0?	

[a]General consistency of predictions with geologic observations for NBR (N), CBR (C) and SBR (S). Symbols: +, observations generally consistent with prediction; 0, observations generally inconsistent with prediction; ~, observations are contradictory or inconclusive; –, insufficient data.

Table 2 Relative velocity of Pacific plate relative to North America and length of plate boundary[a]

Time interval (Ma)	Speed of Pacific plate (mm/yr)[b]	Azimuth of Pacific plate motion[b]	Azimuth of plate boundary[c]	Normal velocity u_0 (mm/yr)[d]	Tangential velocity v_0 (mm/yr)[d]	Average length of plate boundary L (km)[e]
0 (present-day)	46 ± 1	N34W-N38W	N35W-N40W	−5–2	45–47	2600
0–8	52	N36W-N42W	N32W-N42W	−4–9	50–54	2400
8–12	52	N56W-N60W	N32W-N42W	13–24	46–50	2000
12–30	33	N54W-N60W	N30W-N40W	8–17	29–32	500

[a]Velocity estimates are for a point on the plate boundary near the Mendocino triple junction.

[b]From Atwater & Stock (1998), Table 2 and Figure 3, except present day data which are from NUVEL-1A plate model (DeMets et al, 1994).

[c]From Severinghaus & Atwater (1990), Figure 6, Atwater & Stock 1998). In the absence of estimates of uncertainty, an arbitrary range of $\pm 5^\circ$ assumed.

[d]Values calculated from mean values in previous columns using $u_0 = S\sin(\Delta\theta)$ and $v_0 = S\cos(\Delta\theta)$, where S is speed of plate and $\Delta\theta$ is the difference between azimuth of plate motion and azimuth of plate boundary. Plus and minus values indicate the extremes possible, given the uncertainties in S and $\Delta\theta$.

[e]From Figure 6 of Atwater (1970), Figures 10–13 of Severinghaus & Atwater (1990), Figure 2 of Stock & Molnar (1988), and Figure 7 of Atwater & Stock (1998).

have been in contact (Table 2), amounting to some 10° of longitude since 30 Ma (Atwater & Stock 1998, Stock & Molnar 1988). Bohannon and Parsons (1995) have suggested that parts of the North American plate boundary were forced to move with the Pacific plate, perhaps by being tightly coupled to underlying subducted Farallon plate fragments that adopted Pacific velocity once spreading ceased. Broadly distributed extension in the continental interior is postulated to occur wherever the western edge of North America moved with the Pacific plate.

Another proposal is that extension resulted from the instability of the Mendocino triple junction. Because the North America-Pacific plate boundary has been somewhat oblique to the Farallon-North American (Cascadia) plate boundary, the geometrical consequence of assuming all plates are rigid is that as the plates move, a hole must form between them south of the Mendocino triple junction. To avoid surface exposure of asthenosphere, distributed extension of parts of the North American continent has been proposed (Color Plate 2) (Dickinson & Snyder 1979b, Dokka & Ross 1995, Glazner & Bartley 1984, Ingersoll 1982).

Testable consequences of these models generally involve the timing, amount, and distribution of extension (Table 1). The latter two models and the first model (for post-30-Ma time) would predict that initiation of extension should track northward along with the migration of the Mendocino triple junction. While contrary to observations in the NBR, this prediction does seem to hold in the SBR, at least until 20 to 16 Ma (Glazner & Bartley 1984) when extension stopped progressing northwards. The first two models would predict broadly distributed deformation in the continental interior behind much if not all of the Pacific-North American plate boundary, whereas the triple junction instability model would predict more localized deformation around the triple junction. The latter would also predict a southward decrease in displacement of the North American continental edge relative to interior North America, while the other models would not necessarily require this. It should be noted that recent reconstructions of the western boundary of western North America (Atwater & Stock 1998) permit the parallelism of the Pacific-North American and Vancouver-North American plate boundaries when extension in the SBR initiated, meaning that the triple junction was stable and no extension of the continent was required to avoid a hole.

PRESENT-DAY SHEAR Interpretations of geodetic data also link deformation in the continental interior with plate boundary kinematics. VLBI and GPS data (Argus & Gordon 1991, Clark et al 1987, Dixon et al 1995, Minster & Jordan 1987) show that present-day Pacific-North American plate motion is not completely taken up in the San Andreas and California Borderlands regions, so presumably some of it must be taken up in the Basin and Range. However, interpretations of the geodetic data give an ambiguous picture of Basin and Range deformation and its possible kinematic compatibility with plate motion vectors. VLBI data have been interpreted as indicating that the width of the zone of shearing associated with Pacific-North America motion is less than ~450 km (i.e. extending no more than ~225 km to the east of the San Andreas fault), and thus reaches no farther east than the Nevada-California border (Ward 1988). However, recent GPS data indicating ~10 mm yr^{-1} of northwest-trending right lateral shear in the western Basin and Range, compared with negligible shear in the central and eastern Basin and Range (Bennett et al 1998, Dixon et al 1995), might reasonably suggest that the zone of shearing reaches into the western NBR and CBR.

One must be careful to not confuse cause and effect in interpreting the geodetic data. It is easy to conclude that because deformation in the Basin and Range makes up for the shortfall in California of Pacific North-American relative motion, Basin and Range deformation is somehow a result of the plate interaction (Bohannon & Parsons 1995). On the contrary, the necessity that the relative plate motion vector reflect all of the motion of the Pacific plate relative to cratonic North America, regardless of how such motion is distributed, is only a kinematic requirement. As such, it provides no constraint on the origin of the forces driving the deformation.

3.1.2 TESTS OF MODELS INVOKING PLATE BOUNDARY FORCES Close correlations between timing and location of extension within western North America and growth of the transform boundary can support hypotheses that extension is driven by plate boundary forces, but it is also necessary to show that the magnitude of strain rates occurring in the continental interior can be produced by a given velocity or stress condition at the plate boundary.

To estimate the magnitude of strain rates, we treat the lithosphere as a thin viscous layer with a vertically averaged rheology (Bird 1988, Bird 1998, Bird & Piper 1980, England & McKenzie 1982, Sonder et al 1986). This approach is valid when considering deformation whose horizontal extent is much greater than the lithosphere thickness, so it is appropriate for examining the broad characteristics of western United States deformation.

Approximate solutions to the equations of motion for a thin viscous layer allow estimation of strain rates and velocities in the plate interior to be estimated (England et al 1985). Given velocity u_0 acting normal to a plate boundary of length L, the velocity distribution $u(x)$ in the continental interior is approximately $u(x) = u_0 \exp(-\pi\sqrt{n}x/L)$, where x is distance perpendicular to the plate boundary. Similarly, the velocity distribution resulting from a tangential velocity v_0 applied to the boundary is approximately $v(x) = v_0 \exp(-4\pi\sqrt{n}x/L)$ (England et al 1985). The normal strain and shear strain rate distributions $\dot{\varepsilon}_n$ and $\dot{\varepsilon}_s$ are

$$\begin{aligned}\dot{\varepsilon}_n &= \frac{\partial u}{\partial x} \cong \frac{-u_0\sqrt{n}\pi}{2L}\exp\left(\frac{-\sqrt{n}\pi x}{2L}\right)\\ \dot{\varepsilon}_s &\cong \frac{1}{2}\frac{\partial v}{\partial x} \cong \frac{-v_0\sqrt{n}\pi}{L}\exp\left(\frac{-2\sqrt{n}\pi x}{L}\right)\end{aligned} \tag{1}$$

The parameter n is the stress exponent in a power law relationship between stress and strain rate that describes the vertically averaged rheology of the lithosphere. The value of n reflects the relative proportions of brittle and ductile behavior in the lithosphere, with $n = 3$ indicating pure ductile creep and increasing values of n reflecting increasing contributions to lithospheric strength from brittle mechanisms (Sonder & England 1986). A simple rule of thumb for Equation 1 is that shear strain rates will decrease by a factor of 10 over a distance of $L/\pi\sqrt{n}$ perpendicular to the plate boundary, while normal strain rates decrease more slowly, requiring a distance of $4L/\pi\sqrt{n}$.

We use these equations to test whether strain rates in the continental interior could be driven by either transcurrent or divergent motion of the plates off the west edge of North America. Values for the parameters L, u_0, and v_0 in Equation (1) are obtained from plate reconstructions and are listed in Table 2. For n, we use the values 3, 8, and 12, which most likely span the range appropriate for the Basin and Range (Jones et al 1996). Figure 1 shows the resulting strain rate distributions and compares them with estimated and measured rates from geological or geodetic observations (see summary in Section 2).

Calculated extensional strain rates resulting from the westerly motion of the Pacific plate relative to North America are generally less than 10^{-15} s^{-1} (Figure 1*a*–*c*) in the region corresponding to the Basin and Range. These rates are uniformly less than the average strain rates estimated from observations for the CBR and SBR (Figure 1, Section 2.3). Only in the westernmost 100 to 200 km of the NBR during the period 30 to 12 Ma, and across a wider portion of the same region during 12 to 8 Ma, are they comparable with observed values. However, the history of extension in these regions at these times tends to eliminate plate boundary forces as playing a key role in driving extension. Extension is not known in the western part of the NBR during the early stages (pre-25 Ma) of extension, nor did deformation increase significantly between 12 to 8 Ma and slow after 8 Ma, as would be necessary to be consistent with changing plate boundary kinematics.

Likewise, average strain rates in the SBR seem to be much greater than those predicted (Figure 1*a*–*c*), regardless of whether they are driven by motion along the entire length of the plate boundary or along a shorter length associated with local accommodation of the triple-junction instability. We therefore think it unlikely that extension in the SBR was driven solely by plate boundary forces, although they were likely to have been significant particularly in western parts within 200 km of the plate boundary.

However, shear strain rates predicted for the present day (Figure 1*d*) are in accordance with shear strain rates measured along the western side of the NBR and are not much less than rates observed in the western CBR. Thus, at least in the western ~100 to 200 km of the Basin and Range, shearing due to Pacific-North American motion may be influencing active deformation.

Full numerical solution of the thin sheet governing equations indicates essentially the same result as the back-of-the-envelope calculations. Such calculations, run over the ~30 Ma duration of Pacific-North America plate contact, were unable to produce more than ~15 to 20 percent extension across the area corresponding to the Basin and Range—nowhere near the 100 percent indicated by geological observations (Sonder et al 1986).

The possibility of rheological heterogeneity does not alter this conclusion. Several lines of evidence, including the paucity of seismicity (Hill et al 1991), relatively small post-30 Ma strain, low heat flow (Blackwell & Steele 1992), high crustal and mantle seismic velocities (Mooney & Weaver 1989 and

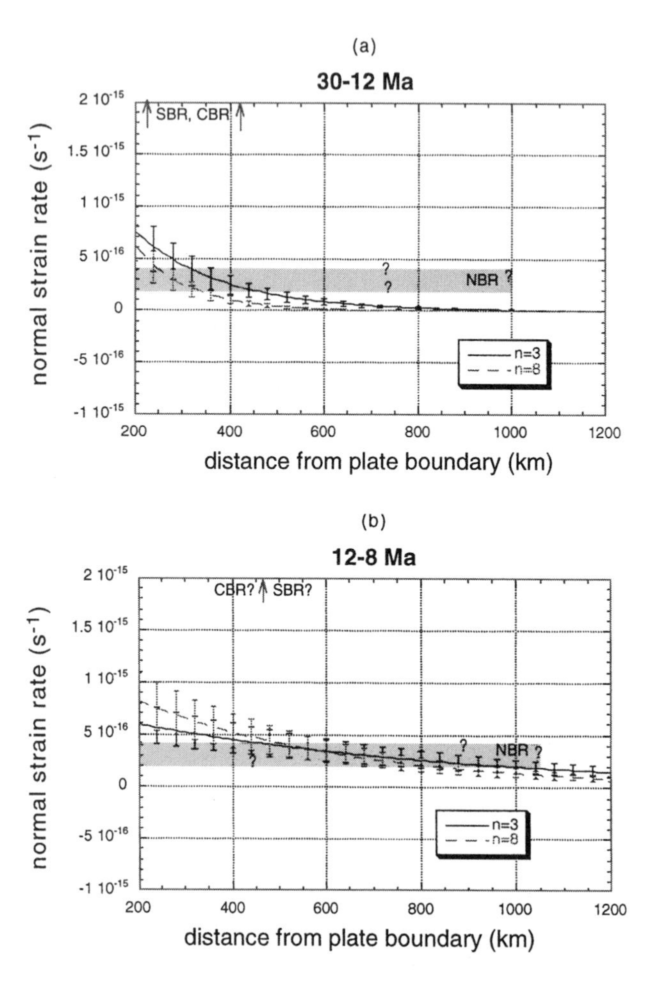

Figure 1 (*a, b, c, d*) Distribution of extensional strain rates in the continental interior resulting from normal velocity applied at the western edge of the North American plate during the period 30–12 Ma (*a*), 12–8 Ma (*b*), or 8–0 Ma (*c*). Shear strain rates resulting from shear velocity applied at the present-day plate boundary (*d*). Error bars indicate range of possible strain rates given the range of plate velocities (Table 2). Calculation based on analytical solutions for deformation of a thin sheet with indicated stress exponent n for vertically averaged power law rheology (from England et al 1985).

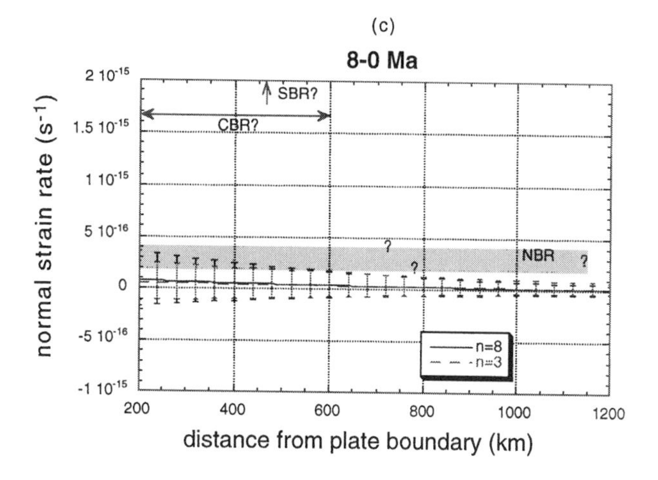

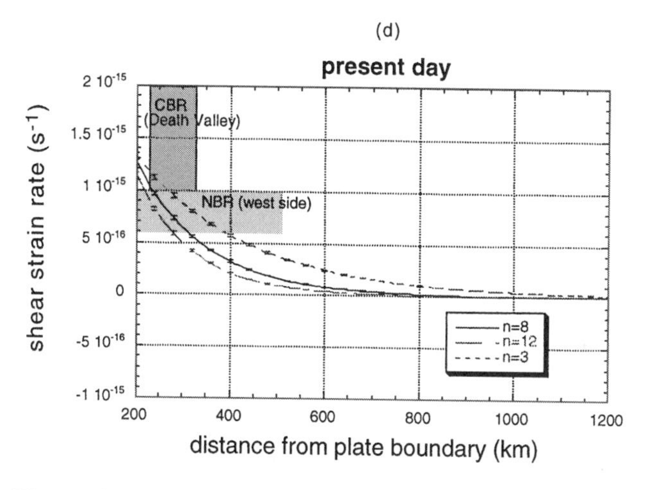

Figure 1 (*Continued*)

references therein; Benz et al 1992) suggest that the Great Valley and possibly the western Sierra Nevada may be mechanically strong. If so, they might be expected to transmit stresses from the continental margin into the Basin and Range. However, numerical solution of the thin sheet equations with a rheologically strong inclusion representing a strong Great Valley/Sierra Nevada block (Figure 2) does not significantly increase shear or extensional strain rates in the Basin and Range.

In summary, it appears that Pacific-North American plate motion produces neither the magnitude nor the rates of extension in the NBR and CBR. At best, plate boundary forces can augment or modify a Basin and Range stress field already extensional in character. The hypothesis of plate boundary-driven extension is also problematic in the NBR because the timing and distribution of deformation do not correlate well with initiation and growth of the Pacific-North American transform boundary. However, it is possible that present-day shear deformation (but not extension) in the western Basin and Range, and past extension in the SBR, may be affected by plate boundary motions.

3.2 *Forces Originating From Sublithospheric Processes*

The general association of mafic or bimodal magmatism with extension in the western United States suggests that the mantle, both lithospheric and sublithospheric, is a crucial component of extensional processes. However, because the mantle is covered by ~30 km of crust, the processes that may be occurring or have occurred in it are only indirectly detectable. This means that mantle processes are considerably more difficult to test than boundary or buoyancy forces acting on the western United States.

Understanding the role of basal processes requires knowing the magnitude of basal stresses produced by these processes, which in turn requires knowledge of mantle viscosities and strain rates. However, mantle viscosities and strain rates are not known to better than an order of magnitude. Therefore we proceed to discuss basal stresses in two ways. First we discuss some simple estimates of the magnitude of basal stresses needed to drive lithospheric deformation at the rates inferred for western North America. Second, we discuss the various models invoking basal stresses that have been proposed to explain extension in western North America, and the observations that may bear on the

→

Figure 2 Numerical solutions for deformation of a thin viscous sheet representing the western United States, showing that a strong Sierra Nevada/Great Valley block does not significantly affect strain rates in the Basin and Range. (*a*) map view of 3 × 1 region; contours are dimensionless horizontal shear strain rate. A shear velocity boundary condition was applied on western ($x = 0$) edge of the solution region, consisting of dimensionless velocity $v' = 0$ for dimensionless coordinate $y' < 0.5$ and $y' > 2.5$, $v' = 1$ for $0.75 < y' < 1.25$, and in the intervening regions $0.5 < y' < 0.75$ and $1.25 < y' < 1.5$, a $\sin^2$ taper was applied, making shear velocities and velocity gradients continuous. The normal component of velocity on this boundary is zero. On the north and south boundaries ($y' = 0$ and $y' = 3$), normal velocity is zero and shear stress is zero; on the eastern boundary ($x' = 1$), which corresponds to stable interior North America, both normal and shear velocity are zero. (*b*) Like (*a*), but with region representing a strong Sierra Nevada/Great Valley block, in which viscosity was set to 10 times that of the surrounding area. Contours are dimensionless shear strain rate. (*c*) Like (*a*), but with normal velocity applied, rather than tangential velocity. (*d*) Like (*c*), but with strong region representing Sierra Nevada/Great Valley block in which viscosity was increased by a factor of 10. In all cases stress exponent $n = 3$. Strain rates may be made dimensional by multiplying by v_0/L, where v_0 is the dimensional magnitude of the boundary velocity (Table 2) and L is the length of the narrow side of the solution region (1000 km).

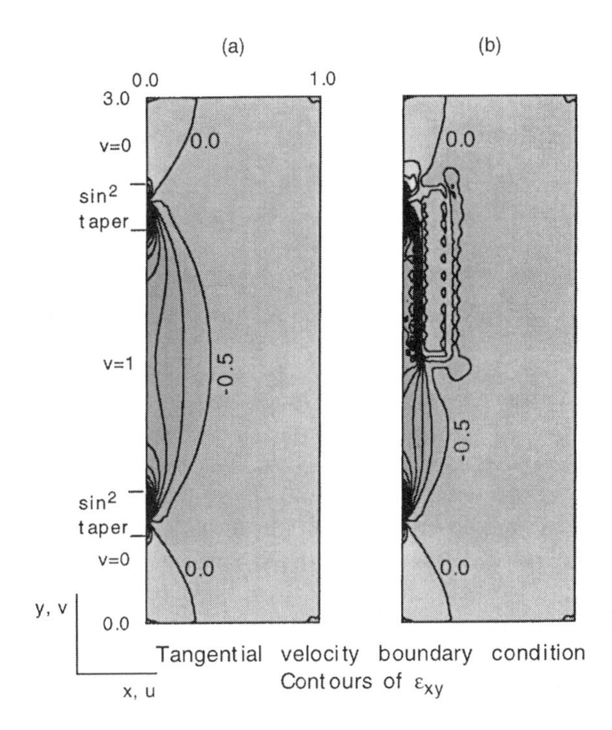

Tangential velocity boundary condition
Contours of ε_{xy}

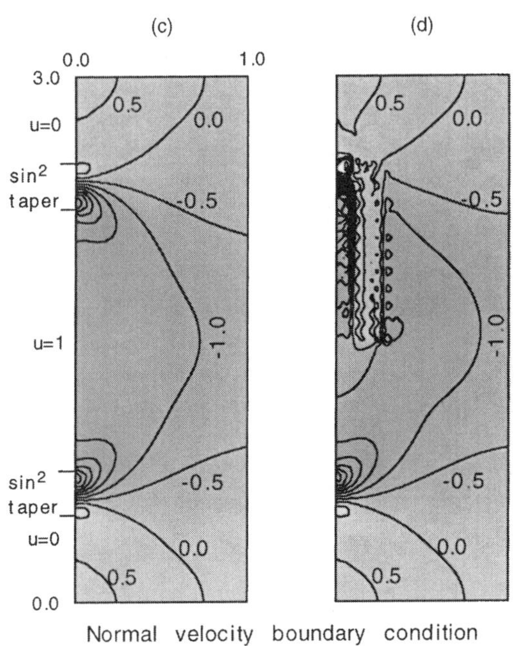

Normal velocity boundary condition
Contours of ε_{xx}

plausibility of such models. For simplicity, we separately consider basal shear stresses and basal normal stresses, although clearly both can and probably do act simultaneously.

3.2.1 MAGNITUDE OF BASAL SHEAR STRESSES Continuing to treat the lithosphere as a thin layer with vertically averaged rheological properties and neglecting any other driving force except basal shear, the horizontal equation of motion can be written (Wdowinski et al 1989) as

$$4L\eta_l \frac{\partial^2 u}{\partial x^2} = -T_b \tag{2}$$

where u is horizontal velocity, x is horizontal distance, T_b is the basal shear stress, L is lithosphere thickness, and lithospheric viscosity η_l is assumed to be Newtonian. If in a deforming region $0 < x < D$ with constant basal shear stress we assume that u is zero at $x = 0$ and horizontal normal strain rate $\dot{\varepsilon}_{xx} = \partial u/\partial x$ decreases to zero at $x = D$, then the horizontally averaged strain rate is $T_b D/8L\eta_l$. To produce an average strain rate of 10^{-15} s^{-1} within a 500-to-1000 km wide region such as the western United States (i.e. $D = 250$–500 km) with lithospheric thickness 100 km and average viscosity of 10^{22}–10^{24} Pa s (England & Houseman 1986, Walcott 1970) would require a basal shear stress greater than ~20 MPa.

This shear stress should be compared with shear stresses inferred to exist in the mantle. Shear stresses calculated from global mantle circulation models (Bai et al 1992; Bird 1998) are less than ~1 MPa. More localized circulation does not seem to produce significantly larger stresses. For example, we may calculate the shear stress due to asthenospheric flow in the corner above a subducting slab (McKenzie 1969). For the region >200 km from the trench that would correspond to the Basin and Range, assuming a mantle with Newtonian viscosity $<10^{20}$ Pa s (Cathles 1975, Hager 1991) results in basal shear stresses less than 10 MPa. This estimate is supported by the results of Wdowinski (Wdowinski & O'Connell 1991, Wdowinski, 1989), who coupled asthenospheric corner flow and deformation of the overlying lithosphere and showed that basal shear stress can enhance extension driven by other forces, but not induce extension by itself. Shallow dips may increase shear stresses, but not sufficiently to produce significant extension. Only if a cold lithospheric slab, with viscosity considerably greater than normal mantle viscosity, directly shears the base of the lithosphere (Bird 1988) are shear stresses likely to be large enough to deform the lithosphere.

3.2.2 MODELS INVOKING BASAL SHEAR STRESSES During the enthusiastic embracing of plate tectonics in the late 1960s and early 1970s, some researchers proposed that the ridge between the Farallon and Pacific plates was subducted and continued to spread (DeLong et al 1978, Dixon & Farrar 1980, Uyeda &

Miyashiro 1974). Such divergent motion was suggested to drive extension in the overriding plate as a consequence of the mantle flow induced by the divergence at the subducted ridge. This idea is no longer taken seriously; not only did the ridge not subduct (Atwater 1970, Atwater & Molnar 1973, Lonsdale 1991, Stock & Molnar 1988), but continued divergence of a ridge following subduction is at odds with current understanding of spreading dynamics (discussion in Severinghaus and Atwater 1990).

Basal shear stresses might also arise from flow induced by the subducting Farallon plate (McKenzie 1969, Tovish et al 1978), as suggested by McKenzie (1978) for extension in the Aegean. As argued above, this mechanism seems to be incapable of generating the shear stresses necessary to generate the magnitude of strain rates observed in the Basin and Range.

3.2.3 MAGNITUDE OF BASAL NORMAL STRESSES Basal normal stresses arise due to mantle density contrasts, either thermal or chemical, which drive upwellings and downwellings. We attempt in this section to constrain the magnitude of basal stresses, but because the effects of basal normal stresses on the lithosphere are virtually indistinguishable from buoyancy forces (discussed in Section 3.3), we leave discussion of their consequences for lithospheric deformation until later.

Like basal shear stresses, the magnitude of basal normal stresses is difficult to determine precisely. An upper bound can be established, however, by recognizing that basal normal stresses will produce dynamic topography (undulations of the earth's surface not isostatically compensated within the lithosphere). The amplitude of dynamic topography is less than 1 km, and possibly less than 200 m (Gurnis 1990, Gurnis 1992, Le Stunff & Richard 1997, Mitrovica et al 1989, Ricard et al 1993). The basal normal stress needed to support this is approximately $\rho_a g Z$ (Houseman & England 1986), where ρ_a is asthenosphere density, g is gravitational acceleration, and Z is the height of the dynamic uplift. Thus 1 km of dynamic uplift implies a basal normal stress of ~30 MPa. Even 150 m of dynamic topography requires a basal normal stress of ~4 MPa. For a uniform lithosphere of uniform viscosity η_l with no other source of deviatoric stress except dynamic topography due to a basal normal stress σ_b, the average deviatoric stress in the lithosphere is approximately $\sigma_b/2$ and the average extensional strain rate is $\sigma_b/4\eta_l$; for $\sigma_b < 30$ MPa and $\eta_l \sim 10^{22}$–10^{24} Pa s, the resulting strain rate is 10^{-17}–10^{-15} s^{-1}. Thus a first-order conclusion is that basal normal stresses may be large enough to drive extension in the overlying lithosphere (Houseman & England 1986).

3.2.4 MODELS INVOKING BASAL NORMAL STRESSES More so than for basal shear stresses, quantitative constraints on basal normal stresses from studies of mantle dynamics are rarely considered in models of western United States extension. In fact, in many models discussed below, the magnitude of basal stresses is never explicit. Rather, many models postulate an ad hoc thermal anomaly, with little or no quantitative consideration of the asthenospheric dynamics that might be required to produce it (Liu & Shen 1998, Sonder et al 1987). Because the connection to mantle dynamics is in many cases so tenuous, tests of these models often involve surface observations of indirect or inferred consequences of processes at the base of the lithosphere, such as the timing and distribution of extension or volcanism, or the presence of isotopic signatures suggesting source regions for melts associated with extension.

SLAB WINDOW A kinematic consequence of cessation of sea-floor spreading as the East Pacific Rise intersected the trench off the west coast of North America could be a hole, or "slab window," in the subducting Farallon plate (Color Plate 2) (Dickinson & Snyder 1979a, Furlong et al 1989, Stewart 1978) which would increase in size with time. Upwelling from the underlying mantle to fill the widening hole between the plates could provide a source of heat to the base of the North American plate, thereby mechanically weakening it (Furlong et al 1989), buoying it up (Axen et al 1993, Glazner & Loomis 1984), and providing a source of basaltic magma for the volcanism associated with extension.

However, the dimensions and shape of the slab window are highly uncertain. The slab window of Dickinson & Snyder (1979a), which was defined by extrapolating in time the positions of the plate edges at the time of subduction and which has been updated by Atwater & Stock (1998), is only one end-member of a series of possible geometries. Because the Farallon plate was already young and warm at the time of subduction, it would quickly become thermally indistinguishable from asthenosphere. Depending on the quantitative criteria one adopts to define "thermally indistinguishable," a range of different scenarios for the size and shape of the slab window results, including a "slab gap" under a large east-west swath of North America where no slab exists at all (Severinghaus & Atwater 1990, Dickinson 1997). The eastern edge of the slab gap is particularly poorly constrained.

Testing a "slab gap" explanation for extension in the Basin and Range is usually attempted by comparing the dimensions of the predicted gap with the distribution of extensional faulting or volcanism (Axen et al 1993, Dickinson & Snyder 1979a, Glazner & Bartley 1984). In some areas, particularly in the SBR between Arizona and Las Vegas, the correlation is qualitatively good: extension and volcanism seem to migrate northward in parallel with the Mendocino fracture zone, which defines the northern boundary of the slab gap/window (Glazner & Bartley 1984). A change in trace element geochemistry of basalts, suggesting a transition from lithospheric sources to asthenospheric sources, occurs in increasingly younger rocks along a south to north traverse from the Death Valley region (CBR) to west-central Nevada (NBR). This change has been linked, with a 2 to 3 Myr delay, to northward growth of the slab gap (Ormerod et al 1988).

There are limits to the quality of these associations: the most recent estimate of the position of the triple junction with time places it slightly south of 35°N at 24 Ma (chron 6c) and near 36°N at 19 Ma (Atwater & Stock 1998), well to the north of coeval extension and volcanism (Axen et al 1993, Glazner & Bartley 1984). The revised plate reconstructions also make the rate of northward migration of the triple junction before about 20 Ma (about 15–25 km/Myr) less than previously inferred. It thus appears that extension and volcanism followed passage of the triple junction by a few million years. Similarly, the revised plate reconstructions indicate that the transition in the western NBR from lithospheric to asthenospheric magma sources postdates the slab edge by about 5 to 10 Myr instead of 2 to 3 Myr as originally suggested (Ormerod et al 1988), with the lag varying inconsistently (11 Myr in the south, but almost none in the north). Growth of a slab gap and emplacement of asthenosphere below the continent should also have a distinct west-to-east migration. Migration of extension and volcanism in this manner has not been documented and instead extension near the west coast of North America appears to be coeval with extension in southern Arizona (Axen et al 1993, Bohannon & Geist 1998, Glazner & Bartley 1984). This latest revision of the triple junction migration (Atwater & Stock 1998) places even greater stress upon the hypothesis that an unstable triple junction caused extension, because the kinematics of this feature require an instantaneous response from the continent.

On a continent-wide basis, growth of slab gaps cannot explain the spatial evolution of Basin and Range extension. For example, it is inconsistent with the southern sweep of magmatism in the NBR prior to 20 Ma (Lipman 1992, Wernicke 1992) and the early initiation of extension in the Pacific Northwest (45–55 Ma). The departure of large-scale extension from near the plate boundary after about 16 Ma complicates any slab-gap inferences because the gap should have the most profound effects near the margin (Dickinson 1997).

Quantitatively testing these models has been difficult because of our present inability to establish a quantitative, physically rigorous link between mantle upwelling and lithospheric tectonics or volcanism. The circulation pattern in the mantle arising from the development of the slab gap is difficult to constrain, due to uncertainties in slab dip, rheological and density structure of the mantle, and geometry of the gap. Therefore only crude estimates of heat supplied to the lithosphere can be made. Further work is needed before these hypotheses can be fully tested in those areas where geological correlations are most promising.

MODELS RELATED TO SLAB DIP Temporal and spatial patterns of volcanism have been interpreted to imply that the Farallon slab subducting under North America gradually shallowed during the early Cenozoic (Laramide time), then began to steepen at ~40 Ma (Coney & Reynolds 1977, Dickinson & Snyder 1978, Keith 1978). The plausibility of such subduction is supported by observations of present-day low-angle Benioff zones under South America (Stauder 1975, Barazangi & Isacks 1976). This low-angle subduction has led to the hypothesis that the Farallon plate directly underlay North America with no intervening asthenospheric wedge (Dickinson & Snyder 1978). Simulations by Bird (1988) suggest that shear stresses imparted by a strong, flat slab under continent would produce crustal thickening over large part of Western North America (i.e. the Laramide orogeny), and then as slab dip increased after 40 Ma, warmer asthenosphere filling in above the slab would cause isostatic uplift and ultimately extension as the lithosphere warms and weakens.

The pattern of post-Laramide volcanism and extension is, in this scenario, controlled by the way in which the slab steepens. An increase in slab dip without concurrent change in strike would produce an east-to-west progression of initiation of extension and volcanism, which is not consistent with observed trends of northward-migrating volcanism in the SBR and southward-migrating extension and volcanism in the NBR; however, slab steepening by downward buckling or double-sided rollback along an east-northeast–trending axis (Humphreys 1995) would be consistent with volcanic trends. Thus, understanding whether removal of a flat slab influenced extension of the Basin and Range depends critically on understanding the mechanics of slabs and mantle circulation.

Independent support for flat subduction comes from determinations of lithospheric thermal gradients using isotopic and fission track cooling ages. These data suggest reduction of thermal gradients in the Basin and Range during the early Cenozoic, which could have resulted from "refrigeration" by a shallow slab (Dumitru et al 1991). Evidence for reduced thermal gradients is stronger to the west, in the Franciscan arc and Sierra Nevada (Dumitru 1988, Dumitru 1990), as might be expected because the slab would have been shallower there, though some controversy exists regarding the magnitude of cooling that occurred (House et al 1997).

On the other hand, isotopic evidence argues against flat subduction, at least to the extreme point that it stripped off the entire continental mantle lithosphere (Bird 1988). Isotope geochemistry of Cenozoic mafic volcanic rocks across much of the Basin and Range (Asmerom et al 1994, Asmerom & Walker 1998, Beard & Johnson 1997, Daley & De Paolo 1992, Farmer et al 1989, Livaccari & Perry 1993), and Mojave Desert (Miller et al 1999) points to a light-rare-earth-element (LREE) enriched source, interpreted to be ancient lithospheric mantle. Younger syn-extensional rocks tend to have asthenospheric sources (higher ε_{Nd}, lower $^{87}Sr/^{86}Sr$, $^{208}Pb/^{204}Pb$, and $^{207}Pb/^{204}Pb$), suggesting thinning of the lithospheric mantle after extension began (Daley & De Paolo 1992, Livaccari & Perry 1993). However, at least some of the lithospheric mantle must have remained to provide the geochemical source for late Cenozoic magmas.

MANTLE CONVECTIVE INSTABILITIES The lithosphere underneath western North America was most likely thickened during the Laramide and Sevier orogenies. Such thickened mantle lithosphere is convectively unstable and it is possible for the lower part to form a blob and detach from the rest of the lithosphere over time scales of a few tens of millions of years (Conrad & Molnar 1997, Houseman et al 1981, Houseman & Molnar 1997, Molnar et al 1998). Before the blob detaches, the lithosphere above it is in a state of horizontal deviatoric compression, but during detachment, warmer asthenosphere will replace the blob, the lithosphere will rapidly uplift and, depending on the amount of uplift, will evolve toward a state of deviatoric tension. Depending on asthenospheric temperature and the thickness of lithosphere removed by the blob, the lithosphere will warm and weaken, and asthenospheric or lithospheric melting may produce basaltic magmas. Following detachment, the driving force for extension is buoyancy stored in the uplifted lithosphere.

In several ways the lithospheric consequences of convective instabilities and removal of a flat slab are similar. Both cause large and rapid increases in surface elevation as the blob or slab detaches from the continental lithosphere. Replacement of either a blob or a slab by warm asthenosphere may result in asthenospheric or lithospheric melting. Both mechanisms may induce spatial migration of extension or volcanism as detachment proceeds. Better understanding of the dynamics of convective instabilities and suducted slabs is needed before such spatial patterns can be used to confidently distinquish between convective instabilities and slab removal.

HOT SPOTS AND MANTLE PLUMES Parsons et al (1994) and Saltus & Thompson (1995) argue that the present elevation of the NBR must be supported by low densities in the asthenosphere, since lithospheric buoyancy is by itself insufficient. They ascribe the buoyant asthenosphere to material emplaced by the head of the Yellowstone plume, and go on to suggest that impingement of the plume head at 16 to 17 Ma on the base of the lithosphere triggered accelerated extension and accompanying basaltic volcanism in a broad (~800 km diameter) region. However, the temporal evolution of tectonism does not strongly support a strong mantle plume influence on most NBR extension. There is no large-scale northeastward migration of extension remotely comparable to the dramatic southward migration of extension observed across the NBR from 55 to ~35 Ma. Indeed, tectonism in Idaho was largely quiescent at a time when it should have been very active given the timing and geometry of a major plume head (Janecke 1992, Janecke 1994). Nor is there a radial expansion of accelerated extension or volcanism, at the scale of the NBR, as might be expected as the rising plume encountered the base of the lithosphere and spread out. Paleoelevations would be extremely useful in testing this model. If a plume of the size proposed above impinged on the lithosphere, surface elevations would have increased significantly and rapidly. The fact that elevations in the NBR in the mid-Tertiary were as high as 2 to 3 km (Gregory-Wodzicki 1997, Wolfe et al 1998, Wolfe et al 1997) would imply even higher elevations at 16 Ma.

More locally, the Yellowstone hot spot has also been linked to the distribution and evolution of extension north and south of the Snake River Plain (Pierce & Morgan 1992, Smith & Braile 1993). Its influence may have been thermal, heating the lithosphere and thereby weakening it (Anders & Sleep 1992) and increasing its buoyancy; or mechanical, increasing forces on the base of the lithosphere (Westaway 1993); or both.

A mantle plume has also been proposed to exist under the SBR (Fitton et al 1991). However, calculated relationships between the chemistry, volume, and temperature of mantle melts suggest that the asthenospheric mantle underneath the SBR was not unusually warm (Bradshaw et al 1993), as would be the case if a plume were present.

Arguments for or against the existence of a plume aside, if a plume were present it could increase driving forces for extension in several ways. First, horizontal asthenospheric flow might exert shear stresses on the base of the lithosphere, although the arguments given in Section 3.2.1 suggest to us that these stresses are not likely to be significant (for an alternative view see Westaway 1993). Second, the plume would warm the lithosphere, thereby decreasing densities and increasing lithospheric buoyancy. Third, its buoyancy could exert a normal stress on the lithosphere, creating dynamic topography and increasing lithospheric buoyancy.

3.3 *Buoyancy Forces*

Discussions abound regarding the role of buoyancy forces in driving continental deformation, in both compressional and extensional regimes (Artyushkov 1973, Bott & Dean 1972, England & Jackson 1989, Fleitout & Froidevaux 1982, Le Pichon 1982, McKenzie 1972, Molnar & Lyon-Caen 1988). Such forces arise from horizontal contrasts in density and as such are intrinsic to the lithosphere rather than imposed from outside. The total buoyancy force (per unit length) acting between two lithospheric columns is the sum over the thickness of the lithosphere of the pressure difference (Figure 3*a*):

$$F_L = \int_{-h}^{L} \Delta P(z)\, dz = \int_{-h}^{L} \left(\int_{-h}^{z} \Delta\rho(z')g\, dz' \right) dz \tag{3}$$

where $\Delta P(z)$ is the pressure difference, $\Delta\rho(z)$ is density contrast, g is gravitational acceleration, z is depth below sea level, h is the elevation of each column, and L is the depth of the base of the lithosphere below sea level. It should be appreciated that F_L may be nonzero even if the two columns are in complete isostatic balance because the condition for isostatic equilibrium requires only that the pressure at $z = L$ be equal and does not constrain the pressure distribution in $z < L$. Indeed, if the two columns are assumed to be in isostatic equilibrium, then

$$F_L = \int_{-h}^{L} \Delta\rho g z\, dz \tag{4}$$

so that F_L takes the form of differences in gravitational potential energy (ΔPE) per unit area. The vertically averaged deviatoric stress exerted by one column on the other can be obtained by dividing F_L by the lithosphere thickness.

To first order (ignoring density differences within crust or mantle), F_L (or ΔPE) depends strongly on crustal thickness (Artyushkov 1973, Molnar & Lyon-Caen 1988). This has led many to view buoyancy forces as arising only from crustal thickness variations, but it is important to recognize (as seen from Equation 3 and Figure 3*b*,*c*) that contributions to F_L may also derive from density variations within crust or mantle, e.g. abnormally warm and buoyant mantle lithosphere. As discussed earlier, basal normal stresses will raise the elevation of the lithosphere and thereby also increase potential energy.

The possible role of buoyancy forces in driving extension in the western United States was first proposed by Molnar and Chen (1983), who suggested that the Basin and Range might represent a more mature version of the Tibetan plateau, which is now extending in an east-west direction. In this view, the rough coincidence of regions of maximum late Mesozoic-early Tertiary crustal thickening in western North America with many Tertiary metamorphic core complexes (Coney & Harms 1984) suggests that buoyancy forces arising from the thickened crust drove subsequent deformation. Additional circumstantial evidence comes from the observation that extension generally occurred perpendicular to the boundaries of thickened crust (Wernicke 1992, Wust 1986). Furthermore, the magnitude of buoyancy forces derived from thickened crust and convectively thinned mantle was sufficiently large to produce the extensional strain observed in western North America, provided that the lithosphere was sufficiently weak (Sonder et al 1987). This weakening could have resulted either

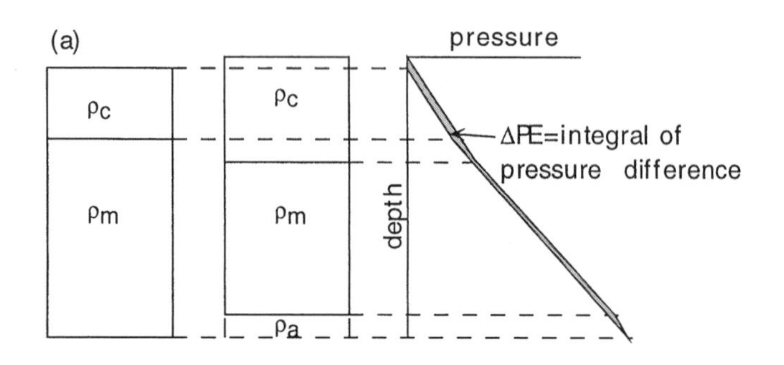
(a)
pressure
ρc
ρc
ρm
ρm
depth
ρa
ΔPE=integral of
pressure difference

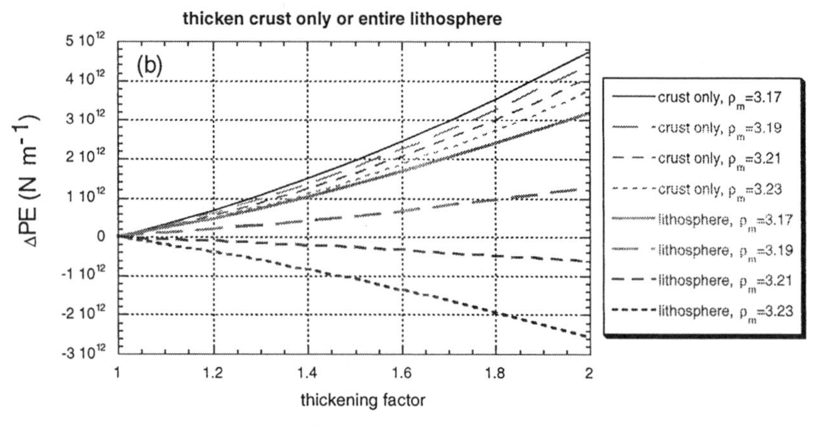
thicken crust only or entire lithosphere
(b)
ΔPE (N m-1)
thickening factor
crust only, ρm=3.17
crust only, ρm=3.19
crust only, ρm=3.21
crust only, ρm=3.23
lithosphere, ρm=3.17
lithosphere, ρm=3.19
lithosphere, ρm=3.21
lithosphere, ρm=3.23

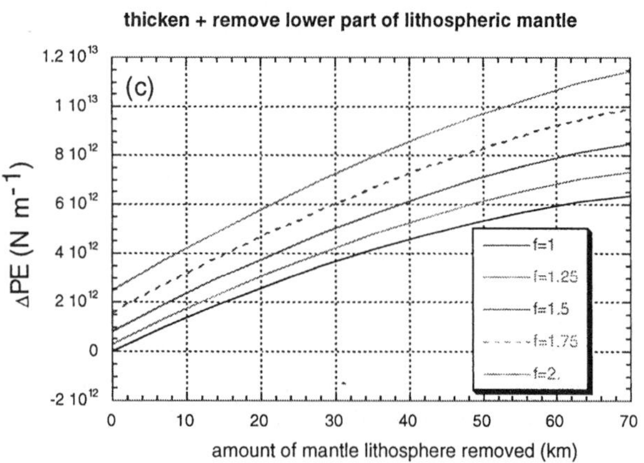
thicken + remove lower part of lithospheric mantle
(c)
ΔPE (N m-1)
amount of mantle lithosphere removed (km)
f=1
f=1.25
f=1.5
f=1.75
f=2.

from lithospheric warming as the lithosphere re-equilibrated following compression or from heat advected upwards in the lithosphere by syn-compressional magmas.

A necessary, but not sufficient, condition that extension was buoyancy driven is that the Basin and Range was high before extension began (Jones et al 1998). Although early paleobotanical evidence has been interpreted to imply that the region was low standing (Axelrod 1966, 1991), recent determinations of paleoaltitude using multivariate analysis of leaf physiognomy to constrain paleoenvironmental variables (Forest et al 1995, Gregory-Wodzicki 1997, Wolfe et al 1998, Wolfe et al 1997) indicate mid-Tertiary elevations of generally 2 to 3 km. These elevations are at least equal to, and in many cases up to 1.5 km greater than, present-day elevations.

Additional support for the importance of buoyancy forces for western United States extension comes from considering present-day tectonics. By converting seismic velocities to densities and calculating the distribution of potential energy, Jones et al (1996) showed that potential energy variations correlate with strain rate variations and that potential energy magnitudes (averaged over regions) are sufficiently large to drive all of the active extension in the western United States (Figure 4). By implication, extensional buoyancy forces should have been even more significant in the past, because crustal thicknesses were up to about twice what they are today.

Although there is strong support for the importance, if not dominance, of buoyancy forces in driving both past and present extension in the western United States, some issues remain. For example, relating potential energy variations to strain rates requires specifying a zero-PE reference state equivalent to the state

←

Figure 3 (*a*, *b*, *c*) Origin of lithospheric buoyancy force (gravitational potential energy) from integral of pressure difference between two isostatically compensated columns of lithosphere (*a*). Buoyancy force derives from differences in lithospheric thickness or density due to crustal or lithospheric thickening (*b*), and/or removal of lithospheric mantle (*c*). In (*b*), four upper curves show potential energy changes with crustal thickening only; lower four curves indicate changes when both mantle and crust are thickened proportionately. Mantle density ρ_m is assumed to decrease linearly with depth z according to $\rho_m = \rho_a[1 + \alpha\gamma(L - z)]$, where ρ_a = asthenosphere density, α = thermal expansion coefficient, γ = mantle thermal gradient, L = lithosphere thickness. The four curves in each set indicate choices of $\alpha\gamma = 1.7, 3.4, 5.1$, and $6.8 \times 10^{-7}\ m^{-1}$, giving average mantle densities of 3.17, 3.19, 3.21, and $3.23 \times 10^3\ kg\ m^{-3}$, respectively. In (*c*), four curves show potential energy changes due to crustal thickening by a factor *f* plus detachment of lowermost mantle lithosphere and replacement by asthenosphere, as might occur due to convective instability or thermal erosion, for example. Lithospheric mantle density is assumed to be constant and equal to $3.3 \times 10^{-3}\ kg\ m^{-3}$. In both (*b*) and (*c*), crust and asthenosphere densities are constant, equaling 2.7 and $3.15 \times 10^3\ kg\ m^{-3}$, respectively; crust and lithosphere thicknesses before thickening are 30 and 100 km, respectively.

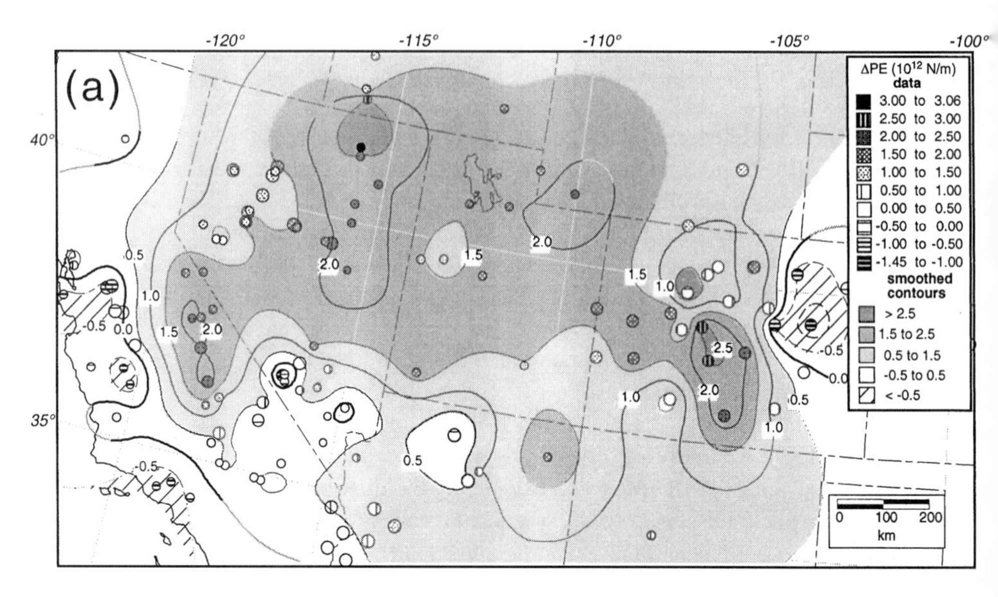

Figure 4 (*a*, *b*) Present-day distribution of gravitational potential energy (ΔPE) in western United States (*a*). Positive (negative) values of ΔPE indicate lithosphere in a state of extension (compression). Comparison (*b*) of observed extensional strain rate in the NBR (*darker gray shading*) with strain rates predicted from the average ΔPE observed in the region (*lighter gray shading*), and estimates of lithospheric strength derived from observations of heat flow, depth to brittle-ductile transition, and crustal thickness. For method of calculation, see Sonder & England (1986). T_M = Moho temperature. Adapted from Figures 1 and 3 of Jones et al (1996).

of zero deviatoric stress (and thus zero strain rate). Considerable ambiguity exists regarding what that reference state should be, and hence there is an uncertainty of $\sim 10^{12}$ N m^{-1} in ΔPE (Coblentz et al 1994, Jones et al 1998, Jones et al 1996). Depending on the rheology, this translates to an uncertainty of as much as an order of magnitude in the rate of deformation that may be produced by buoyancy.

Second, even if buoyancy forces are present, plate kinematics must provide boundary conditions that permit these forces to drive deformation. In other words, plate boundary forces resisting deformation cannot exceed buoyancy forces promoting extension. However, it is not at all clear how the two driving forces interact. The fact that the Pacific plate has been moving away from North America for most of the time since 30 Ma suggests that the boundary conditions permitted extension of the continental interior, although they did not control the distribution of strain rates in the continental interior. It is much less clear how forces at the earlier subduction boundary accommodated extension.

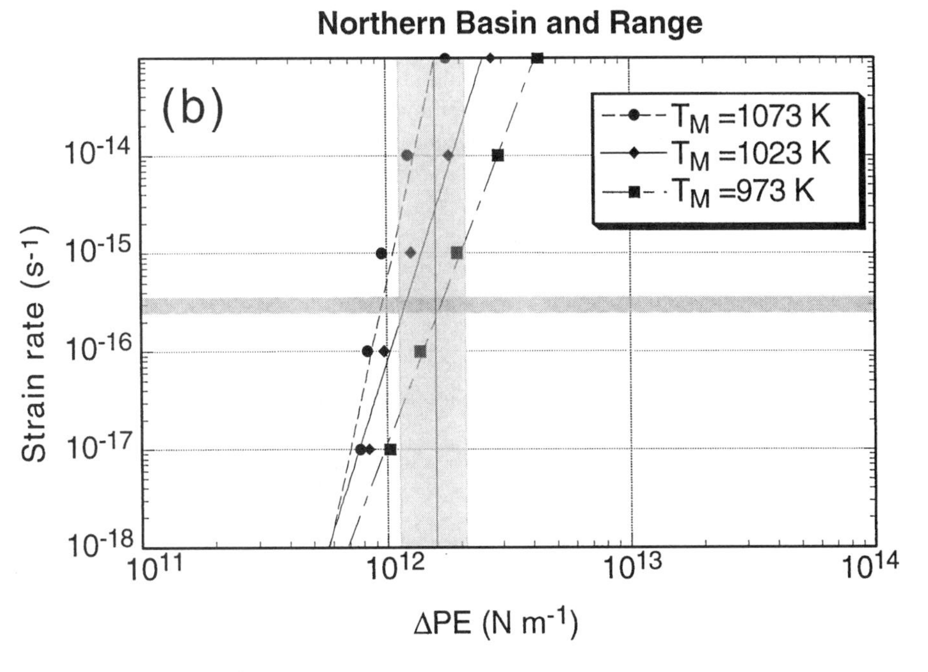

Figure 4 (*Continued*)

Third, although the correlation between potential energy variations and strain rate variations in the western United States is excellent when examined on a province-wide basis (Jones et al 1996), it is less consistent at smaller length scales. Scatter is to be expected at the smallest scales ($<\sim$100 km) where flexural support of topography dominates. At intermediate scales it is likely that other influences, such as rheology, or boundary forces imposed by one province on another due to geometry and directions of potential energy variations, may affect the strain rate field. Such discrepancies may require case-by-case examination (Townsend & Sonder 1997).

Fourth, the source of buoyancy is not well understood. It seems clear that crustal thickness variations alone are not sufficient to produce all of western North American extension (Liu & Shen 1998, Sonder et al 1987). Mantle buoyancy must be invoked; this is most generally taken to be a lithospheric thinning process that results in removal of the cold lower part of the lithospheric mantle and replacement with hot asthenosphere, perhaps by detachment of a blob (Conrad & Molnar 1997, Houseman et al 1981, Houseman & Molnar 1997), by mantle upwelling that thermally erodes the lithosphere (Parsons et al 1994, Pierce & Morgan 1992, Saltus & Thompson 1995), or by steepening or delamination of a shallowly dipping subducting Farallon slab (Bird 1988, Humphreys 1995). However, few quantitative tests have been presented. To resolve the source of buoyancy, we need to better understand controls on subduction angle, when and where slab removal occurred, the resulting mantle circulation, the nature of coupling between slabs and overriding plates, and the behavior of convective instabilities and mantle plumes. Once these are known, it should be possible to calculate how stresses and temperatures at the base of the continental lithosphere should evolve and the consequent changes in lithospheric stress state and rheologic strength.

In the absence of quantitative physical understanding, we are left with testing models by correlating observed events with inferred consequences of processes originating either at the base of the lithosphere or within the lithosphere that increase potential energy. For example, replacement of lithosphere by asthenosphere would cause a corresponding large isostatically driven increase in surface elevation just as extension began. On the other hand, if extension were driven by lithospheric buoyancy developed previously, e.g. by crustal thickening during the Laramide and Sevier orogenies (Sonder et al 1987), the Basin and Range would have been high for some time before the beginning of extension. Thus knowledge of surface elevations throughout the Cenozoic is critical to constraining the causes of extension.

4. OUR BEST GUESS

The discussion above highlights the inability of any single mechanism to explain all the extensional deformation in the western United States. Instead some combination is required. We provide a brief explanation for disregarding certain hypotheses and present a scenario we consider plausible, with the goal of identifying the observations that permit critical evaluation of this and other possibilities. We do this by subregion because it is quite clear that the NBR and SBR share few similarities.

4.1 *Northern Basin and Range*

For this region, we reject boundary forces and basal shears as primary causes of deformation for reasons discussed above. We also doubt that a mantle plume was primarily responsible for extension. This leaves internal body forces and basal normal forces as the most probable driving forces. Crustal body forces produced solely by crustal thickening have sometimes been considered distinct from forces derived from density variations in the mantle. Thus some have viewed extension in the NBR as a two-stage process in which extension driven by crustal body forces was separated from extension driven by emplacement of buoyant mantle by a period of volcanism without extension (Best &

Christiansen 1991, Liu 1996). We reject this in its most extreme form because (*1*) we do not think crustal extension ceased during volcanism, as argued above; (*2*) at a regional scale, crustal extension would necessarily require extension of the mantle lithosphere (otherwise a decollement would have to penetrate the Sierra to the plate boundary at the coast); (*3*) the timing of extension and volcanism from north to south is inconsistent with such a simple story, including nearly coaxial extension around 35 Ma being simultaneously pre-volcanic (southern Nevada), syn-volcanic (Northern Snake Range) and post-volcanic (Idaho), and (*4*) extension continued outside the volcanic belts (Axen et al 1993, Lee 1995).

Crustal potential energy cannot by itself explain the Cenozoic extension because the available potential energy is insufficient to produce the observed magnitude of extension (Liu & Shen 1998, Sonder et al 1987). Even so, crustal buoyancy was not insignificant. The good correlation of thickened crust in the rear of the Sevier thrust belt with later extensional belts argues that crustal body forces played a significant role in driving extension. Thus a combination of potential energy from crustal thickening, removal of mantle lithosphere and/or an attached slab, and asthenospheric upwelling is needed to explain NBR extension.

The fact that Basin and Range extension lies behind that portion of the Mesozoic Cordilleran volcanic arc that shut down in early Cenozoic time argues that this poorly understood event somehow provided a precondition for large-scale extension. Some sort of mantle upwelling following removal of the Farallon slab (Humphreys 1995) is difficult to evaluate, but must be rejected as a sole cause if it is presumed to be associated with generating large silicic volcanic centers, for silicic volcanic activity extended well east of the Basin and Range into areas of little or no extension, e.g. Marysvale and San Juan volcanic fields (Color Plate 1).

Thus we suspect that the lithospheric mantle was convectively destabilized beneath the NBR during the Laramide orogeny, plausibly through cooling or thickening of the mantle lithosphere. As contractional tectonism waned, mantle lithosphere was made buoyant or removed roughly from north to south, increasing the mean elevation of a broad region and beginning the process of warming the lithosphere (Humphreys 1995); this could be associated with volcanism. Extension appeared nearly simultaneously at a given latitude in both the frontal parts of the Sevier thrust belt and the hinterland (Constenius 1996), suggesting that elevation was acquired across the grain of the old Sevier orogen. Where elevation (and thus gravitational potential energy) was acquired as the lithosphere was heated, considerable buoyancy force to drive extension would accompany a reduction of the resisting force from lithospheric strength. This force would act to drive extension within the uplifted area and act on the stronger Sierran

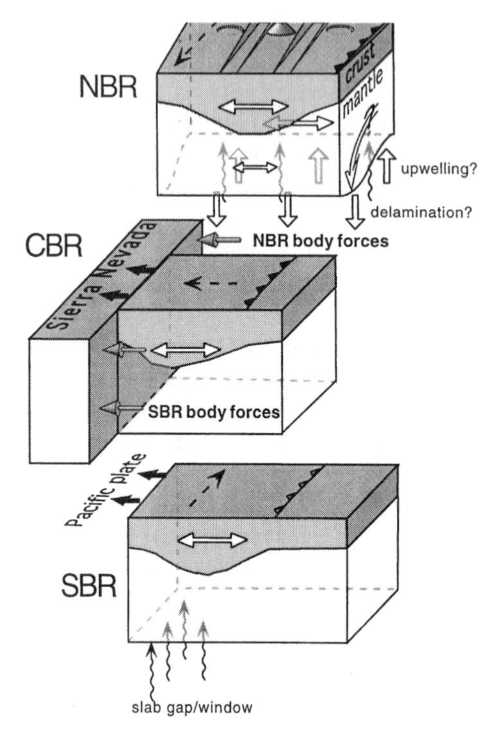

Figure 5 Summary sketch of our scenario for forces driving extension in the western United States (Section 4).

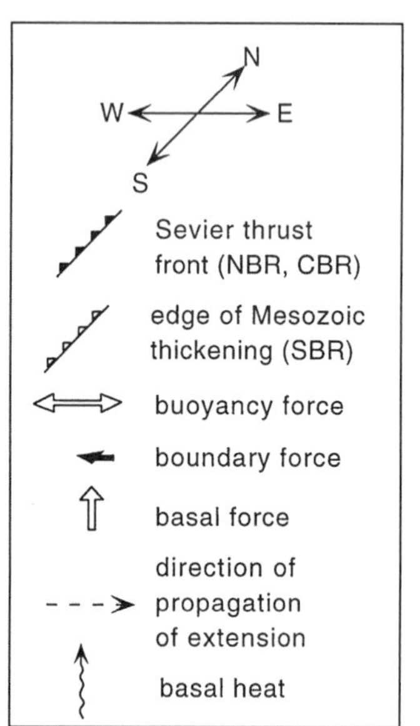

lithosphere to the west, prying it away from North America. This process would help drive extension south of the uplifted, volcanically active region (i.e. in areas with lower potential energy), with displacement decreasing to the south, e.g. the decrease in displacement from the northern through the southern Snake Range to the Stampede detachment (Color Plate 1) as envisioned by Axen et al (1993) and Burke & Axen (1997). This extension would localize in the weakest regions, which would presumably be within the thick pile of sediments west of the Sevier thrust front. To use classic (but imprecise) summary terms, the southward transition from volcanic to nonvolcanic regions represented a change from "active" to somewhat more "passive" extension (Figure 5).

As the mantle lithosphere was altered or removed from north to south, body forces increased. At the same time, volatilized lithospheric mantle was exposed to higher temperatures and lay within a tectonic environment favorable to explosive volcanism. This scenario suggests that elevation of the Sevier hinterland should be diachronous and closely associated with the migration of volcanism. Because the driving forces were largely body forces internal to the crust and derived from the increase in buoyancy in the mantle, careful examination of changes in elevation through time along with the character of the upper mantle will prove most critical in evaluating their presence and magnitude.

4.2 *Southern Basin and Range*

Different associations suggest different causes. The close association of the bulk of extension with the complex evolution of the plate boundary in the SBR (Bohannon & Parsons 1995, Dokka & Ross 1995, Glazner & Bartley 1984) indicates that removal of the subduction zone was a necessary prerequisite for extension. However, this cannot be sufficient. Because spreading ridges still remained offshore as extension initiated, a simple appeal to either divergence of the Pacific from the North American plate or triple junction instability fails if forces and strengths favored continued spreading at the ridge. Bohannon and Parsons (1995) argue that extension in the SBR occurred only inboard of areas where there was direct contact between the Pacific and North American plates, but this seems unlikely given the similar style of extension in the continental borderland directly adjacent to ongoing seafloor spreading (Bohannon & Geist 1998). We suggest that an additional requirement for extension in the SBR is that the gravitational potential energy exceeded that of the spreading centers; thus the region with the greater potential energy and weaker rheology would fail (Dokka & Ross 1995, 1996). The short duration of extension contrasts strongly with the NBR, which was extending behind the Sierran/Great Valley block and over the colder lithosphere being subducted north of the Mendocino triple junction.

A key challenge to the hypothesis that extension in the SBR was strongly influenced by boundary forces is the sudden end of northward migration of extension at the southern end of the Sierra Nevada at ~16 Ma (Dokka & Ross 1995, Glazner & Bartley 1984). Extension ceased to migrate with the triple junction at this time, instead shifting northeastward into the last unextended part of the Mesozoic compressional belt. This shift might be accounted for by the intervention of the strong Sierran/Great Valley block, which had its southern boundary defined by Paleogene time (Wood & Saleeby 1997) and which might have suppressed boundary-driven deformation to its west if the Sierra became strongly coupled to the Pacific plate. However, such a scenario is difficult to envision given the strike-slip deformation occurring between the two and the limited exposure of the Sierra to the Pacific Plate.

4.3 *Central Basin and Range*

Until about 20 Ma, the CBR was the northern part of a substantial unextended area connecting the Sierra Nevada to North America. It lay just south of the edge of massive silicic volcanism that had just swept across the NBR (Best & Christiansen 1991). If our scenario for the NBR is correct, forces derived from buoyancy in the NBR tending to move the Sierra Nevada away from North America and thereby stretch the CBR were resisted by the strength of the CBR and parts of the SBR to the south. As the mantle lithosphere to the north eroded away, as suggested by isotopic evidence in much of the NBR (Daley & De Paolo 1992, Farmer et al 1989), the potential energy of the NBR increased and so the forces acting to extend the CBR increased. Simultaneously the SBR to the south began extending rapidly; thus the region resisting westward motion of the Sierra Nevada became much smaller. Moreover, as the Mendocino triple junction moved north of the southern edge of the Sierra Nevada, there may have been an increased force pulling the Sierra Nevada to the north and west, placing the CBR in additional tension. It is entirely plausible that the CBR had also warmed by this time due to thermal re-equilibration after thickening (Sonder et al 1987) and perhaps removal of the Farallon slab. Thus a large number of plausible reasons exist for the CBR to fail around 15 Ma.

This emphasis on forces external to the CBR seems appropriate. Estimates of gravitational body forces within this area today are low. Because it also appears to have preserved mantle lithosphere (Jones et al 1996, Jones et al 1992) and experienced only minor volcanism in comparison with other areas, there seems to be little evidence that the mantle provided much buoyancy in the past. However, failure of the contractionally thickened crust east of the Sierra indicates that internal forces and strengths have played a role in localizing strain even if they were insufficient to initiate and drive extension. Better knowledge of middle Tertiary elevations, pre-extensional crustal density and thermal structures, and the composition and state of the mantle would all help better constrain the relative contributions of internal and external forces in extending this portion of the Basin and Range.

We envision three extensional systems in the western United States: one exploiting crust and mantle potential energy as the mantle became more buoyant and probably warmer (NBR), one also exploiting crustal potential energy but more in league with plate boundary forces (SBR), and one possibly driven by forces external to itself yet derived, in large part, from the acquisition of body forces elsewhere in the orogen (CBR). Recognizing these differences helps bring into focus the reasons why extensional tectonism occurred in some regions and not others, and why it had the complex temporal evolution we observe. Our scenario is a template that points out the critical need for more complete determinations of paleoelevation, crustal structure, and composition and thermal state of the mantle in order to better understand the forces driving deformation and the fundamental ambiguities that remain in our understanding of the Basin and Range.

ACKNOWLEDGMENTS

This work was partially supported by NSF grants EAR 97-25192 (to LJS) and EAR 97-25755 (to CHJ). Thanks to Greg Houseman for the use of his sybil/basil codes and to Anne Sheehan, Steve Wesnousky, Carl Renshaw, G. Lang Farmer, Peter Molnar, Ken Dueker and Xiahong Feng for commenting on early versions of this paper. Data for Color Plate 3 was provided by CNSS member networks through the CNSS catalog maintained by the Northern California Data Center at U. C. Berkeley.

Literature Cited

Allmendinger RW. 1992. Fold and thrust tectonics of the western United States exclusive of the accreted terranes. In *The Cordilleran Orogen: Conterminous U.S.*, ed. BC Burchfiel, PW Lipman, ML Zoback. pp. 583–607. Geology of North America. Vol. G3. Boulder, Colorado: Geol. Soc. Am.

Anders MH, Sleep NH. 1992. Magmatism and extension: The thermal and mechanical effects of the Yellowstone hotspot. *J. Geophys. Res.* 97:15,379–15,393

Anderson RE. 1971. Thin skin distension in Tertiary rocks of southeastern Nevada. *Geol. Soc. Am. Bull.* 82:43–58

Applegate JDR, Hodges KV. 1995. Mesozoic and Cenozoic extension recorded by metamorphic rocks in the Funeral Mountains, California. *Geol. Soc. Amer. Bull.* 107:1063–76

Argus DF, Gordon RG. 1991. Current Sierra Nevada-North America motion from very long baseline interferometry: Implications for the kinematics of the western United States. *Geology* 19:1085–88

Argus DF, Gordon RG. 1996. Test of the rigid-plate hypothesis and bounds on intraplate deformation using geodetic data from very long baseline interferometry. *J Geophys. Res.* 101:13,555–13,572

Armstrong RL. 1972. Low-angle (denudation) faults, hinterland of the Sevier orogenic belt, eastern Nevada and western Utah. *Geol. Soc. Am. Bull.* 83:1729–54

Armstrong RL, Ward P. 1991. Evolving geographic patterns of Cenozoic magmatism in the North American Cordillera: The temporal and spatial association of magmatism and metamorphic core complexes. *J. Geophys. Res.* 96:13,201–13,224

Artyushkov EV. 1973. Stresses in the lithosphere caused by crustal thickness inhomogeneities. *J. Geophys. Res.* 78:7675–7708

Asmerom Y, Jacobsen SB, Wernicke BP. 1994. Variations in magma source regions during large-scale continental extension, Death Valley region, western United States. *Earth Planet. Sci. Lett.* 125:235–54

Asmerom Y, Walker RJ. 1998. Pb and Os isotopic constraints on the composition and rheology of the lower crust. *Geology* 26:359–62

Atwater T. 1970. Implications of plate tectonics for the Cenozoic tectonic evolution of western North America. *Bull. Geol. Soc. Amer.* 81:3513–36

Atwater T, Molnar P. 1973. Relative motion of the Pacific and North American plates deduced from sea-floor spreading in the Atlantic, Indian, and south Pacific Oceans. In *Conference on Tectonic Problems of the San Andreas Fault System*, ed. RL Kovach, A Nur, 13:136–48. Stanford, CA: Stanford Univ. Publ. Geol. Sci.

Atwater T, Stock J. 1998. Pacific-North America plate tectonics of the Neogene southwestern United States: An update. *International Geology Review* 40:375–402

Axelrod DI. 1966. The Eocene Copper Basin flora of northeastern Nevada. *Univ. Calif. Pubs. Geol. Sci.* 59:125. Berkeley, CA: Univ. Calif. Press

Axelrod DI. 1991. The Middle Miocene Pyramid flora of western Nevada. In *Univ. Calif. Pubs. Geol. Sci.* 137:50. Berkeley, CA: Univ. of Calif. Press

Axen GJ, Taylor WJ, Bartley JM. 1993. Space-time patterns and tectonic controls of Tertiary extension and magmatism in the Great Basin of the western United States. *Geol. Soc. Amer. Bull.* 105:56–76

Bai W, Vigny C, Ricard Y, Froidevaux C. 1992. On the origin of deviatoric stresses in the lithosphere. *J. Geophys. Res.* 97:11,729–11,737

Barazangi M, Isacks BL. 1976. Spatial distribution of earthquakes and subduction of the Nazca plate beneath South America. *Geology* 4:686–92

Beard BL, Johnson CM. 1997. Hafnium isotope evidence for the origin of Cenozoic basaltic lavas from the southwestern United States. *J. Geophys. Res.* 102:20,149–20,178

Beard LS. 1996. Paleogeography of the Horse Spring Formation in relation to the Lake Mead fault system, Virgin Mountains, Nevada and Arizona. In *Reconstructing the history of Basin and Range extension using sedimentology and stratigraphy.*, ed. KK Beratan, 303:1–14. Geological Society of America Special Paper. Boulder, CO: Geol. Soc. Amer.

Bennett RA, Wernicke BP, Davis JL. 1998. Continuous GPS measurements of contemporary deformation across the northern Basin and Range province. *Geophys. Res. Lett.* 25:563–66

Bennett RA, Wernicke BP, Davis JL, Elósegui P, Snow JK, et al. 1997. Global Positioning System constraints on fault slip rates in the Death Valley region, California and Nevada. *Geophys. Res. Lett.* 24:3073–76

Bergquist JR, Nitkiewicz AM. 1982. Geologic map of the Domeland Wilderness and contiguous roadless areas, Kern and Tulare counties, California. *U.S. Geol. Surv. Misc. Field Map.* MF-1395A

Best MG. 1988. Early Miocene change in direction of least principal stress, southwestern United States: Conflicting inferences from dikes and metamorphic core-detachment fault terranes. *Tectonics* 7:249–59

Best MG, Christiansen EH. 1991. Limited extension during peak Tertiary volcanism, Great Basin of Nevada and Utah. *J. Geophys. Res.* 96:13,509–13,528

Bird P. 1988. Formation of the Rocky Mountains, western United States: A continuum computer model. *Science* 239:1501–7

Bird P. 1998. Testing hypotheses on plate-driving mechanisms with global lithosphere models including topography, thermal structure, and faults. *J. Geophys. Res.* 103:10,115–10,129

Bird P, Piper K. 1980. Plane-stress finite element models of tectonic flow in southern California. *Phys. Earth Planet. Inter.* 21:158–75

Blackwell DD, Steele JL. 1991. Geothermal map of North America, Map CSM-007, 4 sheets. Boulder, CO: Geol. Soc. Amer.

Block L, Royden LH. 1990. Core complex geometries and regional scale flow in the lower crust. *Tectonics* 9:557–67

Bogen NL, Schweickert RA. 1985. Magnitude of crustal extension across the northern Basin and Range Province: Constraints from paleomagnetism. *Earth Planet. Sci. Lett.* 75:93–100

Bohannon RG, Geist E. 1998. Upper crustal structure and Neogene tectonic development of the California continental borderland. *Geol. Soc. Amer. Bull.* 110:779–800

Bohannon RG, Parsons T. 1995. Tectonic implications of post-30 Ma Pacific and North American relative plate motions. *Geol. Soc. Amer. Bull.* 107:937–59

Bott MHP, Dean DS. 1972. Stress systems at young continental margins. *Nature* 235:23–25

Bradshaw TK, Hawkesworth CJ, Gallagher K. 1993. Basaltic volcanism in the Southern Basin and Range: no role for a mantle plume. *Earth Planet. Sci. Lett.* 116:45–62

Burchfiel BC, Cowan DS, Davis GA. 1992. Tectonic overview of the Cordilleran orogen in the western United States. In *The Cordilleran Orogen: Conterminous U.S.*, ed. BC Burchfiel, PW Lipman, ML Zoback, *The Geology of North America* G-3:407–79. Boulder, CO: Geol. Soc. Amer.

Burke KJ, Axen GJ. 1997. Structural geometry resulting from episodic extension in the northern Chief Range area, eastern Nevada. In *Geologic studies in the Basin and Range-Colorado Plateau transition in southeastern Nevada, southwestern Utah, and northwestern Arizona, 1995*, ed. F Maldonado, LD Nealey. p. 265–88. *U.S. Geol. Surv. Bull.* Vol. 2153. Reston, VA: U.S. Geol. Surv.

Bursik M, Sieh K. 1989. Range-front faulting and volcanism in the Mono basin, eastern California. *J. Geophys. Res.* 94:15,587–15,609

Camilleri PA, Chamberlain KR. 1997. Mesozoic tectonics and metamorphism in the Pequop Mountains and Wood Hills region, Northeast Nevada; implications for the architecture and evolution of the Sevier Orogen. *Geol. Soc. Amer. Bull.* 109:74–94

Cathles LM. 1975. *The Viscosity of the Earth's Mantle*. Princeton: Princeton University Press. 386 pp.

Christiansen RL, Yeats RS. 1992. Post-Laramide geology of the U. S. Cordilleran region. In *The Cordilleran Orogen: Conterminous United States*, ed. BC Burchfiel, PW Lipman, ML Zoback, G-3:261–406. Boulder, CO: Geol. Soc. Amer.

Clark MM, Yount JC, Vaughan PR, Zepeda RL. 1982. Map showing surface ruptures associated with the Mammoth Lakes, California, earthquakes of May 1980. Miscellaneous

Field Studies Map MF–1396. Reston, VA: U.S. Geol. Surv.

Clark TA, Gordon D, Himwich WE, Ma C, Mallama A, Ryan JW. 1987. Determination of relative site motions in the western United States using Mark II Very Long Baseline Interferometry. *J. Geophys. Res.* 92:12,741–12,750

Coblentz DD, Richardson RM, Sandiford M. 1994. On the gravitational potential of the Earth's lithosphere. *Tectonics* 13:929–45

Coney PJ, Harms TA. 1984. Cordilleran metamorphic core complexes: Cenozoic extensional relics of Mesozoic compresssion. *Geology* 12:550–54

Coney PJ, Reynolds SJ. 1977. Cordilleran Benioff zones. *Nature* 270:403–6

Conrad CP, Molnar P. 1997. The growth of Rayleigh-Taylor-type instabilities in the lithosphere for various rheological and density structures. *Geophys. J. Int.* 129:95–112

Constenius KN. 1996. Late Paleogene extensional collapse of the Cordilleran foreland fold and thrust belt. *Geol. Soc. Amer. Bull.* 108:20–39

Coogan JC, DeCelles PG. 1996. Extensional collapse along the Sevier Desert reflection, northern Sevier Desert basin, Western United States. *Geology* 24:933–36

Daley EE, De Paolo DJ. 1992. Isotopic evidence for lithospheric thinning during extension: Southeastern Great Basin. *Geology* 20:104–8

DeLong SE, Fox PJ, McDowell FW. 1978. Subduction of the Kula Ridge at the Aleutian Trench. *Geol. Soc. Amer. Bull.* 89:83–95

DeMets C, Gordon RG, Argus DF, Stein S. 1994. Effect of recent revisions to the geomagnetic reversal time scale on estimates of current plate motions. *Geophys. Res. Lett.* 21:2191–94

Dickinson WR. 1997. Tectonic implications of Cenozoic volcanism in coastal California. *Geol. Soc. Amer. Bull.* 109:936–54

Dickinson WR, Snyder WS. 1978. Plate tectonics of the Laramide orogeny. In *Laramide Folding Associated with Basement Block Faulting in the Western United States*. ed. V Matthews, Geol. Soc. Amer. Mem. 151:355–66. Boulder, CO: Geol. Soc. Amer.

Dickinson WR, Snyder WS. 1979a. Geometry of subducted slabs related to San Andreas transform. *J. Geol.* 87:609–27

Dickinson WR, Snyder WS. 1979b. Geometry of triple junctions related to San Andreas transform. *J. Geophys. Res.* 84:561–72

Dilles JH, Gans PB. 1995. The chronology of Cenozoic volcanism and deformation in the Yerington area, western Basin and Range and Walker Lane. *Geol. Soc. Amer. Bull.* 107:474–86

Dixon JM, Farrar E. 1980. Ridge subduction, eduction, and the Neogene tectonics of southwestern North America. *Tectonophysics* 67:81–99

Dixon TH, Robaudo S, Lee J, Reheis MC. 1995. Constraints on present-day Basin and Range deformation from space geodesy. *Tectonics* 14:755–72

Dokka RK, Ross TM. 1995. Collapse of southwestern North America and the evolution of early Miocene detachment faults, metamorphic core complexes, the Sierra Nevada orocline, and the San Andreas fault system. *Geology* 23:1075–78

Dokka RK, Ross TM. 1996. Collapse of southwestern North America and the evolution of early Miocene detachment faults, metamorphic core complexes, the Sierra Nevada orocline, and the San Andreas fault system: Reply. *Geology* 24:859–60

Dubiel RF, Potter CJ, Good SC, Snee LW. 1996. Reconstructing an Eocene extensional basin; the White Sage Formation, eastern Great Basin. In *Reconstructing the history of Basin and Range extension using sedimentology and stratigraphy*. ed. KK Beratan, *Special Paper–Geological Society of America*, 303:1–14. Boulder, CO: Geol. Soc. Amer.

Dumitru TA. 1988. Subnormal geothermal gradients in the Great Valley forearc basin, California, during Franciscan subduction: A fission track study. *Tectonics* 7:1201–21

Dumitru TA. 1990. Subnormal Cenozoic geothermal gradients in the extinct Sierra Nevada magmatic arc: Consequences of Laramide and post-Laramide shallow angle subduction. *J. Geophys. Res.* 95:4925–41

Dumitru TA, Gans PB, Foster DA, Miller EA. 1991. Refrigeration of the western Cordilleran lithosphere during Laramide shallow angle subduction. *Geology* 19:1145–48

Eaton GP. 1982. The Basin and Range province: Origin and tectonic significance. *Annu. Rev. Earth Planet. Sci.* 10:409–40

Eddington PK, Smith RB, Renggli C. 1987. Kinematics of Basin and Range intraplate extension. In *Continental Extensional Tectonics*, ed. MP Coward, JF Dewey, PL Hancock. Geol. Soc. Lond. Spec. Pub. 28:371–92. Oxford: Geol. Soc. Lond.

Engebretson DC, Cox A, Gordon RG. 1984. Relative motions between oceanic plates of the Pacific basin. *J. Geophys. Res.* 89:10,291–10,310

England PC, Houseman GA, Sonder LJ. 1985. Length scales for continental deformation in convergent, divergent, and strike-slip environments: Analytical and approximate solutions for a thin viscous sheet model. *J. Geophys. Res.* 90:3551–57

England PC, Jackson J. 1989. Active deformation of the continents. *Annu. Rev. Earth Planet. Sci.* 17:197–226

England PC, Houseman GA. 1986. Finite strain calculations of continental deformation 2. Comparison with the India-Asia collision. *J. Geophys. Res.* 91:3664–76

England PC, McKenzie DP. 1982. A thin viscous shell model for continental deformation. *Geophys. J. R. Astron. Soc.* 70:295–321

Farmer GL, Perry FV, Semken S, Crowe B, Curtis D, DePaolo DJ. 1989. Isotopic evidence on the structure and origin of subcontinental lithospheric mantle in southern Nevada. *J. Geophys. Res.* 94:7885–98

Fitton JG, James D, Leeman WP. 1991. Basic magmatism associated with late Cenozoic extension in the western United States: Compositional variations in space and time. *J. Geophys. Res.* 96:13,693–711

Fleitout L, Froidevaux C. 1982. Tectonics and topography for a lithosphere containing density heterogeneities. *Tectonics* 1:21–56

Forest CE, Molnar P, Emanuel KA. 1995. Paleoaltimetry from energy conservation principles. *Nature* 374:347–50

Frei LS. 1986. Additional paleomagnetic results from the Sierra Nevada: Further constraints on Basin and Range extension and northward displacement in the western United States. *Geol. Soc. Amer. Bull.* 97:840–49

Furlong KP, Hugo WD, Zandt G. 1989. Geometry and evolution of the San Andreas fault zone in northern California. *J. Geophys. Res.* 94:3100–10

Gans PB. 1987. An open-system, two-layer crustal stretching model for the eastern Great Basin. *Tectonics* 6:1–12

Gans PB, Bohrson WA. 1998. Suppression of volcanism during rapid extension in the Basin and Range Province, United States. *Science* 279:66–68

Gans PB, Mahood GA, Schermer E. 1989. Synextensional magmatism in the Basin and Range Province; A case study from the eastern Great Basin. In *Geol. Soc. Am. Spec. Paper*, 233:53. Boulder, CO: Geol. Soc. Amer.

Glazner AF, Bartley JM. 1984. Timing and tectonic setting of Tertiary low-angle normal faulting and associated magmatism in the southwestern United States. *Tectonics* 3:385–96

Glazner AF, Loomis DP. 1984. Effect of subduction of the Mendocino fracture zone on Tertiary sedimentation in Southern California. *Sediment. Geol.* 38:287–303

Gordon D, Ma C, Ryan JW. 1993. Results from the CDP Mobile VLBI Program in the Western United States. In *Contributions of space geodesy to geodynamics; crustal dynamics*, ed. DE Smith, DL Turcotte, *Geodynamics Series*, 23:131–8. Washington, DC: Amer. Geophys. Un.

Greensfelder RW, Kintzer FC, Somerville MR. 1970a. Seismotectonic regionalization of the Great Basin, and comparison of moment rates computed from Holocene strain and historic seismicity. *Geol. Soc. Amer. Bull.* 91:2039–111 (part II)

Greensfelder RW, Kintzer FC, Somerville MR. 1970b. Seismotectonic regionalization of the Great Basin, and comparison of moment rates computed from Holocene strain and historic seismicity: Summary. *Geol. Soc. Amer. Bull.* 91:518–23 (part I)

Gregory-Wodzicki KM. 1997. The late Eocene House Range Flora, Sevier Desert, Utah: Paleoclimate and paleoelevation. *Palaios* 12:552–67

Gurnis M. 1990. Bounds on global dynamic topography from Phanerozoic flooding of continental platforms. *Nature* 344:754–56

Gurnis M. 1992. Rapid continental subsidence following the initiation and evolution of subduction. *Science* 255:1556–58

Hager BH. 1991. Mantle viscosity: A comparison of models from postglacial rebound and from the geoid, plate driving forces, and advected heat flux. In *Glacial Isostasy, Sea-Level and Mantle Rheology*, ed. R Sabadini, K Lambeck, E Boschi, *NATO ASI Series* p. 493–513. Boston: Kluwer Academic Publishers

Hamilton W, Myers WB. 1966. Cenozoic tectonics of the western United States. *Rev. Geophys.* 4:509–49

Hamilton WB. 1987. Crustal extension in the Basin and Range Province, Southwestern United States. In *Continental extensional tectonics*, ed. MP Coward, JF Dewey, PL Hancock. Geol. Soc. London Sp. Pub. 28:155–76. Oxford, UK: Geol. Soc. Lond.

Hill DP, Eaton JP, Ellsworth WL, Cockerham RS, Lester FW, Corbett EG. 1991. The seismotectonic fabric of central California. In *Neotectonics of North America*, ed. DB Slemmons, ER Engdahl, MD Zoback, DD Blackwell, Decade Map Volume 1:107–32. Boulder, CO: Geol. Soc. Amer.

Hodges KV, Walker JD. 1992. Extension in the Cretaceous Sevier orogen, North American Cordillera. *Geol. Soc. Amer. Bull.* 104:560–69

House MA, Wernicke BP, Farley KA, Dumitru TA. 1997. Cenozoic thermal evolution of the central Sierra Nevada from (U-Th)/He thermochronometry. *Earth Planet. Sci. Lett.* 151:167–79

Houseman GA, England PC. 1986. A dynamical model for lithosphere extension and sedimentary basin formation. *J. Geophys. Res.* 91:719–29

Houseman GA, McKenzie DP, Molnar P. 1981. Convective instability of a thickened boundary layer and its relevance for the thermal evolution of continental convergent belts. *J. Geophys. Res.* 86:6115–32

Houseman GA, Molnar P. 1997. Gravitational (Rayleigh-Taylor) instability of a layer with non-linear viscosity and convective thinning of continental lithosphere. *Geophys. J. Int.* 128:125–50

Humphreys ED. 1995. Post-Laramide removal of the Farallon slab, western United States. *Geology* 23:987–90

Ingersoll RV. 1982. Triple-junction instability as cause for late Cenozoic extension and fragmentation of the western United States. *Geology* 10:621–24

Janecke SU. 1992. Kinematics and timing of three superposed extensional systems, east central Idaho; evidence for an Eocene tectonic transition. *Tectonics* 11:1121–38

Janecke SU. 1994. Sedimentation and paleogeography of an Eocene to Oligocene rift zone, Idaho and Montana. *Geol. Soc. Amer. Bull.* 106:1083–95

Jones CH. 1987. Is extension in Death Valley accommodated by thinning of the mantle lithosphere beneath the Sierra Nevada, California? *Tectonics* 6:449–73

Jones CH, Phinney RA. 1998. Constraints on the seismic structure of the lithosphere from teleseismic converted arrivals observed at small arrays in the southern Sierra Nevada and vicinity, California. *J. Geophys. Res.* 103:10,065–90

Jones CH, Sonder LJ, Unruh JR. 1998. Lithospheric gravitational potential energy and past orogenesis: Implications for conditions of initial Basin and Range and Laramide deformation. *Geology* 26:639–42

Jones CH, Unruh JR, Sonder LJ. 1996. The role of gravitational potential energy in active deformation in the southwestern United States. *Nature* 381:37–41

Jones CH, Wernicke BP, Farmer GL, Walker JD, Coleman DS, et al. 1992. Variations across and along a major continental rift: An interdisciplinary study of the Basin and Range Province, western USA. *Tectonophysics* 213:57–96

Jones LM, Dollar RS. 1986. Evidence of basin-and-range extensional tectonics in the Sierra Nevada; the Durrwood Meadows swarm, Tulare County, California (1983–1984). *Bull. Seismol. Soc. Amer.* 76:439–61

Keith SB. 1978. Paleosubduction geometries inferred from Cretaceous and Tertiary magmatic patterns in southwestern United States. *Geology* 6:516–21

Le Pichon X. 1982. Land-locked oceanic basins and continental collision: The eastern Mediterranean as a case example. In *Mountain Building Processes*, ed. KJ Hsü, p. 201–11. New York: Academic Press. 263 pp.

Le Stunff Y, Richard Y. 1997. Partial advection of equidensity surfaces: A solution for the dynamic topography problem? *J. Geophys. Res.* 102:24,655–24,667

Lee J. 1995. Rapid uplift and rotation of mylonitic rocks from beneath a detachment fault; insights from potassium feldspar $^{40}Ar/^{39}Ar$ thermochronology, northern Snake Range, Nevada. *Tectonics* 14:54–77

Lipman PW. 1992. Magmatism in the Cordilleran United States; Progress and problems. In *The Cordilleran Orogen: Conterminous U.S.*, ed. BC Burchfiel, PW Lipman, ML Zoback, *The Geology of North America* G3:481–514. Boulder, CO: Geol. Soc. Amer. 724 pp.

Liu M. 1996. Dynamic interactions between crustal shortening, extension, and magmatism in the North American cordillera. *Pure and Applied Geophysics* 146:447–67

Liu M, Shen Y. 1998. Crustal collapse, mantle upwelling, and Cenozoic extension in the North American Cordillera. *Tectonics* 17:311–21

Livaccari RF. 1979. Late Cenozoic tectonic evolution of the western United States. *Geology* 7:72–75

Livaccari RF, Perry FV. 1993. Isotopic evidence for preservation of Cordilleran lithospheric mantle during the Sevier-Laramide orogeny, western United States. *Geology* 21:719–22

Lonsdale P. 1991. Structural patterns of the Pacific floor offshore of Peninsular California. In *The Gulf and Peninsular Province of the Californias*, ed. JP Dauphin, BRT Simoneit, p. 87–125. Amer. Assoc. Pet. Geol. Mem. Vol. 47. Tulsa, OK: Amer. Assoc. Pet. Geol.

MacCready T, Snoke AW, Wright JE, Howard KA. 1997. Mid-crustal flow during Tertiary extension in the Ruby Mountains core complex, Nevada. *Geol. Soc. Amer. Bull.* 109:1576–94

Martinez LJ, Meertens CJ, Smith RB. 1998. Rapid deformation rates along the Wasatch fault zone, Utah, from first GPS measurements with implications for earthquake hazard. *Geophys. Res. Lett.* 25:567–70

McKenzie DP. 1969. Speculations on the consequences and causes of plate motion. *Geophys. J. R. Astron. Soc.* 18:1–32

McKenzie DP. 1972. Active tectonics of the Mediterranean region. *Geophys. J. R. Astron. Soc.* 30:109–185

McKenzie DP. 1978. Active tectonics of the Alpine-Mediterranean belt: The Aegean Sea and surrounding regions. *Geophys. J. R. Astron. Soc.* 55:217–54

Menard HW. 1960. The east Pacific rise. *Science* 132:1737–46

Miller JS, Glazner AF, Farmer GL, Suayah IB, Keith LA. 1999. Middle Tertiary volcanism across the Mojave Desert and southeastern California. *Geol. Soc. Amer. Bull.* (In press)

Minster JB, Jordan TH. 1987. Vector constraints on western U.S. deformation from space geodesy, neotectonics, and plate motions. *J. Geophys. Res.* 92:4798–804

Mitrovica JX, Beaumont C, Jarvis GT. 1989. Tilting of continental interiors by dynamical effects of subduction. *Tectonics* 8:1079–94

Molnar P, Chen W-P. 1983. Focal depths and fault plane solutions of earthquakes under the Tibetan plateau. *J. Geophys. Res.* 88:1180–96

Molnar P, Houseman GA, Conrad CP. 1998. Rayleigh-Taylor instability and convective thinning of mechanically thickened lithosphere: effects of non-linear viscosity decreasing exponentially with depth and of horizontal shortening of the layer. *Geophys. J. Int.* 133:568–84

Molnar P, Lyon-Caen H. 1988. Some simple physical aspects of the support, structure, and evolution of mountain belts. In *Processes in Continental Lithospheric Deformation*, ed. SP Clarke, BC Burchfiel, J Suppe, *Geol. Soc. Amer. Spec. Pap.* 218:179–207. Boulder, CO: Geol. Soc. Amer. 212 pp.

Mooney WD, Weaver CS. 1989. Regional structure and tectonics of the Pacific Coast states: California, Oregon, and Washington. In *Geophysical Framework of the Continental United States*, ed. LC Pakiser, WD Mooney. *Geol. Soc. Amer. Mem.* 172:129–61. Boulder, CO: Geol. Soc. Amer.

Moore JG, Dodge FCW. 1980. Late Cenozoic volcanic rocks of the southern Sierra Nevada, California; I, Geology and petrology. *Geol. Soc. Amer. Bull.* 91:515–18 (Part I)

Mueller KJ, Cerveny PK, Perkins ME, Snee LW. 1999. Chronology of polyphase extension in the Windermere Hills, NE Nevada. *Geol. Soc. Amer. Bull.* 111:11–27

Mueller KJ, Snoke AW. 1993. Progressive overprinting of normal fault systems and their role in Tertiary exhumation of the East Humboldt-Wood Hills metamorphic complex, northeast Nevada. *Tectonics* 12:361–71

Ormerod DS, Hawkesworth CJ, Rogers NW, Leeman WP, Menzies MA. 1988. Tectonic and magmatic transitions in the Western Great Basin, USA. *Nature* 333:349–53

Parsons T. 1995. The Basin and Range province. In *Continental Rifts: Evolution, Structure, Tectonics*, ed. KH Olsen, *Developments in Geotectonics* 25:277–324. New York: Elsevier

Parsons T, Thompson GA, Sleep NH. 1994. Mantle plume influence on the Neogene uplift and extension of the U.S. western Cordillera? *Geology* 22:83–86

Parsons T, Thompson GA, Smith RP. 1998. More than one way to stretch: A tectonic model for extension along the plume track of the Yellowstone hotspot and adjacent Basin and Range. *Tectonics* 17:221–34

Pierce KL, Morgan LA. 1992. The track of the Yellowstone hot spot: Volcanism, faulting, and uplift. In *Regional Geology of Eastern Idaho and Western Wyoming*, ed. PK Link, MA Kuntz, LB Platt, *Geol. Soc. Amer. Memoir* 179:1–53. Boulder, CO: Geol. Soc. Amer.

Potter CJ, Dubiel RF, Snee LW, Good SC. 1995. Eocene extension of early Eocene lacustrine strata in a complexly deformed Sevier-Laramide hinterland, Northwest Utah and Northeast Nevada. *Geology* 23:181–84

Ricard Y, Richards M, Lithgow BC, Le Stunff Y. 1993. A geodynamic model of mantle density heterogeneity. *J. Geophys. Res.* 98:21,895–21,909

Rosa JWC, Molnar P. 1988. Uncertainties in reconstructions of the Pacific, Farallon, Vancouver, and Kula plates and constraints on the rigidity of the Pacific and Farallon (and Vancouver) plates between 72 and 35 Ma. *J. Geophys. Res.* 93:2997–3008

Saltus RW, Thompson GA. 1995. Why is it downhill from Tonopah to Las Vegas? A case for mantle plume support of the high northern Basin and Range. *Tectonics* 14:1235–44

Savage JC, Lisowski M, Prescott WH, Sanford AR. 1980. Geodetic measurements of horizontal deformation across the Rio Grande rift near Socorro, New Mexico. *J. Geophys. Res.* 85:7215–20

Sawyer DA, Fleck RJ, Lanphere MA, Warren RG, Broxton DE, Hudson MR. 1994. Episodic caldera volcanism in the Miocene southwestern Nevada volcanic field: Revised stratigraphic framework, $^{40}Ar/^{39}Ar$ geochronology, and implications for magmatism and extension. *Geol. Soc. Amer. Bull.* 106:1304–18

Sbar ML, Barazangi M, Dorman J, Scholz C, Smith RB. 1972. Tectonics of the intermountain seismic belt, western United States: microearthquake seismicity and composite fault plane solutions. *Geol. Soc. Amer. Bull.* 83:13–26

Scholz CH, Barazangi M, Sbar ML. 1971. Late Cenozoic evolution of the Great Basin, western United States, as an ensialic interarc basin. *Geol. Soc. Amer. Bull.* 82:2979–90

Severinghaus J, Atwater T. 1990. Cenozoic geometry and thermal state of the subducting slabs beneath western North America. In *Basin and Range Extensional Tectonics Near the Latitude of Las Vegas, Nevada*, ed. BP Wernicke, *Geol. Soc. Amer. Mem.* 176:1–22. Boulder, CO: Geol. Soc. Amer.

Smith RB, Arabasz WJ. 1991. Seismicity of the

intermountain seismic belt. In *Neotectonics of North America*, ed. DB Slemmons, ER Engdahl, MD Zoback, DD Blackwell, *Decade Map* Volume 1:185–228. Boulder, CO: Geol. Soc. Amer.

Smith RB, Braile LW. 1993. Topographical signature, space-time evolution, and physical properties of the Yellowstone-Snake River Plain volcanic system: the Yellowstone hotspot. In *Geology of Wyoming*, ed. AW Snoke, JR Steidtmann, SM Roberts. *Geol. Surv. Wyoming Mem.* 5:694–754. Laramie, WY: Geol. Surv. Wyoming

Smith RB, Sbar M. 1974. Contemporary tectonics and seismicity of the western United States with emphasis on the intermountain seismic belt. *Geol. Soc. Amer. Bull.* 85:1205–18

Sonder LJ, England PC. 1986. Vertical averages of rheology of the continental lithosphere: relation to thin sheet parameters. *Earth Planet. Sci. Lett.* 77:81–90

Sonder LJ, England PC, Houseman GA. 1986. Continuum calculations of continental deformation in transcurrent environments. *J. Geophys. Res.* 91:4797–810

Sonder LJ, England PC, Wernicke BP, Christiansen RL. 1987. A physical model for Cenozoic extension of western North America. In *Continental Extensional Tectonics*, ed. MP Coward, JF Dewey, PL Hancock, Geol. Soc. Lond. Sp. Pub. 28:187–201. Oxford, UK: Geol. Soc. Lond.

Spencer JE, Reynolds SJ. 1991. Tectonics of mid-Tertiary extension along a transect through west central Arizona. *Tectonics* 10:1024–221

Spencer JE, Richard SM, Reynolds SJ, Miller RJ, Shafiqullah M, et al. 1995. Spatial and temporal relationships between mid-Tertiary magmatism and extension in southwestern Arizona. *J. Geophys. Res.* 100:10,321–10,351

Stauder W. 1975. Subduction of the Nazca plate under Peru as evidenced by focal mechanisms and by seismicity. *J. Geophys. Res.* 80:1053–64

Stewart JH. 1978. Basin-range structure in western North America: A review. In *Cenozoic Tectonics and Regional Geophysics of the Western Cordillera*, ed. RB Smith, GP Eaton, *Geol. Soc. Amer. Mem.* 152:1–31. Boulder, CO: Geol. Soc. Amer.

Stock J, Molnar P. 1988. Uncertainties and implications of the late Cretaceous and Tertiary position of North America relative to the Farallon, Kula, and Pacific plates. *Tectonics* 6:1339–84

Thompson GA, Burke DB. 1974. Regional geophysics of the Basin and Range Province. *Annu. Rev. Earth Planet. Sci.* 2:213–38

Tovish A, Schubert G, Luyendyk BP. 1978. Mantle flow pressure and the angle of subduction: Non-Newtonian corner flows. *J. Geophys. Res.* 83:5892–98

Townsend DA, Sonder LJ. 1997. Rheological control of extension in the Rio Grande Rift. *EOS, Fall Mtg. Suppl.* 78:F658–59

Uyeda S, Miyashiro A. 1974. Plate tectonics and the Japanese Islands: A synthesis. *Geol. Soc. Amer. Bull.* 85:1159–70

Walcott RI. 1970. Flexural rigidity, thickness, and viscosity of the lithosphere. *J. Geophys. Res.* 75:3941–54

Wallace RE. 1984. Patterns and timing of late Quaternary faulting in the Great Basin Province and relation to some regional tectonic features. *J. Geophys. Res.* 89:5763–69

Ward PL. 1991. On plate tectonics and the geological evolution of southwestern North America. *J. Geophys. Res.* 96:12,479–12,496

Ward SN. 1988. North America-Pacific plate boundary, an elastic-plastic megashear: Evidence from Very Long Baseline Interferometry. *J. Geophys. Res.* 93:7716–28

Wdowinski S, O'Connell R. 1991. Deformation of the central Andes (15°–27°S) derived from a flow model of subduction zones. *J. Geophys. Res.* 96:12,245–12,255

Wdowinski S, O'Connell RJ, England P. 1989. A continuum model of continental deformation above subduction zones: Application to the Andes and the Aegean. *J. Geophys. Res.* 94:10,331–10,346

Wells ML. 1997. Alternating contraction and extension in the hinterlands of orogenic belts: An example from the Raft River Mountains, Utah. *Geol. Soc. Amer. Bull.* 109:107–26

Wernicke B. 1985. Uniform-sense normal simple shear of the continental lithosphere. *Can. J. Earth Sci.* 22:108–25

Wernicke B. 1992. Cenozoic extensional tectonics of the U. S. Cordillera. In *The Cordilleran Orogen: Conterminous U.S.*, ed. BC Burchfiel, PW Lipman, ML Zoback, *The Geology of North America* G-3:553–81. Boulder, CO: Geol. Soc. Amer.

Wernicke B, Axen GJ, Snow JK. 1988. Basin and Range extensional tectonics at the latitude of Las Vegas, Nevada. *Geol. Soc. Am. Bull.* 100:1738–57

Wernicke B, Getty SR. 1997. Intracrustal subduction and gravity currents in the deep crust: Sm-Nd, Ar-Ar, and thermobarometric constraints from the Skagit gneiss complex, Washington. *Geol. Soc. Amer. Bull.* 109:1149–66

Wernicke B, Snow JK. 1998. Cenozoic tectonism in the central Basin and Range: Motion of the Sierran-Great Valley block. *Int. Geol. Rev.* 40:403–10

Westaway R. 1993. Forces associated with

mantle plumes. *Earth Planet. Sci. Lett.* 119: 331–48

Wilson DS. 1988. Tectonic history of the Juan de Fuca ridge over the last 40 million years. *J. Geophys. Res.* 93:11,863–11,876

Wolfe JA, Forest CE, Molnar P. 1998. Paleobotanical evidence of Eocene and Oligocene paleoaltitudes in midlatitude western North America. *Geol. Soc. Amer. Bull.* 110:664–78

Wolfe JA, Schorn HE, Forest CE, Molnar P. 1997. Paleobotanical evidence for high altitudes in Nevada during the Miocene. *Science* 276:1672–75

Wood DJ, Saleeby JB. 1997. Late Cretaceous-Paleocene extensional collapse and disaggregation of the southernmost Sierra Nevada Batholith. *Int. Geol. Rev.* 39:973–1009

Wust SL. 1986. Regional correlation of extension directions in Cordilleran metamorphic core complexes. *Geology* 14:828–30

Zoback ML, Anderson RE, Thompson GA. 1981. Cainozoic evolution of the state of stress and style of tectonism of the Basin and Range province of the western United States. *Phil. Trans. Roy. Soc. Lond.* A300:407–34

Annu. Rev. Earth Planet. Sci. 1999. 27:463–93

MAJOR PATTERNS IN THE HISTORY OF CARNIVOROUS MAMMALS

Blaire Van Valkenburgh
Department of Organismic Biology, Ecology, and Evolution, University of California, Los Angeles, CA 90095-1606; e-mail: bvanval@ucla.edu

KEY WORDS: carnivores, evolutionary trends, mammalian adaptations, competition, predation

ABSTRACT

The history of carnivorous mammals is characterized by a series of rise-and-fall patterns of diversification in which declining clades are replaced by phylogenetically distinct but functionally similar clades. Seven such examples from the last 46 million years are described for North America and Eurasia. In three of the seven turnover events, competition with replacement taxa may have driven the decline of formerly dominant taxa. In the remaining four this is less likely because inferred functional similarity was minimal during the interval of temporal overlap between clades. However, competition still may have been important in producing the rise-and-fall pattern through suppression of evolution within replacement taxa; as long as the large carnivore ecospace was filled, the radiation of new taxa into that ecospace was limited, only occurring after the extinction of the incumbents. The apparently inevitable decline of incumbent taxa may reflect the tendency for clades of large carnivorous mammals to produce more specialized species as they mature, leading to increased vulnerability to extinction when environments change.

Introduction

Carnivorous mammals include some of the most popular and well-known mammalian species, such as lions, tigers, wolves, and, of course, domestic cats and dogs. Undoubtedly this has encouraged study of their evolutionary history, and they have been the subject of several recent reviews (Martin 1989, Hunt 1996, Werdelin 1996a). Their limited diversity, relative to herbivores, for example, also make them a fairly tractable group to review, and their potential roles as competitors and predators has made them the subject of numerous studies of

0084-6597/99/0515-0463$08.00

possible competition and coevolution in the fossil record (Bakker 1983, Van Valkenburgh 1988, 1991, Janis & Wilhelm 1993, Werdelin 1996b, Werdelin & Turner 1996).

The focus of this review is on carnivorous mammals that eat vertebrates frequently; at a minimum, their diet should be or have been 50 percent vertebrates but it can range upward to 100 percent. In addition, the emphasis is on nonmarine species that are or were jackal size (about 7 kg) and larger. The focus on relatively large species that rely substantially to entirely on vertebrate prey delimits a group of mammals that tend to be highly interactive, both over ecological and evolutionary time (Eaton 1979, Van Valkenburgh 1988, 1995). In extant ecosystems, groups or guilds (Root 1967) of sympatric large carnivores are often characterized by the regular occurrence of interspecific predation and interference competition, despite the fact that carnivore population densities are currently reduced due to human activities (Eaton 1979, Van Valkenburgh 1985, 1995). These negative interactions among species appear to influence the distribution and abundance of the less dominant species, suggesting that selection should favor behavioral and morphological attributes that minimize competition and dangerous interspecific encounters (cf. Van Valkenburgh 1985, 1988; Dayan & Simberloff 1996). Over evolutionary time, intense competition and intraguild predation (cf Polis and Holt 1996) might be capable of driving the replacement of one clade by another.

In the fossil record, the replacement of one clade by another produces a "double-wedge" pattern (Figure 1), in which the rise and fall of an earlier clade is followed by or overlaps the rise of a second clade (Krause 1986, Benton 1987). If the rise of the second clade overlaps the fall of the first clade in time and space, then competition as a cause is implied, and the turnover might be ascribed to "active displacement" or "competitive replacement" (Benton 1987). However, the same pattern could result from chance or an environmental (biotic or physical) change that favored one clade and not the other. If there is little or no temporal overlap between successive clades, it is considered to be "passive replacement" (Benton 1987). In fact, double-wedge patterns do characterize the fossil record of large carnivores (Van Valkenburgh 1991, Werdelin & Solounias 1991, Werdelin 1996a, Werdelin & Turner 1996), and temporal overlap between declining and expanding clades is typical. Whether such reciprocal trends in diversity are due to active displacement as opposed to alternate responses to some external factor, such as climate change, is difficult to assess. However, if double-wedge patterns occur repeatedly in the history of carnivorous mammals and if not all such replacements are associated with significant climate change, the argument for competition (and intraguild predation) as causal agents is strengthened. As I describe here, it more often appears that the previously dominant clade suppressed the diversification of the replacement clade,

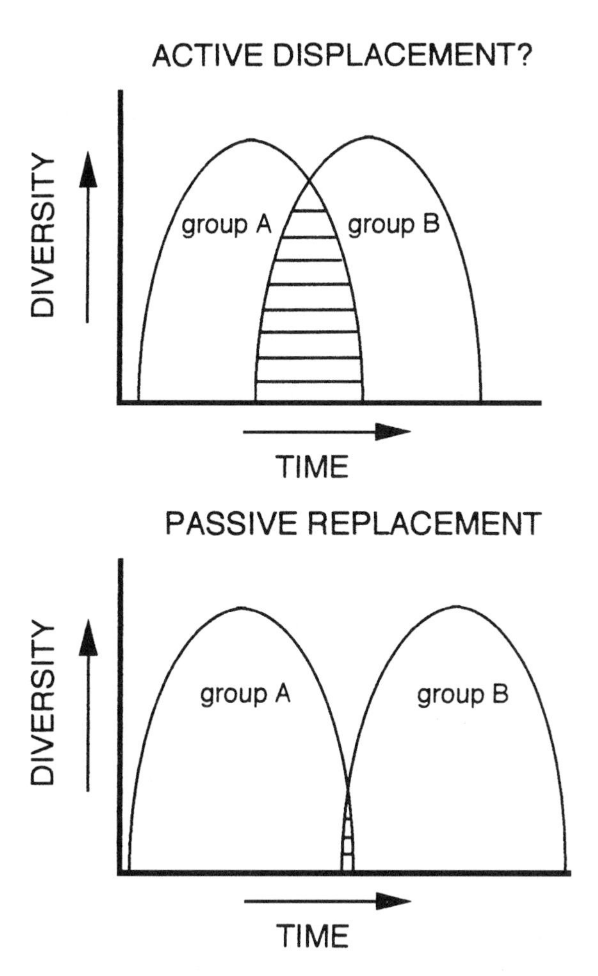

Figure 1 Hypothetical examples of double-wedge patterns of reciprocal diversity trends in two groups of organisms. (*top*) Active displacement or competitive displacement is implied because of the extensive temporal overlap between the two groups. (*bottom*) Passive replacement of group A by group B is implied because of the near absence of temporal overlap.

suggesting a competitive advantage of incumbency (Rosenzweig & McCord 1991) rather than extinction caused by competition.

In addition to competition and intraguild predation, the history of large carnivorous mammals likely has been affected by the evolution of their prey, but the impact seems to have been limited. Although an "arms race" between predator and prey might be expected (e.g. Simpson 1944, Dawkins & Krebs 1979), it has yet to be well documented in the mammalian fossil record. Over the course of the Cenozoic, ungulates have tended to evolve longer, more erect limbs and presumably faster speeds, as the world shifted from the more tropical, closed-habitat world of the Paleogene to the more arid, savannahlike world of the late Cenozoic (Webb 1977, Bakker 1983, Janis & Wilhelm 1993). Interestingly, the limbs of carnivores do not exhibit a parallel trend and instead remain relatively short and flexible until quite late, approximately five million years ago (Bakker 1983, Janis & Wilhelm 1993).

Although carnivorous mammals do not seem to track climate change closely, there are examples where severe temperature declines appear to have altered their evolutionary history. In several instances, sea level falls associated with global cooling exposed land bridges between continents and allowed interchange between formerly isolated flora and fauna (Tedford et al 1987, Webb & Opdyke 1995). It is these sporadic events of interchange that played a more significant role in the history of carnivorous mammals than the long-term cooling and drying trend. As I describe here, in some cases access to new territory with apparently unoccupied ecological space resulted in a rapid adaptive radiation within the immigrant clade. More often, however, the immigrant taxa exist at low diversity in their new continent for some period of time and then replace formerly dominant carnivores as they disappear for reasons that are not usually clear. Although this might be ascribed to competitive replacement of the native by the immigrant, such a conclusion is not often justified because of substantial differences in body size and ecomorphology between the putative competitors.

This review of the major events in the history of carnivorous mammals focuses on the turnover events between "dynasties," that is, the decline of formerly dominant clades and their replacement. Most of the examples will come from North America and Eurasia, because that is where the Cenozoic fossil record is best documented and understood.

The Cast of Characters

Before going further into the history of carnivorous mammals, it is useful to place the key taxa within a phylogenetic framework. Without this, it is difficult to appreciate the magnitude of convergent evolution that has occurred. Most of the species and families of mammals that are mentioned are members of the order Carnivora, and are labeled "carnivorans" to distinguish them from other carnivorous mammals outside the order. The order Carnivora includes both marine species, pinnipeds (seals, sea lions, and walruses), and terrestrial species, fissipeds, which are the subject of this paper. Carnivora is split into two major branches, the Feliformia (such as cats, hyenas, civets, mongooses, and the extinct catlike nimravids) and the Caniformia (such as bears, weasels, skunks, badgers, raccoons, dogs, pinnipeds, and the extinct bear-dogs) (Figure 2 and Table 1) (Flynn et al 1988, Wyss & Flynn 1993). This fundamental

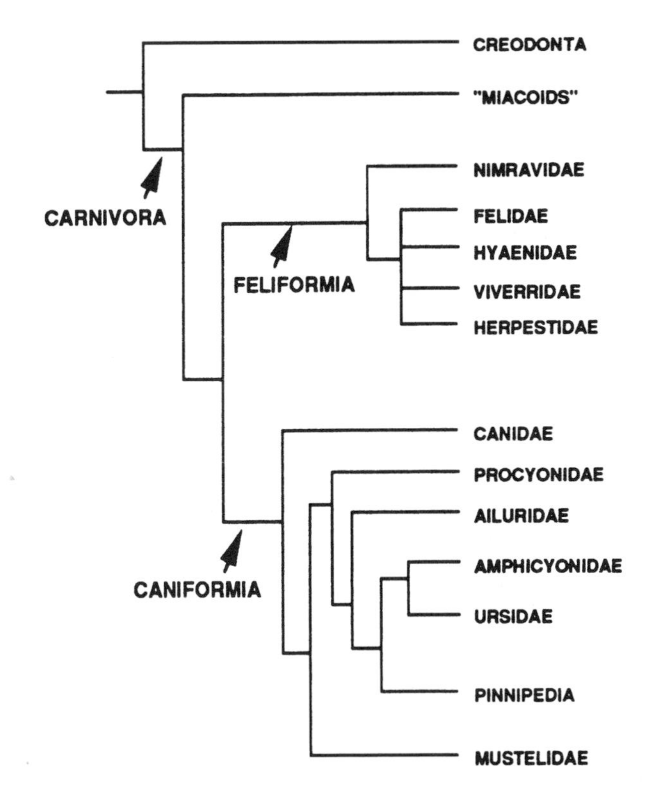

Figure 2 Phylogenetic tree of the order Carnivora. The extinct order Creodonta is the closest sister order, and together they form a group known as the Ferae (Wyss & Flynn 1993). The tree is based on molecular and morphological evidence as presented in Wyss & Flynn (1993).

subdivision was complete by the late Eocene (about 40 MYA) and may have occurred much earlier in the late Cretaceous or earliest Paleocene (70–60 MYA) (Flynn & Galiano 1982, Hunt & Tedford 1993). The sister group relationships within each of the two subdivisions are not fully resolved, but recent molecular and morphological studies have made significant advances. It is now widely accepted that (*a*) pinnipeds are monophyletic, (*b*) mustelids and procyonids are sister taxa, (*c*) the red panda, *Ailurus fulgens*, is a monotypic lineage, and (*d*) mongooses deserve family rank as Herpestidae (Wyss & Flynn 1993, Hunt & Tedford 1993, Vrana et al 1994, Lento et al 1995, Flynn & Nedbal 1998).

In addition to the Carnivora, several extinct groups of carnivorous mammals are relevant. Creodonts and mesonychids were the first large carnivores after the demise of the dinosaurs some 65 MYA, and they had a worldwide distribution,

Table 1 Brief description of carnivorous mammals discussed in this paper

	Extant representatives	Extinct representatives
Family Mesonychidae	—	Terrestrial, doglike omnivores
Order Creodonta		
Limnocyonidae	—	Civetlike creodonts
Oxyaenidae		Short-faced robust creodonts with flexible limbs
Hyaenodontidae	—	Terrestrial doglike creodonts with some cursorial adaptations
Order Carnivora		
Nimravidae	—	Saber-toothed catlike predators
Viverridae	Civets	
Herpestidae	Mongooses	
Hyaenidae	Hyenas, aardwolf	Diverse array of civetlike and wolflike species
Felidae	Felinae: cats	Machairodontinae: saber-toothed cats
Canidae	Caninae: wolflike dogs and foxes	Hesperocyoninae and Borophaginae
Procyonidae	Raccoons, coatis, kinkajous	
Ailuridae	Red panda	
Amphicyonidae		"Bear-dogs:" a diverse group of terrestrial carnivorans, most of which had flexible, fairly robust limbs
Ursidae	Bears	Hemicyoninae: an extinct group of highly carnivorous, fairly cursorial species
Pinnipedia	Seals, sea lions, walruses	
Mustelidae	Weasels, badgers, skunks, and otters	Oligobuninae: medium to large mustelids, some of which were hypercarnivous

exclusive of South America, Australia, and the poles (Bakker 1983, Janis et al 1998a). Creodonts and carnivorans are probably sister taxa (Wyss & Flynn 1993), but mesonychids are more closely related to ungulates and were the probable progenitors of cetaceans (Thewissen 1994, Gingerich et al 1994).

In addition to these two groups, carnivorous marsupials should be mentioned. During South America's long isolation the carnivorous mammal role was filled by a somewhat dog-like group of marsupials known as borhyaenids, which ultimately produced a saber-tooth cat analog, *Thylacosmilus* (Marshall 1976, 1978). The borhyaenids were joined by the occasional large didelphid (possum) as well as giant predatory ground birds, known as phorusrhacids (Marshall 1977). In Australia the typical marsupial carnivores tended to be small and omnivorous, with a few exceptions such as the cat-like wombat relative, *Thylacoleo*, and the dog-like Tasmanian wolf, *Thylacinus* (Archer 1982, Rich 1991). Notably, as in South America the large predator adaptive zone was partially filled by an unexpected group. In this instance giant reptiles (such as lizards and snakes) rather than birds were the dominant large predators for much of Australia's Cenozoic history (Molnar 1991).

The Paleogene

As was true for all mammals, the first major event in the history of carnivorous mammals was the extinction of the dinosaurs at the Cretaceous-Tertiary boundary. The removal of these giants that had dominated the landscape for the first 150 million years of mammalian history resulted in an ecological release for their furry successors. Nevertheless, considerable time passed (approximately 10 million years) before the first large, specialized meat eaters appeared (Gunnell 1998). In the interim, fairly generalized animals such as mesonychids seemed to be the most carnivorous species within the community. Perhaps the first genus that would fit the definition of carnivorous mammal used in this paper is the coyote-sized mesonychid, *Dissacus*, which was present in North America and Europe in the mid-Paleocene, about 62 MYA (Janis et al 1998a). Mesonychids had a somewhat doglike appearance, but a closer look at their teeth and limbs reveals them to be quite different from modern canids. Their limbs were shorter, more robust, and more flexed than typical canids and their teeth were much less specialized for carnivory (Bakker 1983, O'Leary & Rose 1995a, b). The lower tooth row of mesonychids consisted of a series of premolar-like teeth, similar in shape with relatively blunt cusps and weakly developed cutting blades (Figure 3, top). There are no modern analog taxa with this type of dentition, and so it is difficult to infer mesonychid diets with confidence, but they were probably omnivorous. Mesonychids persist at low diversity into the latest Eocene in the New World and slightly later in Asia (O'Leary & Rose 1995a, Archibald 1998). Over their history they tend to become larger and more cursorial, with some genera achieving the size of black bears (Bakker 1983, O'Leary & Rose 1995a).

The mesonychids were joined in the late Paleocene by the earliest known creodonts, described as civetlike in their diets, preying on relatively small mammals, birds, and perhaps arthropods (Gunnell 1998). Clearly, the creodonts had a dentition more specialized for carnivory than the mesonychids. All their lower molars were narrow and bladelike with obvious shearing facets caused by the scissorlike occlusion of lower and upper teeth (Figure 3, middle). The creodonts reached maximum diversity in the early and middle Eocene and produced a diverse array of species including probable bone crackers (e.g. *Patriofelis*), small saber-tooth cat analogs (e.g. *Machaeroides*), and foxlike

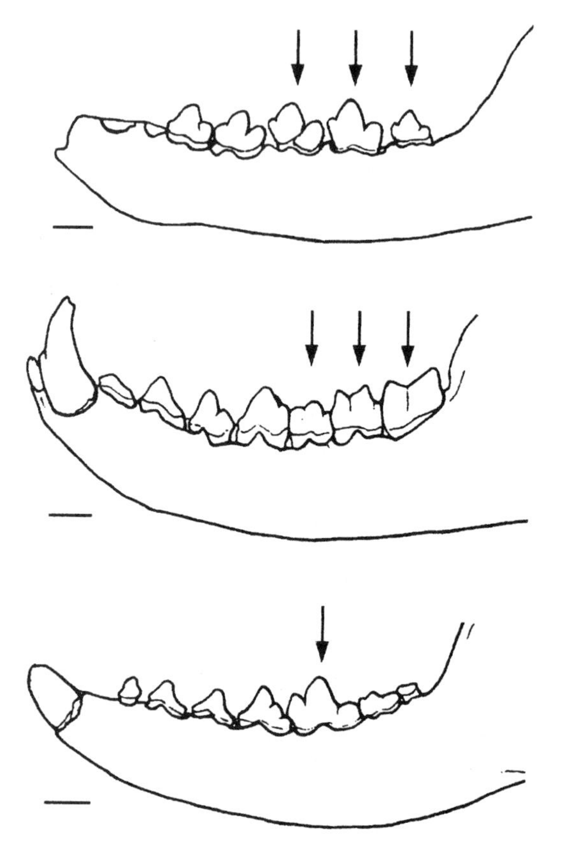

Figure 3 Lateral view of the lower jaws of (*top*) *Dissacus*, a mesonychid, (*middle*) *Hyaenodon crucians*, a creodont, and (*bottom*) *Hesperocyon gregarius*, a canid carnivoran. Arrows indicate molar teeth specialized for cutting. *Dissacus* is from Figure 1 (O'Leary & Rose 1995b); the others are from color transparencies.

forms (e.g. *Limnocyon*). In terms of locomotor adaptations, creodonts fall into two basic types: robust, short-limbed wolverinelike forms (e.g. oxyaenids) and slender-limbed, semicursorial forms (e.g. hyaenodontids) (Mellett 1977, Gunnell and Gingerich 1991, Gebo and Rose 1993, Gunnell 1998). Creodonts and mesonychids coexisted throughout the Eocene, with little or no evidence of competitive displacement (Figure 4).

The first carnivorans appear in the fossil record about 62 MYA, along side the first mesonychids (Flynn 1998, Janis et al 1998a). However, these early carnivorans, called miacoids, were small, weasel- to perhaps fox-size animals, many of whom probably spent part of the day foraging or resting in trees (Heinrich

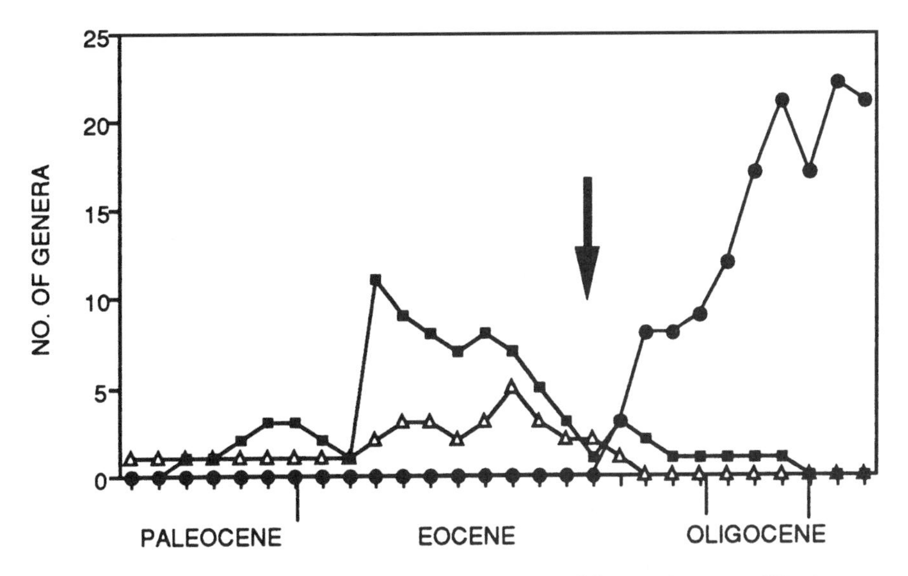

Figure 4 Generic diversity in North American creodonts (*solid squares*), mesonychids, (*open triangles*), and carnivorans (*solid circles*) over time. Arrow indicates approximate time when declining clade(s) diversity is exceeded by that of the replacement clade(s). Unlabeled temporal units correspond to intervals used throughout Janis et al (1998b). (Data are from Gunnell 1998, Archibald 1998, and Flynn 1998.)

and Rose 1995, 1997, Rose 1990). They differ most from the creodonts and mesonychids in their teeth. Miacoids display a key diagnostic feature of the order Carnivora: a single pair of cutting teeth, the carnassials (upper fourth premolar, P4, and lower first molar, m1). This differs from both creodonts, which have two or three pairs of cutting teeth, and mesonychids, which have a series of similar looking premolarized molars (Figure 3).

The significance of the carnivoran configuration is that it seems to have allowed greater evolutionary plasticity in the long term. The molars behind the carnassial remain as less specialized crushing teeth, and the premolars anterior to the carnassial are retained as modified cones for puncturing, holding, or crushing food. By enhancing either the cutting aspects of their teeth (i.e. the carnassials) or the crushing aspects (i.e. the postcarnassial molars), carnivorans can move into hypercarnivorous (meat specialists) or hypocarnivorous (plant matter specialists) niches, respectively. Both the creodonts and the mesonychids modify all their molars for similar function early in their evolutionary history, and this may have constrained them in terms of their adaptive diversity. In particular, the loss of crushing function in creodont molars might have limited them to hypercarnivorous diets. Interestingly, marsupial carnivores (dasyurids) also have all their lower molars modified as cutting blades, and their adaptive diversity is much more limited than that of the Carnivora. Werdelin (1987, 1996a) has argued that the pattern of dental eruption in marsupials, in which molars erupt in sequence over an individual's life span, has acted as a developmental constraint on the evolution of dietary diversity. This is likely true, but even without such an unusual eruption pattern, the examples of the mesonychids and creodonts suggest that the restriction of all molars to a cutting blade function, which involves a significant loss of tooth complexity, may itself act as an evolutionary constraint.

The second major event in the history of carnivorous mammals occurred in the late Eocene, and as was true of the first event, this one affected many mammalian groups. In general, many of the groups that dominated Paleocene and Eocene communities went extinct and were replaced by representatives of orders that survived to the present (Janis 1993). The last 10 million years or so of the Eocene (43–33 MYA) witnessed two relatively rapid and severe cooling events, separated by a slight warming trend (Prothero and Berggren 1992, Prothero 1994), and these shifted the planet from the mostly warm, moist tropical world of the Paleogene toward the cooler, seasonally arid world of the Neogene (Wolfe 1992, Wing 1998). The transition was made in steps, with each temperature drop associated with sea lowering and intercontinental migrations, both between North America and Europe and between Europe and eastern Asia (Webb & Opdyke 1995). North American oxyaenid and limnocyonid creodonts made their last stand in the Uintan (circa 39.5–45.9 MYA), and they were followed by the mesonychids and miacoids in the Duchesnean and early Chadronian (35.5–39.5 MYA) (Gunnell 1998, Flynn 1998, Archibald 1998). Nearly simultaneously with or soon after the loss of these taxa, new predators appeared, including hyaenodontid creodonts, the first canids, amphicyonids, nimravids, and mustelids (Hunt & Tedford 1993, Janis et al 1998a). Although it might seem that the new predators outcompeted the archaic species, the excellent North American record indicates that the decline of the mesonychids and creodonts began well before the first appearance of the replacement taxa (Figure 4). This transition from a mesonychid/diverse creodont/miacoid guild to a predominantly nimravid/hyaenodontid creodont guild presents us with the first double-wedge pattern of turnover.

The causes for this turnover event remain murky. Clearly, it overlapped with a major climatic shift and the extinction of numerous potential prey species such as primitive artiodactyls, condylarths, brontotheres, and plesiadapiform primates (Janis 1993). However, these losses were matched by the arrival of many new taxa, some via migration between continents (Prothero 1994, Webb & Opdyke 1995). Rodents, lagomorphs (rabbits), and various artiodactyls diversified in the late Eocene and should have served as acceptable prey substitutes.

The new predators do not exhibit any obvious morphological advantages (e.g. larger body size, larger brains, better senses, or longer limbs) over those they replaced, except perhaps the presence of a single carnassial pair in the Carnivora. However, that hardly seems likely to have given them a competitive advantage on ecological time scales. Moreover, the wolf-like hyaenodontid creodonts are the largest and most common predators in the late Eocene and early Oligocene, and they bore three carnassial pairs. Thus the cause of the decline of the first dynasty of carnivorous mammals remains a mystery.

The Neogene

The new late Eocene-Oligocene dynasty of predators included an array of species that differed more from one another in feeding and locomotor morphology (and inferred behavior) than the previous dynasty. For the first time there were definitive catlike forms, nimravids, with short snouts, no crushing postcarnassial molars, retractile claws, and elongate upper canine teeth. Although the nimravids were once labeled as true cats (Felidae), aspects of their basicranial anatomy demonstrate that the nimravids belong in their own family (Hunt 1987, Bryant 1991). The convergence in form is remarkable and has been well described previously (Matthew 1910, Martin 1980, Emerson & Radinsky 1980). Nimravids shared the niche of hypercarnivore with the hyaenodontid creodonts, which were more doglike in their post-cranial skeletons but still probably rushed prey from a short distance (Mellett 1977, Van Valkenburgh 1985, Janis & Wilhelm 1993). Also present for the first time at moderate body size were the amphicyonids or bear dogs, an extinct family of caniform carnivorans that reached maximum diversity in the Miocene (Viranta 1996, Hunt 1998b). In the Oligocene the amphicyonids of the Old and New World tended to be coyote size or smaller, with flexible limbs and fairly generalized dentitions, suggesting a moderately omnivorous diet (Van Valkenburgh 1985, Hunt 1998b). The earliest canid, *Hesperocyon*, was present in North America throughout the Oligocene but was a small, civet-like animal that probably climbed trees occasionally and ate a mixed diet of small vertebrates and plant matter (Van Valkenburgh 1985, 1987, Munthe 1998, Wang 1993).

For most of the Oligocene, there was little interchange between the Old and New World despite a sharp temperature drop (and inferred sea lowering) about 33 MYA that clearly increased seasonal aridity in some areas, such as the interior of North America (Hutchison 1996, Prothero 1994, Retallack 1992, Webb and Opdyke 1995). In fact, the faunal composition of central North America was so stable between approximately 37 and 27 MYA that it has been recognized as the "White River Chronofauna" (Emry 1981, Emry et al 1987). Within the North American predator guild, there were numerous species extinctions and replacements, but diversity varied little, and replacement taxa tended to be sister taxa of those they replaced (Van Valkenburgh 1994). The stability of structure of the guild is quite remarkable given that floral and sedimentological evidence indicates a significant vegetational shift toward more savannahlike open habitats (Retallack 1992, Wolfe 1992, Prothero & Heaton 1996). Similar data are not yet available for the Old World, but it appears that faunal evolution was more dynamic. There was a major extinction event among European mammals at approximately 32–33 MYA (the "Grand Coupere"), which was associated with global cooling and the arrival of many immigrant taxa from Asia as a result of the drying of the Turgai Straits (Hooker 1992, Legendre and Hartenberger 1992, Janis 1993, Prothero 1994).

The third major event in the history of carnivorous mammals is the replacement of the Oligocene nimravid/hyaenodontid creodont guild. In North America it was largely replaced by canids, amphicyonids, and ursids, whereas in the Old World it was replaced by hyaenids, amphicyonids, and ursids. In both regions the transition between dynasties is not well documented in the fossil record but is characterized by the absence of either catlike nimravids or true cats. In North America the last Oligocene nimravids disappear at the Oligocene-Miocene boundary, approximately 23 MYA, and catlike species are absent until 17.5 MYA, when a felid immigrated from the Old World (Van Valkenburgh 1991, Hunt 1996). In Europe the timing is not as clear, but there is a significant period of time in the late Oligocene when nimravids are absent and felids have yet to appear (Werdelin 1996a).

During or just prior to this "cat gap," numerous caniform species evolve catlike features indicative of hypercarnivory, such as reduced snouts, somewhat enlarged canines, and fairly extreme reduction of their crushing molars (Van Valkenburgh 1991). In North America the first caniform group of moderate body size to move in the direction of hypercarnivory were the endemic hesperocyonine canids, with three genera (*Parenhydrocyon, Enhydrocyon*, and *Mesocyon*), ranging in size from jackals to small coyotes, appearing in the early Arikareean (circa 28 MYA) (Wang & Tedford 1996, Munthe 1998). Notably, these three evolved alongside the last hyaenodont and the remaining three nimravids, two of which were puma-sized. The small hypercarnivorous canids were soon joined by and ultimately replaced by numerous species from other families which also had evolved more specialized meat-eating teeth and skulls (Van Valkenburgh 1991). These included at least three larger genera of similarly adapted amphicyonids, one endemic (*Daphoenodon*) and two from the Old World (*Temnocyon* and *Mammocyon*), a leopard-sized mustelid (*Megalictis*) as well as two hypercarnivorous bears, the hemicyonines *Cephalogale* and *Phoberocyon* (Hunt & Skolnick 1996, Hunt 1998a, b). The hypercarnivorous amphicyonids and ursids were more cursorial than the canids; all had a digitigrade stance and some were incapable of supination of the forepaw, suggesting

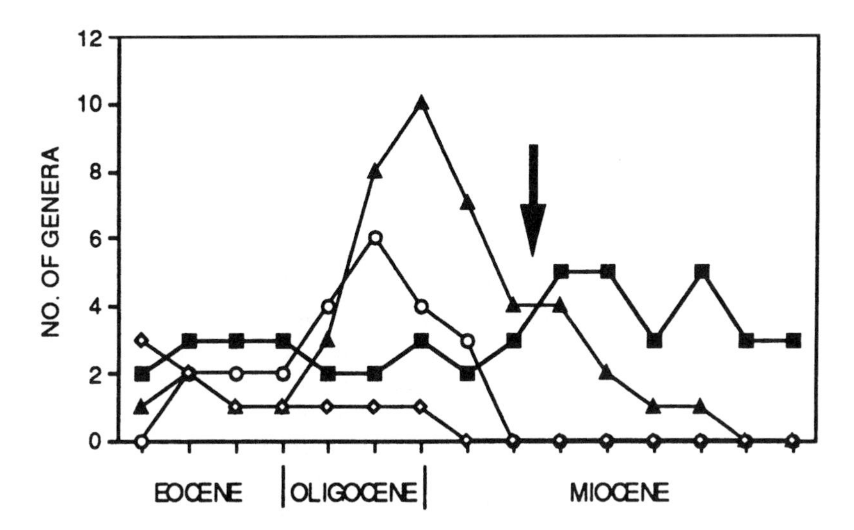

Figure 5 Generic diversity in North American creodonts (*open diamonds*), nimravids (*open circles*), highly carnivorous canids (*solid triangles*), and highly carnivorous amphicyonids and hemicyonine bears (*solid squares*) over time. Arrow indicates approximate time when declining clade(s) diversity is exceeded by that of the replacement clade(s). Unlabeled temporal units correspond to intervals used throughout Janis et al (1998b). Data are from Martin (1998a), Munthe (1998), and Hunt (1998a, 1998b).

specialization for terrestrial locomotion and perhaps greater speed (Hunt 1998a, b). In contrast, the contemporaneous canids tended to be more short-limbed robust forms, some of which likely were capable of climbing (Wang 1993, Munthe 1998). The rise and fall of nimravids and creodonts, followed by hesperocyonine canids, alongside a rise in amphicyonids and ursids presents another example of the double-wedge pattern, with significant temporal overlap between the declining and expanding wedges (Figure 5). Because many of the replacement taxa were similar in body size or larger than those they replaced, competitive displacement might have been responsible, particularly in the decline of the hypercarnivorous canids.

In the Old World, the late Oligocene-early Miocene record is not as good as that of North America, but a similar radiation of nonfeliform hypercarnivores may have occurred. Several of the North American immigrants with hypercarnivorous adaptations were members of Old World subfamilies that have been characterized as specialized meat eaters (e.g. temnocyonine amphicyonids, hemicyonine ursids; Hunt 1998a, 1998b). Consequently, it is likely that the Old World ancestors of the North American immigrants were hypercarnivorous as well. The apparent tendency of taxa to evolve toward hypercarnivory suggests that this is a consistently lucrative niche that is rapidly filled when formerly dominant occupants decline. Interestingly, as will be discussed further below, a similar but less extensive radiation occurred when canids invaded South America approximately 2–3 MYA and found themselves on a cat-free continent (Van Valkenburgh 1991).

The remainder of the Miocene saw a gradual transition in the composition of the predator guild that was not punctuated by any notable events. The catlike feliforms return in the form of nimravids and felids, but their impact is felt gradually. Whereas the early Miocene was characterized by predator guilds dominated by caniform taxa (amphicyonids, ursids, and canids), the late Miocene featured guilds dominated more by feliform taxa (felids and doglike hyaenids in Eurasia; felids and borophagine canids in North America) (Hunt & Tedford 1993, Hunt 1996). There was no single climatic event associated with this transition, and occasional faunal interchange between the continents took place several times over the span of approximately 20 million years (Tedford et al 1987, Webb & Opdyke 1995). Notably, Cenozoic land mammal diversity in North America and western Europe peaked in the mid-Miocene (11–16 MYA) but declined sharply by the latest Miocene (Webb 1983, Stucky 1990, Van Valkenburgh & Janis 1993, Fortelius et al 1996). In Europe, the decline has been ascribed to severe reductions in habitat diversity and increased seasonality as savannahlike conditions came to dominate (Fortelius et al 1996), and similar arguments have been made for North America (Janis 1993, Webb 1983).

There is a double-wedge pattern in the Miocene that is associated with the return of catlike feliforms. Felids show up first in the Old World, approximately 24 MYA in the form of *Proailurus*, a felid the size of a bobcat. (Hunt & Tedford 1993). *Proailurus* is very rare but is succeeded by a more successful genus, *Pseudaeleurus*, which evolved in the Old World at least 20 MYA and immigrated to North America around 18 MYA (Martin 1998b). Nimravid saber-toothed cats reappear as well, first in Eurasia about 18 MYA and subsequently in North America by immigration around 11 MYA. (Martin 1989, 1998a, Werdelin 1996a). In both Eurasian and North American faunas the felids (often saber-toothed forms) increase in both diversity and body size over the course of the Miocene, perhaps displacing, or at least replacing, many of the caniform hypercarnivores (Van Valkenburgh 1991, Werdelin 1996a). These parallel diversifications are accompanied by a radiation of medium- to large-size dog-like predators, hyaenids in the Old World and borophagine canids in North America (Figure 6). The possible victims of the felid, hyaenid, and borophagine radiations seem to have been the hemicyonine bears and amphicyonids, both of which completely disappear by the latest Miocene (Figures 6 and 7). The amphicyonids also may have been affected negatively by the arrival of large

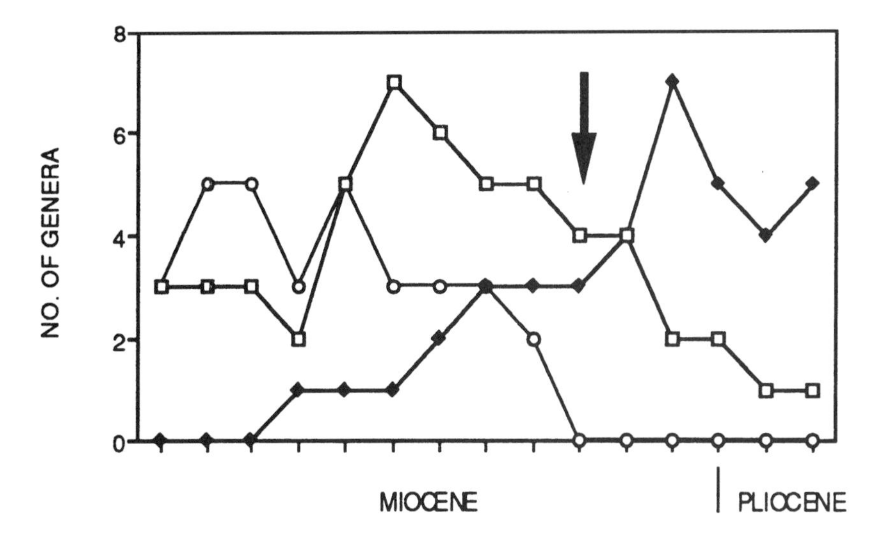

Figure 6 Generic diversity in North American borophagine canids (*open squares*), felids and nimravids, (*solid diamonds*), and amphicyonids and hemicyonine bears (*open circles*) over time. Arrow indicates approximate time when declining clade(s) diversity is exceeded by that of the replacement clade(s). Unlabeled temporal units correspond to intervals used throughout Janis et al (1998b). Data are from Martin (1998a, 1998b), Munthe (1998), and Hunt (1998a, 1998b).

omnivorous agriotherine ursids approximately 12 MYA in Eurasia and then 8 MYA in North America (Viranta 1996, Werdelin 1996a, c). It is interesting that the last amphicyonids, who were jack-of-all-trades massive predators with both blade-like carnassials and large crushing molars, were replaced by taxa, such as the hyaenids, felids, and agriotherine bears, that were more specialized for either carnivory (felids and hyaenids) or omnivory (ursids) (Viranta 1996, Werdelin 1996b).

The next and final major turnover within Holarctic, large mammalian predator guilds occurs relatively rapidly and sets the stage for the predator guilds of the present, which are composed of large felids, wolflike and foxlike canids, bone-cracking hyaenids (in the Old World), and omnivorous ursids. In this case, the turnover was associated with a major extinction event that removed 60 to 70 percent of Eurasian genera and 70 to 80 percent of North American genera (Webb 1983, 1984; Savage & Russell 1983). In both regions the extinctions occur near the Mio-Pliocene boundary (circa 5–8 MYA) and are correlated with a significant climatic event, the desiccation of the Mediterranean Sea (the Messinian salinity crisis) and the associated spread of seasonally arid grasslands in place of more mesic woodlands (Janis 1993).

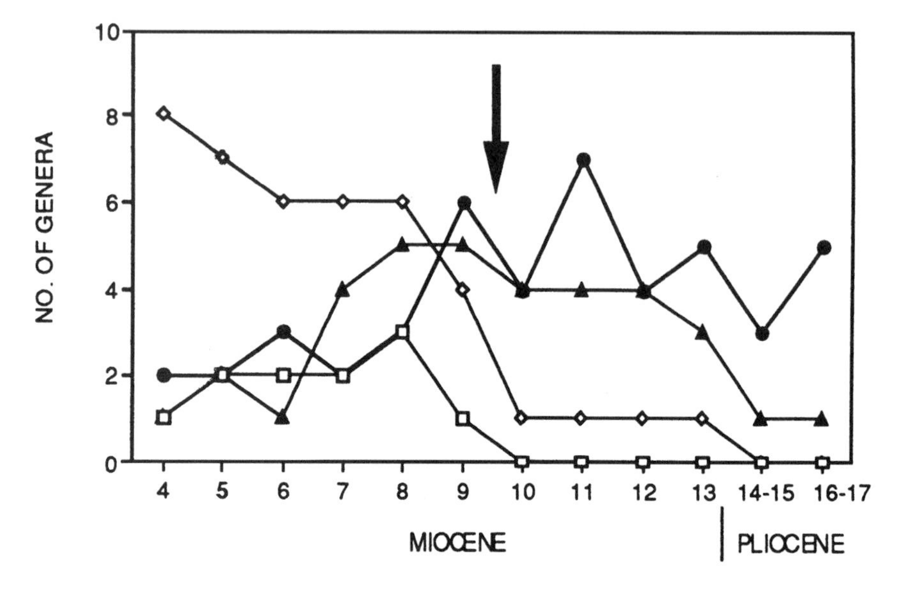

Figure 7 Generic diversity in Eurasian hemicyonine bears (open squares), amphicyonids, (open diamonds), felids and nimravids (solid circles), and non-bone-cracking hyaenids (solid triangles) over time. Arrow indicates approximate time when declining clade(s) diversity is exceeded by that of the replacement clade(s). Temporal units are numbered according to European Neogene Mammal Faunal Zones (MN units); for explanation, see Bernor et al (1996). (Diversity data are from Werdelin & Solounias 1991, 1996; and Werdelin 1996c.)

In Eurasia, the transition to the Pliocene predator guild is marked by a startling decline in hyaenid diversity from 22 species in the Turolian (9–5.3 MYA) to just 5 species in the Ruscinian (5.3–3 MYA) (Figure 7; Werdelin & Turner 1996). The extinctions are almost entirely of the more civet-like and dog-like species rather than the bone crackers (Werdelin & Turner 1996). In North America the borophagine canids undergo a similar drop in diversity from a maximum of five genera in the early late-Miocene to a single genus by latest Pliocene (Munthe 1998).

Again it is the more dog-like ecomorphs that disappear first, leaving only the genera most convergent with bone-cracking hyaenids. Certainly, the fact that bone-cracking species were favored both among the hyaenids and the borophagine canids 5–7 MYA suggests that food (prey) was scarce and the ability to use a carcass more fully was advantageous.

The group that eventually takes over the ecospace formerly held by borophagine dogs and hyaenids is the same in both Eurasia and North America, the modern subfamily of canids, the Caninae (Munthe 1998, Werdelin & Solounias 1991, Werdelin 1996a, 1996c, Werdelin & Turner 1996, Flynn et al 1991).

Canids spent the first 30 million years of their history entirely in North America, finally dispersing in the late Miocene, with a single species, "*Canis*" *cipio*, emerging in the Old World fossil record approximately 7–8 MYA in Spain. This is followed by a gradual diversification of fox-like and wolf-like forms (Werdelin & Turner 1996). In North America, an early fox-like genus of Caninae, *Leptocyon*, coexisted alongside the Borophagine throughout their Miocene heyday, finally beginning to diversify around 7–8 MYA (Hunt 1996, Munthe 1998). So, in both Old and New Worlds, the replacement canids overlap with the taxa they replace, creating another double-wedge pattern (Figure 8). However, the early canines that might have been potential competitors of Eurasian hyaenids or North American borophagines are much smaller in body size and exceedingly rare in the fossil record. In this case, it seems the double wedge is best explained as simple replacement rather than competitive displacement.

Saber-toothed ecomorphs also suffer declines in the latest Miocene but much more so in Eurasia than North America. The nimravids made their final stand approximately 9 MYA in Eurasia and so were already gone by the Mio-Pliocene extinction event (Werdelin 1996a, 1996c). However, there were seven genera of felids in the late Miocene that exhibit elongate, flattened upper canine teeth to varying degrees. That number declined to two in the Pliocene, *Megantereon* and *Homotherium*, while the number of felid genera with more rounded canines (the conical-tooth cats or Felinae) increased from one to three over the same interval (Werdelin & Turner 1996). In North America the picture was different. The last nimravid, *Barbourofelis*, arrived from the Old World in the late mid-Miocene but vanished by 6 MYA, about one million years before the Mio-Pliocene boundary. Across this boundary, there is only a slight decline in felid diversity, from seven to five genera, and both extinctions occur among non–saber-toothed forms (Martin 1998b). The final chapter in the global history of the felids is the dwindling of saber-tooth diversity and the expansion of big cats with more conical canine teeth. Although saber-toothed felids were never very diverse, with most paleocommunities containing two or at most three species, it does seem odd that this successful ecomorph should decline to extinction. The obvious explanation is competition with the larger conical-toothed felines in the genus *Panthera*, but it is not clear what advantage the felines had over the massive, well-equipped saber-toothed forms. An analysis of changes in body size and locomotor adaptations among the potential prey species might provide insight into the causes of the decline of saber-toothed cats.

The Plio-Pleistocene predator guild differed from all previous guilds in that it included a variety of carnivorans that were clearly built for long-distance pursuit. In Eurasia, long-limbed, running hyaenas ("chasmoporthetines") had been present at low diversity since the mid-Miocene, but there were no similarly built carnivorans in North America until the Plio-Pleistocene, when the Caninae

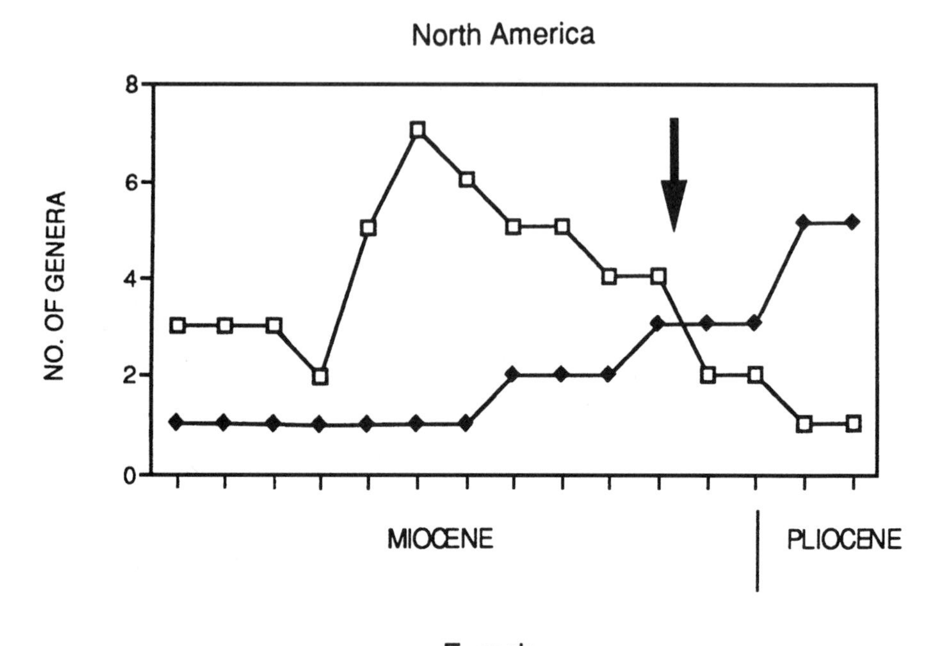
North America
NO. OF GENERA
8
6
4
2
0
MIOCENE
PLIOCENE

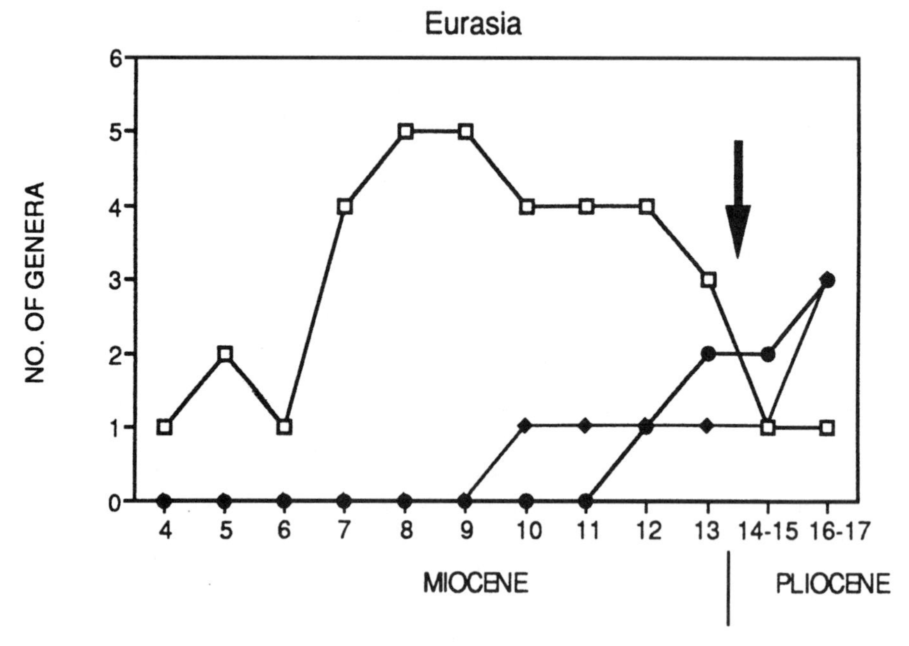
Eurasia
NO. OF GENERA
6
5
4
3
2
1
0
4
5
6
7
8
9
10
11
12
13
14-15
16-17
MIOCENE
PLIOCENE

diversified. The long legs, relatively inflexible forelimbs, and compact paws of the Caninae label them as specialized for terrestrial locomotion, more so than perhaps any previous caniforms (Janis & Wilhelm 1993). In addition to the dogs, cheetahlike felids evolve in parallel in both North America *(Miracinonyx)* and Eurasia *(Acinonyx)*, and a running hyena (*Chasmoporthetes*) migrates from the Old to the New World (Berta 1981, Van Valkenburgh et al 1990, Hunt 1996). Finally, in the Pleistocene of North America, even the ursids produce a long-legged, gigantic short-faced bear *(Arctodus)*. All this provides the large predator guild with a greater diversity of locomotor types than seen previously, running cats, hyaenas, dogs, and bears alongside robust saber-toothed felids and more typical felids.

The apparent increase in the number of cursorial hunters in the Plio-Pleistocene cannot be correlated with a similar pattern among their potential prey. In North America, a study of the evolution of cursorial adaptations among carnivorans and herbivores demonstrated that among herbivores the transition to long, slender limbs and compact, interlocking joints occurred quite early, approximately 20 million years before similarly specialized carnivores appear (Janis & Wilhelm 1993). Apparently, it was not necessary for predators to evolve markedly specialized limbs to keep up with their prey for much of the Cenozoic. It seems that something was different in the Pliocene; perhaps the vegetation structure was significantly more open with more extensive, short grasslands than previously existed. A very open structure makes ambush more difficult and would favor long-distance pursuit abilities among predators. More detailed data on vegetation structure before and after the Mio-Pliocene boundary would help resolve this issue, but such data are difficult to acquire. In addition, better osteological indicators of stamina and running abilities might reveal whether in fact, most pre-Pliocene predators were incapable of long-distance pursuit, as has been suggested (Bakker 1983, Janis & Wilhelm 1993).

Beyond North America and Eurasia

South America, Africa, and Australia existed as isolated continents to varying degrees over the Cenozoic with Africa having the most contact with a larger

←

Figure 8 (*top*) Generic diversity in North American borophagine canids (*open squares*) and canine canids (*solid diamonds*). Unlabeled temporal units correspond to intervals used throughout Janis et al (1998b). (Data are from Munthe 1998.) (*bottom*) Generic diversity in European bone-cracking hyaenids (solid diamonds), non-bone-cracking hyaenids (open squares), and canids (*solid circles*). Arrow indicates approximate time when declining clade(s) diversity is exceeded by that of the replacement clade(s). Temporal units for Eurasia are numbered according to European Neogene Mammal Faunal Zones (MN units). For explanation see Bernor et al (1996). (Data are from Werdelin & Solounias 1991, Werdelin 1996c, Werdelin & Turner 1996.)

landmass via a land bridge, South America somewhat less, and Australia none. Both Africa and South America made important contributions to Cenozoic carnivore history, especially in the Plio-Pleistocene.

The African land mammal record is interrupted by numerous gaps that make it difficult to reconstruct the history of connections to Eurasia. Nevertheless, it is clear that the Neogene sequence of carnivorous mammals in Africa was not too different from that of Eurasia; creodonts and nimravids were replaced by amphicyonids, felids, and hyaenids in the Miocene and then by canids, felids, and hyaenids in the Plio-Pleistocene (Savage & Russell 1983). In several instances, predator clades make their final appearance in Africa after having gone extinct elsewhere, suggesting that Africa served as a refuge from Eurasian conditions (changing climate and competitive milieu). This was true for the hyaenodontid creodonts, amphicyonids, and the doglike hyaenids (Werdelin & Solounias 1991, Viranta 1996, Gunnell 1998, Werdelin & Turner 1996). Perhaps Africa's most significant contribution to the history of carnivorous mammals is that at present it contains the most diverse large predator guild and therefore has served as a model for much of the work on paleoecology of carnivore guilds (e.g. Van Valkenburgh 1987, 1988, Werdelin & Turner 1996, Lewis 1997). It appears likely that three of the major players in both modern and Pleistocene predator guilds evolved in Africa: the lion, leopard, and spotted hyena (Turner 1990). Pleistocene lions are known from every continent except Australia, making them the most widespread carnivoran species to have ever existed (if they are all the same species) (Kurtén 1968, Kurtén and Anderson 1980).

The most significant event in the history of carnivorous mammals in South America was the opening of a substantial land bridge to Central and North America approximately 2-3 MYA. Just prior to the emergence of the Panamanian isthmus, the South American predator community consisted of one bear-like procyonid carnivoran, three species of carnivorous didelphid marsupials (one of which was the size of a coyote), and a gigantic, predaceous ground bird (Marshall 1977). With the possible exception of the rare ground bird, none of these species was a specialized hypercarnivore. Consequently, carnivorans from the north were entering a cat-free world, and not surprisingly, there was a radiation of hypercarnivorous forms within one of the first families to arrive, the canids. Over the interval from 2.5 MYA to 10 KYA, 16 new species of canids appeared, seven of which were adapted for hypercarnivory (Berta 1988, Van Valkenburgh 1991). Felids appear in the South American fossil record at the same time as canids but apparently do not diversify to the same degree, with only three new species found in the Plio-Pleistocene record (Berta and Marshall 1978). The reasons for this are unclear; it is possible that the early record is incomplete and canids arrived much earlier to South America and won the advantage of incumbency. What seems clear is that the radiation of South American canids is an excellent example of adaptive radiation in the face of ecological opportunity. Even today, despite late Pleistocene extinctions, the canid fauna of South America is the most diverse of any continent.

Discussion

The dominant feature of the history of carnivorous mammals is the repeated occurrence of double-wedge patterns of replacement in which formerly dominant clades are replaced by clades that were previously absent or at low diversity. In many cases, the replacement taxa are similar in morphology to those they replace, thus providing multiple examples of functional convergence among carnivores over the course of the Cenozoic.

Seven examples of double-wedge patterns were provided (Table 2), here listed in chronological order, by the declining or outgoing dynasty: (*a*) Eocene mesonychid/diverse creodont/miacoid; (*b*) Oligocene nimravid/hyaenodontid creodont (Eurasia and North America); (*c*) early Miocene hesperocyonine canid/amphicyonid/hemicyonine ursid (North America) and hyaenid/amphicyonid/hemicyonine ursid (Eurasia); and (*d*) late Miocene borophagine canid/felid (North America) and dog-like hyaenid/felid (Eurasia). The demise of the last two dynasties occurred near the Mio-Pliocene boundary and was followed by the establishment of similar dynasties in both Eurasia and North America that included canine canids, conical-toothed and saber-toothed felids, and bone-cracking hyaenids (in Eurasia only).

The causes of these turnovers are unlikely to be the same in each instance. Double-wedge patterns in which the expanding clade temporally and spatially overlaps the declining clade suggest the possibility of competitive displacement. Such an overlap existed in most or all of the examples mentioned, and given the incompleteness of the fossil record the observed time span of overlap is a minimum value. Consequently, in all cases, the overlap was likely of sufficient duration to have allowed competition to drive the decline of the formerly dominant group. However, even when there is substantial temporal and spatial overlap, it is critical to examine whether the morphology and body size of the newcomers are consistent with the notion that they might have outcompeted the incumbents (cf Krause 1986, Maas et al 1988). For carnivorous mammals the key parameters are body size, diet, and locomotor mode, all of which can be deduced for extinct species from morphology (cf. Van Valkenburgh 1987, 1988, 1990, Lewis 1997). If the newcomers differ greatly from the incumbents in inferred body size, diet, and locomotor mode, a conclusion of competitive displacement is not well justified. Moreover, when possible, the relative abundance of both newcomers and incumbents should be examined. Ideally, if competition were important, the fossil record would record a decline in the

Table 2 Turnover events with declining and replacement taxa shown

Location and approximate time of turnover	Declining taxa	Replacement taxa	Competition as cause of turnover[a]
1) North America and Eurasia: late Eocene–Oligocene, 46–35 MYA	Mesonychids, non-hyaenodontid creodonts, miacoid carnivorans	Hyaenodontid creodonts, canids, amphicyonids, nimravids	Doubtful
2) North America: late Oligocene–early Miocene, 30–20 MYA	Hyaenodontid creodonts, nimravids	Hesperocyonine canids, amphicyonids, hemicyonine ursids	Possible
3) Eurasia: late Oligocene–early Miocene, 30–20 MYA	Hyaenodontid creodonts, nimravids	Amphicyonids, hemicyonine ursids, hyaenids	Possible
4) North America: mid-late Miocene, 13–6 MYA	Amphicyonids, hemicyonine ursids	Borophagine canids, felids	Possible
5) Eurasia: mid-late Miocene, 12.5–9.5 MYA	Amphicyonids, hemicyonine ursids	Doglike hyaenids, felids	Possible
6) North America: late Miocene–Pliocene, 5–2 MYA	Borophagine canids	Canine canids, conical-toothed and saber-toothed felids	Doubtful
7) Eurasia: late Miocene–Pliocene, 7–4 MYA	Doglike hyaenids	Canine canids, conical-toothed and saber-toothed felids, bone-cracking hyaenids	Doubtful

[a]Competition is not considered to have been important as a cause of turnover in cases where temporal overlap is minimal and/or replacement taxa differ significantly in body mass or inferred diet from the declining taxa. See text for further explanation.

numbers of individuals of incumbents along side a rise in those of newcomers. However, the fossil record of carnivores tends to be patchy, and it is unlikely that changes in abundance would be clear.

In three of the examples, the decline of the Eocene and the two late Miocene dynasties, competition does not seem likely to have been the driving force. In the Eocene example the decline of the mesonychids and creodonts was already well underway by the time most replacement taxa appeared. In the late Miocene, the replacement clade, the Caninae, overlapped with the incumbents, borophagine and hyaenids, but these early canines were smaller and much rarer than the incumbents. The major diversification of the Caninae took place after the extinction of the groups they replaced, implying an opportunistic replacement rather than active displacement. Interestingly, in all three of these examples the turnover is associated with significant extinction and turnover among other land mammals, as well as a major temperature drop and increasing seasonality.

In the remaining three double-wedge examples, the case for competitive displacement is stronger but not certain. In North America the hypercarnivorous canids and amphicyonids that replaced the late Oligocene hypercarnivorous nimravids and hyaenodonts were of moderate size in the early Arikareean, when all four groups were present. Similarly, the gradual turnover in the mid-Miocene of Eurasia involved reciprocal diversity trends between similarly sized amphicyonids and dog-like hyaenids, both of which had dentitions indicative of moderate carnivory (Viranta 1996; Werdelin 1996a,1996c; Werdelin & Solounias 1996). The parallel turnover that occurred in North America was also between taxa of similar size and diet, amphicyonids and borophagine canids. None of these three examples is associated with such severe climate shifts as those in the late Eocene and late Miocene.

One interesting pattern emerging from this review is that each predator dynasty tends to be composed of a few subfamilies of carnivores, such as borophagine canids, hemicyonine ursids, and hyaenodontid creodonts. As a result, these subfamilies share similar histories of a fairly rapid rise in diversity followed by a decline to extinction, with their heyday of high diversity having lasted perhaps 10 to 15 million years. Interestingly, such short life spans do not seem to characterize the subfamilies of most of the smaller carnivorans. The fossil record of small carnivorans is more difficult to reconstruct than that of larger species because of the limited amount of material and taxonomic uncertainties caused by the fragmentary nature of many specimens. However, it is possible to compare subfamily durations of large and small carnivorans from a molecular-based phylogeny.

In a tree based on hybridization of unique sequence DNA (Wayne et al 1989) the splitting of families within the Carnivora is shown to have occurred nearly simultaneously, with the exception of the Canidae, which split some 10 million years earlier (Figure 9). This is in partial agreement with the fossil record; the earliest canids, mustelids, ursids, nimravids, and amphicyonids all appear approximately 37-40 MYA (Janis et al 1998a).

Subsequent to these basal divisions on the DNA hybridization tree, the families vary in their branching patterns. Among the larger, more carnivorous families (canids, felids, and hyaenids), all the extant species appear to have diverged less than 10 MYA. Not surprisingly, all the extant taxa within each of these families belong to the same subfamily. In contrast, among the smaller and more omnivorous forms, the splits tend to be much deeper in the tree, and these deep splits often correspond to subfamily (or proposed family) subdivisions. For example, the raccoon (*Procyon lotor*) and red panda (*Ailurus fulgens*)

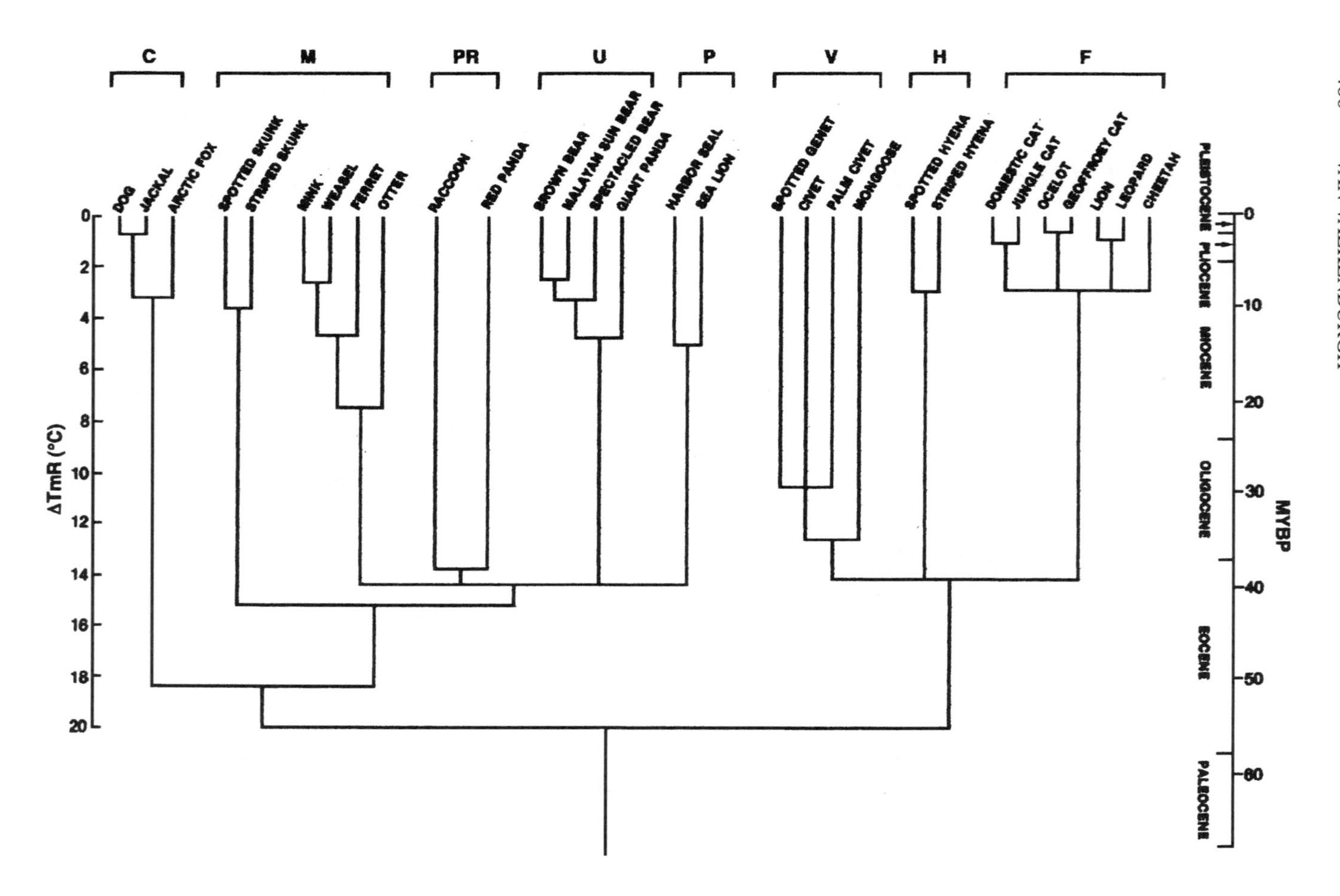
C
M
PR
U
P
V
H
F
DOG
JACKAL
ARCTIC FOX
SPOTTED SKUNK
STRIPED SKUNK
MINK
WEASEL
FERRET
OTTER
RACCOON
RED PANDA
BROWN BEAR
MALAYAN SUN BEAR
SPECTACLED BEAR
GIANT PANDA
HARBOR SEAL
SEA LION
SPOTTED GENET
CIVET
PALM CIVET
MONGOOSE
SPOTTED HYENA
STRIPED HYENA
DOMESTIC CAT
JUNGLE CAT
OCELOT
GEOFFROEY CAT
LION
LEOPARD
CHEETAH
ΔTmR (°C)
0
2
4
6
8
10
12
14
16
18
20
MYBP
0
10
20
30
40
50
60
PLEISTOCENE
PLIOCENE
MIOCENE
OLIGOCENE
EOCENE
PALEOCENE

are now considered to belong to different families (Flynn and Nedbal 1998). Among the four species grouped as viverrids in this tree, mongooses are usually given family rather than subfamily status, as the Herpestidae (Flynn & Nedbal 1998). The deep splits are most extreme among the Viverridae but are also apparent within the Mustelidae and less so among the Ursidae. This reflects the fact that each of these three families is today represented by two or more subfamilies.

The fact that extant canids, felids, and hyaenids include but one subfamily each reflects their dynamic history as members of the large predator guild. Previous subfamilies have gone extinct, such as the borophagine canids, machairodont (saber-toothed) felids, and doglike hyenas. The alternative pattern of persistence of subfamilies among the smaller and/or more omnivorous carnivorans is likely due to at least three factors. First, as smaller taxa, they tend to exist at higher diversity, and therefore entire subfamilies are unlikely to disappear. Second, as smaller, more omnivorous forms, they are less specialized, and this also might favor persistence when environments change. Third, levels of interspecific competition are not likely to be as high as that within the large predator guild, and consequently, displacement is less likely. On the other hand, clades that evolve into the large predator adaptive zone are likely to find that greater body size and dental specialization are advantageous in the short run, despite the consequent loss of evolutionary versatility and species diversity. Natural selection is not prescient, and adaptations for current conditions may prove detrimental when the environment changes, particularly if those adaptations involve significant specialization. Evolution within the large carnivore niche may behave like a ratchet, with interspecific competition favoring increasing body mass and dental specialization, which together are likely to make the species more vulnerable to extinction if the environment changes. In relation to smaller species, large carnivores exist at lower population densities, making them more vulnerable to extinction, and tend to disperse widely, discouraging genetic isolation and thereby speciation (Stanley 1979, p. 278). Consequently, as predator clades age, they produce fewer new taxa, and eventually, extinction rate exceeds speciation rate, sending them into a decline.

The idea that evolution might proceed in a ratchetlike manner is usually considered with reference to the advantage of sex. The hypothesis, known as

←

Figure 9 Phenetic tree of carnivoran relationships based on thermal stability of DNA hybrids from Wayne et al (1989). *C*, Canidae; *M*, Mustelidae; *PR*, Procyonidae and Ailuridae; *U*, Ursidae; *P*, Pinnipedia; *V*, Viverridae and Herpestidae; *H*, Hyaenidae; and *F*, Felidae. The timescale was based on a divergence time of approximately 40 MYA for all the carnivoran families except the Canidae. (From Wayne et al 1989, published with permission of Cornell University Press.)

Muller's ratchet, states that populations of clonal organisms will accumulate deleterious mutations over time due to drift and the fact that back mutations are extremely rare (Futuyma 1998 p. 610). Consequently, the population becomes dominated by more and more inferior genotypes over time, and these are less likely to persist through an environmental perturbation. Recombination, or sex, tends to rid the population of deleterious mutations and thereby provides a long-term advantage for survival of the lineage. The parallel ratchet for the carnivores is a macroevolutionary one, in which new species are budded off rather than cloned. In the carnivore example, the ratchet works by the loss of generalized features (small body mass, versatile dentition) as speciation tends to produce more specialized forms. Given that evolutionary reversals are uncommon, clades that have become dominated by specialized forms are hypothesized to be at a disadvantage during an environmental perturbation (but see Armbruster & Baldwin 1998).

In support of this hypothesis, a tendency for mean body size to increase within subfamilies and/or families of carnivores has been reported for North American amphicyonids (Hunt 1998b) and hyaenodontids (Mellett 1977), as well as Eurasian felids and hyaenids (Werdelin & Solounias 1991, Werdelin 1996a). Similarly, the last representatives of subfamilies or families are often described as exhibiting relatively specialized dentitions for their clade: for example, amphicyonids (Viranta 1996, Hunt 1998b), nimravids (Martin 1998a), North American hyaenodonts (Mellett 1977), and borophagine dogs (Munthe 1998). Rates of extinction and speciation over the life span of each subfamily have yet to be quantified; it would be especially useful to document whether extinction rate increased as each clade aged and the replacement taxa appear.

Conclusion

The key to understanding the history of carnivorous mammals over evolutionary time lies in fundamental characteristics of the large predator adaptive zone that have changed little over the last 65 million years. First, there are a fairly limited number of ways to hunt, kill, and consume prey, and consequently, sympatric predators have tended to diverge along the same lines, no matter where or when they lived. There are bone crackers, meat specialists, and omnivores. Moreover, unlike the situation for herbivores, the material properties of the food of carnivores did not change over the course of the Cenozoic. Although the prey may have acquired longer legs and tougher teeth, they were still composed of skin, muscle, bone, and viscera, and so their predators continued to evolve the blades, daggers, and hammers to eat them. On different continents and at different times the same kinds of carnivores appear, often bearing remarkably similar adaptations despite independent evolutionary histories, such as saber-like canine teeth and bone-cracking cheek teeth (Werdelin 1996a, Martin 1989).

A second feature of the large predator adaptive zone that has greatly influenced the evolutionary history of carnivores is the proclivity for interspecific competition and intraguild predation. Interspecific competition tends to be more intense among large carnivores because prey are often difficult to capture and can represent a sizable quantity of food that is worthy of stealing and defending. In addition, competition appears to be a motive for much intraguild predation because the victim often is not eaten (Cooper 1991, Johnson et al 1996). In many ecosystems today, interspecific relations among sympatric predators play a primary role in their distribution and abundance (Laurenson 1994, Johnson et al 1996, Creel & Creel 1996, Mills & Gorman 1997, White & Garrott 1997), so much so that for some species, such as the cheetah and wild dog, conservation may be dependent on density reduction among other larger species (Creel & Creel 1998, Laurenson 1994). Given such strong levels of interaction, it is not surprising to see parallels on the evolutionary time scale.

Literature Cited

Archer M, ed. 1982. *Carnivorous Marsupials*. Mosman, N.S.W.: R. Zool. Soc. N.S.W.

Archibald D. 1998. Archaic ungulates ("Condylarthra"). See Janis et al 1998b, p. 292–331

Armbruster WS, Baldwin BG. 1998. Switch from specialized to generalized pollination. *Nature* 394:632

Bakker RT. 1983. The deer flees, the wolf pursues: incongruencies in predator-preycoevolution. In *Coevolution*, ed. DJ Futuyma, M Slatkin, p. 350–82

Benton MJ. 1987. Progress and competition in macroevolution. *Biol. Rev.* 62:305–38

Bernor RL, Fahlbusch V, Mittmann H-W, eds. 1996. *The Evolution of Western Eurasian Neogene Mammal Faunas*. New York: Columbia Univ. Press

Berta A. 1981. The Plio-Pleistocene hyaena *Chasmoporthetes ossifraga* from Florida. *J. Vertebr. Paleontol.* 1:341–56

Berta A. 1988. Quaternary evolution and biogeography of the large South American Canidae (Mammalia: Carnivora). *Univ. Calif. Publ. Geol. Sci.* 132:1–149

Berta A, Marshall LG. 1978. *South American Carnivora. Fossilium Catalogus I: Animalia, pars 125*. The Hague: W. Junk

Bryant HN. 1991. Phylogenetic relationships and systematics of the Nimravidae (Carnivora). *J. Mammal.* 72:56–78

Cooper SM. 1991. Optimal hunting group size: the need for lions to defend their kills against loss to spotted hyaenas. *Afr J. Ecol.* 29:130–36

Creel S, Creel NM. 1996. Limitation of African wild dogs by competition with other carnivores. *Conserv. Biol.* 10:526–38

Creel S, Creel NM. 1998. Six ecological factors that may limit African wild dogs, *Lycaon pictus*. *Anim. Conserv.* 1:1–9

Dawkins R, Krebs JR. 1979. Arms races between and within species. *Proc. R. Soc. London Ser. B.* 205:489–511

Dayan T, Simberloff D. 1996. Patterns of size separation in carnivore communities. In *Carnivore Behavior, Ecology and Evolution*, ed. JL Gittleman, p. 243–66. Ithaca, NY: Cornell Univ. Press

Eaton RL. 1979. Interference competition among carnivores: a model for the evolution of social behavior. *Carnivore* 2:9–16

Emerson SB, Radinsky LB. 1980. Functional analysis of sabertooth cranial morphology. *Paleobiology* 6:259–312

Emry RJ. 1981. Additions to the mammalian fauna of the type Duchesnean, with

comments on the status of the Duchesnean "age". *J. Paleontol.* 55:563–70
Emry RJ, Bjork PR, Russell LS. 1987. The Chadronian, Orellan, and Whitneyan land mammal ages. In *Cenozoic Mammals of North America, Geochronology and Biostratigraphy*, ed. MO Woodburne, p. 118–52. Berkeley: Univ. Calif. Press
Flynn JJ. 1998. Early Cenozoic Carnivora ("Miacoidea"). See Janis et al 1998b, p. 110–23
Flynn JJ, Galiano H. 1982. Phylogeny of early Tertiary Carnivora, with a description of a new species of *Parictis* from the middle Eocene of northwestern Wyoming. *Am. Mus. Novit.* 2725:1–64
Flynn JJ, Nedbal MA. 1998. Phylogeny of the Carnivora (Mammalia): congruence vs. incompatibility among multiple data sets. *Mol. Phylogenet. Evol.* 9:414–26
Flynn JJ, Neff NA, Tedford RH. 1988. Phylogeny of the Carnivora. In *The Phylogeny and Classification of the Tetrapods.* vol. 2: *Mammals*, ed. MJ Benton, p. 73–115. Syst. Assoc. Spec. Vol. 35B. Oxford, UK: Clarendon
Flynn LJ, Tedford RH, Qiu ZX. 1991. Enrichment and stability in the Pliocene mammalian fauna of north China. *Paleobiology* 17:246–65
Fortelius M, Werdelin L, Andrews P, Bernor RL, Humphrey L, et al. 1996. Provinciality, diversity, turnover and paleoecology in land mammals faunas of the later Miocene of western Eurasia. See Bernor et al 1996 p. 414–48
Futuyma DJ. 1998. *Evolutionary Biology*, 3rd ed. Sunderland, MA: Sinauer
Gebo DL, Rose KD. 1993. Skeletal morphology and locomotor adaptation in *Prolimnocyon atavus*, an early Eocene hyaenodontid creodont. *J. Vertebr. Paleontol.* 13:125–44
Gingerich PD, Raza SM, Arif M, Anwar M, Zhou XY. 1994. New whale from the Eocene of Pakistan and the origin of cetacean swimming. *Nature* 368:844–47
Gunnell GF. 1998. Creodonta. See Janis et al 1998b, p. 91–109
Gunnell GF, Gingerich PD. 1991. Systematics and evolution of late Paleocene and early Eocene Oxyaenidae (Mammalia, Creodonta) in the Clarks Fork Basin, Wyoming. *Contrib. Mus. Paleontol. Univ. Mich.* 28:141–79
Heinrich RE, Rose KD. 1995. Partial skeleton of the primitive carnivoran *Miacis petilus* from the early Eocene of Wyoming. *J. Mammal.* 76:148–62
Heinrich RE, Rose KD. 1997. Postcranial morphology and locomotor behaviour of two early Eocene miacoid carnivorans, *Vulpavus* and *Didymictis*. *Palaeontology* 23:279–305
Hooker JJ. 1992. British mammalian paleocommunities across the Eocene-Oligocene transition and their environmental implications. See Prothero & Berggren 1992, p. 494–515
Hunt RM. 1987. Evolution of the aeleuroid Carnivora: significance of auditory structure in the nimravid cat *Dinictis*. *Am. Mus. Novit.* 2886:1–74
Hunt RM. 1996. Biogeography of the order Carnivora. In *Carnivore Behavior, Ecology and Evolution*, ed. JL Gittleman, p. 485–541. Ithaca, NY: Cornell Univ. Press
Hunt RM. 1998a. Ursidae. See Janis et al 1998b, p. 174–95
Hunt RM. 1998b. Amphicyonidae. See Janis et al 1998b, p. 196–227
Hunt RM, Skolnick R. 1996. The giant mustelid *Megalictis* from the early Miocene carnivore dens at Agate Fossil Beds National Monument, Nebraska: earliest evidence of dimorphism in New World Mustelidae (Carnivora, Mammalia). *Contrib. Geol. Univ. Wyo.* 31:35–48
Hunt RM, Tedford RH. 1993. Phylogenetic relationships within the aeluroid Carnivora and implications of their temporal and geographic distribution. In *Mammal Phylogeny*, eds. FS Szalay, MJ Novacek, MC McKenna, p. 53–73. New York: Springer-Verlag
Hutchison JH. 1996. Testudines. See Prothero & Emry 1996, p. 337–53
Janis CM. 1993. Tertiary mammal evolution in the context of changing climates, vegetation, and tectonic events. *Annu. Rev. Earth Planet. Sci.* 24:467–500
Janis CM, Baskin JA, Berta A, Flynn JJ, Gunnell GF, et al. 1998a. Carnivora. See Janis et al 1998b, p. 73–90
Janis CM, Scott KM, Jacobs LL, eds. 1998b. *Evolution of Tertiary Mammals of North America*. New York: Cambridge Univ. Press
Janis CM, Wilhelm PB. 1993. Were there mammalian pursuit predators in the Tertiary? Dances with wolf avatars. *J. Mammal. Evol.* 1:103–25
Johnson W, Fuller TK, Franklin WL. 1996. Sympatry in canids: a review and assessment. In *Carnivore Behavior, Ecology and Evolution*, ed. JL Gittleman, p. 189–218. Ithaca, NY: Cornell Univ. Press
Krause DW. 1986. Competitive exclusion and taxonomic displacement in the fossil record: the case of rodents and multituberculates in North America. In *Vertebrates, Phylogeny, and Philosophy*, ed. KM Flanagan, JA Lillegraven, p. 95–118. *Univ. Wyo. Spec. Pap.* 3:119–30
Kurtén B. 1968. *Pleistocene Mammals of Europe*. Chicago: Aldine
Kurtén B, Anderson E. 1980. *Pleistocene Mammals of North America*. New York: Columbia Univ. Press

Laurenson K. 1994. High juvenile mortality in cheetahs (*Acinonyx jubatus*) and its consequences for maternal care. *J. Zool.* 234:387–408

Laurenson K. 1995. Implications of high offspring mortality for cheetah population dynamics. In *Serengeti II: Dynamics, Management, and Conservation of an Ecosystem,* ed. ARE Sinclair, P Arcese, p. 385–99. Chicago: Univ. Chicago Press

Legendre S, Hartenberger J-L. 1992. The evolution of mammalian faunas in Europe during the Eocene and Oligocene. In Prothero & Berggren 1992, p. 516–28

Lento GM, Hickson RE, Chambers GK, Penny D. 1995. Use of spectral analysis to test hypotheses on the origin of pinnipeds. *Mol. Biol. Evol.* 12:28–52

Lewis ME. 1997. Carnivoran paleoguilds of Africa: implications for hominid food procurement strategies. *J. Human Evol.* 32:257–88

Maas MC, Krause DW, Strait SG. 1988. The decline and extinction of Plesiadapiforms (Mammalia: Primates) in North America: displacement or replacement. *Paleobiology* 14:410–31

Marker-Kraus L, Kraus D. 1997. Conservation of strategies for the long-term survival of the cheetah *Acinonyx jubatus* by the Cheetah Conservation Fund, Windhoek. *Int. Zoo Yearb.* 35:59–66

Marshall LG. 1976. Evolution of the family Thylacosmilidae, fossil marsupial sabertooths of South America. *PaleoBios* 23:1–20

Marshall LG. 1977. Evolution of the carnivorous adaptive zone in South America. In *Major Patterns in Vertebrate Evolution*, ed. MK Hecht, PC Goody, BM Hecht, p. 709–21. New York: Plenum

Marshall LG. 1978. Evolution of the Borhyaenidae, extinct South American predaceous marsupials. *Univ. Calif. Publ. Geol. Sci.* 117:1–89

Martin LD. 1980. Functional morphology and the evolution of cats. *Trans. Nebr. Acad. Sci.* 8:141–54

Martin LD. 1989. Fossil history of the terrestrial Carnivora. In *Carnivore Behavior, Ecology and Evolution*, ed. JL Gittleman, p. 536–68. Ithaca, NY: Cornell Univ. Press

Martin LD. 1998a. Nimravidae. See Janis et al 1998b, p. 228–35

Martin LD. 1998b. Felidae. See Janis et al 1998b, p. 236–42

Matthew WD. 1910. The phylogeny of the Felidae. *Bull. Am. Mus. Nat. Hist.* 28:289–316

Mellett JS. 1977. Paleobiology of North American *Hyaenodon* (Mammalia: Creodonta). *Contrib. Vertebr. Evol.* 1:1–134

Mills MGL, Gorman ML. 1997. Factors affecting the density and distribution of wild dogs in the Kruger National Park. *Conserv. Biol.* 11:1397–406

Molnar RE. 1991. Fossil reptiles in Australia. In *Vertebrate Palaeontology of Australasia*, ed. P Vickers-Rich, JM Monaghan, RF Baird, TH Rich, p. 605–702. Melbourne: Monash Univ. Publ.

Munthe K. 1998. Canidae. See Janis et al 1998b, p. 124–43

O'Leary MA, Rose KD. 1995a. Postcranial skeleton of the early Eocene mesonychid *Pachyaena* (Mammalia, Mesonychia). *J. Vertebr. Paleontol.* 15:401–30

O'Leary MA, Rose KD. 1995b. New mesonychian dentitions from the Paleocene and Eocene of the Bighorn Basin, Wyoming. *Ann. Carnegie Mus.* 64:147–72

Polis GA, Holt RD. 1996. Intraguild predation–the dynamics of complex trophic interactions. *Trends Ecol. Evol.* 7:151–54

Prothero DR. 1994. The late Eocene-Oligocene extinctions. *Annu. Rev. Earth Planet. Sci.* 22:145–65

Prothero DR, Berggren WA, eds. 1992. *Eocene-Oligocene Climatic and Biotic Evolution.* Princeton, NJ: Princeton Univ. Press

Prothero DR, Emry RG, eds. 1996. *The Terrestrial Eocene-Oligocene Transition in North America.* New York: Cambridge Univ. Press

Prothero DR, Heaton TH. 1996. Faunal stability during the early Oligocene climatic crash. *Palaeogeogr. Palaeoclimatol. Palaeoecol.* 127:257–83

Retallack GJ. 1992. Paleosols and changes in climate and vegetation across the Eocene/Oligocene boundary. See Prothero & Berggren 1992, p. 382–98

Rich TH. 1991. The history of mammals in Terra Australis. In *Vertebrate Palaeontology of Australasia*, ed. P Vickers-Rich, JM Monaghan, RF Baird, TH Rich, p. 893–1070. Melbourne: Monash Univ. Publ.

Root RB. 1967. The niche exploitation pattern of the blue-gray gnatcatcher. *Ecol. Monogr.* 37:317–50

Rose KD. 1990. Postcranial skeletal remains and adaptations in early Eocene mammals from the Willwood Formation, Bighorn Basin, Wyoming. In *Dawn of the Age of Mammals in the Northern Part of the Rocky Mountain Interior of North America*, ed. TM Bown, KD Rose. *Spec. Pap. Geol. Soc. Am.* 243:107–34

Rosenzweig ML, McCord RD. 1991. Incumbent replacement: evidence for long-term evolutionary progress. *Paleobiology* 17:202–13

Savage DE, Russell DE. 1983. *Mammalian Paleofaunas of the World.* Reading, MA: Addison-Wesley

Simpson GG. 1944. *Tempo and Mode of Evolution*. New York: Columbia Univ. Press

Stanley SM. 1979. *Macroevolution: Pattern and Process*. San Francisco: WH Freeman

Stucky RK. 1990. Evolution of land mammal diversity in North America during the Cenozoic. In *Current Mammalogy*, vol. 2, ed. HH Genoways, p. 375–432. New York: Plenum

Tedford RH, Galusha T, Skinner MF, Taylor BE, Fields RW, et al. 1987. Faunal succession and biochronology of the Arikareean through Hemphillian interval (late Oligocene through earliest Pliocene Epochs) in North America. In *Cenozoic Mammals of North America, Geochronology and Biostratigraphy*, ed. MO Woodburne, p. 152–210. Berkeley: Univ. Calif. Press

Thewissen JGM. 1994. Phylogenetic aspects of cetacean origins: a morphological perspective. *J. Mammal. Evol.* 2:157–84

Turner A. 1990. The evolution of the guild of larger terrestrial carnivores during the Plio-Pleistocene in Africa. *Geobios* 23:349–68

Van Valkenburgh B. 1985. Locomotor diversity within past and present guilds of large predatory mammals. *Paleobiology* 11:406–28

Van Valkenburgh B. 1987. Skeletal indicators of locomotor behavior in living and extinct carnivores. *J. Vertebr. Paleontol.* 7:162–82

Van Valkenburgh B. 1988. Trophic diversity within past and present guilds of large predatory mammals. *Paleobiology* 14:156–73

Van Valkenburgh B. 1990. Skeletal and dental predictors of body mass in carnivores. In *Body Size in Mammalian Paleobiology*, ed. J Damuth, BJ MacFadden, p. 181–206. New York: Cambridge Univ. Press

Van Valkenburgh B. 1991. Iterative evolution of hypercarnivory in canids (Mammalia: Canidae): evolutionary interactions among sympatric predators. *Paleobiology* 17:340–62

Van Valkenburgh B. 1994. Extinction and replacement among predatory mammals in the North American late Eocene-Oligocene: tracking a guild over twelve million years. *Hist. Biol.* 8:1–22

Van Valkenburgh B. 1995. Tracking ecology over geologic time: evolution within guilds of vertebrates. *Trends Ecol. Evol.* 10:71–76

Van Valkenburgh B, Grady F, Kurtén B. 1990. The Plio-Pleistocene cheetah-like cat *Miracinonyx inexpectatus* of North America. *J. Vertebr. Paleontol.* 10:434–54

Van Valkenburgh B, Janis CM. 1993. Historical diversity patterns in large mammalian herbivores and carnivores. In *Species Diversity in Ecological Communities: Historical and Geographical Perspectives*, eds. R Ricklefs, D Shluter, p. 330–40. Chicago: Univ. Chicago Press

Viranta S. 1996. European Miocene Amphicyonidae–taxonomy, systematics, and ecology. *Acta Zool. Fenn.* 204:1–61

Vrana PB, Milinkovitch MC, Powell JR, Wheeler WC. 1994. Higher level relationships of the arctoid Carnivora based on sequence data and "total evidence". *Mol. Phylogenet. Evol.* 3:47–58

Wang X. 1993. Transformation from plantigrady to digitigrady: functional morphology of locomotion in *Hesperocyon* (Canidae: Carnivora). *Am. Mus. Novit.* 3069:1–23

Wang X, Tedford RH. 1996. Canidae. See Prothero & Emry 1996, p. 433–52

Wayne RK, Benveniste RE, Janczewski DN, O'Brien SJ. 1989. Molecular and biochemical evolution of the Carnivora. In *Carnivore Behavior, Ecology and Evolution*, ed. JL Gittleman, p. 465–94. Ithaca, NY: Cornell Univ. Press

Webb SD. 1977. A history of savannah vertebrates in the New World. Part 1: North America. *Annu. Rev. Ecol. Syst.* 8:355–80

Webb SD. 1983. The rise and fall of the late Miocene ungulate fauna in North America. In *Coevolution*, ed. MD Nitecki, p. 267–306. Chicago: Univ. Chicago Press

Webb SD. 1984. Ten million years of mammal extinctions in North America. In *Quaternary Extinctions: a Prehistoric Revolution*, eds. PS Martin, RG Klein, p. 189–210. Tucson: Univ. Ariz. Press

Webb SD, Opdyke ND. 1995. Global climatic influences on Cenozoic land mammal faunas. In *Effects of Past Global Change on Life*, ed. Board on Earth Sciences and Resources, National Research Council, p. 184–208. Washington, DC: Natl. Acad. Sci. USA

Werdelin L. 1987. Jaw geometry and molar morphology in marsupial carnivores: analysis of a constraint and its evolutionary consequences. *Paleobiology* 13:342–50

Werdelin L. 1996a. Carnivoran ecomorphology: a phylogenetic perspective. In *Carnivore Behavior, Ecology and Evolution*, ed. JL Gittleman, p. 582–624. Ithaca, NY: Cornell Univ. Press

Werdelin L. 1996b. Community-wide character displacement in Miocene hyaenas. *Lethaia* 29:97–106

Werdelin L. 1996c. Carnivores, exclusive of Hyaenidae, from the later Miocene of Europe and western Asia. See Bernor et al 1996, p. 271–89

Werdelin L, Solounias N. 1991. The Hyaenidae: taxonomy, systematics and evolution. *Foss. Strata* 30:1–104

Werdelin L, Solounias N. 1996. The evolutionary history of hyaenas in Europe and western Asia during the Miocene. See Bernor et al 1996, p. 290–306

Werdelin L, Turner A. 1996. Turnover in the guild of larger carnivores in Eurasia across the Miocene-Pliocene boundary. *Acta Zool. Cracov.* 39:585–92

White PJ, Garrott RA. 1997. Factors regulating kit fox populations. *Can. J. Zool.* 75:1982–88

Wing SL. 1998. Tertiary vegetation of North America as a context for mammalian evolution. See Janis et al 1998b, p. 37–65

Wolfe JA. 1992. Climatic, floristic, and vegetational changes near the Eocene/Oligocene boundary in North America. See Prothero & Berggren 1992, p. 421–36

Wyss A, Flynn JJ. 1993. A phylogenetic analysis and definition of the Carnivora. In *Mammal Phylogeny*, eds. FS Szalay, MJ Novacek, MC McKenna, p. 32–52. New York: Springer-Verlag

SUBJECT INDEX

A

B

D

E

O

P

R

S

W

CUMULATIVE INDEXES

CONTRIBUTING AUTHORS, VOLUMES 17–27

CHAPTER TITLES, VOLUMES 17–27

PALEONTOLOGY, STRATIGRAPHY, AND SEDIMENTOLOGY

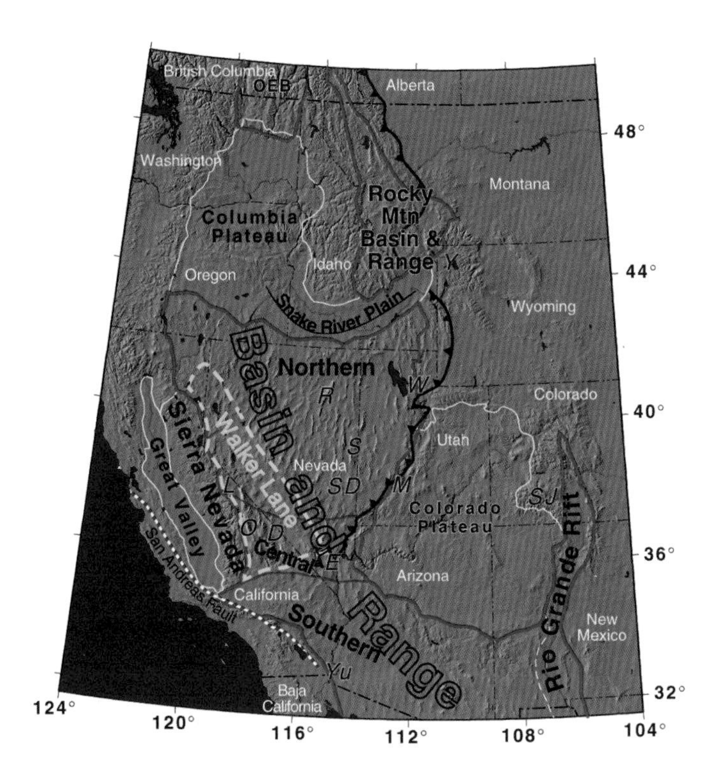

Color Plate 1 General index map indicating extent of Tertiary extensional provinces *(thick red outlines)* in the western United States and localities mentioned in the text. *L,* Long Valley; *O,* Owens Valley; *D,* Death Valley; *E,* Eldorado Mountains; *R,*Ruby Mountains; *OEB,* Omineca extensional belt; *S,* Snake Range; *SJ,* San Juan volcanic field; *M,* Marysvale volcanic field; *SD,* Stampede detachment; *W,* Wasatch fault; *Y,* Yellowstone; *Yu,* Yuma. *Black line with teeth* shows easternmost extent of Sevier-age thrusting.

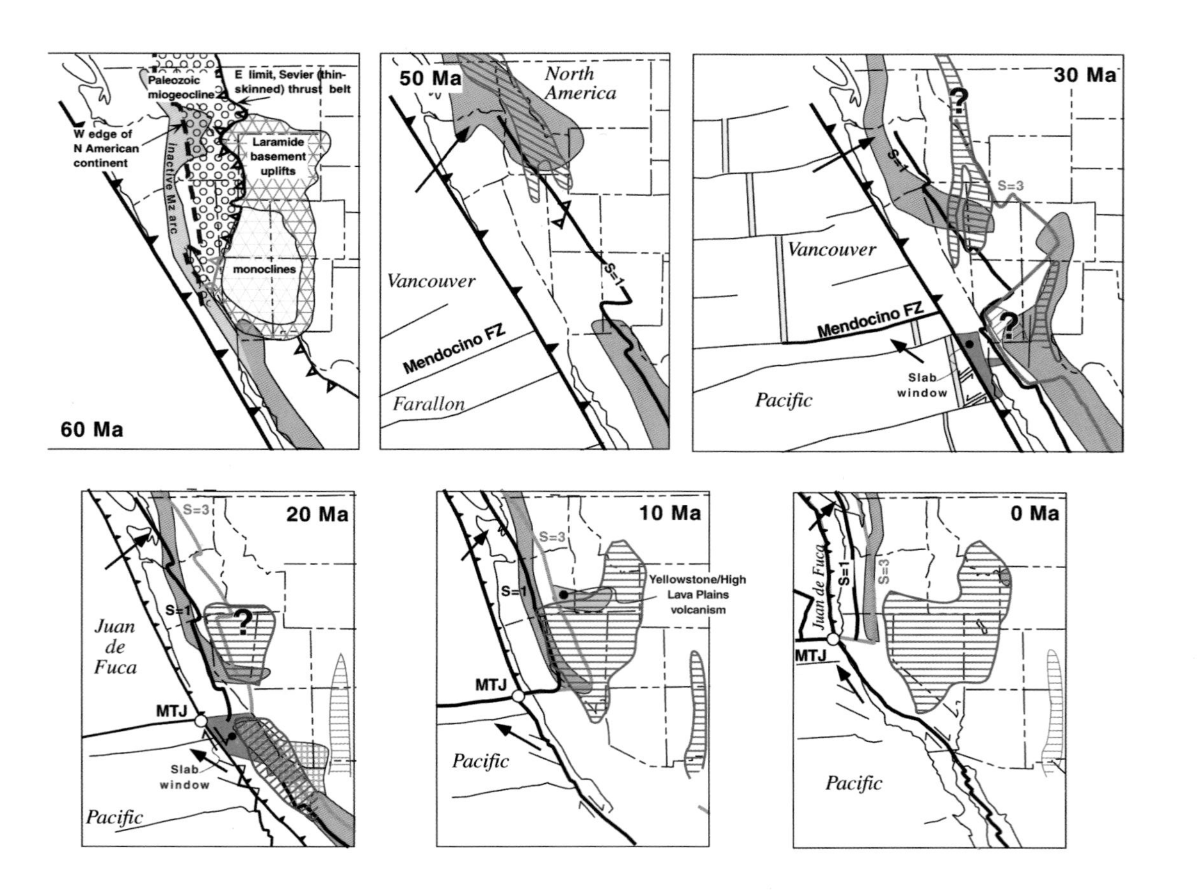
60 Ma
Paleozoic miogeocline
E limit, Sevier (thin-skinned) thrust belt
W edge of N American continent
inactive Mz arc
Laramide basement uplifts
monoclines
50 Ma
North America
S=1
Vancouver
Mendocino FZ
Farallon
30 Ma
S=1
S=3
Vancouver
Mendocino FZ
Slab window
Pacific
20 Ma
S=3
S=1
Juan de Fuca
MTJ
Slab window
Pacific
10 Ma
S=3
S=1
Yellowstone/High Lava Plains volcanism
MTJ
Pacific
0 Ma
Juan de Fuca
S=1
S=3
MTJ
Pacific

Color Plate 2 Tectonic setting relevant to the Cenozoic evolution of the western United States. In the 60 Ma panel, *orange areas* indicate active magmatic arc; *light orange areas* indicate late Mesozoic magmatic arcs inactive by 60 Ma. Active edge of foreland fold-and-thrust (Sevier belt) shown with black line with teeth; Sevier hinterland area > 100--150 km west of the line. *Grayed sections* indicate portions inactive by 60 Ma. Edge of North American continent *(dashed line)* generally defined by western limit of rocks with initial 87Sr/86Sr > 0.706. In 50 Ma and younger panels, *diagonally hatched regions* show areas of active extension; hatch orientation is approximately parallel to extension direction. Major volcanic fields (mostly rhyolitic/andesitic) shown in *orange. Arrows* indicate plate velocities relative to North American plate. S=1 and S=3 curves (from Severinghaus & Atwater 1990) indicate plausible range in extent of Vancouver/Farallon slab underneath North America. Velocities and plate configurations from Severinghaus & Atwater (1990), Atwater & Stock (1998), Stock & Molnar (1998), and Wilson (1988). Geology from Axen et al (1993), Christiansen & Yeats (1992), Janecke (1994), Wernicke (1992) and references therein. *FZ,* fracture zone; *MTJ,* Mendocino triple junction.

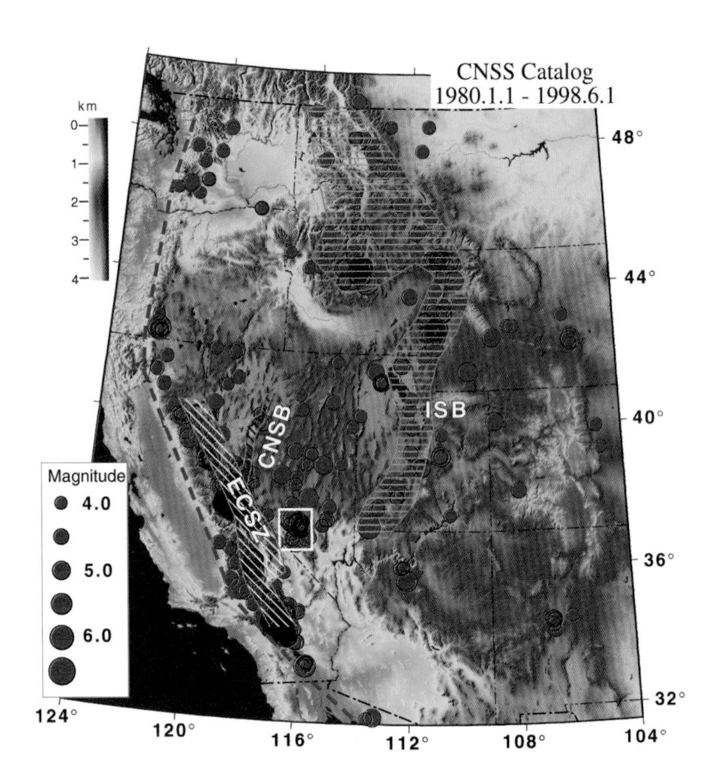

Color Plate 3 Seismicity (M≥4) since 1980 in the western United States, from the council of the National Seismic System combined catalog (http://quake.geo.berkeley.edu/cnss/). *CNSB,* Central Nevada Seismic Belt; *ECSZ,* Eastern California Seismic Zone; *ISB,* Intermountain Seismic Belt. *Box* indicates Nevada Test Site area, where significant seismicity is related to nuclear testing.

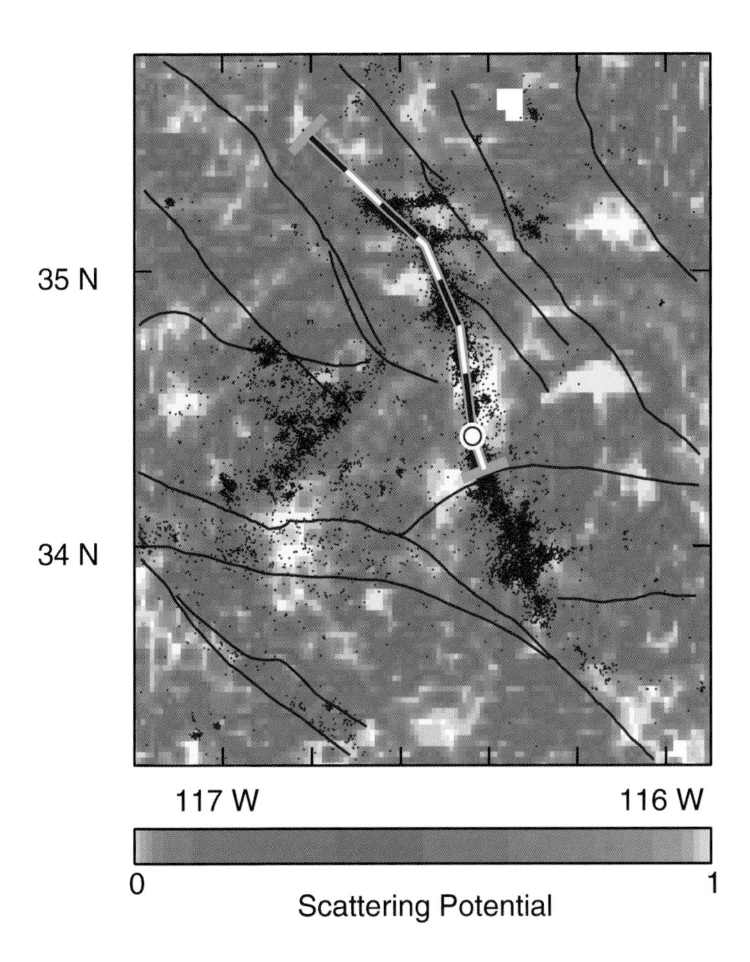

Color Plate 1 Scattering potential of the upper crust surrounding the 1992 Landers earthquake *(epicenter marked by bull's-eye)*. ML ≥ 2.0 aftershock seismicity (*black dots*) clusters in areas of highest scattering potential. *Green and black line* marks fault line used in Figure 4.

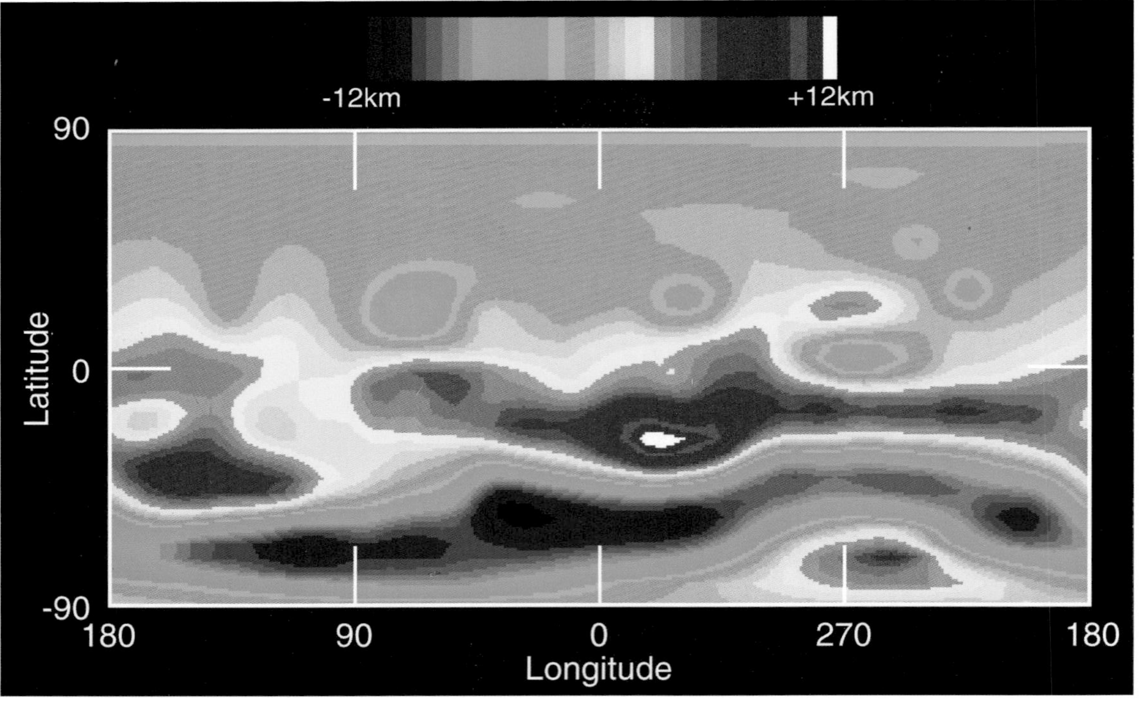

Color Plate 1 Fifty-six WFPC2 images from 1994 combined with seventy-eight images from the favorable 1996 opposition of Vesta create a topographic map. The conversion scale from color to elevation is included in the figure. The most remarkable feature is the large crater near the south pole. This crater may have been the source of a number of smaller asteroids as well as a class of igneous meteorites. The figure is reprinted with permission from the American Association for the Advancement of Science (Thomas et al 1997a).